EXAM
#3

2 Probs = 10 pts
2 Probs 20 pts

Covers
  CH 3&4
  excepting 4-6

Mon
12-1-86

Final Exam
  Wed Dec 10
  2 hrs
  5 probs
  Topics
  Mohrs Circle
  indeterminate Prob
  elastic Curve
  Rotation
  Shear-Moment Diagrams & equations
  Columns

MECHANICS OF ENGINEERING MATERIALS

William H. Bowes

Professor of Civil Engineering
Carleton University
Ottawa, Ontario

Leslie T. Russell

Professor of Mechanical Engineering
Technical University of Nova Scotia
Halifax, Nova Scotia

Gerhard T. Suter

Professor of Civil Engineering
Carleton University
Ottawa, Ontario

John Wiley & Sons

New York • Chichester • Brisbane • Toronto • Singapore

PRODUCTION SUPERVISOR: CINDY STEIN-LAPIDUS
SPONSORING EDITOR: WILLIAM STENQUIST
MANUSCRIPT EDITOR: DEBORAH HERBERT
TEXT DESIGN: LORETTA SARACINO-MAROTTO
COVER PHOTOGRAPHY: PAUL SILVERMAN

Library of Congress Cataloging in Publication Data:

Bowes, William H.
  Mechanics of engineering materials.

  Includes indexes.
  1. Strength of materials.   I. Russell, Leslie T.
II. Suter, Gerhard T.   III. Title.
TA405.B68 1983   620.1'1   83-12341
ISBN 0-471-86145-6

Printed in the United States of America

10 9 8 7 6 5 4 3 2 1

A NOTE REGARDING UNITS

This textbook has been written in such a way that the user can elect to work completely in Imperial units or in SI units. To provide this flexibility:

- Each example problem is worked in both sets of units.
- Imperial units are presented in parentheses alongside SI units, and shading is employed for Imperial units to allow ready recognition.
- All problems at the end of each chapter are defined in both sets of units.
- Tables of material and structural properties are printed twice with a separate appendix for each set of units.

This arrangement will enable the reader to use the entire textbook with ease in either set of units.

GENERAL PREFACE

During many years of teaching, we have found that students seldom acquire an understanding of the mechanics of engineering materials by studying the subject as it is presented in conventional textbooks. At the beginning of their study, students are frequently confused by a diagram, showing a cross section that cuts through a stressed body of general shape, that is supposed to illustrate the general state of stress on a plane. They are equally confounded by Mohr's circle of stress when their understanding of stress—particularly shear stress—is not yet well developed. This conventional approach may be likened to a science student starting with studies of the general theory of relativity in the hope that it will hold the key to all other physics. In reality, students acquire an understanding of a subject by gradually adding to their stock of familiar topics and slowly pushing back the frontier of the unknown.

Our observations of the learning process have prompted us to write a textbook that treats the topics in an order that is logical from the point of view of the student. For example, stress is first treated in axially loaded members, then in bending, and later in shells. Shear stress is introduced through bolted connections. Only after a good working knowledge of stress has been achieved is the question of stress on inclined planes presented. We do not have a chapter dealing with stress and another dealing with strain. These two topics are inseparable and, hence, are treated together. Furthermore, they permeate the whole subject of engineering materials and consequently appear in all chapters.

The subject herein is presented as a tool for use in engineering design and emphasis is placed on the design process. The practical side of design is shown wherever possible, and there are frequent reminders that safety is a prime consideration in all engineering design.

To help the student acquire an intuitive feeling for the proportions and load-carrying capacity of real members and structures, we have chosen only realistic sizes and loads for the structures that are given in examples and problems. The materials specified are chosen from those that are commonly used in engineering practice, and the stresses to which they are subjected are realistic.

We appreciate the fact that the conversion of American industry from Imperial to SI units will take place over a long transition period. During the transition the educational system will gradually change to keep in step with the requirements of industry. Many educators will prefer to work entirely in Imperial units until the changeover and then to use SI units exclusively. To aid instructors in making the transition, *the book has been written in such a manner that the user can work entirely in either set of units* without stumbling over the other set. To accomplish this, all example problems and tables of properties are presented twice, once in each set of units. Similarly, all assignment problems are given in both sets of units. Consequently, users will be able to make the changeover when required without referring to another textbook. The user also has the option of working with a mixture of units in any proportion that is deemed appropriate.

Problems are provided at the end of each chapter. To help synchronize the problems with the text, problem numbers are given in the text at the point where the student should be ready to solve those particular problems. In addition to the regular problems, at the end of each chapter there are a few more challenging problems presented. These will require some original thought on the part of the student, but the initiative required is not unlike that demanded in the everyday practice of engineering.

This book evolved over several years from course-notes written for a two-semester course. The course was required for civil, electrical, and mechanical engineering, and this is reflected in the wide variety of topics covered. The topics have been grouped and presented in such a way that material of limited interest can be omitted. For all disciplines, Chapters 1 through 6 and the first part of Chapter 7 contain essential, basic material. The remainder of the book contains topics that are of an elective nature, and instructors will find it possible to choose chapters and parts of chapters to meet almost any need. For example, for a mechanical class it might be beneficial to omit Beams in Bending II, Columns II, and Energy Methods and instead to concentrate on Pressure Vessels and Design for Cyclic Load. It may seem students in such a class have bought a book more complex than necessary. However, it is important that students be made aware of the limits of their knowledge so that they will apply not, for example, the bending formulas from Beams in Bending I to unsymmetrical sections. Readers would be forewarned about the hazards of such a practice by a casual examination of Beams in Bending II. There are similar benefits to be gained by having material, which is not covered in a formal course, readily available for future reference in a familiar textbook.

Although the first half of the book contains core material, some topics can be omitted without destroying continuity. For example, it would be possible to omit any or all of the following Sections. Thermal Stresses, Singularity Functions, Moment-Area Method, Deflection of Long-radius Curved Beams, and Failure Theories.

It is our feeling that we have provided a level of flexibility that will enable instructors to meet their particular needs whether the class is composed of students specializing in the mechanics of materials or those with only a very general interest in the subject.

Finally, we would like to thank the following people for their constructive criticism and suggestions while reviewing the text: William W. Bradley, Clarkson College; Michael E. James, Jr., Texas A & M University; Robert A. Lucas, Lehigh University; James F. McDonough, University of Cincinnati; J. J. Salinas, Carleton University; and J. R. Snowden, University of Louisville.

William H. Bowes
Leslie T. Russell
Gerhard T. Suter

STUDENT PREFACE

Your everyday life brings you into contact with many machine parts and structural members that have been designed to perform with adequate strength and stiffness. You might wonder just how a satisfactory design has been achieved.

While your knowledge of statics enables you to solve for forces, moments, and torques acting on members, you are not yet able to analyze a member to determine if it is overloaded or to design it for a specified strength and stiffness. This, then, is where a course on the mechanics of engineering materials and this textbook come into play. Through such a course and text you will learn to analyze and design members subjected to common loadings such as tension, compression, shear, torsion, or pressure. You will develop the capability of determining if a particular shaft, beam, or column can safely carry the loads that are imposed on it. Alternatively, based on limiting stresses, you will be able to design such members.

When a member is being loaded or stressed, it is also being deformed or strained. To be acceptable a particular member must not only be strong enough to sustain the loads that act upon it, but it must also be stiff enough so that it does not deform unduly.

In the mechanics of engineering materials we are therefore always interested in ensuring that both strength and stiffness are adequate. This book treats both requirements in depth and in this way lays the foundation for you to become a professional who can analyze and design members. This foundation is of great value to all engineering disciplines because adequate strength and stiffness are controlling factors in engineering. Only with such a foundation can a designer fulfill his obligation to society to provide safe and serviceable members and structures.

W. H. B.
L. T. R.
G. T. S.

CONTENTS

1. **INTRODUCTION TO MECHANICS OF ENGINEERING MATERIALS**

   1-1 Introduction 2
   1-2 Design of Axially Loaded Members 3
   1-3 Design of Bolted End Connections for Axially Loaded Members 10
   1-4 Design of Welded End Connections for Axially Loaded Members 13
   1-5 Shear and Bending Moment Diagrams 17
   1-6 Bending Stresses in a Beam—Approximate Method 28
   1-7 Stresses Due to Pressurized Fluid in a Thin-Walled Vessel 38
   1-8 Stress Due to Axial Force and Torque on a Thin-Walled Vessel 44
   1-9 Allowable Stress 47
   1-10 Stress–Strain Diagrams 53
   1-11 Stress Concentration 58
   1-12 Deformation of Truss Members 62
   1-13 Introduction to Statically Indeterminate Systems 63
   1-14 Thermal Stresses 70

2. **BEAMS IN BENDING I**

   2-1 Introduction 91
   2-2 Stress-Moment Relationship in Bending 92
   2-3 Moment of Inertia of Area 100
   2-4 Design of Beams 107
   2-5 Eccentric Axial Loads 108
   2-6 Shearing Stress in Beams 112
   2-7 Curved Beams 125
   2-8 Summary 136

3. **DEFLECTIONS DUE TO BENDING**

   3-1 Introduction 151
   3-2 Beam Flexure Equation 151
   3-3 Statically Indeterminate Beams 157
   3-4 Singularity Functions 165
   3-5 Superposition 172
   3-6 Moment-Area Method 175

3-7   Deflection of Long-Radius Curved Beams   179
3-8   Summary   185

4. COLUMNS I

4-1   Introduction   195
4-2   Euler's Buckling Load   196
4-3   Energy Considerations in Buckling   203
4-4   Slenderness Ratio   204
4-5   Column Design   206
4-6   End Conditions   211

5. ANALYSIS OF STRESS

5-1   Introduction   224
5-2   Stress on an Inclined Plane in a Uniaxial Stress Field   224
5-3   Stresses on an Inclined Plane in a General Plane Stress Field   228
5-4   Mohr's Circle and Principal Stress Equations   238
5-5   Stresses in Three Dimensions   244
5-6   Failure Theories   246

6. STRESS–STRAIN RELATIONSHIPS

6-1   Introduction   269
6-2   Strains from Known Stresses   269
6-3   Stresses from Known Strains   275
6-4   Stresses from Strain Gage Data   281
6-5   Principal Strains Using Strain Gage Rosettes   285
6-6   Summary   289

7. TORSION

7-1   Introduction   295
7-2   Torsional Stress in a Cylindrical Rod   295
7-3   Torsional Bars and Helical Springs   308
7-4   Membrane Analogy   314
7-5   Stresses in Noncircular Sections   318
7-6   Thin-Walled Irregular Sections   322
7-7   Torsion in the Inelastic Range   326
7-8   Summary   334

8. BEAMS IN BENDING II

8-1   Introduction   341
8-2   Inclined Bending of Beams Having Symmetrical Cross Sections   342

8-3 Beams Having Nonsymmetrical Cross Sections   345

8-4 Properties of Cross Sections: Moment of Inertia, Product of Inertia, and Mohr's Circle of Inertia   348

8-5 Shear Center   358

8-6 Local Buckling   365

8-7 Composite Beams   368

8-8 Nonlinear Bending   374

8-9 Bending into the Inelastic Range   379

8-10 Plastic Design   386

9. COLUMNS II

9-1 Introduction   401

9-2 Partial End Constraints   401

9-3 Eccentrically Loaded Columns   412

9-4 Beam Columns   422

9-5 Complex Buckling Modes   435

9-6 Column Design   436

10. ENERGY METHODS

10-1 Introduction   445

10-2 Strain Energy in a Stretched Spring   445

10-3 Strain Energy Density   447

10-4 Strain Energy in Prismatic Members   449

10-5 The Direct Energy Method   453

10-6 Minimum Potential Energy   464

10-7 The Rayleigh-Ritz Method   473

10-8 Castigliano's First Theorem   478

10-9 Castigliano's Second Theorem   491

11. PRESSURE VESSELS

11-1 Introduction   523

11-2 General Formula for Shell Stress   524

11-3 Membrane Shell Theory and Membrane Analogy   531

11-4 Flat Plates Supporting a Uniform Pressure   533

12. DESIGN FOR CYCLIC LOAD

12-1 Introduction   542

12-2 The Nature of Fatigue Failure   543

12-3 Relationship Between Strength and Number of Cycles   544

12-4 Strength Reduction Factor: Reliability   545

12-5 Mathematical Relationship Between Strength and Number of Cycles   549

12-6 Strength Reduction Factor: Mean Stress   550

12-7   Strength Reduction Factor: Size   556
12-8   Strength Reduction Factor: Finish   557
12-9   Strength Reduction Factor: Concentration of Stress   561
12-10   Strength Reduction Factor: Other Conditions   566
12-11   Summary of Equations and Strength Reduction
           Factors   567
12-12   Loads of Varying Amplitude   572
12-13   Summary   573

APPENDIX A TABLES (SI UNITS)   578

APPENDIX B TABLES (IMPERIAL UNITS)   592

PHOTO CREDITS   605

INDEX   607

INTRODUCTION TO MECHANICS OF ENGINEERING MATERIALS

1-1 INTRODUCTION

Mechanics of Materials deals with the response of various bodies, usually called *members*, to applied forces. In Mechanics of Engineering Materials the members have shapes that either exist in actual structures or are being considered for their suitability as parts of proposed engineering structures. The materials in the members have properties that are characteristic of commonly used engineering materials such as steel, aluminum, concrete, and wood.

The forces, usually called loads, may come from a wide variety of sources such as:

- Gravity acting on manufactured products piled on a warehouse floor.
- Water pressure acting on a sluice gate.
- Vehicles moving across a bridge.
- Temperature changes.
- The flow of air across an aircraft wing.
- The pressure of fluid inside a pressure vessel.
- The weight of a container being hoisted by a crane.

The shapes of the members are also numerous and varied such as:

- A cable supporting a portion of a stadium roof.
- A tube serving as a television antenna mast.
- A connecting rod in an automobile.
- A high-voltage transmission line.
- A cantilever beam supporting a building canopy.
- A vessel containing high-pressure steam.
- A shaft subjected to torque in a transmission drive train.

As you can see already from the variety of materials, forces, and shapes mentioned, Mechanics of Engineering Materials is of interest to all fields of engineering. The engineer uses the principles of Mechanics of Materials to determine if the material properties and the dimensions of a member are adequate to ensure that it can carry its loads safely and without excessive distortion. In general, then, we are interested in both the *safe load* that a member can carry and the associated *deformation*. Engineering design would be a simple process if the designer could take into consideration the loads and the mechanical properties of the materials, manipulate an equation, and arrive at suitable dimensions. Design is seldom that simple. Usually, on the basis of experience, the designer selects a trial member and then does an analysis to see if that member meets the specified requirements. Frequently, it does not and then a new trial member is selected and the analysis repeated. This *design cycle* continues until a satisfactory solution is obtained. The number of cycles required to find an acceptable design diminishes as the designer gains experience. In this book the analysis part of the design cycle will receive the most attention, but you will also be given an opportunity to gain some experience with the complete design cycle.

The subject matter of Chapter 1 is so arranged that you will quickly encounter a large variety of practical engineering problems. Although it might appear to you that the topics dealt with in this chapter have little in common, they provide an opportunity to use your knowledge of statics to develop some new principles. The application of these principles to practical engineering problems provides an exciting and challenging introduction to Mechanics of Engineering Materials.

1-2 DESIGN OF AXIALLY LOADED MEMBERS

To give you some insight into the design cycle, an extremely simple member will be dealt with first. That member, shown in Fig. 1-1a, is a prismatic bar with a force, P, acting along its longitudinal axis in the direction such that it tends to elongate the bar. Such a force is referred to as an *axial tensile load*, and we can readily imagine it trying to pull the fibers apart and to cause failure on a transverse plane. It is safe to assume that all fibers of the bar, in regions remote from the point of application of the load, are being pulled apart with the same load intensity. With this assumption, the load intensity or *stress* is uniform on a transverse plane and is given by

$$\sigma = \frac{P}{A}$$

(1-1a)

When P is in newtons and A is in square metres, stress, σ, is in newtons per square metre (N/m^2), which is by definition pascals (Pa).

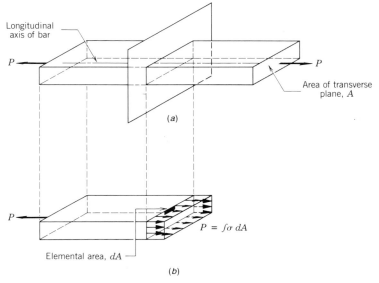

FIG. 1-1 Axially loaded member.

Since a newton is a very small force and a square metre is a large area, the stresses encountered in engineering applications are usually quite large, frequently being more than 100 000 000 Pa. To avoid such unwieldy numbers, engineers use the megapascal as the unit of stress, which results in stresses of the order of 100 megapascals (MPa). Stress in megapascals is equal to force in newtons divided by area in square millimetres. These are practical working units, since cross sections are frequently dimensioned in millimetres.

In Imperial units, when P is in pounds and A is in square inches, (1-1a) gives stress in pounds per square inch (psi). Since the pound is a rather small unit for expressing forces of the magnitude encountered in many applications, the values of P and σ are often five-digit numbers. To reduce the number of digits that must be manipulated, it is common practice to use the kilopound, abbreviated as kip, for the unit of force and the kilopound per square inch, abbreviated as ksi, for the unit of stress. Then, with force in kip and area in square inches, (1-1a) gives stress in ksi.

Before leaving Fig. 1-1a, let us have a closer look at the action inside the bar. From our experience in statics we know that if we cut the bar of Fig. 1-1a at any cross section, a force equal to P must be applied on that section in order to maintain equilibrium. This *internal* force P is made up of the summation of the fiber forces that act on the section under study. When dealing with a summation process, it is useful to involve calculus; hence we can say that the internal force P is made up of the summation of minute forces which act with stress σ on elemental areas, dA. We can then write

$$P = \int \sigma \, dA \qquad \textbf{(1-1b)}$$

where the integral sign refers to summation over the whole cross section. Since the stress is constant in the simple axial tensile loading case of the prismatic member in Fig. 1-1, we get

$$P = \sigma \int dA = \sigma A$$

which is in agreement with (1-1a). Note that if the load in Fig. 1-1 had been applied at the edge of the member instead of along its longitudinal axis, the stress would not be uniform and we would have to determine the relationship between σ and the location of dA before being able to carry out the integration of (1-1b).

Determining the location of the longitudinal axis for the member shown in Fig. 1-1 poses no problem. However, when the cross section is not symmetrical, the location is not so obvious. In that case, if the stress is to be uniformly distributed, the longitudinal load must act along the line that joins the centroids of the cross sections of the member. The line joining the centroids is known as the *centroidal axis*.

For a given axial load and given dimensions, the stress can be calculated from (1-1a) and compared with the stress that can be safely carried by the material. The safe stress, known as the design stress or *allowable stress*, is determined by tests performed on material made to the same specifications as the part being considered. A safety factor, frequently imposed by a legally established code, is applied to the strength, as determined by tests, to give the allowable stress.

The allowable stress, σ_a, is given by

$$\sigma_a = \frac{\sigma_f}{n} \qquad (1\text{-}2)$$

where σ_f is the stress at which the material fails (failure to be defined later) and n is the *safety factor*.

Before approving trial dimensions, the designer makes certain that the design is safe by determining that the inequality

$$\sigma_a \geqslant \sigma$$

or

$$\sigma_a \geqslant \frac{P}{A} \qquad (1\text{-}3a)$$

is satisfied. The inequality is usually more convenient in the form

$$A \geqslant \frac{P}{\sigma_a} \qquad (1\text{-}3b)$$

It might at first seem that the designer would always dimension the cross section so that the stress would exactly equal the allowable stress. However, it may be very costly to produce parts that have nonstandard sizes, so it is usually more economical to waste some material by selecting the next larger standard size above that required by the allowable stress. Departure from standard sizes is justified in cases where the penalty for excess weight is very severe, as in aircraft or space-ship design. The following design example will illustrate the design approach to an axially loaded member.

Example 1-1

Design members *A-B*, *B-C*, and *B-D* of the structure shown in Fig. 1-2a. For tension members use the allowable stress given in Table A of Appendix A at the end of the book for low-strength structural steel and select rods from Table D. For the compression member use an allowable compressive stress of 60 MPa and select a standard pipe from Table C.

Note: The allowable compressive stress is not arrived at by simple means. The design of compressive members is quite complex and will be dealt with in Chapter 4. In the meantime, reasonable values of allowable compressive stress will be given where required in the problems.

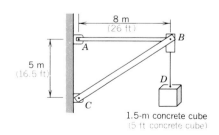

FIG. 1-2*a*

Solution

Weight of Supported Load.   Density of concrete $= 2400 \dfrac{kg}{m^3}$ (Table A)

Volume of cube $= 1.5^3 m^3 = 3.375 \ m^3$

Mass of cube $= 2400 \times 3.375 \dfrac{kg}{m^3} m^3$

$= 8100 \ kg$

Weight of cube $= 8100 \times 9.81 \ kg \dfrac{N}{kg}$ (see Table I)

$= 79\ 500 \ N$

$= 79.5 \ kN$

The *forces in members* are shown in Fig. 1-2*b*.

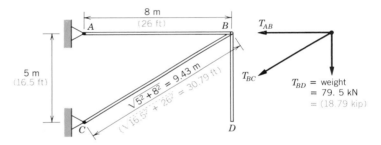

FIG. 1-2*b*

$$\sum F_y = 0$$

$$-\frac{5}{9.43} T_{B\text{-}C} - 79.5 = 0$$

$$T_{B\text{-}C} = -\frac{9.43}{5} \times 79.5 \frac{m}{m} \cdot kN$$

$$= -150 \ kN$$

$$= \textbf{150 kN (compression)}$$

$$\sum F_x = 0$$

$$-T_{A\text{-}B} - \frac{8}{9.43} T_{B\text{-}C} = 0$$

$$T_{A\text{-}B} = -\frac{8}{9.43} T_{B\text{-}C}$$

$$= -\frac{8}{9.43}(-150)\frac{m}{m}\cdot kN$$

$$= \textbf{127 kN (tension)}$$

Design of Member B-D

$$P = T_{B\text{-}D} = 79.5 \text{ kN} \qquad \text{(tension)}$$

$$\sigma_a = 140\,\frac{N}{mm^2} \qquad \text{(Table A)}$$

From (1-3b),

$$A \geqslant \frac{P}{\sigma_a} = \frac{79.5\times10^3}{140}\frac{N}{N/mm^2} = 568 \text{ mm}^2$$

From Table D, a 30-mm rod will provide 707 mm^2.

Use 30-mm rod

Design of Member A-B

$$P = T_{A\text{-}B} = 127 \text{ kN} \qquad \text{(tension)}$$

$$\sigma_a = 140\,\frac{N}{mm^2} \qquad \text{(Table A)}$$

From (1-3b),

$$A \geqslant \frac{P}{\sigma_a} = \frac{127\times10^3}{140}\frac{N}{N/mm^2} = 907 \text{ mm}^2$$

From Table D, a 35-mm rod will provide 962 mm^2.

Use 35-mm rod

Design of Member B-C

$$P = T_{B\text{-}C} = 150 \text{ kN} \qquad \text{(compression)}$$

$$\sigma_a = 60\,\frac{N}{mm^2} \qquad \text{(given)}$$

From (1-3b),

$$A \geqslant \frac{P}{\sigma_a} = \frac{150\times10^3}{60}\frac{N}{N/mm^2} = 2500 \text{ mm}^2$$

From Table C, a 125-mm pipe will provide 2773 mm^2.

Use a 125-mm pipe

Note that the standard-size sections provide an excess of material above the minimum required for a safe structure. In members *A-B*, *B-C* and *B-D* the excess material is 6%, 10% and 20% respectively. Under most conditions it would not be economical to produce non-standard sizes that are closer to the minimum requirements.

Example 1-1

Design members *A-B*, *B-C*, and *B-D* of the structure shown in Fig. 1-2a. For tension members use the allowable stress given in Table A of Appendix A at the end of the book for low-strength structural steel and select rods from Table D. For the compression member use an allowable compressive stress of 8.7 ksi and select a standard pipe from Table C.

Note: The allowable compressive stress is not arrived at by simple means. The design of compressive members is quite complex and will be dealt with in Chapter 4. In the meantime, reasonable values of allowable compressive stress will be given where required in the problems.

Solution

Weight of Supported Load.   Density of concrete $= 0.087 \dfrac{\text{lb}}{\text{in.}^3}$ (Table A)

Volume of cube $= 5^3 \text{ft}^3 = 125 \text{ ft}^3$

Mass of cube $= 0.087 \times 125 \times 12^3 \dfrac{\text{lb}}{\text{in.}^3} \text{ft}^3 \dfrac{\text{in.}^3}{\text{ft}^3}$

$$= 18\ 780 \text{ lb}$$

Weight of cube $= \dfrac{\text{lb mass}}{g} \times a = \dfrac{18\ 790}{32.2} \times 32.2$

$$= 18\ 790 \text{ lb} = 18.79 \text{ kip}$$

The *forces in members* are shown in Fig. 1-2b.

$$\sum F_y = 0$$

$$-\frac{16.5}{30.79} T_{B\text{-}C} - 18.79 = 0$$

$$T_{B\text{-}C} = -\frac{30.79}{16.5} \times 18.79 \frac{\text{ft}}{\text{ft}} \cdot \text{kip}$$

$$= -35.06 \text{ kip}$$

$$= \textbf{35.06 kip (compression)}$$

$$\sum F_x = 0$$

$$-T_{A\text{-}B} - \frac{26}{30.79} T_{B\text{-}C} = 0$$

$$T_{A\text{-}B} = -\frac{26}{30.79} T_{B\text{-}C}$$

$$= -\frac{26}{30.79}(-35.06)\frac{ft}{ft}\cdot kip$$

$$= \textbf{29.60 kip (tension)}$$

Design of Member B-D

$$P = T_{B\text{-}D} = 18.79 \text{ kip} \qquad \text{(tension)}$$

$$\sigma_a = 20 \text{ ksi} \qquad \text{(Table A)}$$

From (1-3b),

$$A \geqslant \frac{P}{\sigma_a} = \frac{18.79}{20}\frac{kip}{kip/in.^2} = 0.939 \text{ in.}^2$$

From Table D, a $1\frac{1}{4}$-in. rod will provide 1.23 in.^{2}.

Use $1\frac{1}{4}$-in. rod

Design of Member A-B

$$P = T_{A\text{-}B} = 29.60 \text{ kip} \qquad \text{(tension)}$$

$$\sigma_a = 20 \text{ ksi} \qquad \text{(Table A)}$$

From (1-3b),

$$A \geqslant \frac{P}{\sigma_a} = \frac{29.6}{20}\frac{kip}{kip/in.^2} = 1.48 \text{ in.}^2$$

From Table D, a $1\frac{1}{2}$-in. rod will provide 1.77 in.^{2}.

Use $1\frac{1}{2}$-in. rod

Design of Member B-C

$$P = T_{B\text{-}C} = 35.06 \text{ kip} \qquad \text{(compression)}$$

$$\sigma_a = 8.7 \text{ ksi} \qquad \text{(given)}$$

From (1-3b),

$$A \geqslant \frac{P}{\sigma_a} = \frac{35.06}{8.7}\frac{kip}{kip/in.^2} = 4.03 \text{ in.}^2$$

From Table C, a 5-in. pipe will provide 4.30 in.^{2}.

Use a 5-in. pipe

Note that the standard-size sections provide an excess of material above the minimum required for a safe structure. In members *A-B*, *B-C* and *B-D* the excess material is 16%, 6% and 24% respectively. Under most conditions it would not be economical to produce non-standard sizes that are closer to the minimum requirements.

Problems 1-1 to 1-6

1-3 DESIGN OF BOLTED END CONNECTIONS FOR AXIALLY LOADED MEMBERS

Trusses are structures having members and loads so arranged that the members are subjected to axial forces only. A typical truss is shown in Fig. 1-3a. The members and their end connections are usually arranged so that the stress is uniformly distributed over the cross section of each member. One way to accomplish this is by using a pin to connect converging members; this method was, in fact, used for many years. When the pin is properly located, and if there is no friction, the stress will

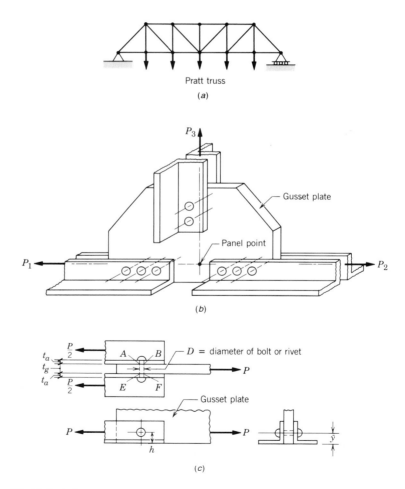

Pratt truss

(a)

(b)

(c)

FIG. 1-3 Typical truss connection.

be uniform. Although the real connection today may be made by means of several bolts or rivets, our analysis of forces in truss members is still based on the simplified single-pin connection; this simplification is acceptable because the resultant forces are known to be substantially correct.

The members of trusses frequently consist of pairs of angles joined through a *gusset plate* as shown in Fig. 1-3b. This makes a very convenient system for connecting the members that converge on a *panel point*, which is the imaginary point through which the axes of the members pass. It is the location that the center of the pin would occupy if the truss were pin-connected.

If the connection between the pair of angles and the gusset plate required only one bolt, the locating dimension *h* in Fig. 1-3c would be made equal to $\bar{y}$, the dimension locating the centroid of the cross-sectional area of the angles. If that value of *h* is used at both ends of the member, the stress throughout most of the member will be *uniform*. When more than one bolt is required, which is the usual case, the location of the bolt row at height *h*, equal to $\bar{y}$, does not necessarily ensure that the stress in the member will be uniform, but in practice the row is so located and the stress is assumed to be uniform.

The fact that some metal is removed in drilling the bolt hole does not weaken the member if it is in compression. However, in the design of tension members the reduced cross-sectional area, known as the *net section*, and shown in Fig. 1-4, must be used in determining the acceptability of the member. The stress on the net section is not uniform, being much larger than the average stress near the hole and smaller than the average stress in the extreme fibers. The amount of non-uniformity can only be determined by very advanced methods of stress analysis. However, for reasons that will be explained later in this Chapter, designers normally do not need to be concerned about this non-uniformity.

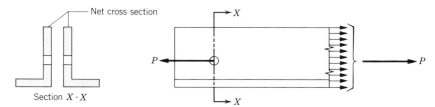

FIG. 1-4

As part of the design of a bolted (or riveted) joint, the diameter and number of bolts must be determined. For a single bolt as shown in Fig. 1-3c the most likely mode of failure, if the bolt is too small, is by sliding along planes *A-B* and *E-F*, that is, by sliding or shearing across two areas of the bolt cross section. This is indicated by saying that the bolt is in *double shear*. It is customary to assume that the *shearing stress*, τ, is uniform on both of these areas. The stress is

$$\tau = \frac{P}{2\frac{\pi}{4}D^2} = \frac{2P}{\pi D^2}$$

If there are N bolts and the *allowable shear stress* is τ_a, the design will be acceptable if the inequality

$$\boxed{\tau_a \geqslant \frac{2P}{N\pi D^2}} \tag{1-4}$$

is satisfied.

The shear stress in the bolt is by no means uniform on the cross section but this is taken into account in establishing the safety factor for parts subjected to shear.

It is important that we be aware of the difference in the nature of the stress in the bolt and the stress in an axially loaded member such as that shown in Fig. 1-1. Under an axial load the force is normal to the plane on which failure might be expected to occur. Consequently, that stress is designated as *normal stress*. In contrast, the bolt is subjected to a force that tends to slide material on one side of a plane across the material on the other side. The force is tangential to the plane, and we might be inclined to use the term *tangential stress*. However, in practice the term *shear stress* is always used. Also it is customary to shorten *normal stress* to *stress*.

When you encounter *shear stress* and *stress*, each term should bring a different image to mind. The stresses are quite different and material reacts differently to them; a particular material may be weak with respect to one kind of stress and strong with respect to the other. Much later we will treat stress more generally and will find that certain relationships exist between them; in the meantime, you should recognize the difference and keep these stresses in their proper places.

Another mode of failure that must be considered in the design of a bolted connection would occur if the *bearing pressure* at the interface between either the bolt and the angle, or the bolt and the gusset plate, is too large. The pressure causes stresses, of the same magnitude as the pressure, in the fibers at and near the surface. If this stress is excessive, failure will occur by crushing of the material in the region of high bearing pressure. The strength of material in crushing is usually very high so that the *allowable bearing stress*, σ_{ab}, is usually larger than the other allowable stresses. The pressure at the bolt-angle interface is given, on rather uncertain theoretical grounds, as

$$p = \frac{P}{2t_a D}$$

where t_a is the thickness of the angle. Hence, for N bolts, a safe design requires that

$$\boxed{\sigma_{ab} \geqslant \frac{P}{2Nt_a D}} \tag{1-5a}$$

is satisfied. If the bolt-gusset interface is to be safe, it is required that

$$\sigma_{ab} \geqslant \frac{P}{Nt_g D}$$

(1-5b)

where t_g is the thickness of the gusset plate. Both (1-5a) and (1-5b) must be satisfied for the design to be acceptable.

Problems 1-7 to 1-11

1-4 DESIGN OF WELDED END CONNECTIONS FOR AXIALLY LOADED MEMBERS

Angles and gusset plates in a truss structure are frequently connected by welding as shown in Fig. 1-5. If the stress is uniformly distributed over the cross sections of the angles, the force P will act at the centroid of the cross-sectional area located by $\bar{y}$. By knowing $\bar{y}$ and the depth of the angle, P can be broken into two statically equivalent loads, one in line with the upper pair of welds and one in line with the lower pair of welds. These loads can be considered to be imposed on their associated welds. For angles placed as in Fig. 1-5, the lower weld will carry the larger load and hence must be stronger. The extra strength is obtained by making the lower weld longer than the more lightly loaded upper weld. The weld metal is subjected to shear and must be designed to carry the load without exceeding the allowable stress on the plane of maximum shear

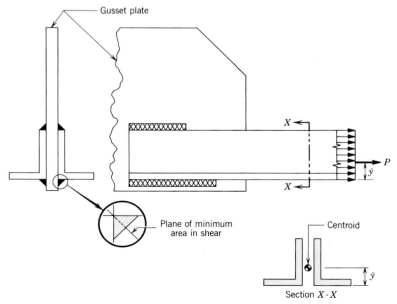

FIG. 1-5 Welded truss connection.

stress. The maximum shear stress will occur on the plane of minimum area. For the type of weld shown in Fig. 1-5, known as a *fillet weld*, the plane of minimum area is at 45° to the surface of the gusset.

Example 1-2

A member, consisting of a pair of $75 \times 75 \times 8$ angles, carries a tensile load of 300 kN. The angles are connected to a gusset plate by 6-mm fillet welds as shown in Fig. 1-6. The allowable shear stress for the weld metal is 110 MPa. Specify the lengths of the welds required for a safe connection.

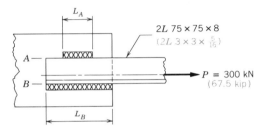

FIG. 1-6 Angles welded to a gusset plate.

Solution.   We will first resolve the load on one angle (300/2 = 150 kN) into statically equivalent loads in line with the welds by writing moment equations according to Fig. 1-7.

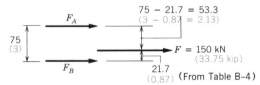

FIG. 1-7

$$F_A \times 75 = 150 \times 21.7$$

$$F_A = \frac{150 \times 21.7}{75} \frac{\text{kN} \cdot \text{mm}}{\text{mm}} = 43.4 \text{ kN}$$

$$F_B \times 75 = 150 \times 53.3$$

$$F_B = \frac{150 \times 53.3}{75} \frac{\text{kN} \cdot \text{mm}}{\text{mm}} = 106.6 \text{ kN}$$

The lateral dimension of the minimum cross section of the welds is

$$6 \times \sin 45° = 4.24 \text{ mm}   \text{(see Fig. 1-5)}$$

The areas in shear are then

$$4.24 \times L_A \qquad \text{for the weld at } A$$

and

$$4.24 \times L_B \qquad \text{for the weld at } B$$

The shearing stress

$$\frac{F_A}{4.24 \times L_A} \qquad \text{at } A$$

and

$$\frac{F_B}{4.24 \times L_B} \qquad \text{at } B$$

must be less than the allowable stress, 110 N/mm^2.
For the weld at A to be safe

$$\frac{F_A}{4.24 \times L_A} \leqslant 110$$

$$4.24 \times L_A \times 110 \geqslant F_A = 43.4 \times 10^3$$

$$L_A \geqslant \frac{43.4 \times 10^3}{4.24 \times 110} \frac{\text{N}}{\text{mm} \cdot \text{N/mm}^2} = 93.0 \text{ mm}$$

Use $L_A = 100$ mm

For the weld at B to be safe

$$\frac{F_B}{4.24 \times L_B} \leqslant 110$$

$$4.24 \times L_B \times 110 \geqslant F_B = 106.6 \times 10^3$$

$$L_B \geqslant \frac{106.6 \times 10^3}{4.24 \times 110} = 228.6 \text{ mm}$$

Use $L_B = 230$ mm

Note: The weld at B is quite long and would require a rather large gusset plate. The weld length could be reduced by specifying a wider fillet weld, but this is limited by the thickness of the angles (8.0 mm). At this stage in a design the engineer would probably reconsider the selection of the angles and perhaps, if all other factors are equal, choose angles having shorter but thicker legs, thus making it possible to use wider and shorter welds and consequently more compact connections. In this case a pair of $L65 \times 65 \times 10$ with 8-mm welds might be considered.

Example 1-2

A member, consisting of a pair of $3 \times 3 \times \frac{5}{16}$ angles, carries a tensile load of 67.5 kip. The angles are connected to a gusset plate by $\frac{1}{4}$-in. fillet welds as shown in Fig. 1-6. The allowable shear stress for the weld metal is 16 ksi. Specify the lengths of the welds required for a safe connection.

Solution.   We will first resolve the load on one angle (67.5/2 = 33.75 kip) into statically equivalent loads in line with the welds by writing moment equations according to Fig. 1-7.

$$F_A \times 3 = 33.75 \times 0.87$$

$$F_A = \frac{33.75 \times 0.87}{3} \frac{\text{kip} \cdot \text{in.}}{\text{in.}} = 9.79 \text{ kip}$$

$$F_B \times 3 = 33.75 \times 2.13$$

$$F_B = \frac{33.75 \times 2.13}{3} \frac{\text{kip} \cdot \text{in.}}{\text{in.}} = 23.96 \text{ kip}$$

The lateral dimension of the minimum cross section of the welds is

$$0.25 \times \sin 45° = 0.177 \text{ in.} \qquad \text{(see Fig. 1-5)}$$

The areas in shear are then

$$0.177 \times L_A \qquad \text{for the weld at } A$$

and

$$0.177 \times L_B \qquad \text{for the weld at } B$$

The shearing stress

$$\frac{F_A}{0.177 \times L_A} \qquad \text{at } A$$

and

$$\frac{F_B}{0.177 \times L_B} \qquad \text{at } B$$

must be less than the allowable stress, 16 ksi.
For the weld at A to be safe

$$\frac{F_A}{0.177 \times L_A} \leqslant 16$$

$$0.177 \times L_A \times 16 \geqslant F_A = 9.79$$

$$L_A \geqslant \frac{9.79}{0.177 \times 16} \frac{\text{kip}}{\text{in.} \cdot \text{kip/in.}^2} = 3.46 \text{ in.}$$

Use L_A = 3.5 in.

For the weld at B to be safe

$$\frac{F_B}{0.177 \times L_B} \leqslant 16$$

$$0.177 \times L_B \times 16 \geqslant F_B = 23.96$$

$$L_B \geq \frac{23.96}{0.177 \times 16} \frac{\text{kip}}{\text{in.} \cdot \text{kip/in.}^2} = 8.46 \text{ in.}$$

Use $L_B = 8.5$ in.

Note: The weld at B is quite long and would require a rather large gusset plate. The weld length could be reduced by specifying a wider fillet weld, but this is limited by the thickness of the angles ($\frac{5}{16}$ in.). At this stage in a design the engineer would probably reconsider the selection of the angles and perhaps, if all other factors are equal, choose angles having shorter but thicker legs thus making it possible to use wider and shorter welds and consequently more compact connections. In this case a pair of $L2\frac{1}{2} \times 2\frac{1}{2} \times \frac{3}{8}$ with $\frac{5}{16}$-in. welds might be considered.

Problems 1-12 to 1-16

1-5 SHEAR AND BENDING MOMENT DIAGRAMS

Many machine parts and structural members support loads that act normal to the axis of the member. Everyone has experience with this class of load. A person standing on the end of a diving board exerts a lateral force on the board and subjects the board to bending. Members that carry lateral loads and, consequently, bending loads are referred to as *beams*. The amount of the bending load is expressed as the *bending moment*.

The bending moment in a beam is analogous to the longitudinal load in an axially loaded member. However, bending moment usually varies from section to section along the beam so that the description of the load is not nearly as simple as it was in the case of the axially loaded member. The bending moment is best presented by means of a bending moment diagram in which the bending moment at any section along the beam is represented by a vector drawn at that section and at right angles to the beam. The envelope formed by the ends of the vectors constitutes a bending moment diagram. A typical bending moment diagram is shown in Fig. 1-14d. We will attach a sign to bending moments according to the following convention, which is illustrated in Fig. 1-8:

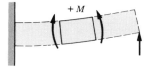

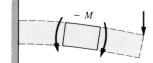

FIG. 1-8 Sign convention for bending moment.

A bending moment that produces compressive stresses in the upper fibers (at this stage intuition derived from experience in bending common items such as pencils, erasers, rulers, etc., must be relied upon to determine where the compressive stresses are located) *is defined as positive*, and *bending moment that produces compressive stresses in the lower fibers of a beam is negative.*

Beginners in this subject frequently try to relate the sign of a bending moment to the sign convention for the moment of a force, that is, to define the sign in terms of clockwise or counterclockwise. Such attempts invariably lead to frustration and a return to the definition given above.

Example 1-3

Draw a bending moment diagram for the beam shown in Fig. 1-9a.

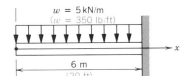

FIG. 1-9a **Cantilever beam.**

FIG. 1-9b **Free-body diagram.**

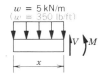

FIG. 1-9c **Bending moment diagram.**

   Solution.   Take a section at a distance x from the left end of the beam and the part to the left of the section as the free body. Applying the internal reactions and external loads gives the free-body diagram shown in Fig. 1-9b.

   Note that the direction of the moment is consistent with the assumption that the bending moment is positive. The magnitude and sign of the moment must be such that the moment equilibrium equation is satisfied. Also note that the free body could not be in equilibrium without V, which is the force tending to cause the material on one side of the section to slide over the material on the other side. This force can be readily recognized as a shear force.

   The moment equilibrium equation

$$\sum M = 0$$

with counterclockwise moments as positive, and moments about any point in the section plane, gives

$$\frac{wx^2}{2} + M = 0$$

hence

$$M = \frac{-wx^2}{2}$$

Plotting this *bending moment equation* gives the bending moment diagram of Fig. 1-9c.

Example 1-3

Draw a bending moment diagram for the beam shown in Fig. 1-9a.

   Solution.   Take a section at a distance x from the left end of the beam and the part to the left of the section as the free body. Applying the internal reactions and external loads gives the free-body diagram shown in Fig. 1-9b.

   Note that the direction of the moment is consistent with the assumption that the bending moment is positive. The magnitude and sign of the moment must be such that the moment equilibrium equation is satisfied. Also note that the free body could not be in equilibrium without V, which is the force tending to cause the material on one side of the section to slide over the material on the other side. This force can be readily recognized as a shear force.

The moment equilibrium equation

$$\sum M = 0$$

with counterclockwise moments as positive, and moments about any point in the section plane, gives

$$\frac{wx^2}{2} + M = 0$$

hence

$$M = \frac{-wx^2}{2}$$

Plotting this *bending moment equation* gives the bending moment diagram of Fig. 1-9c.

Problems 1-17 to 1-19

In Example 1-3 it was found that as well as a bending moment on a typical section, it was necessary to make provision for a shearing force, V. This was perhaps not expected, as we are not often made aware of the existence of shear and have probably never observed failure from excessive shearing force. Nevertheless, failures do occur from shear, as we will see later, and shear must be taken into account in beam design. In designing a beam to withstand shear, the shear force diagram is usually drawn. As in the case of bending moment, a sign convention is required, and again one that is related to the usual force convention is impractical. The standard sign convention for shear, V, is indicated in Fig. 1-10 and can be stated as follows:

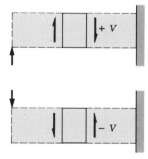

FIG. 1-10 Sign convention for shear force.

A shearing force that tends to slide the portion of a beam to the left of the section upward relative to the portion to the right, is positive.

A shearing force that tends to slide the portion of a beam to the left of the section downward relative to the portion to the right, is negative.

The definitions of shear and bending moment require reference to direction given by "right" and "upward" which apply when the member is horizontal. For vertical or inclined members these directions are often meaningless. To resolve this difficulty, consider a set of rectangular coordinate axes that are rotated in either direction until the x-axis is parallel to the member. "Right" is then taken as being in the direction of the positive x-axis and "upward" in the direction of the positive y-axis.

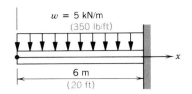

FIG. 1-11*a* Cantilever beam.

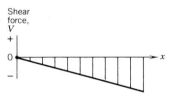

FIG. 1-11*b* Free-body diagram.

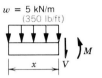

FIG. 1-11*c* Shear diagram.

Example 1-4

Draw a shear force diagram for the beam loaded as shown in Fig. 1-11*a*.

Solution. Taking a section at a distance *x* from the left end, we obtain the free body of Fig. 1-11*b*.

Note that the direction of *V* is consistent with the assumption of positive shear on the cross section.

Using the equilibrium equation

$$\sum F_y = 0$$

gives

$$-wx - V = 0$$

hence the *shear equation*, $V = -wx$, which is represented graphically in Fig. 1-11*c*.

Example 1-4

Draw a shear force diagram for the beam loaded as shown in Fig. 1-11*a*.

Solution. Taking a section at a distance *x* from the left end, we obtain the free body of Fig. 1-11*b*.

Note that the direction of *V* is consistent with the assumption of positive shear on the cross section.

Using the equilibrium equation

$$\sum F_y = 0$$

gives

$$-wx - V = 0$$

hence the *shear equation*, $V = -wx$, which is represented graphically in Fig. 1-11*c*.

Problems 1-20 to 1-23

Although diagrams showing shear and bending moment are normally used for the design of beams, occasionally the same information is required in the form of shear and bending moment equations. The equations can be derived by taking a typical section between loads and satisfying the equilibrium conditions. An equation derived in this way is applicable only between concentrated loads; hence several equations will be required for a beam carrying a number of loads. Care must be taken to guard against using an equation outside its range of applicability. The following example illustrates the derivation of shear and bending moment equations.

Example 1-5a

For the beam shown in Fig. 1-12a, write the shear and bending moment equations for the entire beam and draw the shear and bending moment diagrams.

Solution. From the free-body diagram for the beam given in Fig. 1-12b, the reactions R_L and R_R are found as follows:

$$\sum M_L = 0 \qquad -\left(10 \times 2 \times \frac{2}{2}\right) + 3R_R - 4.5 \times 5 = 0$$

$$R_R = \frac{4.5 \times 5 + 10 \times 2 \times 1}{3}$$

$$= 14.17 \text{ kN}$$

$$\sum F_y = 0 \qquad -10 \times 2 + R_L + R_R - 5 = 0$$

$$R_L = 25 - R_R = 10.83 \text{ kN}$$

Since there are two discontinuities in the system of loads, there are three regions for which separate shear and bending moment equations must be found.

Region 1 $\quad 0 \leqslant x \leqslant 2.$ Using as a free body a portion of the beam within the uniformly loaded region and assuming positive shear and bending moment at the cut section as shown in Fig. 1-12c, we find by using the equilibrium equations

$$\sum F_y = 0$$

$$10.83 - 10x - V = 0$$

shear equation: $V = 10.83 - 10x$

$$\sum M = 0$$

$$-10.83x + 10(x)\frac{x}{2} + M = 0$$

moment equation; $M = \dfrac{-10x^2}{2} + 10.83x$

$$= -5x^2 + 10.83x$$

Region 2 $\quad 2 < x < 3.$ Equilibrium equations for the free body in Fig. 1-12d give:

$$\sum F_y = 0$$

$$10.83 - 10 \times 2 - V = 0$$

shear equation; $V = -20 + 10.83 = -9.17 \text{ kN}$

$$\sum M = 0$$

$$-10.83x + 10 \times 2\left(x - \frac{2}{2}\right) + M = 0$$

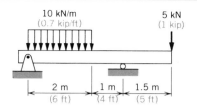

FIG. 1-12a

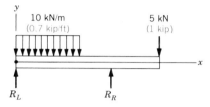

FIG. 1-12b

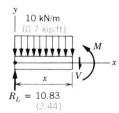

FIG. 1-12c

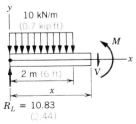

FIG. 1-12d

moment equation; $M = -20(x-1) + 10.83x$

$$= -20x + 20 + 10.83x$$

$$= -9.17x + 20$$

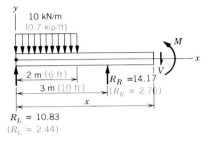

FIG. 1-12e

Region 3 $3 < x < 4.5$. Equilibrium equations for the free body in Fig. 1-12e give:

$$\sum F_y = 0$$

$$10.83 - 10 \times 2 + 14.17 - V = 0$$

shear equation; $V = +5$

$$\sum M = 0$$

$$-10.83x + 10(2)(x-1) - 14.17(x-3) + M = 0$$

$$-10.83x + 20x - 20 - 14.17x + 42.51 + M = 0$$

moment equation; $M = 5x - 22.51$

By using these shear and bending moment equations, the shear force and bending moment diagrams can now be drawn as shown in Fig. 1-12f.

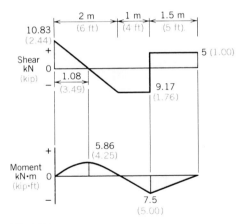

FIG. 1-12f

Example 1-5a

For the beam shown in Fig. 1-12a, write the shear and bending moment equations for the entire beam and draw the shear and bending moment diagrams.

Solution.   From the free-body diagram for the beam given in Fig. 1-12*b*, the reactions R_L and R_R are found as follows:

$$\sum M_L = 0 \qquad -\left(0.7 \times 6 \times \frac{6}{2}\right) + 10R_R - 15 \times 1 = 0$$

$$R_R = \frac{15 \times 1 + 0.7 \times 6 \times 3}{10}$$

$$= 2.76 \text{ kip}$$

$$\sum F_y = 0 \qquad -0.7 \times 6 + R_L + R_R - 1 = 0$$

$$R_L = 5.2 - R_R = 2.44 \text{ kip}$$

Since there are two discontinuities in the system of loads, there are three regions for which separate shear and bending moment equations must be found.

Region 1 $0 \leqslant x \leqslant 6$.   Using as a free body a portion of the beam within the uniformly loaded region and assuming positive shear and bending moment at the cut section as shown in Fig. 1-12*c*, we find by using the equilibrium equations

$$\sum F_y = 0$$

$$2.44 - 0.7x - V = 0$$

$$\text{shear equation: } V = 2.44 - 0.7x$$

$$\sum M = 0$$

$$-2.44x + 0.7(x)\frac{x}{2} + M = 0$$

$$\text{moment equation; } M = -\frac{0.7x^2}{2} + 2.44x$$

$$= -0.35x^2 + 2.44x$$

Region 2 $6 < x < 10$.   Equilibrium equations for the free body in Fig. 1-12*d* give:

$$\sum F_y = 0$$

$$2.44 - 0.7 \times 6 - V = 0$$

$$\text{shear equation; } V = -4.2 + 2.44 = -1.76 \text{ kip}$$

$$\sum M = 0$$

$$-2.44x + 0.7 \times 6\left(x - \frac{6}{2}\right) + M = 0$$

$$\text{moment equation; } M = -4.2(x - 3) + 2.44x$$

$$= -4.2x + 12.6 + 2.44x$$

$$= -1.76x + 12.6$$

Region 3 $10 < x < 15$. Equilibrium equations for the free body in Fig. 1-12e give:

$$\sum F_y = 0$$

$$2.44 - 0.7 \times 6 + 2.76 - V = 0$$

$$\text{shear equation; } V = 1$$

$$\sum M = 0$$

$$-2.44x + 0.7(6)(x-3) - 2.76(x-10) + M = 0$$

$$-2.44x + 4.2x - 12.6 - 2.76x + 27.6 + M = 0$$

$$\text{moment equation; } M = 1x - 15$$

Problems 1-24 to 1-30

When drawing shear and bending moment diagrams, great care must be taken to avoid mistakes that, like any error in engineering, can be disastrous. After gaining a substantial body of experience in this work, it becomes repetitive and most designers look for a simplified routine with fewer chances of making mistakes. Fortunately, the determination of shear on any section can be reduced to the simple algorithm:

The shear on any section can be determined (with the correct sign) *by summing the upward forces to the left of the section* (downward forces are included but with a negative sign).

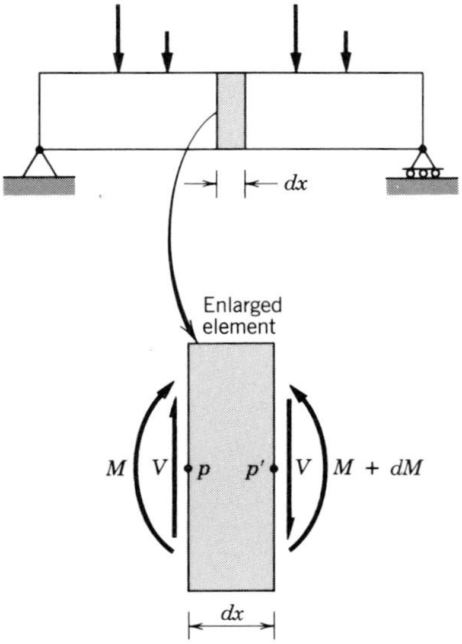

It follows from this that the shear remains constant in regions of the beam where there is no load, and that there is a step in the diagram at each concentrated load. The step has the direction and magnitude of the load.

Using these observations will greatly facilitate the construction of shear diagrams. We will now develop an equally useful principle for bending moments.

Consider an element chosen from any beam as shown in Fig. 1-13. All forces acting on this free body are shown. Note that provision is made for a change in moment between the two sections whereas the shear force is known to be the same on both vertical planes because no external vertical load acts on the element.

Taking moments about any convenient point, such as p or p', leads to the equilibrium equation:

$$-M - V\,dx + M + dM = 0$$

hence

FIG. 1-13 Shear and bending on a typical element.

$$dM = V\,dx \qquad \textbf{(1-6a)}$$

and

$$\boxed{\frac{dM}{dx} = V}$$ **(1-6b)**

Equation 1-6b may be stated:

The slope of the bending moment diagram at any section is equal to the value of the shear at that section.

This observation is occasionally useful, but a far more useful statement of the same basic principle can be obtained through integrating (1-6a). Consider a section of the beam at $x = x_1$, where the bending moment is M_1, and another section at any distance to the right, where $x = x_2$ and the bending moment is M_2. By integrating (1-6a) between corresponding limits, we get

$$\int_{M_1}^{M_2} dM = \int_{x_1}^{x_2} V\, dx$$

$$\boxed{M_2 - M_1 = \int_{x_1}^{x_2} V\, dx}$$ **(1-6c)**

The integral in (1-6c) is the area under the shear diagram between sections 1 and 2. Equation 1-6c can be stated:

The change in the bending moment in going from any section to a section further to the right is equal to the area under the shear diagram between the sections.

Using this statement, and the previously given one pertaining to the shear force, reduces the drawing of shear and bending moment diagrams to a routine.

Such routines should not be used until a thorough understanding of the topic has been acquired by working exercises from first principles. When an adequate level of understanding has been reached, the routine procedure is to be preferred because it allows the designer to concentrate on producing mistake-free results.

Example 1-5b

Draw shear and bending moment diagrams for the beam given in Example 1-5a.

Solution. This solution is not typical in that all steps will be presented. After some experience has been gained, you will do most of the calculations mentally or entirely within your calculator. When you have reached that stage, the diagrams can be drawn directly without any intermediate work appearing on paper.

Since the distributed load leads to a curved bending moment diagram, we will evaluate at stations located 0.5 m apart in the region of the distributed load and draw a smooth curve between the stations as shown in Fig. 1-14.

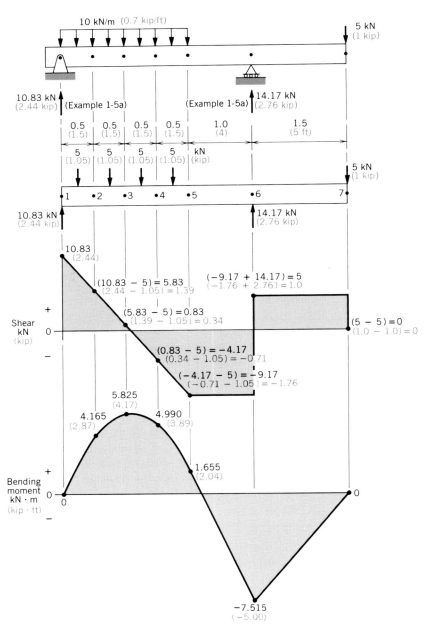

FIG. 1-14

$$M_1 = 0 \qquad \text{(pivoted end)}$$

From (1-6c),

$$M_2 = M_1 + \frac{10.83 + 5.83}{2} \times 0.5 = 4.165 \text{ kN} \cdot \text{m}$$

$$M_3 = M_2 + \frac{5.83 + 0.83}{2} \times 0.5 = 5.825 \text{ kN} \cdot \text{m}$$

$$M_4 = M_3 + \frac{0.83 - 4.17}{2} \times 0.5 = 4.99 \text{ kN} \cdot \text{m}$$

$$M_5 = M_4 + \frac{-4.17 - 9.17}{2} \times 0.5 = 1.655 \text{ kN} \cdot \text{m}$$

$$M_6 = M_5 + (-9.17 \times 1.0) = -7.515 \text{ kN} \cdot \text{m}$$

$$M_7 = M_6 + (5.0 \times 1.5) = -0.015 \text{ kN} \cdot \text{m} \qquad \text{(round-off error)}$$

Example 1-5b

Draw shear and bending moment diagrams for the beam given in Example 1-5a.

Solution.   This solution is not typical in that all steps will be presented. After some experience has been gained, you will do most of the calculations mentally or entirely within your calculator. When you have reached that stage, the diagrams can be drawn directly without any intermediate work appearing on paper.

Since the distributed load leads to a curved bending moment diagram, we will evaluate at stations located 1.625 ft apart in the region of the distributed load and draw a smooth curve between the stations as shown in Fig. 1-14.

$$M_1 = 0 \qquad \text{(pivoted end)}$$

From (1-6c),

$$M_2 = M_1 + \frac{2.44 + 1.39}{2} \times 1.5 = 2.87 \text{ kip} \cdot \text{ft}$$

$$M_3 = M_2 + \frac{1.39 + 0.34}{2} \times 1.5 = 4.17 \text{ kip} \cdot \text{ft}$$

$$M_4 = M_3 + \frac{0.34 - 0.71}{2} \times 1.5 = 3.89 \text{ kip} \cdot \text{ft}$$

$$M_5 = M_4 + \frac{-0.71 - 1.76}{2} \times 1.5 = 2.04 \text{ kip} \cdot \text{ft}$$

$$M_6 = M_5 + (-1.76 \times 4) = -5.00 \text{ kip} \cdot \text{ft}$$

$$M_7 = M_6 + (1.0 \times 5) = 0 \text{ kip} \cdot \text{ft}$$

Problems 1-31 to 1-41

1-6 BENDING STRESSES IN A BEAM — APPROXIMATE METHOD

The following example will lead to some important conclusions. You are advised to attempt to solve the problems presented in the example before examining the given solution because successful completion of the problems will lead to discoveries that will be much more meaningful than if the same conclusions are found in the given solution.

Example 1-6a

Draw shear and bending moment diagrams for a simply supported beam subjected to four point loads as shown in Fig. 1-15a.

Solution. From the free-body diagram in Fig. 1-15b we can solve for the end reactions.

$$\sum M_B = 0$$

$$- R_L \times 40 + 350 \times 32 + 350 \times 24 + 350 \times 16 + 450 \times 8 = 0$$

$$R_L = 720 \text{ kN}$$

$$\sum F_y = 0$$

$$720 - 350 - 350 - 350 - 450 + R_R = 0$$

$$R_R = 780 \text{ kN}$$

350 kN
(80 kip)

350 kN
(80 kip)

350 kN
(80 kip)

450 kN
(100 kip)

5 @ 8m = 40 m (5 @ 26 ft = 130 ft)

FIG. 1-15a

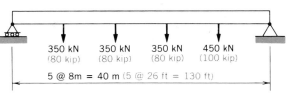

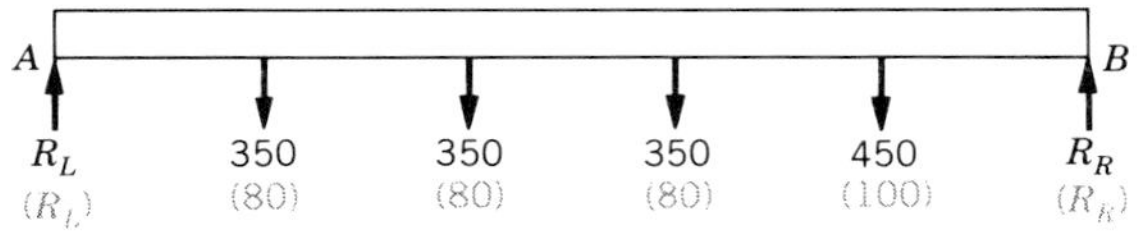

A B

R_L 350   350   350   450 R_R
(R_L) (80)   (80)   (80)   (100) (R_R)

FIG. 1-15b

We now readily obtain the shear force and bending moment diagrams depicted in Figs. 1-15c and d. This requires repeatedly summing forces to the left of the section to get the shear values and then using areas on the shear diagram to get bending moments. The details of these calculations are not shown.

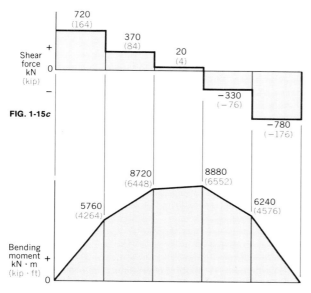

FIG. 1-15c

FIG. 1-15d

Example 1-6a

Draw shear and bending moment diagrams for a simply supported beam subjected to four point loads as shown in Fig. 1-15a.

Solution. From the free-body diagram in Fig. 1-15b we can solve for the end reactions.

$$\sum M_B = 0$$

$$-R_L \times 130 + 80 \times 104 + 80 \times 78 + 80 \times 52 + 100 \times 26 = 0$$

$$R_L = 164 \text{ kip}$$

$$\sum F_y = 0$$

$$164 - 80 - 80 - 80 - 100 + R_R = 0$$

$$R_R = 176 \text{ kip}$$

We now readily obtain the shear force and bending moment diagrams depicted in Figs. 1-15c and d. This requires repeatedly summing forces to the left of the section to get the shear values and then using areas on the shear diagram to get bending moments. The details of these calculations are not shown.

Example 1-6b

Determine the forces in the members of the truss given in Fig. 1-16a. Take advantage of the diagrams in Example 1-6a to reduce the work to a minimum.

How would you describe the function of the chords (i.e., the upper and lower horizontal members) and of the diagonals in terms of the loads (i.e., the shearing load and the bending load)?

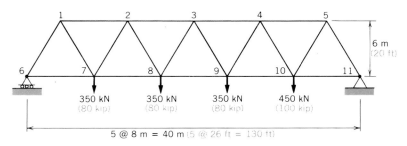

FIG. 1-16a

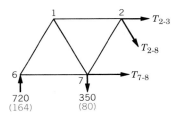

FIG. 1-16b

Solution.   We will start by using the method of sections to determine the force in 2-8, a typical diagonal member. Passing a section plane to cut 2-8 gives the free body of Fig. 1-16b.

From the equilibrium equation $\sum F_y = 0$

$$720 - 350 - T_{2\text{-}8 \text{ vertical}} = 0$$

or

$$T_{2\text{-}8 \text{ vertical}} = 720 - 350$$

The right side of the equation is easy enough to evaluate but, before doing this, note that it is the sum of the upward forces to the left of the section plane which, by the rule given earlier, is just the shear. This can be taken from the shear diagram at any point in the region 12 to 16 m from the left end. The advantage of using the shear diagram rather than subtracting 350 from 720 is dubious, but once we have noted the principle involved we can quickly evaluate the forces in the other diagonal members. Using the dimensions of the structure, we can relate the vertical component of force in the member to the total force as indicated, for instance, for member 2-8 in Fig. 1-16c.

$$\frac{T_{2\text{-}8 \text{ vertical}}}{T_{2\text{-}8}} = \frac{6m}{7.21\ m}$$

$$T_{2\text{-}8} = \frac{7.21}{6} T_{2\text{-}8 \text{ vertical}}$$

$$= 1.202\, T_{2\text{-}8 \text{ vertical}}$$

$$= 1.202 \times (\text{shear in region 2-8})$$

$$= 1.202 \times 370 = 445\ \text{kN}$$

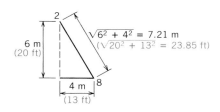

FIG. 1-16c

Since all diagonal members in this truss have the same slope, we can write for any general diagonal member *M-N*

$$T_{M-N} = \pm 1.202 \times (\text{shear in region } M\text{-}N)$$

and from this calculate with ease the forces in the members. This will give the correct values for forces but not necessarily the correct sense of the force. We would like to have a plus sign associated with force if it is tensile and a minus sign if it is compressive. The recommended method for determining the sign is by imagining that the member in question is removed and then, by looking at the direction of the shear loads on the region of the member, determining whether the points, that were formerly connected by the member, move toward one another or move apart. If they tend to separate, the member is in tension; if they tend to come together, the member is in compression.

Using member 2-8 to illustrate, we will consider the region of the structure consisting of 2-8 and adjacent members as shown in Fig. 1-16d. Since the shear diagram is positive, the shear loads will be in the directions shown. Under these loads, if member 2-8 were absent, the pin-jointed assembly would be unable to carry the loads and would begin to deform to a shape such as that indicated by the broken lines in Fig. 1-16e. Pins 2 and 8 are obviously moving away from one another as the structure collapses. The member 2-8, which is necessary to prevent the collapse, must then keep pins 2 and 8 from separating and in so doing is subjected to a tensile load.

The previous discussion pertains to diagonal members and the shear that they carry. We now look for a similar method for determining the forces in the chords. Let us first try to find the force in member 7-8 by the conventional procedure. The section and free body of Fig. 1-16b will be employed again.

The best technique, since $T_{2\text{-}3}$ and $T_{2\text{-}8}$ are not now being evaluated, is to write the equilibrium equation for moments about panel point 2. That is,

$$\sum M_2 = 0$$

$$T_{7\text{-}8} \times 6 - 720 \times 12 + 350 \times 4 = 0$$

$$T_{7\text{-}8} \times 6 = 720 \times 12 - 350 \times 4$$

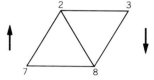

FIG. 1-16d

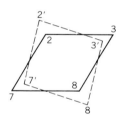

FIG. 1-16e

An inspection of the right side of this equation reveals that it is the bending moment at a point 12 m from the left end, which is the location of panel point 2. Once this is recognized, the right-hand side can be determined from the bending moment diagram, in this case by interpolating between 5760 and 8720 to get 7240 kN·m. The equilibrium equation then gives

$$T_{7\text{-}8} = \frac{7240}{6} = 1207 \text{ kN}$$

A more intuitive approach leads to the same results, is much less time-consuming, and tends to give a better understanding of the function of the chord members. This approach requires that we visualize the effect of removing the chord member in which the force is to be determined. Considering the structure of Fig. 1-16a with member 7-8 removed, we arrive at Fig. 1-16f. It is quite evident that the left portion of the structure and the right portion, which are now connected only by a pin at 2, will collapse, with the portions rotating relative to one another about pin 2. If the pin were imagined to be badly rusted, and hence not able to act as a hinge, it would be subjected to the bending moment that is found in the bending moment diagram directly below point 2; in this case 7240 kN·m.

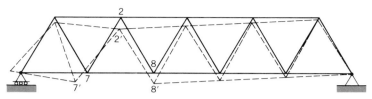

FIG. 1-16f

However, in practice, rotation is not prevented by the pin but rather by force exerted by member 7-8. The perpendicular distance from the line of action of the force to the center of the pin is 6 m. Consequently, to resist the moment of 7240 kN·m a force of $\frac{7240}{6} = 1207$ kN must be exerted by the member 7-8.

To determine the sense of the force, we inspect the motion of the end points when the member is removed. In this case, when member 7-8 is removed, the positive bending moment would cause displacement that would move 7 and 8 apart. The member must prevent this motion by holding the points together; it is thus in tension.

Once this point of view is understood, the calculation of the chord member forces is simple. Let us consider one more member, 2-3, to confirm the method. If member 2-3 were removed, the structure would collapse by the two ends rotating about panel point 8 with a moment of 8720 kN·m. To prevent the collapse, a force on 2 and 3 of $\frac{8720}{6} = 1453$ kN is required. In the imaginary collapse the points 2 and 3 would approach one another. The member prevents collapse by holding the points apart. Therefore, member 2-3 carries a compressive load of 1453 kN.

By imagining all members of the chords removed successively, the forces can be readily calculated and the results indicated in Fig. 1-16g can be obtained.

It will be noted in the foregoing that only shear was used in determining the forces in the diagonals and only bending moment was used in determining the forces in the chords. It might, therefore, be concluded that *the diagonals carry the shear load and the chords carry the bending load.*

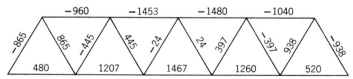

FIG. 1-16g Member forces (kN).

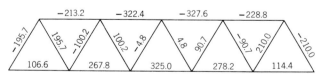

FIG. 1-16g Member forces (kip).

Example 1-6b

Determine the forces in the members of the truss given in Fig. 1-16a. Take advantage of the diagrams in Example 1-6a to reduce the work to a minimum.

How would you describe the function of the chords (i.e., the upper and lower horizontal members) and of the diagonals in terms of the loads (i.e., the shearing load and the bending load)?

Solution.   We will start by using the method of sections to determine the force in 2-8, a typical diagonal member. Passing a section plane to cut 2-8 gives the free body of Fig. 1-16*b*.

From the equilibrium equation $\sum F_y = 0$

$$164 - 80 - T_{\text{2-8 vertical}} = 0$$

or

$$T_{\text{2-8 vertical}} = 164 - 80$$

The right side of the equation is easy enough to evaluate but, before doing this, note that it is the sum of the upward forces to the left of the section plane which, by the rule given earlier, is just the shear. This can be taken from the shear diagram at any point in the region 39 to 52 ft from the left end. The advantage of using the shear diagram rather than subtracting 80 from 164 is dubious, but once we have noted the principle involved we can quickly evaluate the forces in the other diagonal members. Using the dimensions of the structure, we can relate the vertical component of force in the member to the total force as indicated, for instance, for member 2-8 in Fig. 1-16*c*.

$$\frac{T_{\text{2-8 vertical}}}{T_{\text{2-8}}} = \frac{20 \text{ ft}}{23.85 \text{ ft}}$$

$$T_{\text{2-8}} = \frac{23.85}{20} T_{\text{2-8 vertical}}$$

$$= 1.193 \, T_{\text{2-8 vertical}}$$

$$= 1.193 \times (\text{shear in region 2-8})$$

$$= 1.193 \times 84 = 100.2 \text{ kip}$$

Since all diagonal members in this truss have the same slope, we can write for any general diagonal member *M-N*

$$T_{M\text{-}N} = \pm 1.193 \times (\text{shear in region } M\text{-}N)$$

and from this calculate with ease the forces in the members. This will give the correct values for forces but not necessarily the correct sense of the force. We would like to have a plus sign associated with the force if it is tensile and a minus sign if it is compressive. The recommended method for determining the sign is by imagining that the member in question is removed and then, by looking at the direction of the shear loads on the region of the member, determining whether the points, that were formerly connected by the member, move toward one another or move apart. If they tend to separate, the member is in tension; if they tend to come together, the member is in compression.

Using member 2-8 to illustrate, we will consider the region of the structure consisting of 2-8 and adjacent members as shown in Fig. 1-16*d*. Since the shear diagram is positive, the shear loads will be in the directions shown. Under these loads, if member 2-8 were absent, the pin-jointed assembly would be unable to carry the loads and would begin to deform to a shape such as that indicated by the broken lines in Fig. 1-16*e*. Pins 2 and 8 are obviously moving away from one-another as the structure collapses. The member 2-8, which is necessary to prevent the collapse, must then keep pins 2 and 8 from separating and in so doing is subjected to a tensile load.

The previous discussion pertains to diagonal members and the shear that they carry. We now look for a similar method for determining the forces in the chords. Let us first try to find the force in member 7-8 by the conventional procedure. The section and free body of Fig. 1-16b will be employed again.

The best technique, since T_{2-3} and T_{2-8} are not now being evaluated, is to write the equilibrium equation for moments about panel point 2. That is,

$$\sum M_2 = 0$$

$$T_{7-8} \times 20 - 164 \times 39 + 80 \times 13 = 0$$

$$T_{7-8} \times 20 = 164 \times 39 - 80 \times 13$$

An inspection of the right side of this equation reveals that it is the bending moment at a point 39 ft from the left end, which is the location of panel point 2. Once this is recognized, the right-hand side can be determined from the bending moment diagram, in this case by interpolating between 4264 and 6448 to get 5356 kip·ft. The equilibrium equation then gives

$$T_{7-8} = \frac{5356}{20} = 267.8 \text{ kip}$$

A more intuitive approach leads to the same results, is much less time-consuming, and tends to give a better understanding of the function of the chord members. This approach requires that we visualize the effect of removing the chord member in which the force is to be determined. Considering the structure of Fig. 1-16a with member 7-8 removed, we arrive at Fig. 1-16f. It is quite evident that the left portion of the structure and the right portion, which are now connected only by a pin at 2, will collapse, with the portions rotating relative to one-another about pin 2. If the pin were imagined to be badly rusted, and hence not able to act as a hinge, it would be subjected to the bending moment that is found in the bending moment diagram directly below point 2; in this case 5356 kip·ft.

However, in practice, rotation is not inhibited by the pin but rather by force exerted by member 7-8. The perpendicular distance from the line of action of the force to the center of the pin is 20 ft. Consequently, to resist the moment of 5356 kip·ft a force of $\frac{5356}{20} = 267.8$ kip must be exerted by the member 7-8.

To determine the sense of the force, we inspect the motion of the end points when the member is removed. In this case, when member 7-8 is removed, the positive bending moment would cause displacement that would move 7 and 8 apart. The member must prevent this motion by holding the points together; it is thus in tension.

Once this point of view is understood, the calculation of the chord member forces is simple. Let us consider one more member, 2-3, to confirm the method. If member 2-3 were removed, the structure would collapse by the two ends rotating about panel point 8 with a moment of 6448 kip·ft. To prevent the collapse, a force on 2 and 3 of $\frac{6448}{20} = 322.4$ kip is required. In the imaginary collapse the points 2 and 3 would approach one another. The member prevents collapse by holding the points apart. Therefore, member 2-3 carries a compressive load of 322.4 kip.

By imagining all members of the chords removed successively, the forces can be readily calculated and the results indicated in Fig. 1-16g can be obtained.

It will be noted in the foregoing that only shear was used in determining the forces in the diagonals and only bending moment was used in determining the forces in the chords. It might, therefore, be concluded that *the diagonals carry the shear load and the chords carry the bending load.*

Problems 1-42 to 1-44

Example 1-7

Select pairs of equal leg angles from Table B-4 to safely carry the loads on the members of the truss in Example 1-6b assuming:

1. The allowable stress in tension is 140 MPa.
2. The cross section of tensile members is not reduced by bolt holes.
3. The allowable stress in compression is 80 MPa.

Solution. Dividing the force in each member by the appropriate allowable stress gives the required minimum areas. These areas are given in Fig. 1-17*a*. Standard-size angles, chosen from Table B-4, such that a pair provides at least the required area, are indicated in Fig. 1-17*b*.

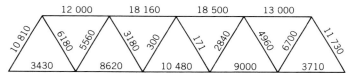

FIG. 1-17a Required minimum areas (mm²).

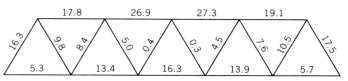

FIG. 1-17a Required minimum areas (in.²).

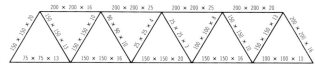

FIG. 1-17b Standard size pairs of angles (mm).

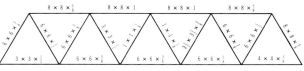

FIG. 1-17b Standard size pairs of angles (in.).

In practice it might be considered that the structure contains too many different angle sizes. Reducing the number of sizes would lead to stresses in some members being considerably below the specified allowable stresses.

Example 1-7

Select pairs of equal leg angles from Table B-4 to safely carry the loads on the members of the truss in Example 1-6b assuming:

1. The allowable stress in tension is 20 ksi.
2. The cross section of tensile members is not reduced by bolt holes.
3. The allowable stress in compression is 12 ksi.

Solution. Dividing the force in each member by the appropriate allowable stress gives the required minimum areas. These areas are given in Fig. 1-17a. Standard-size angles, chosen from Table B-4, such that a pair provides at least the required area, are indicated in Fig. 1-17b.

In practice it might be considered that the structure contains too many different angle sizes. Reducing the number of sizes would lead to stresses in some members being considerably below the specified allowable stresses.

Problems 1-45 to 1-47

Recognizing that the bending load on a truss is carried by the upper and lower chords, and that the shear load is carried by the members that connect the chords, provides us with a very simple and reliable means for analyzing forces and stresses in truss members. The standard W-shape or S-shape listed in Tables B-1 and B-2 can be looked upon as being similar to a truss; there is a concentration of material in the upper and lower flanges, which seem analogous to chords; and the flanges are connected by a web, which seems analogous to the members connecting the chords in truss. There are obvious differences, but if these are ignored and the conclusions reached for the truss are applied to the W-shape or S-shape, we would say that *in a W-shape or S-shape the bending load is carried by the flanges and the shear load by the web*. This conclusion was reached by the application of a principle, which is valid for a truss, but only approximate for a W-shape or S-shape. In most cases it gives results that although not exact, are reasonably accurate; therefore, it is a useful but *approximate* method. The following example illustrates the determination of stresses in an S-shape by this method.

Example 1-8

For the beam consisting of an S-shape loaded as shown in Fig. 1-18a, determine by the approximate method the maximum tensile stress in the flange and the maximum shear stress in the web.

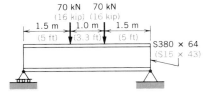

FIG. 1-18a

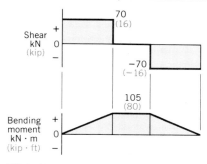

FIG. 1-18b

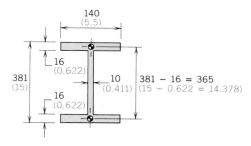

FIG. 1-18c

Solution.   We first determine the shear force and bending moment diagrams as shown in Fig. 1-18b in order to see where the maximum loads occur. The dimensions of the cross section given in Fig. 1-18c were taken from Table B-2. They permit us to calculate the flange and web areas as follows:

$$\text{flange area, } A_F = 140 \times 16 = 2240 \text{ mm}^2$$

$$\text{web area, } A_W = (381 - 2 \times 16) \times 10 = 3490 \text{ mm}^2$$

In the region of maximum bending moment:

$$\text{force in flange} = \frac{105}{365 \times 10^{-3}} \frac{\text{kN} \cdot \text{m}}{\text{m}} = 288 \text{ kN} = 288\,000 \text{ N}$$

$$\text{stress in flange} = \frac{\text{force in flange}}{A_F} = \frac{288\,000}{2240} \frac{\text{N}}{\text{mm}^2} = \textbf{128 MPa}$$

In regions of maximum shear:

$$\text{shearing force} = 70 \text{ kN} = 70\,000 \text{ N}$$

$$\text{shear stress in web} = \frac{\text{shear force}}{A_W} = \frac{70\,000}{3490} \frac{\text{N}}{\text{mm}^2} = \textbf{20.1 MPa}$$

Later, when exact methods have been developed we will be able to check the accuracy of these approximate stresses. The values in the example will be found to be in error by 20% on the safe side for the bending stress and 5% on the unsafe side for the shearing stress. The error is dependent upon the shape of the cross section, but for most sections the results are accurate enough for a *first approximation*. The calculations by this method are much simpler than those of the exact method; hence engineers frequently use the approximate method in the preliminary stage of design when precise answers are not required.

Example 1-8

For the beam consisting of an S-shape loaded as shown in Fig. 1-18a, determine by the approximate method the maximum tensile stress in the flange and the maximum shear stress in the web.

Solution. We first determine the shear force and bending moment diagrams as shown in Fig. 1-18b in order to see where the maximum loads occur. The dimensions of the cross section given in Fig. 1-18c were taken from Table B-2. They permit us to calculate the flange and web areas as follows:

$$\text{flange area, } A_F = 5.5 \times 0.622 = 3.42 \text{ in.}^2$$

$$\text{web area, } A_W = (15 - 2 \times 0.622) \times 0.411 = 5.65 \text{ in.}^2$$

In the region of maximum bending moment:

$$\text{force in flange} = \frac{80}{14.378/12} \frac{\text{kip} \cdot \text{ft}}{\text{ft}} = 66.77 \text{ kip}$$

$$\text{stress in flange} = \frac{\text{force in flange}}{A_F} = \frac{66.77}{3.42} \frac{\text{kip}}{\text{in.}^2} = \textbf{19.52 ksi}$$

In regions of maximum shear:

$$\text{shearing force} = 16 \text{ kip}$$

$$\text{shear stress in web} = \frac{\text{shear force}}{A_W} = \frac{16}{5.65} \frac{\text{kip}}{\text{in.}^2} = \textbf{2.83 ksi}$$

Later when exact methods have been developed, we will be able to check the accuracy of these approximate stresses. The values in the example will be found to be in error by about 20% on the safe side for the bending stress and about 5% on the unsafe side for the shearing stress. The error is dependent upon the shape of the cross section, but for most sections the results are accurate enough for a *first approximation.* The calculations by this method are much simpler than those of the exact method; hence engineers frequently use the approximate method in the preliminary stage of design when precise answers are not required.

Problems 1-48 to 1-53

1-7 STRESSES DUE TO PRESSURIZED FLUID IN A THIN-WALLED VESSEL

We now consider the stresses in a simple pressure vessel where fluid under pressure acts on the inner surface of the vessel wall. The solution to this problem for a vessel that has a perfectly general shape is too complex to be treated at this stage. Only the simplest of shapes will be treated here, that is, the thin-walled circular cylinder, such as the central region of the vessel shown in Fig. 1-19a.

Intuition tells us that the fluid pressing on the inner surface will tend to make the vessel expand in all directions. The tendency for the diameter, and hence the circumference, to increase will stretch the fibers in the circumferential direction and cause a *circumferential stress* or

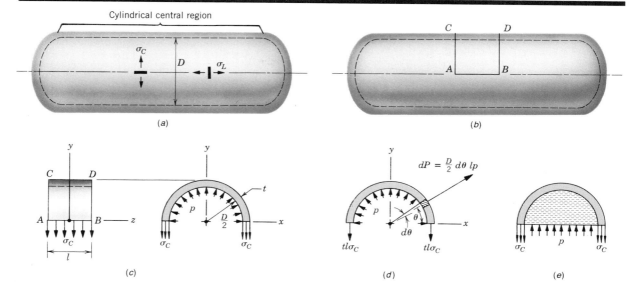

FIG. 1-19 Stresses in a cylindrical pressure vessel.

hoop stress, σ_c. Fluid acting on the end caps will tend to push the ends apart and will put longitudinal stress, σ_L, into the fibers of the cylindrical portion of the vessel. Relationships between pressure, p, the dimensions of the vessel, and stress will now be established partly because the stress formulas are of engineering interest, but also to develop some principles that have wider application.

Repeating the technique that we have used to determine other stresses, we will section the vessel by a plane normal to the stress that is required. In this case σ_c will be determined by using the cutting plane A-B in Fig. 1-19b. Further cutting must be done to isolate a free body. Many planes could be tried; some would not lead to a solution while others would be successful. The planes A-C and B-D will lead to a solution by fairly straightforward processes.

The free body will be taken as the part of the vessel between these planes with loads imposed on it by the fluid and the stress on the cut sections as shown in Fig. 1-19c. Certain assumptions must be made with respect to the stresses:

1. The stress does not vary from point to point around the circumference.
2. The stress does not vary from point to point along the length of the cylinder.
3. The stress does not vary from point to point through the thickness of the wall.

Assumption 1 is reasonable in the light of the axial symmetry of the system. Assumption 2 follows from an assumption that the action in all transverse planes is identical, provided that the planes are not near the end caps; hence the circumferential stress should not vary from point to point on A-B. We shall accept assumption 3 without proof. Studies, by more advanced methods, show that the stress varies through the wall thickness but becomes nearly constant as the ratio of wall thickness to diameter becomes small. Consequently, the theory being developed is *thin-wall theory*, and we are in effect determining the average circumferential stress. The actual peak stress will differ from the average by less than 5.2% when the thickness-to-diameter ratio is less than $\frac{1}{20}$. A great many vessels have a thickness ratio less than $\frac{1}{20}$, and in these cases design based on the average stress is not seriously in error. Furthermore, the safety factor can be raised slightly to compensate for the error.

The free-body diagram, omitting forces that have no y-component, is shown in Fig. 1-19c. The force resulting from pressure acting on the curved surface must be obtained by considering the force on an incremental area as shown in Fig. 1-19d, and integrating. From $\sum F_y = 0$

$$-tl\sigma_c - tl\sigma_c + \int_0^\pi dP \sin \theta = 0$$

$$2tl\sigma_c = \int_0^\pi \frac{D}{2}\, d\theta \; lp \sin \theta = \frac{D}{2} pl \int_0^\pi \sin \theta \; d\theta$$

$$= \frac{D}{2} pl[-\cos \theta]_0^\pi = \frac{D}{2} pl[-(-1) - (-1)] = Dpl$$

$$\sigma_c = \frac{Dpl}{2tl}$$

$$\boxed{\sigma_c = \frac{Dp}{2t}} \qquad \text{Failure more likely} \qquad \textbf{(1-7)}$$

A more direct way to solve this problem is to cut the vessel as before, but to include the fluid, as shown in Fig. 1-19e, as part of the free body. The equilibrium equation $\sum F_y = 0$ then becomes

$$-tl\sigma_c + Dlp - tl\rho_c = 0$$

Thus $\sigma_c = Dp/2t$, which agrees with the previous solution but requires far less manipulation.

Another method will be presented because it will introduce ideas that will be required later in the solution to the general shell problem.

In this case, the element shown in Fig. 1-20a is taken from the free body in Fig. 1-19c. The forces acting on the element are shown in Fig. 1-20b.

Substituting into the equilibrium equation

$$\sum F_{x'} = 0$$

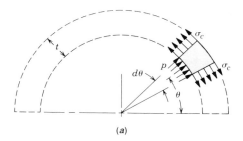

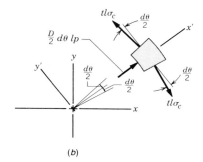

(b)

FIG. 1-20

gives

$$\frac{D}{2} d\theta \ lp - tl\sigma_c \sin\frac{d\theta}{2} - tl\sigma_c \sin\frac{d\theta}{2} = 0$$

$$2tl\sigma_c \sin\frac{d\theta}{2} = \frac{D}{2} d\theta \ lp$$

For infinitesimal angles the sine and the angle are equal. Therefore,

$$2tl\sigma_c \frac{d\theta}{2} = \frac{D}{2} d\theta \ lp$$

$$\sigma_c \equiv \frac{(D/2) \ d\theta \ lp}{2tl(d\theta/2)}$$

$$\sigma_c = \frac{Dp}{2t}$$

which is in agreement with the solution obtained by the other methods.

As noted earlier, stresses in the axial direction should also be expected to exist in a pressure vessel. Again, to find stress in a given direction we take a section normal to that direction, in this case a transverse section such as that shown in Fig. 1-21a.

The inclusion of the fluid with the shell on one side of the cutting plane leads to a much more direct solution than if the fluid is not included. The free body is then shown in Fig. 1-21b. Note that the longitudinal stress

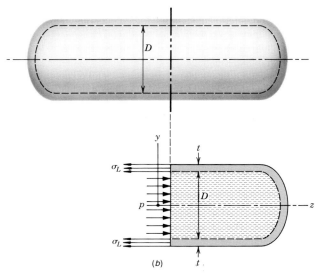

FIG. 1-21 Longitudinal stress in a cylindrical pressure vessel.

is assumed to be constant through the thickness of the wall. At this stage such an assumption must be made in order to solve the problem, and constant stress seems by intuition to be a reasonable assumption. More advanced methods of analysis would show this to be a correct assumption.

From the equilibrium equation

$$\sum F_z = 0$$

we get

$$-\sigma_L \pi (D+t)t + p\frac{\pi}{4}D^2 = 0$$

$$\sigma_L = \frac{D^2}{4(D+t)t}p \qquad \textbf{(1-8a)}$$

When t is small relative to D, we can replace $(D+t)$ with D and simplify the equation to

$$\boxed{\sigma_L = \frac{Dp}{4t}} \qquad \sigma_c = 2\sigma_L \qquad \textbf{(1-8b)}$$

This approximate equation for a thin-walled cylindrical pressure vessel errs on the safe side in that it indicates a higher stress than would be given by the exact equation (1-8a). The approximate equation (1-8b) when compared with (1-7) shows that *in a cylindrical shell under pressure the axial stress is approximately half of the circumferential stress*. This conclusion is useful in the design of pressure vessels made by welding

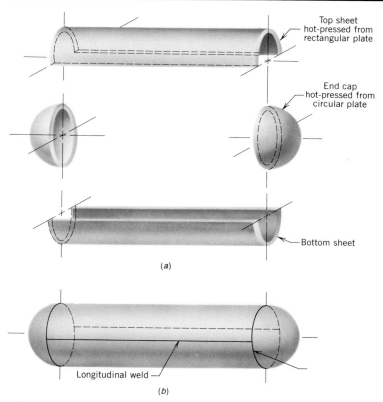

FIG. 1-22 Pressure vessel construction. (*a*) Parts of a pressure vessel. (*b*) Assembled pressure vessel.

together parts such as those shown in Fig. 1-22. It is common practice to press or roll plates into semicircular sheets and weld two such sheets together by longitudinal seams to form the cylindrical portion of a vessel. *End caps* are formed by hot-pressing circular plates. The profile of the end caps may be elliptical, circular (so that the caps are hemispheres) or composed of a short radius curve near the edge and a long radius curve near the center. In the latter case the shell is composed of a portion of a torus near the edge and a portion of a sphere in the central region, thus forming the commonly used torispherical head.

The end caps, when attached to the cylindrical portion by circumferential welds, complete the pressure vessel. Since the stress that tends to pull the longitudinal welds apart is twice that to which the circumferential welds are subjected, the longitudinal welds must be of the best possible quality. In practice these welds are usually made by procedures which ensure a joint that is as strong as the original plate: This is expressed by saying that the weld has an *efficiency* of 100%. The circumferential welds are subjected to only half the stress in the longitudinal welds and consequently do not require the same quality of workmanship or the same rigorous inspection. They could have an efficiency as low as 50% and still be theoretically acceptable. However, in practice a somewhat better quality circumferential weld is usually called for by the specifications.

Problems 1-54 to 1-58

1-8 STRESS DUE TO AXIAL FORCE AND TORQUE ON A THIN-WALLED VESSEL

Oil and gas transmission lines, steam and water lines in power plants, and pipelines in chemical processing plants are usually of such proportions that they may be classed as thin-walled vessels and the formulas that have been developed can be used to determine stresses due to internal fluid pressure. However, the design of these important elements is much more complex than the formulas would suggest. There are significant loads, other than those imposed by the fluid, from the weight of the pipe and contents, from the expansion or contraction of the pipes with change in temperature, and from other sources. These loads may impose on a section of the pipe (1) an axial force, (2) a torque about the pipe axis, (3) a bending moment, or (4) any combination of force, torque, and moment. When bending moment or torque is applied to an elbow, the determination of the stress is a very difficult procedure and is usually attempted only by specialists in piping design. Axial load and torque load, when applied to a straight section of pipe, are within our ability to analyze and will be dealt with next. We will find the results useful in future developments, but we must realize that actual pipeline design is often quite complex.

An axial force applied to a straight portion of a pipe, as in Fig. 1-23a, constitutes a system having complete axial symmetry. We should then expect a tensile stress in the axial direction that does not vary from point to point around the circumference of the pipe. We will assume that the stress does not vary through the thickness of the pipe wall, an assumption that happens to be valid even when the wall is thick. The free body in Fig. 1-23b gives the equilibrium equation

$$-\sigma_L \pi (D + t)\, t + F = 0$$

or

$$\sigma_L = \frac{F}{\pi (D + t)\, t} \qquad \textbf{(1-9a)}$$

If the wall is thin relative to the diameter, there is only a small error introduced by dropping the t in $(D + t)$ to give:

$$\sigma_L = \frac{F}{\pi D t} \qquad \textbf{(1-9b)}$$

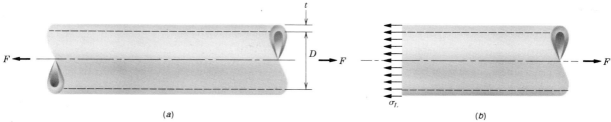

FIG. 1-23 Thin-walled vessel subjected to axial load.

It is interesting to note, in passing, that (1-9b) will become (1-8b) if the axial force is taken as the force resulting from the fluid pressure acting on the cross-sectional area of the fluid.

When a torque load is applied by a couple in the plane that is normal to the pipe centerline, the system again has axial symmetry. If we consider a plane such as A-B in Fig. 1-24a, the applied torque will cause the fibers on one side of A-B to tend to slide over the fibers on the other side of the plane, thus inducing shearing stress, τ. We will assume that τ

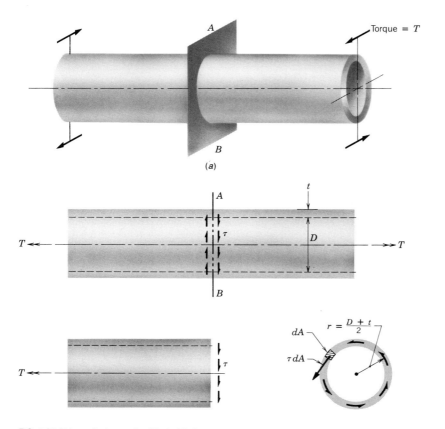

FIG. 1-24 Thin-walled vessel subjected to torque.

is uniform around the circumference on the basis of polar symmetry and that τ is constant across the wall thickness. The latter assumption is based on intuition and will prove to be reasonably accurate as long as the ratio of wall thickness to pipe diameter is small, which is the case for the proportions of most pipes.

With the assumption of uniform shearing stress we shall use the free body of Fig. 1-24c to evaluate the stress. This is done by taking moments about the axis of the pipe. The stress on an incremental area dA gives a tangential force $\tau\,dA$ acting at the centroid of the cross-hatched area. The actual distance to the centroid is difficult to express but can be given to good approximation by treating the area as a rectangle. This approximation becomes more accurate as the thickness to diameter ratio decreases. The moment due to stress acting on dA is then given by

$$dm = \tau\,dA\,\frac{D+t}{2}$$

and the total moment by

$$M = \int \tau\,dA\,\frac{D+t}{2} = \tau A\,\frac{D+t}{2} = \tau \pi (D+t)\,t\left(\frac{D+t}{2}\right)$$

Substituting into the equilibrium equation

$$\sum M = 0$$

gives

$$-T + \tau\,\frac{\pi (D+t)^2 t}{2} = 0$$

$$\tau = \frac{2T}{\pi (D+t)^2 t} \qquad\qquad \textbf{(1-10a)}$$

This is already an approximate equation based on small thickness-to-diameter ratios. Carrying the approximation a step further, we drop the t from $(D+t)$ to obtain

$$\tau = \frac{2T}{\pi D^2 t} \qquad\qquad \textbf{(1-10b)}$$

Because of the approximations, (1-10b) will be 5.5% in error for a thickness-to-diameter ratio of 1:20. The equation gives a stress that is too high and, hence, errs on the safe side.

Problems 1-59 to 1-62

The answer that you have given to part (d) of Prob. 1-62 may not seem entirely satisfactory. At this stage it is impossible to give an answer on any sound theoretical basis. The question was asked in order to encourage you to think about the effect of several stresses acting simultaneously. The question arises: Can the material be safely subjected to the three stresses applied simultaneously just because each stress alone was safe, or do the stresses combine in some way to give a stress, perhaps in an entirely new direction, that is excessive? We cannot answer this question now; it will provide motivation for studies that will be made in Chapter 5.

1-9 ALLOWABLE STRESS

We have been working with allowable stresses that were given without explanation other than that they were safe. It is time now to reexamine these stresses to see how they were determined.

A test for the safety of a structural member might be done by loading the proposed member to failure to see if its capacity exceeds that which we calculate will be imposed on it in service. If we are thinking of a tension member for a truss, it might be tested in a simple test rig such as that of Fig. 1-25. By tightening the nut, we stretch the member so that we are not really loading the member but rather elongating it. Let us assume that by measuring the torque necessary to turn the nut, we are capable of calculating the force induced in the member by virtue of the imposed elongation. A test would consist of turning the nut until the member broke, thus determining the failure load of the member. If the measured failure load was equal to or greater than the working load for the member, we might conclude that it would have been a satisfactory member if we had not destroyed it. Another member for the structure would then have to be made identical to the test member. The dimensions as well as the material would have to be the same. By ensuring that the chemical content of the material, metallurgical processes, and all dimensions were reproduced, we could be reasonably certain that the replacement part would also be satisfactory. Because of some uncertainty in the reproducibility of all the parameters, we would want the test failure load to be larger than the expected load on the actual member. Also, the expected load is usually an estimated value and, therefore, is not known precisely. Because of all these uncertainties we insist that the failure load be greater than the calculated design load by a factor that is called either the factor of safety, or the *safety factor*.

It would, of course, be impractical to make full-size tests on all the components of a structure or machine. In practice we determine the

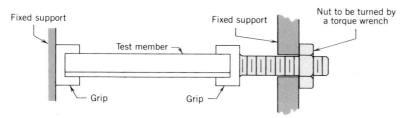

FIG. 1-25 Simplified tension testing device.

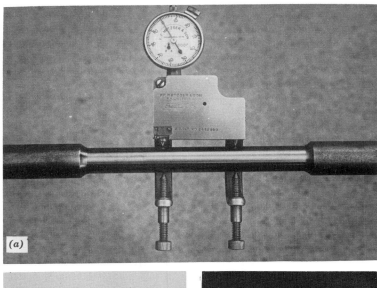

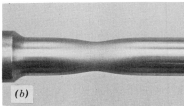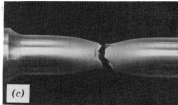

FIG. 1-26 Tensile test specimen. (*a*) Mechanical extensometer clamped to a tensile test specimen. (*b*) Necking of tensile test specimen. (*c*) Ductile tensile test specimen after failure.

strength of a material by testing a small specimen such as that shown in Fig. 1-26. The material in the test specimen is made to the same specifications as the material that is to be used in the structure, and we assume that both materials have the same strength when expressed as a stress. In actual fact a part larger than the test specimen is likely to fail at a slightly lower stress; allowance for this is made in establishing the value of the safety factor. Test specimens are stretched in a testing machine such as that shown in Fig. 1-27, which has the basic elements of our simple apparatus in Fig. 1-25 but has a force-measuring system that is much superior. Most testing machines do, in fact, stretch the material and determine the required force, rather than load the specimen with a force and accept the change in length that results. We usually say that we are "loading" the specimen, which is not strictly true. However, this error in terminology is accepted as part of the engineering language.

In discussing strength tests, we made reference to failure that would normally be expected to mean that the specimen had broken. However, if we do a more complete test, we will discover reasons for taking a wider definition of failure. Let us consider that we have instruments capable of measuring the change in length of a typical portion of the specimen as it

FIG. 1-27 Testing machine set up for a tensile test.

is stretched. An instrument that will measure that change is shown in Fig. 1-26a and is referred to as a mechanical *extensometer*. A mechanical extensometer measures change in length by mechanical rather than by optical or electrical means. In the case of the extensometer shown in Fig. 1-26a, the instrument is clamped to the specimen and the change in length, Δl, is measured between two points on the shank of the specimen. The change in length is amplified mechanically through a set of levers and indicated on a dial gauge that is capable of detecting a change as small as 0.001 mm (0.00004 in.). The two points on the shank, such as A and C in Fig. 1-28, between which the measurements are made as the test proceeds denote the *gage length*, l. We are especially interested in the change in length divided by the gage length and use the word *strain*, and the symbol ε, to refer to this ratio. Thus

$$\varepsilon = \frac{\Delta l}{l}$$

(1-11a)

Then strain is the deformation per unit of length and if Δl and l are given in the same units, the strain is dimensionless. However, a strain, which can be stated simply as a pure number, 0.0015 for example, is more likely to be expressed by engineers as 0.0015 mm/mm (in./in.) or as 0.15%.

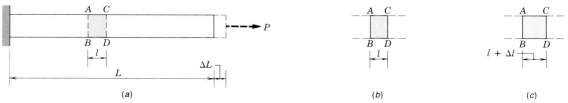

FIG. 1-28 (*a*) Typical element. (*b*) Element before straining. (*c*) Element after straining.

If the full length of the member in Fig. 1-28 is taken into consideration, the strain equation can be written

$$\varepsilon = \frac{\Delta L}{L} \tag{1-11b}$$

In practice we often need to know the change in length of a member when the strain is known. Then the best form of the equation is

$$\Delta L = \varepsilon L \tag{1-11c}$$

If the strain varies along the length, as it would in a tapered member, the "gage length" must be reduced to infinitesimal size. Then

$$\varepsilon = \frac{d\Delta l}{dl} \tag{1-11d}$$

or

$$d\Delta l = \varepsilon \, dl \tag{1-11e}$$

The change in length of a member having a varying strain requires that the strain be integrated in

$$\Delta L = \int_0^L \varepsilon \, dL \tag{1-11f}$$

Let us now return to the tensile specimen of Fig. 1-26a. In order to better understand the behavior of a material, we would conduct a test by stretching the specimen in small steps, taking length and force measurements at each step. The data thus obtained will reveal much about the material if it is plotted on a diagram similar to Fig. 1-29. Although the shape of the curve in Fig. 1-29 is typical of a low-carbon steel, the following discussion applies to many engineering materials.

During the test whenever the change in length is measured the corresponding force is also measured. From these measurements the strain and its related stress are calculated. A point corresponding to this pair of values is then plotted. If several pairs of measurements are taken and plotted, we find that the points lie on a straight line, as in Fig. 1-29a. If, at any step, we reverse the procedure by reducing the load and taking new pairs of readings, we find that these points also fall substantially on the original straight line. If we release the specimen, we find that it returns to its original dimensions and we say that the material has been subjected to *elastic strain* and that it behaves in a *linearly elastic* manner.

Let us imagine that we again load the specimen and plot as before to get Fig. 1-29b. In the early stages of loading, the stress–strain curve is the same as that of the original test. However, as we continue to deform the specimen we will notice a change: the points begin to depart from the straight line. If we reverse the loading, we find a set of points on a straight line that is parallel to, but offset from, the original straight line. When there is no longer any force in the member, there remains a small deformation, which is referred to as *permanent set*. The material may be said to have been loaded into the *inelastic* or *plastic* region. In Fig. 1-29c, where stretching continues beyond that of the previous curve, the

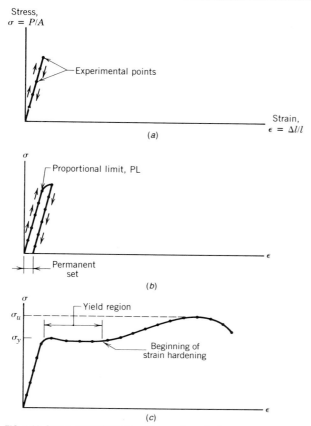

FIG. 1-29 Stress-strain behavior of low-carbon steel.

rounding of the curve previously noted continues into a horizontal region known as the yield region. While yielding, the specimen undergoes large *inelastic strains* at a constant stress; this stress is referred to as *yield stress*, σ_y. Note that for the low-carbon steel specimen the stress at the *proportional limit*, that is, where the curve begins to depart from the straight line, and the yield stress occur very close to each other; however, this is not the case for all materials.

We can see from Fig. 1-29c that after considerable yielding the stress begins to rise again. This is attributed to *strain hardening*, which, characterized by a change in the metallurgical nature of the material, continues until a peak is reached. Following this the load, and stress, decrease rapidly, and shortly thereafter the specimen breaks.

The stress at the peak of the curve is referred to as the *ultimate stress* or *ultimate strength*, σ_u.

If we watch the specimen carefully while the load is being applied, we will probably not be aware of any change taking place as long as the material is in the elastic range. During yielding the elongation can be detected visually; there is a very obvious reduction in diameter, or *necking*, taking place as shown in Fig. 1-26*b*. A view after failure is presented in Fig. 1-26*c*.

It might be argued that at any stage in the test the stress should be calculated by dividing the force by the cross-sectional area that then exists. This would give a *true stress*, but the determination of the area at each stage in loading would be very time-consuming. Consequently, it is standard practice to use the original area for all stress calculations; the value thus obtained is properly designated as *engineering stress* but is usually referred to simply as *stress*.

Materials that exhibit large deformation before failure are described as being *ductile*. The large deformation of these materials serves to give warning of impending failure when a part is grossly overloaded. Ductile materials are also better for absorbing energy because the area under the curve is a measure of the *energy* required to cause failure and the largest part of the area is under the plastic portion of the curve. Consequently, ductile materials are preferred where there is impact loading. Also, it is essential that ductile materials be selected for parts that are to be cold-formed, since the cold-forming process stretches the material severely.

A measure of the ductility of a material is given by the permanent elongation in the test specimen after failure expressed as a percentage of gage length. Another commonly used measure of ductility is the reduction in the cross-sectional area of the specimen after failure expressed as a percentage of the original area. To give you an example of the ductility of low-carbon steel in tension, a typical stress-strain curve is shown in Fig. 1-30. Note the very small strain of 0.14% or 0.0014 mm/mm (in./in.) at commencement of yielding and the 16-fold increase in strain to 2.2% while the stress remains constant at the value of the yield stress. The final strain of 22% indicates a very ductile material; the strain being 160 times the maximum elastic strain.

We are now ready to reconsider the basis for choosing an allowable stress. Since a material can be loaded to its ultimate stress without

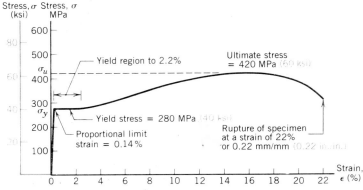

FIG. 1-30 Typical stress-strain diagram for low-carbon steel in tension.

breaking, it would seem that this is a natural basis for arriving at an allowable stress. However, a stress near the ultimate will be accompanied by large deformation, which by itself may be a kind of failure. For example, if the crankshaft of an internal combustion engine is stressed near the ultimate, the deformation of the shaft is almost certain to cause a greater misalignment of parts than can be tolerated. In fact, any strain beyond yield would probably be excessive in such an application. Also, a bridge or a building might be safe with stresses near the ultimate, but the attendant strain would lead to large deformation that could easily be sufficient to alarm the user of the structure. In these cases it might be said that the part "failed" when the stress reached the yield point. For these cases we should apply the safety factor to the yield stress in order to arrive at the allowable stress, and the safety factor is stated as a certain number (e.g., 2.0) based on the yield stress. In other applications it may be more reasonable to base the allowable stress on the ultimate. The safety factor would then be specified as a certain number (e.g., 4.0) based on the ultimate strength.

Problems 1-63 to 1-69

1-10 STRESS–STRAIN DIAGRAMS

Members of the structures that have been considered up to this point have been shaped and loaded in such a manner that stress was induced in one direction only. Being stressed in one direction, the material is said to be in a state of *uniaxial stress*. We considered the strain in the same direction and found it useful to plot the relationship between stress and strain as determined by tests. Under these conditions there is no lateral or transverse stress, but there is a *transverse strain* that we have not yet considered. A member of a truss subjected to an axial tensile load will not only lengthen, as expected, but also its lateral dimensions will decrease slightly due to transverse strain. This contraction occurs in all materials; you have probably observed it in an elastic band, where the width of the rubber decreases noticeably as the band is stretched in tension. For the present we will disregard the lateral strain, as it is of no significance in the design of axially loaded members; it will be dealt with later when plates are considered.

A number of typical stress–strain curves for various materials under uniaxial tensile load are given in Fig. 1-31. For the low-carbon steel of curve A, we have already observed that there is a large amount of strain before the material breaks and it is therefore classified as a *ductile* material. We have also noted that *ductility* is a useful property of any engineering material because the large deformation of a structure often provides a warning that collapse is imminent. Not all materials have this property. For example, the stress–strain curve for cast iron is more likely to resemble curve B. In this curve a straight-line portion, where the strain is proportional to the stress, again exists, but the large plastic deformation is absent. The cast iron, or any other material with a curve similar to B, is described as being *brittle*.

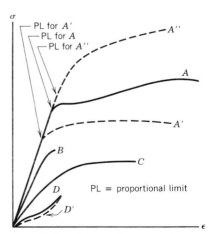

FIG. 1-31 Typical stress-strain curves for engineering materials.

Tests on aluminum and aluminum alloys give curves similar to *C*. In this curve the straight-line portion is quite small and there is an appreciable amount of ductility.

All the curves discussed so far display a linear portion even though the proportional limit in some cases was quite low. As pointed out earlier with reference to Fig. 1-29, the curves up to the proportional limit display the *linearly elastic* behavior of the material. This behavior applies to the material while it is being both loaded and unloaded and, hence, the loading and unloading curves coincide along a straight line. Although most engineering materials behave in this way, a few materials such as rubber and neoprene follow a curve similar to *D* on being loaded and *D'* on being unloaded. The behavior is still elastic, in that the unloaded material returns to its original size, but the response is nonlinear. We refer to these materials as being *nonlinearly elastic*.

The curves of Fig. 1-31 could be used to compare the strengths of the various materials. Generalized conclusions should not be drawn about the relative strengths of steel, cast iron, and aluminum from the given curves, as change in chemical composition and heat treatment can alter the curves considerably. For example, a change in composition of the steel of curve *A* could alter the relationship to that of curve *A'* or *A''*. The properties of the three steels are quite different, but they all have one property in common: the slope of the straight-line portion of the curves. The numerical value of this slope is known as *Young's modulus** or the *modulus of elasticity*. It is represented by the symbol *E* and is the same for all three steels in spite of the fact that the proportional limits are different. The value of *E* can be determined experimentally by measuring the stress and strain for any point on the straight-line portion of the curve and substituting into

$$E = \frac{\sigma}{\varepsilon} \qquad \textbf{(1-12a)}$$

The expression for direct proportionality between stress and strain at stresses below the proportional limit is generally referred to as *Hooke's Law* because it was first annunciated by the English scientist Robert Hooke in 1676. Since σ is a fairly large number such as 200 MPa (29 ksi) and ε is a very small number such as 0.001 mm/mm (in./in.), *E* is always an extremely large number, for example, 200 000 MPa (29 000 ksi) for steel. Because strain is a pure number, the units of *E* are those of stress; however, in trying to gain an understanding of *E*, you should not make the mistake of viewing *E* as a stress. It is perhaps best to think of *E* as nothing more than a measure of the *stiffness* of a material. The larger the *E* value, the stiffer the material. From the curve in Fig. 1-31 the relative slopes, and hence the relative stiffnesses, are obvious; it can be concluded that cast iron (*B*) is a stiffer material than aluminum (*C*) and that steel (*A*) is the stiffest of the three metals.

Values of Young's modulus for many engineering materials are given in Table A and this enables us to relate stress to strain so that, if the

---

*The modulus is named after the English scientist Thomas Young (1773–1829) who studied the elastic behavior of bars and provided a definition of the modulus of elasticity.

stress in a given material is known, the strain can be calculated. For this purpose (1-12a) is more convenient in the form

$$\varepsilon = \frac{\sigma}{E}$$

(1-12b)

or if strain is known, stress can be calculated by

$$\sigma = E\,\varepsilon$$

(1-12c)

In using these equations, it should always be remembered that they are applicable only to the straight-line portion of the stress-strain curve, and that if values beyond the proportional limit are used, the results are meaningless. For points on the stress-strain curve beyond the proportional limit, the *tangent modulus* and *secant modulus* as indicated in Fig. 1-32 are used as a measure of stiffness of the material. The *tangent modulus*, E_t, is defined as the slope of the stress strain curve at a particular stress level. In the case of an elastic material below the proportional limit it is obvious that E_t equals E and that, for the low-carbon specimen of Fig. 1-30, E_t changes from E to zero at the yield stress. The *secant modulus*, E_s, is the ratio of stress to strain at any point on the stress–strain curve. Again, for the case of an elastic material below the proportional limit, E_s would simply be equal to E.

We have already mentioned, in connection with the stress-strain curves of Fig. 1-31, that many materials do not exhibit a well-defined yield point. For these materials it is customary to arrive at a pseudo-yield stress by means of the *0.2% offset method* illustrated in Fig. 1-32b. At a strain of 0.2%, which represents a relatively large strain when viewed in the light of the associated elastic deformations, a line is drawn parallel to the straight-line portion of the initial stress-strain curve. The stress at the intersection with the stress-strain curve, point A, is then considered to be the yield stress of the material.

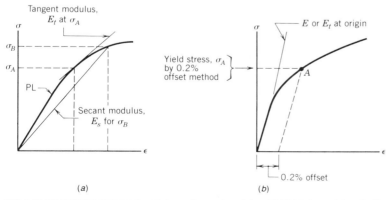

FIG. 1-32 (*a*) Meaning of tangent modulus and secant modulus. (*b*) Yield stress determination by 0.2% offset method.

Some materials, when tested in compression, reveal strengths that are different from those in tension. For example, concrete, masonry, and cast iron are stronger in compression than in tension. However, if a short specimen of steel is tested in compression, its yield stress is found to be very close to its yield stress in tension. If we plotted compressive test results on Fig. 1-31, negative strain coupled with negative stress would give points in the third quadrant and we would again find the curves to be straight lines near the origin. For each material we would find the straight line in compression to be the extension of its straight line in tension. Hence *Young's modulus in compression and tension is the same* and equations (1-12) are valid for compression as well as tension up to the respective proportional limits.

If the compressive test on steel had been made with a long test specimen, a different and false value for the yield would have been indicated. The specimen, if examined carefully, would reveal a different mode of failure; it would have failed by *buckling* in a manner similar to the member in Fig. 4-1. This mode of failure together with the definition of long and short members will be reconsidered in Chapter 4, where buckling is treated rigorously.

Throughout all our discussions on the stress–strain behavior of various materials, we have assumed that the material is *homogeneous*, that is, it displays the same properties at all points within the material. Such an assumption is reasonable for most engineering materials, and even such a relatively nonhomogeneous, or *heterogeneous*, material as concrete is considered to be homogeneous for design purposes. One other property of material needs to be mentioned at this time. Unless otherwise stated, we will assume that all engineering materials are *isotropic*; that is, the behavior of a specimen of material does not depend upon the direction of the axis of the specimen relative to the piece of material from which it is cut. A notable exception to this is timber, where we might intuitively realize from observing the growth rings of a tree that the material has different properties in the longitudinal, transverse, and radial directions. Such a material is classified as *anisotropic*.

With our knowledge of stress and strain it is now useful to derive a general formula for the deformation of axially loaded members of constant cross section as shown in Fig. 1-33. From (1-11c)

$$\Delta L = \varepsilon L$$

But strain, ε, is related to stress through E by

$$\varepsilon = \frac{\sigma}{E} \quad \text{and} \quad \sigma = \frac{P}{A}$$

Hence

$$\Delta L = \frac{\sigma}{E} L = \frac{PL}{AE}$$

$$\boxed{\Delta L = \frac{PL}{AE}} \tag{1-13}$$

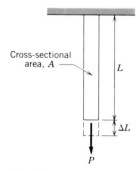

Cross-sectional area, A

L

ΔL

P

FIG. 1-33

Example 1-9

A truss member, 8 m long, is made of $2 - 150 \times 150 \times 16$ low-strength steel angles. Determine the change in length of the member when a tensile load of 1260 kN is applied.

Solution

$$A \text{ (one angle)} = 4540 \text{ mm}^2 \qquad \text{(Table B-4)}$$

$$A = 2 \times 4540 \text{ mm}^2 = 9080 \text{ mm}^2$$

$$\text{stress}, \sigma = \frac{P}{A} = \frac{1260 \times 10^3}{9080} \quad \frac{\text{N}}{\text{mm}^2} = 139 \text{ MPa}$$

$$E = 200\,000 \text{ MPa} \qquad \text{(Table A)}$$

Since the actual stress (139 MPa) is less than the yield stress (230 MPa), equation (1-12) is valid. From (1-12b),

$$\varepsilon = \frac{\sigma}{E} = \frac{139}{200\,000} \quad \frac{\text{MPa}}{\text{MPa}} = 0.000695 \text{ mm/mm}$$

From (1-11c) change in length,

$$\Delta L = L\varepsilon = 8 \times 10^3 \times 0.000695 \text{ mm} \cdot \text{mm/mm} = 5.56 \text{ mm}$$

Alternatively, substitution into (1-13) gives directly

$$\Delta L = \frac{1260 \times 10^3}{9080} \frac{8 \times 10^3}{200\,000} \quad \frac{\text{N}}{\text{mm}^2} \cdot \frac{\text{mm}}{\text{N/mm}^2}$$

$$= 5.56 \text{ mm}$$

Example 1-9

A truss member, 26 ft long, is made of $2 - 6 \times 6 \times 5/8$ low-strength steel angles. Determine the change in length of the member when a tensile load of 285 kip is applied.

Solution

$$A \text{ (one angle)} = 7.11 \text{ in.}^2 \qquad \text{(Table B-4)}$$

$$A = 2 \times 7.11 \text{ in.}^2 = 14.22 \text{ in.}^2$$

$$\text{stress}, \sigma = \frac{P}{A} = \frac{285}{14.22} \quad \frac{\text{kip}}{\text{in.}^2} = 20.04 \text{ ksi}$$

$$E = 29,000 \text{ ksi} \qquad \text{(Table A)}$$

Since the actual stress (20.04 ksi) is less than the yield stress (33 ksi), equation (1-12) is valid. From (1-12b),

$$\varepsilon = \frac{\sigma}{E} = \frac{20.04}{29,000} \quad \frac{\text{ksi}}{\text{ksi}} = 0.000691 \text{ in./in.}$$

From (1-11c) change in length,

$$\Delta L = L\varepsilon = 26 \times 12 \times 0.000691 \text{ in} \cdot \text{in./in.} = 0.216 \text{ in.}$$

Alternatively, substitution into (1-13) gives directly

$$\Delta L = \frac{285}{14.22} \frac{26 \times 12}{29 \times 10^3} \frac{\text{kip}}{\text{in.}^2} \cdot \frac{\text{in.}}{\text{kip/in.}^2}$$

$$= 0.216 \text{ in.}$$

Problems 1-70 to 1-77

1-11 STRESS CONCENTRATION

The design of a simple member, such as that shown in Fig. 1-1, subjected to an axial tensile load, is based on the comparison of the stress in the member with the strength of the material determined by testing. In both cases the stress is assumed to be *uniformly* distributed over the cross section and experience has shown that this assumption leads to safe designs.

When the member being designed has a hole, as illustrated in Fig. 1-4, or a notch or other feature that reduces the cross section, this is taken into account by using the *net* cross section in (1-1a) to determine the stress. However, the assumption underlying (1-1a), that is, that the stress is uniformly distributed, is not valid for the net section. If we were capable of measuring the stresses on the net section, we would find that (1-1a) gives the *average stress* and that some of the fibers are subjected to stresses that are considerably higher than the average. In the light of this nonuniformity of stress, we must reconsider the safety of members designed by (1-1a), (1-2), and (1-3). We will treat the distribution of *nonuniform stress* in order to determine the variation between the *peak stress* and the *average stress* and decide how the design procedure is to be modified to make allowance for the peak stress. To accomplish this, we will consider a very simple case and through it introduce some new principles.

Consider a plate under tensile load *P* as in Fig. 1-34. For simplicity let us take the load *P* as being a uniformly distributed traction over the ends of the plate. With this traction there is no reason to expect the stresses to be anything but uniform in all the fibers of the plate; in fact, they would prove to be uniform if we had the means to measure them.

Consider now the same load acting on a plate with a small hole in the center as in Fig. 1-35. We recognize that the critical cross section is the one through the hole, since, out of all the sections that we might consider, it has the smallest cross-sectional area. Using that cross-sectional area in (1-1a) gives a stress that we will designate as σ_0 which is assumed to be uniformly distributed and is, in fact, the *average* stress on the *net* cross section. This stress is shown in broken lines in Fig. 1-35. If we had a method of determining the stress distribution on the net cross section, we would find that it is nonuniform and the variation would follow a curve similar to that shown in solid lines in Fig. 1-35.

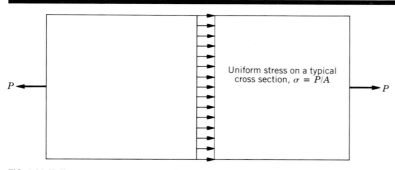

FIG. 1-34 Uniform stress distribution in a plate subjected to concentric tension.

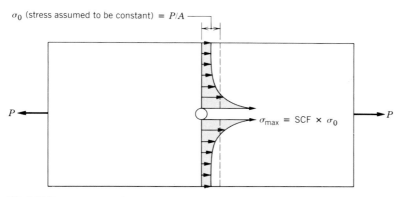

FIG. 1-35 Stress concentration due to circular hole in plate.

It is evident from the curves that (1-1a) gives a stress that is far short of the actual maximum stress, σ_{max}. A convenient quantity in stress calculations is the ratio of σ_{max} to σ_0, which is designated the *stress concentration factor* or SCF. Thus

$$SCF = \frac{\sigma_{max}}{\sigma_0} \tag{1-14a}$$

or rearranging

$$\sigma_{max} = SCF \times \sigma_0 \tag{1-14b}$$

The latter is a convenient form of the equation for stress evaluation. We can calculate σ_0 without difficulty; however, it is impractical for most engineers to determine the SCF. Since this is a frequently occurring problem, researchers have put much effort into determining stress concentration factors for *elastic* material of various shapes and proportions.*

*The results of stress research are reported in technical papers and books such as *Formulas for Stress and Strain* by Roark and Young (McGraw-Hill, New York, 1975).

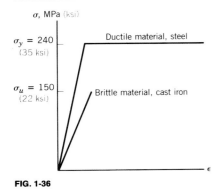

FIG. 1-36

It is interesting to note that for the case given in Fig. 1-35, if the diameter of the hole is *small* relative to the length and width of the plate, the SCF is 3.0. This is a value that designers remember and is useful in that it indicates the magnitude of the error that can be introduced by using average stress when the maximum is required.

To appreciate the importance of the SCF in design, consider two rectangular plates subjected to tensile loading. The first plate is made of *brittle* material, such as gray cast iron, while the second plate is made of a *ductile* material such as steel. For the two materials we will assume the stress-strain curves shown in Fig. 1-36. Note that for both materials we are assuming a somewhat *idealized* stress-strain relationship when compared to the curves A and B in Fig. 1-31. For the brittle cast iron, Fig. 1-36 shows linear behavior right up to failure at the ultimate stress, σ_u. For the ductile steel Fig. 1-36 omits the strain-hardening effect. Such a degree of idealization is often assumed by designers and, therefore, represents practice. The idealized behavior of the ductile material is referred to as *elastic-perfectly plastic*.

We will now return to the SCF problem and deal with the first plate consisting of *brittle* cast iron. In Fig. 1-34 let us imagine that the load starts at zero and is gradually increased. As the load increases, the fibers are subjected to an increasing stress and strain, which are related at any instant by a point on the curve in Fig. 1-36. Throughout most of the loading period no noticeable change in the material would occur. However, the load will eventually reach the level where the stress will exceed the material strength, σ_u. A crack will start to open and, because the material has virtually no ductility, will rapidly spread normal to the direction of the stress until the part has failed completely. In this case failure would occur suddenly and without warning when the stress reaches 150 MPa.

Now consider that the cast iron plate has a small hole drilled near the center as in Fig. 1-35. Again the load is gradually applied and the stress and strain are related as in Fig. 1-36. However, this time all points are not stressed equally; the stresses at the edge of the hole being three times the average stress.

When the maximum stress reaches 150 MPa, the fibers at the edge of the hole will have been loaded and deformed to the end of the stress-strain curve and failure will be imminent. Only a small amount of material is on the verge of failure, but once a crack starts the sharp point at the leading edge of the crack causes a very high stress concentration which, in a brittle material, promotes the rapid spread of the crack. Hence this piece of cast iron fails at an average stress of 50 MPa and thus fails at a load which is only *one-third* of the load that caused the unperforated plate to fail.

Let us now look at the effect of the SCF on a plate made of *ductile* steel. When the load on the steel plate is small, the stress distribution on the net cross section will again be as shown in Fig. 1-35. That is, the maximum stress will be three times the average stress on the net section. The stress relationships will continue to be as shown as long as all the particles of the material are stressed to levels below the proportional limit and the stress concentration factor will continue to be 3.0.

Hence nothing unusual happens before the maximum stress reaches the yield stress and the stress distribution in the plate is of the form of curve *A* in Fig. 1-37. When the maximum stress has reached the yield stress of 240 MPa, the average stress is 80 MPa and it might seem that the limit for the load has been reached. However, the fact that the material is ductile now begins to have an influence.

As the load increases further, the plate will stretch more, and the fibers will have larger strains and, in general, larger stresses. The great bulk of the material is stressed well below yield and hence continues to increase in stress as the load grows.

As the load increases beyond that which stressed the fibers near the edge of the hole to the yield point, those fibers are further strained, but the stress remains constant at 240 MPa. *Plastic* flow takes place in these fibers, but the great bulk of the plate remains elastic and the stress in the elastic regions continues to increase with the increased load. The stress distribution curve is now of the form of curve *B* in Fig. 1-37. The stress concentration factor, which was based on an *elastic analysis*, is no longer 3.

Further increase in load will cause the plastic region to expand beyond the boundary shown in Fig. 1-37 and the stress distribution to change to that given by curve *C*. It is evident that in the limit the whole net section will become a plastic region and be stressed to 240 MPa, the yield stress. At this limit the fibers at the edge of the hole will be subjected to large strains, but, for the amount of ductility normally specified in engineering applications, will be far short of the strain required to rupture the material.

This example shows us that a ductile part having a discontinuity that introduces a stress concentration can carry a load up to the yield stress multiplied by the net cross-sectional area. Consequently, the *stress concentration factor is of no significance in the design of a ductile part.* This explains why we were able to select angles for truss structures using the net section; disregarding the nonuniformity of stress on the net section.

We have seen that stress is raised, or concentrated, in the region where there is a change in cross section. In the cases cited, a circular hole caused the change and hence the stress concentration. A stress

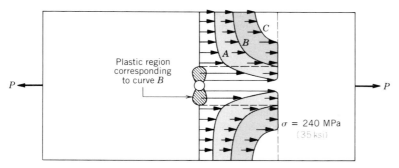

FIG. 1-37 Plastic flow in a plate of ductile material such as mild steel.

concentration factor of 3.0 was given for a small hole in a uniformly loaded plate. If the hole had been elliptical rather than circular, the factor would have been larger or smaller depending on whether the long axis of the ellipse was normal or parallel to the direction of the load.

If the hole had been square, or of any shape having sharp corners, the SCF would have been larger than 3.0 and such a hole would greatly weaken a part made of brittle material.

A change in the stress distribution could also be caused by edge notches in the plate. The notches could be circular arcs, V-shaped, cracks, or other shapes. All would have different SCFs, but, in general, the sharper the profile at the bottom of the notch the larger the SCF.

We have seen that the stress concentration does not influence the design of parts made of ductile materials. Nevertheless, designers wherever possible avoid shapes that cause concentration and when large SCFs are unavoidable the careful designer will compensate by reducing the allowable stress by an amount based on experience.

When the load is applied and released cyclically the SCF has a significant influence on strength of both brittle and ductile materials. However, the effect of *cyclic load* is a whole new subject in itself and is dealt with in Chapter 12. If you look ahead to Chapter 12, you will find graphs that give stress concentration factors for a variety of shapes that are encountered frequently in machine design. In these graphs the SCF is designated K_t and is referred to as the theoretical stress concentration factor.

1-12 DEFORMATION OF TRUSS MEMBERS

The numerical values used in Example 1-9 are typical of those found in actual structures; hence the change in length of 5.56 mm (0.22 in.) is typical of the change in length of an actual member. It would be reasonable to wonder if there is any need for doing this calculation when it leads to such a small value. The change in length of a single isolated member is seldom of interest, but when all members of a structure such as that shown in Fig. 1-38, are considered, the changes have a cumulative effect so that, although no member in this truss changes length by more than 4.5 mm (0.18 in.) the deflection at the center panel point of the top chord is 51 mm (2.0 in.). This may not seem to be a very large value, but a very small deflection is quite noticeable in such an application and a sagging bridge makes the user feel quite insecure. In the actual design of this truss the deflection of all upper-chord panel points would be calculated and the structure built with its top chord initially curved upward so that under load the deflection would cause the chord either to straighten or to retain some of its initial camber. The determination of deflection for such a truss is a lengthy process requiring special techniques that will be developed in later chapters.

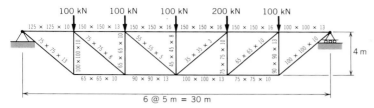

FIG. 1-38 SI dimensions and loads for a typical truss.

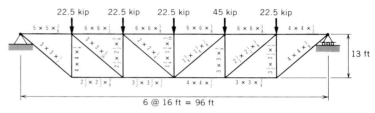

FIG. 1-38 Imperial dimensions and loads for a typical truss.

Problems 1-78 to 1-81

1-13 INTRODUCTION TO STATICALLY INDETERMINATE SYSTEMS

If we are required to find the reactions for the beam in Fig. 1-39a, there are two unknown quantities and two equations available. The equation $\sum F_x = 0$ would be trivial and contribute nothing to the solution, leaving us with only $\sum F_y = 0$ and $\sum M_p = 0$. Since the number of unknowns is the same as the number of equations of statics, there is then no difficulty in solving for the reactions and hence the problem is said to be *statically determinate*.

Providing another support in Fig. 1-39b at C adds another unknown, and plainly two equations are insufficient to solve for three unknowns. We might attempt to generate another equation by writing a second moment equation with respect to a point other than p that was used in the first moment equation. This would give three equations in three unknowns, and we might now expect a solution. However, no amount of mathematical manipulation will produce a solution, the difficulty being that the third equation is in reality nothing more than the original moment equation plus a constant times the force equilibrium equation. Since the additional moment equation is not independent, the matrix of the coefficients is singular and the reactions cannot be determined. This case is described as being *statically indeterminate* because the equations of statics are not sufficient to determine the required unknowns.

In the case just presented, the system became indeterminate when one support was provided in excess of the number required to maintain equilibrium; we then have one *redundant support*. Incidentally, if one of the supports under the beam in Fig. 1-39a had been removed, the structure would have moved under load and we would have turned the structure into a *mechanism*. Indeterminate systems can also result from too many internal structural members being provided. For example, in the truss shown in Fig. 1-16a, if a member were added to connect point 3

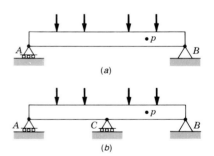

FIG. 1-39 Beam with redundant support.

to point 7, the force in that *redundant member*, and some others, could not be determined by the equations of statics alone. As in all statically indeterminate cases, the equations of equilibrium are still valid but are not numerous enough to permit a complete solution. To obtain a solution, additional *independent* equations are generated by consideration of *displacement* in one form or another.

The structure in the Example that follows is indeterminate and is solved in a very simple manner by taking displacement into account. More advanced indeterminate problems will be solved in Chapters 3 and 10.

Example 1-10

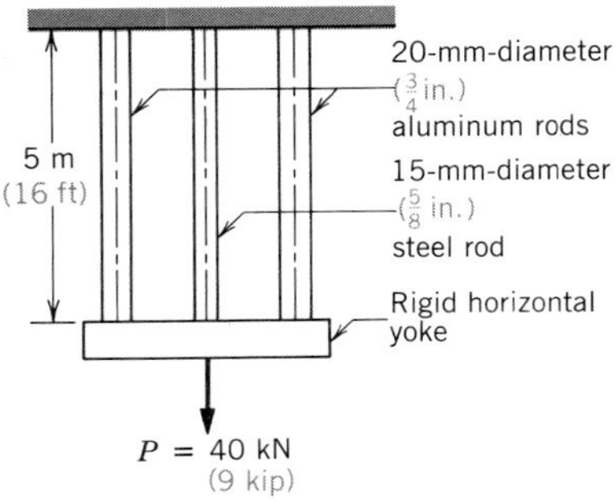

The rods in Fig. 1-40a have their upper ends securely embedded in an unyielding support and their lower ends fastened to an inflexible horizontal yoke. Initially, there is no stress in the rods.

Determine the stresses in the rods when a vertical downward load of 40 kN is applied.

Solution. To determine the stresses in the rods, we must first obtain the forces P_a and P_s indicated in Fig. 1-40b. From the equilibrium equation $\sum F_y = 0$,

$$P_a + P_s + P_a - P = 0$$

$$2P_a + P_s = P = 40 \text{ kN} \tag{1}$$

One more simultaneous equation is required, but all other equilibrium equations are either trivial or are a repetition of equation (1). We must, therefore, resort to consideration of displacement in order to generate an independent equation. In this case the displacement is the downward motion of the yoke, δ, which is equal to the elongation in each of the rods.

From (1-13)

$$\Delta L = \frac{PL}{AE}$$

or

$$P = \frac{AE}{L} \Delta L = \frac{AE}{L} \delta$$

For each aluminum rod

$$P_a = \frac{A_a E_a}{L} \delta \tag{2}$$

and for the steel rod

$$P_s = \frac{A_s E_s}{L} \delta \tag{3}$$

Substituting into equation (1), we find that

$$2 \frac{A_a E_a}{L} \delta + \frac{A_s E_s}{L} \delta = P$$

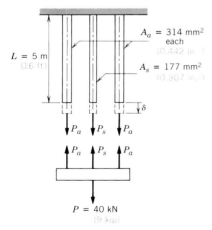

FIG. 1-40a

20-mm-diameter $\left(\frac{3}{4} \text{ in.}\right)$ aluminum rods

15-mm-diameter $\left(\frac{5}{8} \text{ in.}\right)$ steel rod

Rigid horizontal yoke

$P = 40 \text{ kN}$ (9 kip)

5 m (16 ft)

$A_a = 314 \text{ mm}^2$ each (0.442 in.²)

$A_s = 177 \text{ mm}^2$ (0.307 in.²)

$L = 5 \text{ m}$ (16 ft)

P_a P_s P_a

$P = 40 \text{ kN}$ (9 kip)

FIG. 1-40b

$$\delta = \frac{PL}{2A_aE_a + A_sE_s}$$

$$= \frac{40 \times 10^3 \times 5 \times 10^3}{2 \times 314 \times 70\,000 + 177 \times 200\,000} \, \frac{\text{N} \cdot \text{mm}}{\text{mm}^2 \cdot \text{N/mm}^2}$$

$$= 2.52 \text{ mm}$$

From (2),

$$\sigma_a = \frac{P_a}{A_a} = \frac{E_a}{L}\delta = \frac{70\,000 \times 2.52}{5000} = 35.3 \, \frac{\text{N}}{\text{mm}^2}$$

From (3),

$$\sigma_s = \frac{P_s}{A_s} = \frac{E_s}{L}\delta = \frac{200\,000 \times 2.52}{5000} = 100.8 \, \frac{\text{N}}{\text{mm}^2}$$

Example 1-10

The rods in Fig. 1-40a have their upper ends securely embedded in an unyielding support and their lower ends fastened to an inflexible horizontal yoke. Initially, there is no stress in the rods.

Determine the stresses in the rods when a vertical downward load of 9 kip is applied.

Solution. To determine the stresses in the rods, we must first obtain the forces P_a and P_s indicated in Fig. 1-40b. From the equilibrium equation $\sum F_y = 0$,

$$P_a + P_s + P_a - P = 0$$

$$2P_a + P_s = P = 9 \text{ kip} \tag{1}$$

One more simultaneous equation is required, but all other equilibrium equations are either trivial or are a repetition of equation (1). We must, therefore, resort to consideration of displacement in order to generate an independent equation. In this case the displacement is the downward motion of the yoke, δ, which is equal to the elongation in each of the rods.

From (1-13)

$$\Delta L = \frac{PL}{AE}$$

or

$$P = \frac{AE}{L}\Delta L = \frac{AE}{L}\delta$$

For each aluminum rod

$$P_a = \frac{A_aE_a}{L}\delta \tag{2}$$

and for the steel rod

$$P_s = \frac{A_s E_s}{L} \delta \qquad (3)$$

Substituting into equation (1), we find that

$$2 \frac{A_a E_a}{L} \delta + \frac{A_s E_s}{L} \delta = P$$

$$\delta = \frac{PL}{2 A_a E_a + A_s E_s}$$

$$= \frac{9 \times 16 \times 12}{2 \times 0.442 \times 10,000 + 0.307 \times 29,000} \frac{\text{kip} \cdot \text{in.}}{\text{in.}^2 \cdot \text{ksi}}$$

$$= 0.097 \text{ in.}$$

From (2),

$$\sigma_a = \frac{P_a}{A_a} = \frac{E_a}{L} \delta = \frac{10,000 \times 0.097}{16 \times 12} = 5.05 \text{ ksi}$$

From (3),

$$\sigma_s = \frac{P_s}{A_s} = \frac{E_s}{L} \delta = \frac{29,000 \times 0.097}{16 \times 12} = 14.65 \text{ ksi}$$

A second indeterminate problem will further demonstrate the use of displacement to obtain a solution where the equations of statics do not suffice.

Example 1-11

For the statically indeterminate steel structure shown in Fig. 1-41a, calculate the stresses in the members.

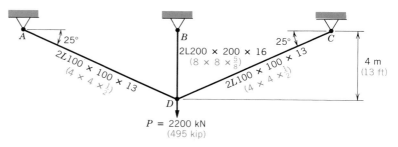

FIG. 1-41a

Solution.   The lengths of the members are given in Fig. 1-41*b*.

$$A_1 = 2 \times 2430 \qquad \text{(Table B-4)} \qquad A_2 = 2 \times 6140 \qquad \text{(Table B-4)}$$

$$= 4860 \text{ mm}^2 \qquad\qquad\qquad = 12\,280 \text{ mm}^2$$

$$E_1 = E_2 = E = 200\,000 \frac{\text{N}}{\text{mm}^2} \qquad \text{(Table A)}$$

From symmetry it is evident that point *D* will move vertically when the load is applied. The first objective will be to calculate, δ, the downward displacement at *D* shown in Fig. 1-41*c*.

$$\varepsilon_1 = \frac{\Delta L_1}{L_1} \qquad\qquad\qquad\qquad \varepsilon_2 = \frac{\Delta L_2}{L_2}$$

$$\sigma_1 = E\varepsilon_1 = \frac{E \,\Delta L_1}{L_1} = \frac{E \sin 25^\circ}{L_1} \delta \quad \textbf{(1)} \qquad \sigma_2 = E\varepsilon_2 = \frac{E \,\Delta L_2}{L_2} = \frac{E}{L_2} \delta \qquad \textbf{(2)}$$

$$T_1 = \sigma_1 A_1 = \frac{E A_1 \sin 25^\circ}{L_1} \delta \quad \textbf{(1.1)} \qquad T_2 = \sigma_2 A_2 = \frac{E A_2}{L_2} \delta \qquad \textbf{(2.1)}$$

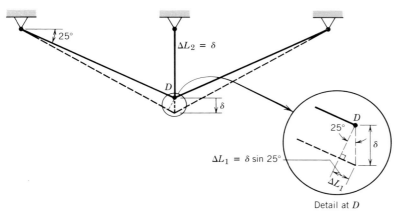

FIG. 1-41*b*

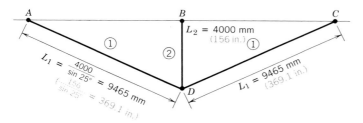

Detail at *D*

FIG. 1-41*c*

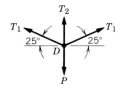

FIG. 1-41d

Forces acting on point D are indicated in Fig. 1-41d.

$$\sum F_y = 0$$

$$T_1 \sin 25° + T_2 + T_1 \sin 25° - P = 0$$

$$2 \sin 25° T_1 + T_2 = P \qquad (3)$$

Substituting the tension forces given in (1.1) and (2.1) into (3) gives an equilibrium equation in one unknown, δ.

$$2 \sin 25° \frac{EA_1 \sin 25°}{L_1} \delta + \frac{EA_2}{L_2} \delta = P$$

Solving for δ, we obtain

$$\delta = \frac{P}{\dfrac{2 \sin^2 25° EA_1}{L_1} + \dfrac{EA_2}{L_2}} \qquad (4)$$

$$= \frac{2\,200\,000}{\dfrac{2 \sin^2 25° \times 200\,000 \times 4860}{9465} + \dfrac{200\,000 \times 21\,280}{4000}} \frac{\text{N}}{\dfrac{\text{N}}{\text{mm}^2} \cdot \dfrac{\text{mm}^2}{\text{mm}} + \dfrac{\text{N}}{\text{mm}^2} \cdot \dfrac{\text{mm}^2}{\text{mm}}}$$

$$= 3.14 \text{ mm}$$

Substituting into (1) and (2) gives the required stresses:

$$\sigma_1 = \frac{200\,000 \sin 25°}{9465} 3.14 \frac{\text{N}}{\text{mm}^2} \cdot \frac{\text{mm}}{\text{mm}}$$

$$= 28 \frac{\text{N}}{\text{mm}^2}$$

$$\sigma_2 = \frac{200\,000}{4000} 3.14 \frac{\text{N}}{\text{mm}^2} \cdot \frac{\text{mm}}{\text{mm}}$$

$$= 157 \frac{\text{N}}{\text{mm}^2}$$

As a matter of interest we calculate the force in the vertical member

$$T_2 = \sigma_2 A_2 = 157 \times 12\,280 \frac{\text{N}}{\text{mm}^2} \cdot \text{mm}^2$$

$$= 1930 \times 10^3 \text{ N} = 1930 \text{ kN}$$

Note that the stress in the vertical member is more than five times the stress in the inclined members and that the bulk of the load (88%) is carried by the vertical member. This illustrates the principle that when there are several routes by which a load may be carried to the supports, the stiffest route carries more load than those that are more flexible. This is a characteristic of indeterminate systems.

The above problem could also have been solved by substituting δ as given in (4) directly into (1) and (2) without evaluating the displacement. With that substitution it would have been found that E could be canceled out of the stress formulas. This means that the value of E is not significant in the determination of stress. Consequently, the stresses and loads would remain unchanged if members of the same dimensions, but different Young's modulus, were substituted for the steel angles. Of course, the stresses in the new members would have to stay within the elastic range for the analysis to remain valid.

Example 1-11

For the statically indeterminate steel structure shown in Fig. 1-41a, calculate the stresses in the members.

Solution. The lengths of the members are given in Fig. 1-41b.

$$A_1 = 2 \times 3.75 \quad \text{(Table B-4)} \qquad A_2 = 2 \times 9.61 \quad \text{(Table B-4)}$$

$$= 7.50 \text{ in.}^2 \qquad\qquad = 19.22 \text{ in.}^2$$

$$E_1 = E_2 = E = 29,000 \text{ ksi} \qquad \text{(Table A)}$$

From symmetry it is evident that point D will move vertically when the load is applied. The first objective will be to calculate, δ, the downward displacement at D shown in Fig. 1-41c.

$$\varepsilon_1 = \frac{\Delta L_1}{L_1} \qquad\qquad\qquad \varepsilon_2 = \frac{\Delta L_2}{L_2}$$

$$\sigma_1 = E\varepsilon_1 = \frac{E\,\Delta L_1}{L_1} = \frac{E\sin 25°}{L_1}\delta \quad \textbf{(1)} \qquad \sigma_2 = E\varepsilon_2 = \frac{E\,\Delta L_2}{L_2} = \frac{E}{L_2}\delta \quad \textbf{(2)}$$

$$T_1 = \sigma_1 A_1 = \frac{EA_1\sin 25°}{L_1}\delta \quad \textbf{(1.1)} \qquad T_2 = \sigma_2 A_2 = \frac{EA_2}{L_2}\delta \quad \textbf{(2.1)}$$

Forces acting on point D are indicated in Fig. 1-41d.

$$\sum F_y = 0$$

$$T_1 \sin 25° + T_2 + T_1 \sin 25° - P = 0$$

$$2 \sin 25° \, T_1 + T_2 = P \qquad\qquad \textbf{(3)}$$

Substituting the tension forces given in (1.1) and (2.1) into (3) gives an equilibrium equation in one unknown, δ.

$$2 \sin 25° \frac{EA_1\sin 25°}{L_1}\delta + \frac{EA_2}{L_2}\delta = P$$

Solving for δ, we obtain

$$\delta = \frac{P}{\dfrac{2\sin^2 25° EA_1}{L_1} + \dfrac{EA_2}{L_2}} \qquad\qquad \textbf{(4)}$$

$$= \frac{495}{\dfrac{2\sin^2 25° \times 29,000 \times 7.5}{369.1} + \dfrac{29,000 \times 19.22}{156}} \quad \frac{\text{kip}}{\dfrac{\text{kip}}{\text{in.}^2} \times \dfrac{\text{in.}^2}{\text{in.}} + \dfrac{\text{kip}}{\text{in.}^2} \times \dfrac{\text{in.}^2}{\text{in.}}}$$

$$= 0.13 \text{ in.}$$

Substituting into (1) and (2) gives the required stresses:

$$\sigma_1 = \frac{29,000 \sin 25°}{369.1} 0.13 \frac{\text{kip}}{\text{in.}^2} \cdot \frac{\text{in.}}{\text{in.}}$$

$$= 4.32 \text{ ksi}$$

$$\sigma_2 = \frac{29,000}{156} 0.13 \frac{\text{kip}}{\text{in.}^2} \cdot \frac{\text{in.}}{\text{in.}}$$

$$= 24.17 \text{ ksi}$$

As a matter of interest we calculate the force in the vertical member

$$T_2 = \sigma_2 A_2 = 24.17 \times 19.22 \frac{\text{kip}}{\text{in.}^2} \text{ in.}^2$$

$$= 464.5 \text{ kip}$$

Note that the stress in the vertical member is more than five times the stress in the inclined members and that the bulk of the load (94%) is carried by the vertical member. This illustrates the principle that when there are several routes by which a load may be carried to the supports, the stiffest route carries more load than those that are more flexible. This is a characteristic of indeterminate systems.

The above problem could also have been solved by substituting δ as given in (4) directly into (1) and (2) without evaluating the displacement. With that substitution it would have been found that E could be canceled out of the stress formulas. This means that the value of E is not significant in the determination of stress. Consequently, the stresses and loads would remain unchanged if members of the same dimensions, but different Young's modulus, were substituted for the steel angles. Of course, the stresses in the new members would have to stay within the elastic range for the analysis to remain valid.

Problems 1-82 to 1-91

1-14 THERMAL STRESSES

When the temperature of a body changes, the particles expand or contract according to well-known laws of physics. The *thermal strain* in unconstrained particles is given by

$$\boxed{\varepsilon_T = \alpha \, \Delta T} \tag{1-15}$$

where α is the coefficient of thermal expansion and ΔT is the change in temperature. This strain would cause a dimension L in the material to change by

$$\boxed{\Delta L = L\varepsilon_T = L\alpha \, \Delta T} \tag{1-16}$$

When the particles are not constrained, the thermal strains cause no stresses, but when all, or part, of the body is confined so that it cannot change size freely, a stress known as *thermal stress*, σ_T, will result. Values for α for various engineering materials are listed in Table A in the Appendix.

Example 1-12

The steel rod shown in Fig. 1-42a is built into a structure so that its ends are embedded in rigid supports. When the temperature is 20°C, the ends are secured and there is no stress in the bar. Determine the stress when the temperature drops to −35°C. Assume that the supports do not move as a result of temperature change or force.

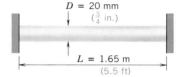

$D = 20$ mm
($\frac{3}{4}$ in.)

$L = 1.65$ m
(5.5 ft)

FIG. 1-42a

Solution. Imagine that the rod is cut while at the higher temperature and still stress-free. The cut can be made in any location but in this case is made at the right support as shown in Fig. 1-42b. Now when the temperature drops there is no constraint, and no stress, and a gap opens at the right end as the rod becomes shorter. If we consider the bar with a load P applied to the end, the bar will elongate an amount given by ΔL, as indicated in Fig. 1-42c.

We will adjust P so that ΔL equals the amount of the gap that appeared in the previous step; that is, P must satisfy

$L\alpha\,\Delta T$

FIG. 1-42b

$$\frac{PL}{AE} = L\alpha\,\Delta T$$

Then the required P is

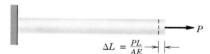

P

$\Delta L = \frac{PL}{AE}$

FIG. 1-42c

$$P = \frac{AEL\alpha\,\Delta T}{L} = AE\alpha\,\Delta T$$

If we superimpose the last two diagrams, we see that P given above is the force that the right support must exert on the bar if the temperature drops while the bar is attached to the rigid supports. This gives a thermal stress

$$\sigma_T = \frac{P}{A} = \frac{AE\alpha\,\Delta T}{A} = E\alpha\,\Delta T = 200\,000 \times 12 \times 10^{-6} \times 55 = 132 \text{ MPa}$$

Note that the thermal stress is independent of the dimensions of the bar so that, for equal temperature change, the same stress would exist in a small machine bolt as a long railroad track if both are fixed without provision for expansion. In this illustration a stress almost equal to the allowable stress for steel was induced by temperature change alone. This would leave a very small margin for other loads that the bar might be required to carry. Also, note that the temperature range was not uncommon for outdoors in a temperate climate. A range of 500°C or more is quite common for pipelines in steam power plants. Consequently, most steam lines must be provided with expansion joints to eliminate, or at least alleviate, thermal stresses.

The simple equation obtained above,

$$\sigma_T = E\alpha\,\Delta T \qquad (1\text{-}17)$$

can be used for identical cases, but care must be taken not to misapply it. For example, it will give meaningless results if applied to Problem 1-95. In that problem the solution can only be obtained by using the principles that were illustrated in Example 1-11.

Example 1-12

The steel rod shown in Fig. 1-42a is built into a structure so that its ends are embedded in rigid supports. When the temperature is 70°F, the ends are secured and there is no stress in the bar. Determine the stress when the temperature drops to −30°F. Assume that the supports do not move as a result of temperature change or force.

Solution. Imagine that the rod is cut while at the higher temperature and still stress-free. The cut can be made in any location but in this case is made at the right support as shown in Fig. 1-42b. Now when the temperature drops there is no constraint, and no stress, and a gap opens at the right end as the rod becomes shorter. If we consider the bar with a load P applied to the end, the bar will elongate an amount given by ΔL, as indicated in Fig. 1-42c.

We will adjust P so that ΔL equals the amount of the gap that appeared in the previous step, that is, P must satisfy

$$\frac{PL}{AE} = L\alpha\,\Delta T$$

Then the required P is

$$P = \frac{AEL\alpha\,\Delta T}{L} = AE\alpha\,\Delta T$$

If we superimpose the last two diagrams, we see that P given above is the force that the right support must exert on the bar if the temperature drops while the bar is attached to the rigid supports. This gives a thermal stress

$$\sigma_T = \frac{P}{A} = \frac{AE\alpha\,\Delta T}{A} = E\alpha\,\Delta T = 29 \times 10^3 \times 6.5 \times 10^{-6} \times 100 = 18.9 \text{ ksi}$$

Note that the thermal stress is independent of the dimensions of the bar so that, for equal temperature change, the same stress would exist in a small machine bolt as a long railroad track if both are fixed without provision for expansion. In this illustration a stress almost equal to the allowable stress for steel was induced by temperature change alone. This would leave a very small margin for other loads that the bar might be required to carry. Also, note that the temperature range was not uncommon for outdoors in a temperate climate. A range of 900°F or more is quite common for pipelines in steam power plants. Consequently, most steam lines must be provided with expansion joints to eliminate, or at least alleviate, thermal stresses.

The simple equation obtained above,

$$\sigma_T = E\alpha\,\Delta T \tag{1-17}$$

can be used for identical cases, but care must be taken not to misapply it. For example, it will give meaningless results if applied to Problem 1-95. In that problem the solution can only be obtained by using the principles that were illustrated in Example 1-11.

Problems 1-92 to 1-99

Thermal stress frequently results from external constraints preventing the fibers of a material from expanding or contracting freely when the temperature changes. Constraint can also be imposed by one portion of the material on another portion when there is a differential expansion within the material, usually due to nonuniform temperature. This is best illustrated by imagining a large flat plate that is initially stress-free. If a flame is directed at one region of the plate, say the center, that region will attempt to expand but will be partially constrained by the surrounding cooler material. The central region will be in compression and the surrounding region will be subjected to a complex stress pattern. Even for such a simple model as the plate, thermal stresses are not easily calculated. They are extremely difficult to determine in many practical problems such as the cylinder-head of an internal combustion engine. Here metal having a very intricate shape is heated on the surfaces of the combustion chamber while other surfaces are chilled by the cooling fluid. The temperature distribution is hard to evaluate and the resultant thermal stresses are even more difficult to determine; hence any calculated results can be expected to be rough approximations only. This is often the case with thermal stresses.

From the magnitudes of the stresses encountered in the problems, where temperature changes were quite moderate, it is evident that very large *residual stresses* could exist in castings when some parts are cooled more rapidly and solidify before other parts. The design of castings, such as those required for engine blocks and heads, is often governed by consideration of thermal stress, residual stress, and cracking.

Problem 1-100

PROBLEMS

1-1 For the structure shown in the illustration:

    a. Select standard steel rods from Table D for the three members. Use the allowable stress given in Table A for low-strength structural steel.

    b. How much do the rods weigh?

    c. As well as carrying the ingot, the rods must carry their own weight. In part (a) the weight of the rods was not taken into account. Was a serious error introduced?

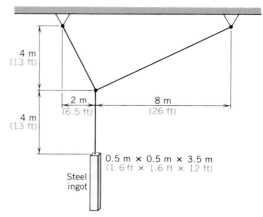

PROBLEM 1-1

1-2 Select a pair of angles from Table B-4 for member A-B. The material is low-strength structural steel. Use the allowable stress given in Table A.

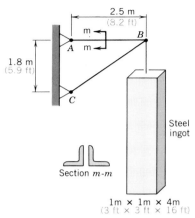

PROBLEM 1-2

1-3 Select pairs of angles from Table B-4 for A-B and B-C. The allowable stress is 140 MPa (20 ksi) in tension and 60 MPa (9 ksi) in compression.

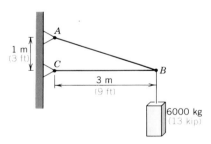

PROBLEM 1-3

1-4 The allowable stress in compression is 70 MPa (10 ksi). Select standard pipes from Table C for members A-B and A-C of the given crane.

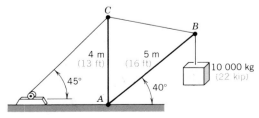

PROBLEM 1-4

1-5 For the simple pin-jointed truss select C shapes from Table B-3 for each of the members A-B, B-C, B-D, A-D, and D-C if the allowable compressive stress is 65 MPa (9.4 ksi) and the allowable tensile stress is 150 MPa (22 ksi).

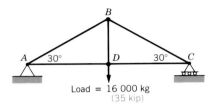

PROBLEM 1-5

1-6 A hoist was constructed from Douglas fir as shown. Member A-B was designed with an allowable stress of 9.0 MPa (1.3 ksi), and a 100 × 150 mm (4 × 6 in.) cross section was required. What size of low-strength, structural steel rod may be substituted for A-B in order to maintain the same lifting capacity of the hoist? Use the allowable stress given in Table A.

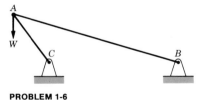

PROBLEM 1-6

1-7 A member consisting of a pair of 55 × 55 × 6 (2 × 2 × $\frac{1}{4}$) angles carries an axial tensile load of 80 kN (18 kip). The angles are attached to a 10-mm (0.4-in.) gusset plate by a row of 2-20 mm (0.8 in.) bolts. Calculate

a. The maximum tensile stress in the angles.
b. The shearing stress in the bolts.
c. The bearing stress between the bolts and the angles.
d. The bearing stress between the bolts and the gusset.

1-8 How many 25-mm (1-in.) bolts are required to connect a pair of 65 × 65 × 10 ($2\frac{1}{2} \times 2\frac{1}{2} \times \frac{3}{8}$) angles carrying a load of 180 kN (40 kip) to a 16-mm (0.63-in.) gusset plate? Allowable stresses are $\tau_a = 90$ MPa (13 ksi) and $\sigma_{ab} = 260$ MPa (38 ksi).

1-9 A lap joint is to be made as shown. What load, *P*, can be safely carried by the joint if one 30-mm (1.18-in.) bolt is used to join the bars? Assume the following allowable stresses:

$$\sigma_a(\text{tension}) = 160 \text{ MPa (23 ksi)}$$

$$\tau_a = 90 \text{ MPa (13 ksi)}$$

$$\sigma_{ab} = 250 \text{ MPa (36 ksi)}$$

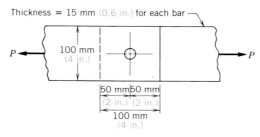

PROBLEM 1-9

1-10 If the steel bars in Prob. 1-9 are joined by two 20-mm (0.79-in.)-diameter bolts as shown, by how much (%) will the safe load change?

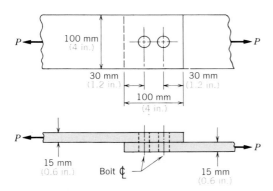

PROBLEM 1-10

1-11 Design the bolted connection choosing both plate thickness and bolt diameter if the basic configuration is as shown and the connection is to carry a load *P* = 175 kN (39 kip). Allowable stresses are

$$\sigma_a(\text{tension}) = 140 \text{ MPa (20 ksi)}$$

$$\sigma_{ab} = 260 \text{ MPa (38 ksi)}$$

$$\tau_a = 100 \text{ MPa (14 ksi)}$$

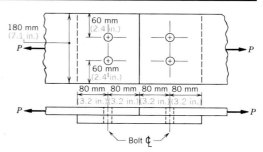

PROBLEM 1-11

1-12 A pair of 125 × 90 × 10 (5 × 3½ × ⅜) angles are connected to a gusset plate by 8-mm (0.32-in.) fillet welds as shown in the sketch. For an applied axial load, *P*, of 600 kN (135 kip) and an allowable shearing stress in the welds of 110 MPa (16 ksi), check the welds for safety.

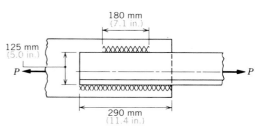

PROBLEM 1-12

1-13 For the welded connection shown in the illustration determine the safe load that can be carried if the allowable stress in the 10-mm (0.39-in.) fillet welds is 100 MPa (14 ksi).

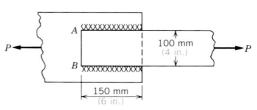

PROBLEM 1-13

1-14 If the weld in Prob. 1-13 is continued along the edge *A-B*, by how much can *P* be increased?

1-15 For the connection shown in the illustration, determine the length, L, of 10-mm (0.39-in.) fillet welds to be applied along A-B and C-D so that the allowable stress in the bar is reached when the stress in the weld is τ_a. This is known as a balanced design. Assume

$$\sigma_a\text{(tension in bar)} = 140 \text{ MPa (20 ksi)}$$

$$\tau_a\text{(weld)} = \ 90 \text{ MPa (12 ksi)}$$

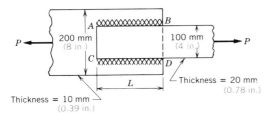

PROBLEM 1-15

1-16 A $200 \times 200 \times 10$ ($8 \times 8 \times \frac{3}{8}$) angle is to be welded to the plate with 8-mm (0.32-in.) fillet welds as shown. Find L such that the resultant restraining force due to the three welds, A, B, and C when loaded to the allowable stress is in direct alignment with the centroidal axis of the angle. With this length of weld, determine the safe load, P. Assume that $\tau_a = 100$ MPa (14 ksi) in the weld.

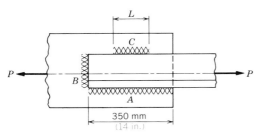

PROBLEM 1-16

1-17 Determine the bending moment equation for each region of the beam and draw the bending moment diagram.

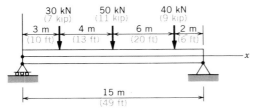

PROBLEM 1-17

1-18 Determine the bending moment equation for each region of the beam and draw the bending moment diagram.

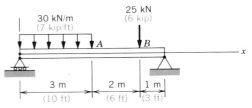

PROBLEM 1-18

1-19 Determine the bending moment equation for each region of the beam and draw the bending moment diagram.

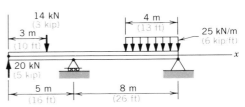

PROBLEM 1-19

1-20 Determine the shear equation for each region of the beam and draw the shear diagram.

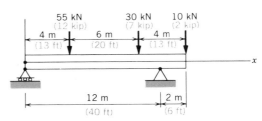

PROBLEM 1-20

1-21 Determine the shear equation for each region of the beam and draw the shear diagram.

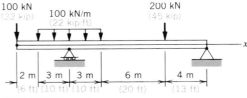

PROBLEM 1-21

1-22 Determine the shear equation for each region of the beam and draw the shear diagram.

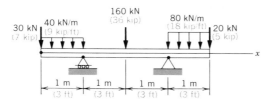

PROBLEM 1-22

1-23 Determine the shear equation for each region of the beam and draw the shear diagram.

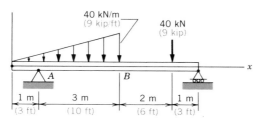

PROBLEM 1-23

1-24 to 1-30 For the given beams derive the shear and bending moment equations and sketch the shear and bending moment diagrams. Identify the magnitude and location of the maximum values for both the shear force and the bending moment.

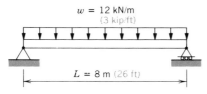

PROBLEM 1-24

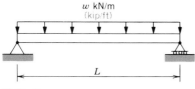

PROBLEM 1-25

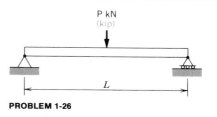

PROBLEM 1-26

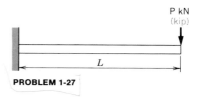

PROBLEM 1-27

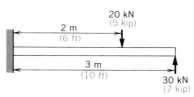

PROBLEM 1-28

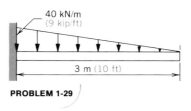

PROBLEM 1-29

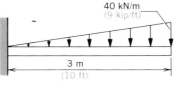

PROBLEM 1-30

1-31 to 1-41   Draw shear and bending moment diagrams for the given beams. Determine the location and the magnitude of the largest bending moment.

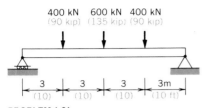

PROBLEM 1-31

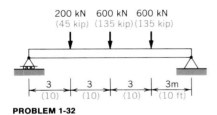

PROBLEM 1-32

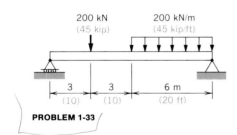

PROBLEM 1-33

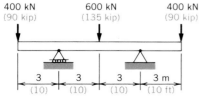

PROBLEM 1-34

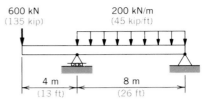

PROBLEM 1-35

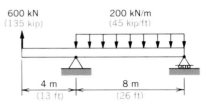

PROBLEM 1-36

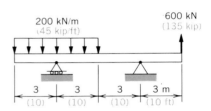

PROBLEM 1-37

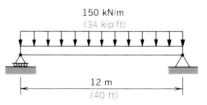

PROBLEM 1-38

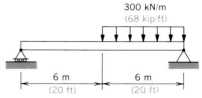

PROBLEM 1-39

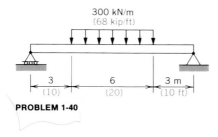

PROBLEM 1-40

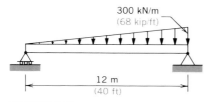

PROBLEM 1-41

1-42 to 1-44   Use shear and bending moment diagrams to determine the forces in the members of the given truss. Report your answers by a sketch similar to Fig. 1-16g.

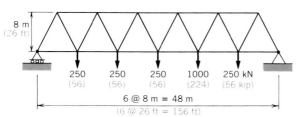

PROBLEM 1-42

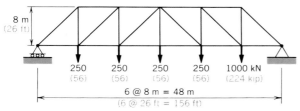

PROBLEM 1-43

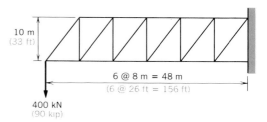

PROBLEM 1-44

1-45   Using an allowable stress of 140 MPa (20 ksi) in tension and 90 MPa (13 ksi) in compression, select pairs of angles from Table B-4 for the members of the structure in Problem 1-44.

1-46   Using an allowable stress of 140 MPa (20 ksi) in tension and 80 MPa (11 ksi) in compression, select pairs of angles from Table B-5 for the members of the given structure.

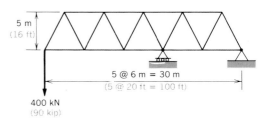

PROBLEM 1-46

1-47   Using an allowable stress of 140 MPa (20 ksi) in tension and 85 MPa (12 ksi) in compression, select pairs of angles from Table B-4 for the members of the given structure.

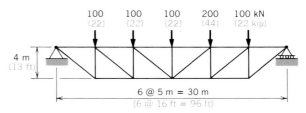

PROBLEM 1-47

1-48 to 1-51   In each of the following four problems use the principles demonstrated in Example 1-8 to:

a. Determine the normal stress at *A* and the shearing stress at *B*.
b. If the material is low-strength structural steel and the allowable stresses are those given in Table A, is the beam safe?

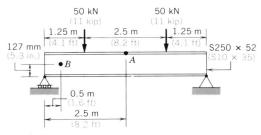

PROBLEM 1-48

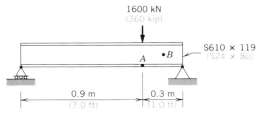

PROBLEM 1-49

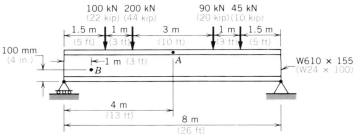

PROBLEM 1-50

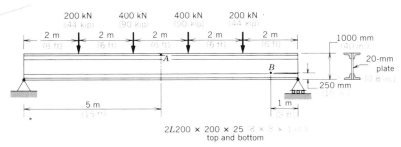

PROBLEM 1-51

1-52 Determine:

a. The maximum stress in the upper flange.
b. The maximum stress in the lower flange.
c. The maximum shearing stress in the web.

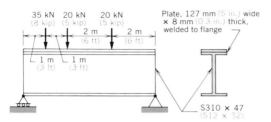

PROBLEM 1-52

1-53 A member is fabricated by welding together two *C*-shapes and a plate as shown.

a. If the allowable shear stress is 70 MPa (10 ksi), determine the required thickness of the vertical plate.
b. Determine the maximum tensile stress in the lower channel.

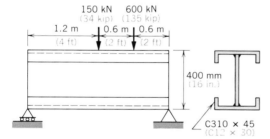

PROBLEM 1-53

1-54 to 1-56 The given pressure vessel has an inside diameter of 2.5 m (8.2 ft) and a thickness of 12 mm (0.472 in.). The material has an allowable stress of 165 MPa (24 ksi).

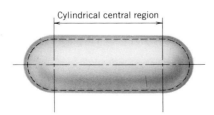

PROBLEMS 1-54 to 1-56

1-54 What internal pressure can the vessel safely carry if all welds are 100% efficient?

1-55 What internal pressure can the vessel safely carry if the longitudinal welds are 80% efficient and the circumferential welds 60% efficient?

1-56 What internal pressure can the vessel safely carry if the longitudinal welds are 100% efficient and the circumferential welds 40% efficient?

1-57 A vessel similar to that of Prob. 1-54 has an inside diameter of 3 m (118 in.) and a thickness of 20 mm (0.787 in.). The allowable stress in tension is 95 MPa (14 ksi). The vessel contains fluid at a pressure of 1000 kPa (145 psi) and the efficiencies of the longitudinal and circumferential seams are 85 and 50%, respectively. Check the vessel for safety.

1-58 The stress equations (1-7) and (1-8a) were developed for pressure vessels. For a pipe the same element could be chosen and (1-7) arrived at again. However, the ends of a pipeline are seldom capped in a manner similar to the pressure vessel; in fact, pipes usually terminate by being connected to some massive, immovable object such as a pump, storage tank, or steam boiler. The free body can then not be Fig. 1-21*b*.

If a pipe is straight and runs from a fixed unit, such as a pump, to another fixed unit, say, a storage tank, the longitudinal stresses are very difficult to calculate and may be very large. However, most pipelines consist of straight lengths of pipe joined by elbows. By taking as a free body two straight lengths and an elbow with deflection angle, θ, and assuming that there are no bending stresses, show that (1-8a) is valid for the pipe.

1-59 to 1-61 A piece of steel pipe has an inside diameter of 34.5 mm (1.36 in.). Determine the required wall thickness for the conditions specified in each problem.

1-59 The allowable stress in tension is 100 MPa (14.5 ksi), and the pipe contains fluid at a pressure of 9000 kPa (1300 psi).

1-60 The allowable stress in tension is 100 MPa (14.5 ksi), and the pipe is subjected to an axial load of 20 kN (4.5 kip).

1-61 The allowable stress in shear is 50 MPa (7.25 ksi), and the pipe is subjected to torque of 1.0 kN·m (740 lb·ft).

1-62 A pipeline is made of steel having allowable stresses 200 MPa (29 ksi) in tension and 120 MPa (17 ksi) in shear. The pipe wall is 3 mm (0.118 in.) thick and the outside diameter is 102 mm (4.0 in.).

  a. Determine the pressure that can be safely carried by the pipe.

  b. If an axial tensile force is applied along with the pressure load determined in (a), what force can be safely applied?

  c. If a torque load alone acts on the pipe, what load can be safely applied?

  d. If the loads determined in (b) and (c) are applied simultaneously, will the pipe be safe? (*Note:* Give part (d) some thought even if you do not feel competent to answer the question.)

1-63 A pair of low-strength structural steel equal leg angles must carry an axial load of 1.85 MN (416 kip). The code that governs the design requires a safety factor of 1.65 on the yield. Select the angles.

1-64 A brake rod is required to carry a tensile force of 15 kN (3.4 kip). The rod is made of AISI 1090 and has a diameter of 8 mm (0.315 in.).

  a. What is the safety factor, based on the yield?

  b. Determine the diameter of a 6061-T6 aluminum alloy replacement rod for a safety factor equal to that of the original design.

1-65 Repeat Prob. 1-64 with the safety factor based on the ultimate.

1-66 The brake drum is 240 mm (9.45 in.) in diameter and has a torque of 1.92 kN·m (1420 lb·ft) applied to it by the shaft. A force P of 1.00 kN (0.225 kip) is required to stop the drum. Determine the safety factor in the brake band based on

  a. The yield.

  b. The ultimate.

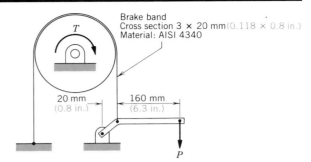

PROBLEM 1-66

1-67 What was the safety factor used for the design of the pin at A in the bell-crank mechanism shown below? The pin is made from AISI 1020 steel and design was based on yield stress. The yield stress in shear is 0.6 times the tensile yield stress. The pin diameter, d, is 15 mm (0.6 in.).

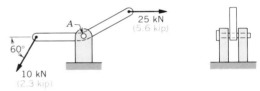

PROBLEM 1-67

1-68 A boiler drum with an inside diameter of 1500 mm (60 in.) contains superheated steam at 370°C (700°F) and 8000 kPa (1160 psi). At that temperature the steel that is to be used for the drum has a yield stress of 175 MPa (25 ksi) and an ultimate stress of 320 MPa (46 ksi). All welds are 100% efficient.

    The code specifies that the vessel must have a safety factor of at least 1.5 on yield and at least 3.0 on ultimate.

    Determine the minimum safe plate thickness.

1-69 A hollow steel shaft turns at 2500 rpm and transmits 2000 kW (2680 horsepower). The outside diameter is 75 mm (3 in.) and the material is AISI 1090 steel. Determine the maximum allowable inside diameter if the safety factor is to be 1.50 based on yield. Assume that the yield stress in shear is 60% of the yield stress in tension.

1-70 A 200-mm (8-in.)-long rubber band on a slingshot is being stretched by 45 mm (1.8 in.). What is the strain in the rubber?

1-71 The steel tension member shown in the figure has a tapering circular cross section. Calculate its elongation under the load of 60 kN (14 kip). (*Note:* Because the shaft has a nonuniform cross section, you will have to determine the deformation of an elemental slice and then integrate the deformation over the length of the tapering shaft.)

$D_2 = 25$ mm
(1.0 in.)

300 mm
(12 in.)

$D_1 = 15$ mm
(0.6 in.)

60 kN
(14 kip)

PROBLEM 1-71

1-72 The following load versus deformation data pertain to a tensile test of 35, grade 400 (#11, grade 60), reinforcing bar. Plot the stress–strain curve for the bar if the extensometer readings refer to a gage length of 100 mm (4 in.).

Load		Extensometer Reading	
(kN)	(kip)	(mm)	(10^{-3} in.)
40	(9.0)	0.020	(0.8)
80	(18.0)	0.041	(1.6)
120	(27.0)	0.060	(2.4)
160	(36.0)	0.079	(3.2)
200	(45.0)	0.101	(4.0)
240	(54.0)	0.121	(4.8)
280	(62.9)	0.141	(5.6)
300	(67.4)	0.151	(6.0)
320	(71.9)	0.165	(6.6)
340	(76.4)	0.180	(7.2)
360	(80.9)	0.203	(8.1)
380	(85.4)	0.237	(9.5)
400	(89.9)	0.275	(11.0)
420	(94.4)	0.334	(13.4)
440	(98.9)	0.426	(17.0)
460	(103.4)	0.580	(23.2)
480	(107.9)	End of data acquisition	

1-73 For the reinforcing bar of Prob. 1-72, verify from the stress–strain curve that E agrees reasonably well with the published value for steel and the yield stress is equal to or exceeds the specified strength of grade 400 (60) steel.

1-74 The following load-deformation data were obtained in a tensile test of an aluminum alloy specimen. The flat bar specimen had a cross section of 20 by 4 mm (0.787 by 0.157 in.) and the extensometer readings were taken over a gage length of 50 mm (2 in.). Plot the stress–strain curve and determine

a. The proportional limit.
b. The modulus of elasticity, E. How well does your value compare with the published E?
c. The yield stress by the 0.2% offset method.

Load		Extensometer Reading	
(kN)	(kip)	(mm)	(10^{-3} in.)
1.6	(0.36)	0.014	(0.56)
3.2	(0.72)	0.029	(1.16)
4.8	(1.08)	0.043	(1.72)
6.4	(1.44)	0.058	(2.32)
8.0	(1.80)	0.072	(2.88)
9.6	(2.16)	0.086	(3.44)
11.2	(2.52)	0.101	(4.04)
12.8	(2.88)	0.120	(4.80)
14.4	(3.24)	0.140	(5.60)
16.0	(3.60)	0.171	(6.84)
17.6	(3.96)	0.209	(8.36)
19.2	(4.32)	0.257	(10.28)
20.8	(4.68)	0.312	(12.48)
22.4	(5.04)	0.382	(15.28)
24.0	(5.40)	0.475	(19.00)
25.6	(5.75)	0.591	(23.64)
27.2	(6.11)	Readings discontinued	

To give you some idea about the ultimate strength and ductility of this aluminum alloy specimen, note that it reached an ultimate stress of 340 MPa (49.3 ksi) and an elongation, after failure, of 7.1 percent.

1-75 From the stress–strain curve of Prob. 1-74 determine the secant modulus and tangent modulus at 200 MPa (29 ksi).

1-76 The following stress–strain data refer to a high-strength concrete cylinder that was tested in compression. Carefully plot the entire stress–strain curve and, in comparison with the stress–strain behavior of metals you have met before, comment on the shape of the curve and the ductility of concrete.

Stress		Strain
(MPa)	(ksi)	(%)
4	(0.58)	0.010
8	(1.16)	0.022
12	(1.74)	0.034
16	(2.32)	0.047
20	(2.90)	0.062
24	(3.48)	0.078
28	(4.06)	0.096
32	(4.64)	0.119
34	(4.93)	0.134
36	(5.22)	0.156
37	(5.36)	0.183
36	(5.22)	0.220
34	(4.93)	0.250
32	(4.64)	0.272
30	(4.35)	0.288
28	(4.06)	0.304
26	(3.77)	Virtual disintegration of specimen

1-77 From the concrete stress–strain curve of Prob. 1-76, determine

a. The initial tangent modulus at the origin.
b. The tangent modulus and secant modulus at a stress of 20 MPa (2.9 ksi).

Do your values appear reasonable in comparison with published values?

1-78 For the truss given in Fig. 1-38, determine the change in length of each of the members in the upper chord.

1-79 How much travel must the roller at the right support of the truss in Fig. 1-38 be capable of accommodating?

1-80 For the truss dealt with in Examples 1-6 and 1-7, verify the forces of all bottom chord members and determine the change in length of each of these members.

1-81 Determine the change in length of the vertical members of the truss shown in Fig. 1-38.

1-82 A standard 250-mm (10-in.) pipe is filled with low-strength concrete and flat, rigid plates are placed over the ends. If an axial compressive load of 1750 kN (400 kip) is applied, calculate the stress in the concrete.

1-83 Determine the stress in the steel and in the low-strength concrete when an axial compressive load of 3600 kN (800 kip) is applied to the rigid top plate.

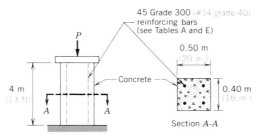

PROBLEM 1-83

1-84 Calculate the allowable axial compressive load if grade 350 (50) reinforcing bars and high-strength concrete are used. Take allowable stresses from Table A.

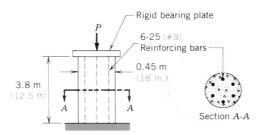

PROBLEM 1-84

1-85 A high-strength concrete pier 400-mm (16-in.) square is reinforced with 4-20 (#6), grade 400 (60), reinforcing bars, one near each corner of the concrete cross section. If the pier is loaded through a rigid steel plate to ensure equal shortening of the concrete and steel, what load can be carried safely by the pier? See Table A for allowable stresses.

1-86 Using allowable stresses given in Table A, determine the safe load, P, for the material combination:

   a. Low-strength concrete—Grade 300 (40) steel
   b. Low-strength concrete—Grade 400 (60) steel
   c. High-strength concrete—Grade 300 (40) steel
   d. High-strength concrete—Grade 400 (60) steel

Since materials having higher strength are usually more costly, is it always good practice to use the highest-strength materials?

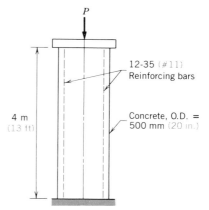

PROBLEM 1-86

1-87 To increase the load-carrying capacity of a short Douglas fir post, the member is reinforced with steel plates on all four sides as shown in the cross section. If the allowable stress in the wood is 6 MPa (0.9 ksi) and in the steel 140 MPa (20 ksi), what axial compressive load can be carried by the post? Assume that the materials are securely fastened together and shorten the same amount.

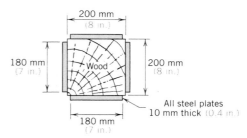

PROBLEM 1-87

1-88 The timber post of Prob. 1-87 is 600 mm (23.6 in.) long and the steel plates are 0.3 mm (0.0118 in.) shorter than the post. What is the allowable axial load on the reinforced post? Assume that relative axial motion between the steel and the wood can occur.

1-89 Determine the stress in the steel and in the aluminum due to a downward force of 30 kN (7 kip) at *P*.

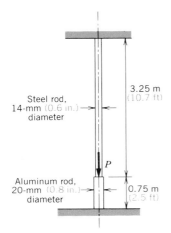

3.25 m
(10.7 ft)

Steel rod,
14-mm (0.6 in.)
diameter

P

Aluminum rod,
20-mm (0.8 in.)
diameter

0.75 m
(2.5 ft)

PROBLEM 1-89

1-90 A solid steel rod is surrounded by an aluminum tube as shown in the figure. A rigid bearing block is supported on the aluminum tube, and a gap of 0.2 mm (0.008 in.) exists between the block and the top of the steel rod. Determine the permissible load, *P*, if the maximum allowable stress is 140 MPa (20 ksi) in the steel and 100 MPa (14 ksi) in the aluminum.

P

0.2 mm
(0.008 in.)

Aluminum
tube
OD = 45 mm
(1.77 in.)
ID = 40 mm
(1.58 in.)

25 mm
(1.00 in.)

500 mm
(20 in.)

Steel

PROBLEM 1-90

1-91 Determine the stresses in the steel and aluminum rods due to the load *P*.

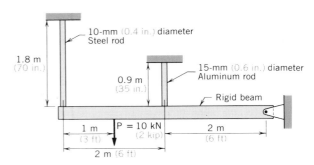

10-mm (0.4 in.) diameter
Steel rod

1.8 m
(70 in.)

0.9 m
(35 in.)

15-mm (0.6 in.) diameter
Aluminum rod

Rigid beam

1 m
(3 ft)

P = 10 kN
(2 kip)

2 m
(6 ft)

2 m (6 ft)

PROBLEM 1-91

1-92 a. Determine the change in length of the members in the upper chord of the truss shown in Fig. 1-38 due to the given loads.
  b. The roller at the right support must be capable of accommodating how much travel?
  c. If the bridge is erected in the summer at 25°C (77°F) and in winter experiences −40°C (−40°F), how much travel must the roller be capable of accommodating?

1-93 A 16-mm (0.630-in.) magnesium alloy rod is stress-free when securely attached to the unyielding supports as shown in the figure. What is the allowable temperature rise in the rod in order not to exceed a stress of 35 MPa (5 ksi) in compression?

300 mm
(12 in.)

PROBLEM 1-93

1-94 After the rod in Prob. 1-93 has been attached to the supports, the supports move 1 mm (0.0394 in.) away from one another. Determine the allowable temperature rise.

1-95 If the steel bar is unstressed at 20°C (68°F), what is the maximum stress at −35°C (−31°F)?

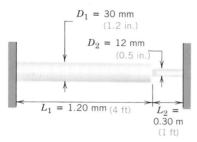

$D_1 = 30$ mm
(1.2 in.)

$D_2 = 12$ mm
(0.5 in.)

$L_1 = 1.20$ mm (4 ft) $L_2 = 0.30$ m
(1 ft)

PROBLEM 1-95

1-96 A hollow steel cylinder of cross-sectional area 2000 mm² (3.10 in.²) concentrically surrounds a solid aluminum cylinder of cross-sectional area 6000 mm² (9.30 in.²). Both cylinders have the same length of 500 mm (20 in.) before the rigid block weighing 200 kN (45.0 kip) is applied at 20°C (68°F) as shown in the figure. Determine the temperature rise required for the total load to be carried by the aluminum.

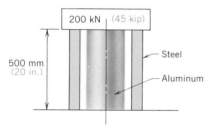

200 kN   (45 kip)

500 mm
(20 in.)

— Steel

— Aluminum

PROBLEM 1-96

1-97 Determine the stress in the rod in Example 1-12 if the right support is not perfectly rigid but moves 0.01 mm (0.00175 in.) for each kN (kip) of force applied by the rod.

1-98 For the assembly of Prob. 1-96, how much load is carried by each cylinder at 60°C (140°F)?

1-99 For the structure of Prob. 1-91, consider that the aluminum rod is well insulated and hence held at a constant temperature while the temperature of the steel rod increases. If the allowable stress in the aluminum rod is 120 MPa (17 ksi), by how many degrees can the temperature of the steel rod increase without exceeding the allowable stress in the aluminum rod?

1-100 A long S310 × 47 (S12 × 32) member is heated so that the web is at 70°C (158°F) and the flanges are at 120°C (248°F). Determine the thermal stresses.

Advanced Problems

1-A Assume that the truss shown in Fig. 1-38 was erected in the fall at a temperature of 10°C (50°F) and the contractor failed to provide a roller for temperature movements; that is, both ends were anchored securely. Determine the thermal stresses in the members of the upper chord of the truss

a. In summer at a temperature of 30°C (86°F).
b. In winter at a temperature of −30°C (−22°F).

1-B The structure is designed to support, by a cable attached to B, a 1.6-m (63-in.) cube of concrete.

   a. Calculate the force in each member.
   b. Determine the change in length of each member.
   c. Calculate the angles at A and C after the load is applied.
   d. Recalculate the forces in the members.
   e. How much error (in percent) would be made by ignoring the change in shape of the structure due to the load?

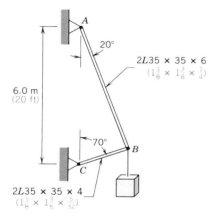

PROBLEM 1-B

1-C For the reinforced concrete member shown in the figure, assume allowable stresses of 140 MPa (20 ksi) for steel and 16 MPa (2.3 ksi) for concrete.

   a. Determine the maximum allowable load P.
   b. In order to carry the load determined in part (a), what values must the dimensions x and y have?

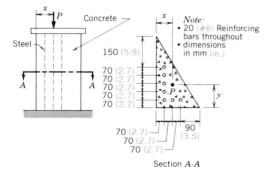

PROBLEM 1-C

1-D A reinforced concrete member has a 300 × 400 mm (12 × 16 in.) cross section and is reinforced by 8-25 (#8) reinforcing bars.

A low-strength concrete is used and grade 300 (40) steel. If the factor of safety for both materials is 2.5 on ultimate strength, calculate the safe compressive load for the member.

How much gain in strength could be achieved by using a grade 400 (60) steel, which is stronger and also more costly? Note that the grade number refers to the yield strength of the reinforcing steel.

1-E Determine the stresses in the rods of the structure given in Example 1-10 if the given load is applied and the temperature is increased by 40°C (72°F).

1-F A pin-jointed structure is composed of two steel bars attached to immovable supports. Before the load is applied, the bars are unstressed and horizontal as shown.

   a. Determine the load P required to displace the center pin 240 mm (9.45 in.) downward. Determine the stress in the bars.
   b. If the bars are high-strength structural steel, determine the safe load.
   c. Determine the stress in the bars when the load is 2000 kN (450 kip).

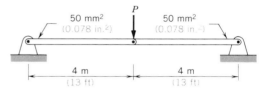

PROBLEM 1-F

1-G A tube, made of aluminum alloy 7075-T6, has an outside diameter of 201.26 mm (7.924 in.) and a wall thickness of 6 mm (0.236 in.). A tube made of AISI 4063 has an inside diameter of 200.00 mm (7.874 in.) and a wall thickness of 2 mm (0.0787 in.). Both tubes are 500 mm (20 in.) long.

By heating the steel tube and cooling the aluminum tube, the inside diameter of the steel tube can be made to exceed the outside diameter of the aluminum tube. In this state the aluminum tube is quickly slipped inside the steel tube. When the assembly returns to ambient temperature a shrink fit has been achieved.

Determine the stress in the steel and in the aluminum when the assembly is at ambient temperature.

1-H The aluminum tube in Prob. 1-G has an outside diameter of 200.00 mm (7.874 in.). All other dimensions remain as given in Prob. 1-G.

The aluminum tube is placed inside the steel tube and the assembly is then heated 200°C (360°F) above ambient temperature. Determine the stress in the steel and in the aluminum while the assembly is at the elevated temperature.

1-I If the coefficient of friction between the tube surfaces in Prob. 1-H is 0.15, determine the force required to slide one tube over the other longitudinally while they are at the elevated temperature.

2

BEAMS IN BENDING I

2-1 INTRODUCTION

Beams, that is members that are subjected to lateral loads and hence to bending, have cross sections that differ greatly from those of tension or compression members. We have already seen that angles in pairs are very convenient for the members of a truss structure, but such an arrangement would not be used for beams because the material would be distributed in a manner that is inefficient for carrying bending loads. The shapes of some commonly used beams are shown in Fig. 2-1. For the standard steel beam or so-called *S-shape* of Fig. 2-1a, the material is concentrated in an upper and lower *flange* with a thinner *web* connecting the flanges. We have already dealt with this type of member in a superficial manner and have calculated the stresses by applying, without adequate justification, some rules that we had discovered for trusses. In this chapter we will develop a new theory of bending that will be more accurate than the approximate method of Chapter 1.

The wide-flange beam section or so-called *W-shape* of Fig. 2-1b has the material arranged in two flanges and a web as in the S-shape. However, compared to the S-shape the proportions are different in that the width of the flange is larger relative to the depth of the member. Although both the S-shape and the W-shape are well proportioned to act as beams, the large flange width of the W-shape permits that section also to be used very effectively for columns.

The S-shapes and W-shapes are manufactured as rolled steel sections. When the bending moments are too large to be carried by the largest available rolled section, the basic section can be reinforced by *cover plates* as shown in Fig. 2-1c. The plates may run the full length of the beam or be placed only in the region of highest bending moment. The cover plates must be secured to the rolled section by rivets, bolts, or welds to ensure that all elements act together as a single unit. The selection of rivet size and spacing, or the design of the welds, poses some interesting engineering problems. The analysis and design of a riveted cover plate will be illustrated toward the end of this chapter.

When the bending moment is too large to be carried by the largest available rolled section with cover plates, it is necessary to fabricate a section as a built-up member. In practice this is done by welding or riveting angles and plates together to form a *plate girder* such as that shown in Fig. 2-1d. The decision as to the number, size, and method of attachment of cover plates will be made by the designer on the basis of the bending moment diagram, sizes of available materials, and other factors, including cost.

In actual practice, rivets have been almost entirely replaced by welds (for shop-fabricated parts) and by bolts (for field assembly). We will treat riveted cover plates and riveted plate girders because these designs serve as a useful introduction to similar welded members.

In many applications concrete has advantages over steel as a material for beams. We might, as beginners, make the error of merely copying the general form that seemed satisfactory for steel members with larger thicknesses to compensate for the lower strength of concrete. However, while the strength of steel is substantially the same in both

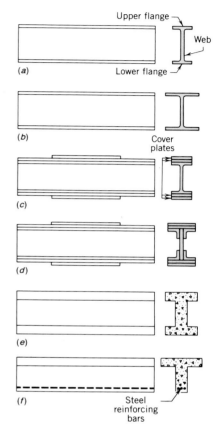

FIG. 2-1 Beam cross sections. (*a*) S-shape. (*b*) W-shape. (*c*) S-shape with cover plates. (*d*) Plate girder. (*e*) Hypothetical concrete beam. (*f*) Reinforced concrete beam.

tension and compression, the strength of concrete in tension is so low that the allowable stress in tension is usually taken as zero or near zero. This means that if the concrete beam in Fig. 2-1e is used in an application where the ends are simply supported and there are downward loads near the center, there will be tension in the concrete of the lower flange. If the load is applied gradually, the beam will fail at a very low load by vertical cracks developing in the lower flange near midspan. To make this a practical member, it should be reinforced by placing steel reinforcing bars in the lower flange to carry the tensile load. Since the concrete in the lower flange cannot be used to carry tensile load, the lower flange can be reduced in size and the beam will evolve into a *reinforced concrete T-beam* such as that shown in Fig. 2-1f.

With this background for beams in bending we are now ready to establish by rigorous methods the relationship between bending moment and stress. Such a relationship is useful because it will permit us to analyze a beam under load to determine the actual stress and compare it with the allowable stresses stipulated in the applicable code. In arriving at a moment–stress relationship, we will find, as expected, that the cross section will have an influence on the relationship, but the significant property of the cross section is one that we have not yet encountered.

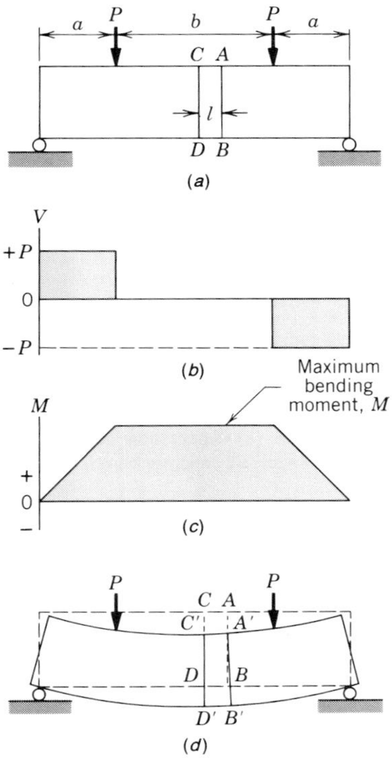

2-2 STRESS-MOMENT RELATIONSHIP IN BENDING

In order to derive an equation for stress in the fibers of a cross section subjected to bending, we will consider a beam as shown in Fig. 2-2a. The objective will be to determine the stresses in the fibers at the Section A-B. This section is somewhat special in that it is in a region where there is no shear and, consequently, no difference between the moments on neighboring sections. This special arrangement will make the analysis that follows much simpler than it would be if the section were in the region near either end of the beam where there is shear on the section. The solution to the general case would contain additional terms that would tend to obscure the meaningful part of the analysis and, at any rate, would finally be eliminated from the stress equation. Consequently, obtaining the solution by this special case is preferable at this time.

In the statically indeterminate cases encountered in Chapter 1, it was found that displacement, deflection, or strain had to be taken into account in order to obtain a solution. The present problem is similar in that no solution can be obtained without making use of the deformation of the material. Figure 2-2d shows the shape of the beam before loading, in broken lines, and its deflected shape after loading. The displacements are much exaggerated in the diagram because actual displacements that occur in real structures are difficult to detect by eye and would not be visible in a beam such as that of the illustration. We will focus our attention on planes A-B and C-D, which are plane sections, by definition,

FIG. 2-2 (a) Beam and loads. (b) Shear diagram. (c) Bending momemt diagram. (d) Deflected beam.

before deformation. After deformation the particles that were located on *A-B* and *C-D* will move to new locations such as *A'-B'* and *C'-D'* in Fig. 2-2*d* and Fig. 2-3. There is no way that we could predict the shape of the *A'-B'* and *C'-D'* sections; yet without some knowledge of the shapes we cannot proceed with this analysis. Therefore, we will make the simplest possible assumption, that is, that the *transverse sections that were initially plane will remain plane during the deformation*. In the diagrams the straight lines *A-B* and *C-D*, which are edge views of planes, are assumed to remain straight lines *A'-B'* and *C'-D'*. To simplify the analysis, we will superimpose the two views of the element so that the plane *C'-D'* coincides with *C-D* as in Fig. 2-3. The edges *A'-C'* and *B'-D'* should show a slight curvature, but no significant error is introduced by treating them as straight lines. Having superimposed *C'-D'* on *C-D*, we have no choice but to allow *A'-B'* to fall where it may; of course, it will not coincide with *A-B* but will be rotated through a small angle, α. Since intuition leads us to expect tension in the lower fibers and compression in the upper fibers, and hence lengthening of *B-D* and shortening of *A-C*, we are not surprised to find that the side *A'-B'* is rotated as shown and that there is a region of axially stretched fibers at the bottom of the section and shortened fibers at the top. Between these regions there is a location where neither stretching nor shortening takes place. This location is indicated by *N* in Fig. 2-3. The fibers located on the plane *S-N* are neither strained nor stressed in tension or compression during the loading of the beam. In other words, the fibers are neutral and are said to be located on the neutral plane or *neutral surface*. Material on any line that lies in plane *S-N* and is normal to the plane of Fig. 2-3 is unstressed. Such a line is referred to as a *neutral axis*. The neutral axis that shows as the point *N* is of interest, since it constitutes the axis about which the section *A-B* appears to rotate.

At this stage we are likely to have lost sight of our objective, which was to establish a relationship between bending moment and stress, preferably in the form of an equation by which stress can be evaluated from any given bending moment. Instead of trying to achieve this objective directly, it is profitable to pursue another objective first: to *find the moment required to rotate one face of an element*, such as *A-B* in Fig. 2-3, *through an angle, α (in radians), when the other end of the element is fixed*. This intermediate goal will take us a long way toward solving the stress–moment relationship.

Consider an element such as that in Fig. 2-4 taken from a beam. We can no longer ignore the shape of the cross section, which in practice may be a rectangle, S-shape, W-shape, or some other shape. In order not to limit the generality of this analysis, we will not specify the shape of the cross section but will make it as general as possible. We will place only one restriction on the cross-sectional area: It will have an axis of symmetry located as shown in Fig. 2-4*b*.

The correct location for the neutral axis is not known at this stage; hence its position is chosen arbitrarily. The analysis is greatly simplified by making the *x*-axis coincide with the neutral axis, as in Fig. 2-4*b*.

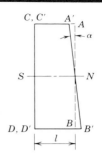

FIG. 2-3

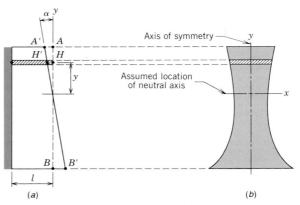

FIG. 2-4 (*a*) Beam element. (*b*) End view of element.

Consider the typical lamina *G-H*, which has a length *l* before deformation. After deformation, the point *H* will move to *H'* through a distance $\alpha \times y$, α always being a small angle. This will strain the fibers in the lamina an amount given by

$$\varepsilon = \frac{\Delta l}{l} \qquad \text{from (1-11a)}$$

$$= \frac{\alpha y}{l}$$

For the directions indicated in Fig. 2-4, it is obvious that the strain is compressive. If we want the equation to indicate the sense of the strain, it needs a negative sign for compression and then becomes

$$\varepsilon = -\frac{\alpha y}{l} \tag{2-1}$$

Since *A'-B'* is a straight line, (2-1) applies at all locations on the cross section.

If we restrict our study to material that is not strained beyond the proportional limit, stress is related to strain by

$$\sigma = E\varepsilon \qquad \text{from (1-12c)}$$

$$= -\frac{E\alpha}{l} y \tag{2-2}$$

Below the neutral axis, *y* is negative and (2-2) gives a positive value for stress, which indicates tension. Above the neutral axis the equation gives a compressive stress. Thus the equation shows that the stress varies linearly from a maximum tensile stress in the bottom fibers, through zero at the neutral axis, to a maximum compressive stress at the top. The resultant stress distribution is shown in Fig. 2-5*a*.

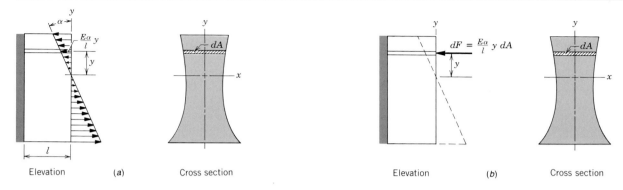

FIG. 2-5 (*a*) **Stress distribution.** (*b*) **Force due to stress on a typical area.**

Problems 2-1 to 2-5 (*Note:* Do these problems on the basis of your present knowledge. Working the solutions will reinforce and extend your understanding of the fundamentals of bending.)

On a typical element of area, dA in Fig. 2-5*b*, located by dimension y, the stress acting on the area gives a force

$$dF = \sigma \, dA$$

Substituting for stress from (2-2), we obtain

$$dF = -\frac{E\alpha}{l} \, y \, dA \qquad \textbf{(2-3)}$$

acting as shown in Fig. 2-5*b*, where the vector pointing toward the left conveys the same information as the negative sign in (2-3).

In the original problem, the beam section was subjected to pure bending without any overall tensile or compressive load. If the same state is to exist in this analysis, the total normal force on the cross section must be zero, that is,

$$\int dF = 0$$

or

$$\int \frac{E\alpha}{l} \, y \, dA = 0$$

Taking the constants outside the integral gives

$$\frac{E\alpha}{l} \int y \, dA = 0 \qquad \textbf{(2-4)}$$

This will be satisfied if either E or α are zero; however, a zero E would imply a material that had no resistance to deformation and would be of no interest as a load-carrying material, while a zero α corresponds to no rotation and, hence, to a trivial problem. Then, realistically, (2-4) can be satisfied only if

$$\int y \, dA = 0 \qquad (2\text{-}5)$$

This integral occurred when treating properties of areas in basic mechanics. It is the moment of area about the x-axis and was used to determine the location of the centroid of an area. It will be recalled that when the moment of an area about an axis was zero, the centroid of the area was located on the axis. Since condition (2-5) calls for a zero moment of area about the x-axis, the x-axis, which was selected to coincide with the neutral axis, must pass through the centroid of the cross-sectional area. Recall that (2-5) arose from the requirement that the net force on the cross section be zero; hence, *if there is no axial load on a beam, the neutral axis will automatically position itself to pass through the centroid of the cross-sectional area.*

Nothing in the derivation of (2-2) and (2-3) is invalidated by the discovery that the distance designated as y must be taken from an axis through the centroid of the area so that we will make use of these equations but will henceforth always position the x-axis to pass through the centroid.

In order to deform the element in the assumed manner, that is, with plane sections remaining plane and one section rotating through α, we must apply stress to the surface as shown in Fig. 2-5a or apply to a typical element of cross-sectional area the force dF as shown in Fig. 2-5b. This requires an increment of moment

$$dM = -y \, dF$$

$$= y \, \frac{E\alpha}{l} \, y \, dA \qquad (2\text{-}6)$$

$$= \frac{E\alpha}{l} \, y^2 \, dA$$

The total moment to deform all elements of area that make up the whole cross section will be given by integrating (2-6) over the cross-sectional area. The total moment is then

$$M = \frac{E\alpha}{l} \int y^2 \, dA \qquad (2\text{-}7)$$

We have now arrived at our intermediate objective; that is, we have an equation that gives the moment required to rotate one end of a beam element of length l through an angle α. Except for the integral, the equation is straightforward. The integral, however, is not always easy to evaluate and is usually represented by the symbol I and evaluated as a separate problem. This is done by changing (2-7) to

$$M = \frac{E\alpha}{l} \, I \qquad (2\text{-}8)$$

where

$$I = \int y^2 \, dA \qquad \text{(2-9a)}$$

In the study of dynamics an integral similar to (2-9a) is encountered. It arises in connection with the problem of finding the moment necessary to overcome the inertia of a rotating body. With some justification, that integral is called the moment of inertia. The integral for I in (2-9a) has nothing to do with inertia, but because of its similarity to the integral in the dynamics problem, it is also given the name *moment of inertia*. It is more properly called the *second moment of the area*, but engineers seem to prefer *moment of inertia*.

When the applied moment is known and α is to be determined, a more convenient arrangement of (2-8) is

$$\alpha = \frac{Ml}{EI} \qquad \text{(2-10)}$$

The angle-moment relationship, in this form, will be used later in Chapter 3 to determine the bending deflection of beams. It is immediately useful for substitution into (2-2) to give

$$\sigma = -\frac{E}{l}\left(\frac{Ml}{EI}\right)y$$

$$\sigma = -\frac{My}{I} \qquad \text{(2-11)}$$

This equation gives stress on a cross section subjected to a bending moment, M, at any point that is located a distance y from the neutral axis (which passes through the centroid of the cross-sectional area). The value of I is required for substitution into the equation and may be obtained by the integration of (2-9a). You can learn much about the process of integration by working exercises in moments of inertia, but when beams consisting of standard rolled steel sections are used, the values of I are taken from tables of properties of standard sections (see Tables B-1 to B-5). It is important that you gain complete mastery of the techniques required for evaluating (2-9a) in spite of the fact that in engineering practice the moment of inertia is seldom determined by integration. This mastery is essential because it lets you develop an intuitive feeling for the important part that I plays in bending stresses. It will also help you appreciate why the commonly used beam cross sections have a large amount of material placed in the flanges as far as possible away from the neutral surface. Another reason for becoming thoroughly familiar with I lies in the fact that, although technologists or draftsmen frequently select standard sections from tables, you must be able to check their work and also carry out the determination of I for nonstandard sections. This example of the role of technologist and engineer is typical of all practice: The routine operations performed by the technologist must be understood by the engineer, but the engineer must have a sufficiently extensive understanding to take over and solve problems when routine methods do not apply. This is the justification for keeping engineers on staff and paying them more than technologists.

Example 2-1

For the timber beam loaded as shown in Fig. 2-6a, calculate the bending stress at point P.

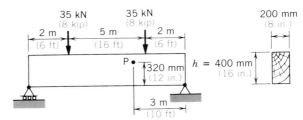

FIG. 2-6a

Solution. To determine the bending stress at P according to (2-11), we must obtain values for M, y, and I. From the symmetry of the loading pattern, the point P is located in the constant bending moment region between the two point loads. The bending moment is equal to

$$M = 35 \text{ kN} \times 2 \text{ m} = 70 \text{ kN} \cdot \text{m} = 70 \times 10^6 \text{ N} \cdot \text{mm}$$

The location of point P with respect to the neutral axis is determined by

$$y = 320 - \frac{h}{2} = 320 - 200 = 120 \text{ mm}$$

It follows from the symmetry of the cross section that the neutral axis is located at middepth of the section.

To determine the moment of inertia, I, consider integration for the cross section shown in Fig. 2-6b.

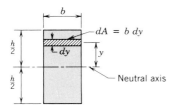

FIG. 2-6b

$$I = \int y^2 \, dA = \int_{-h/2}^{h/2} y^2 b \, dy = b \int_{-h/2}^{h/2} y^2 \, dy$$

$$= b \left[\frac{y^3}{3} \right]_{-h/2}^{h/2} = \frac{2b}{3} \left[\frac{h^3}{8} \right] = \frac{bh^3}{12}$$

This general moment of inertia equation for a rectangular cross section, which agrees with Case 4 of Table H, can now be evaluated for the dimensions of our problem as

$$I = \frac{bh^3}{12} = \frac{200 \times (400)^3}{12} \text{ mm} \cdot \text{mm}^3 = 1.067 \times 10^9 \text{ mm}^4$$

We can finally obtain the stress at point P as

$$\sigma = -\frac{My}{I} = -\frac{70 \times 10^6 \times 120}{1.067 \times 10^9} \frac{\text{N} \cdot \text{mm} \cdot \text{mm}}{\text{mm}^4}$$

$$= -7.87 \frac{\text{N}}{\text{mm}^2} = -\textbf{7.87 MPa}$$

The negative sign indicates that the stress at point P is compressive; intuitively, this is to be expected from the bending action of the beam.

Example 2-1

For the timber beam loaded as shown in Fig. 2-6a, calculate the bending stress at point P.

   Solution.   To determine the bending stress at P according to (2-11), we must obtain values for M, y, and I. From the symmetry of the loading pattern, the point P is located in the constant bending moment region between the two point loads. The bending moment is equal to

$$M = 8 \text{ kip} \times 6 \text{ ft} = 48 \text{ kip} \cdot \text{ft}$$

The location of point P with respect to the neutral axis is determined by

$$y = 12 - \frac{h}{2} = 12 - 8 = 4 \text{ in.}$$

It follows from the symmetry of the cross section that the neutral axis is located at middepth of the section.

   To determine the moment of inertia, I, consider integration for the cross section shown in Fig. 2-6b.

$$I = \int y^2 \, dA = \int_{-h/2}^{h/2} y^2 b \, dy = b \int_{-h/2}^{h/2} y^2 \, dy$$

$$= b \left[\frac{y^3}{3} \right]_{-h/2}^{h/2} = \frac{2b}{3} \left[\frac{h^3}{8} \right] = \frac{bh^3}{12}$$

This general moment of inertia equation for a rectangular cross section, which agrees with Case 4 of Table H, can now be evaluated for the dimensions of our problem as

$$I = \frac{bh^3}{12} = \frac{8 \times (16)^3}{12} \text{ in.}^4 = 2.73 \times 10^3 \text{ in.}^4$$

We can finally obtain the stress at point P as

$$\sigma = -\frac{My}{I} = -\frac{48 \times 12 \times 4}{2.73 \times 10^3} \frac{\text{kip} \cdot \text{ft} \cdot \text{in.}/\text{ft} \cdot \text{in.}}{\text{in.}^4}$$

$$= -0.84 \text{ kip/in.}^2 = -\textbf{0.84 ksi}$$

The negative sign indicates that the stress at point P is compressive; intuitively, this is to be expected from the bending action of the beam.

The stress equation (2-11) was derived for a region of a beam that had a constant bending moment or, saying the same thing in different form, zero shear. The assumption that sections that were plane before loading remained plane after loading was essential to the solution and is a valid assumption for a region of zero shear. A more general analysis, applicable to regions of nonzero shear and varying bending moment, will show that the assumption is not strictly valid but will, after much mathematical manipulation, produce the same stress equation. Such an analysis, at this stage, would not contribute to your understanding of beams and is better left to studies at a much more advanced level.

Bending stress at any point in a cross section can be found by (2-11), but usually the designer is interested in the maximum stress on the cross section. This is obtained by making y the greatest value that it can have while still indicating a point within the boundaries of the cross section. We use the symbol c to represent the distance to the extreme fibers. Thus the greatest stress on a cross section is given by

$$\sigma = -\frac{Mc}{I} \qquad \textbf{(2-12)}$$

where c is the *distance from the neutral axis to the extreme fiber*. When a beam lacks symmetry about the neutral axis, the distance to the extreme fiber in the tension field may be different from that of the compression field and it may be necessary, especially if the material strength in tension is not the same as in compression, to calculate two bending stresses.

Equation (2-12) will give the correct sign for stress ($+$ for tension and $-$ for compression) provided that the correct signs for M and c are entered. However, engineers seldom use this feature, preferring instead to get the numerical value from the equation and then determine the nature of the stress by inspection. When this practice is followed, the negative sign in the equation is redundant; consequently, most engineers prefer the equation

$$\boxed{\sigma = \frac{Mc}{I}} \qquad \textbf{(2-13)}$$

Problems 2-6 to 2-8

2-3 MOMENT OF INERTIA OF AREA

The value of the moment of inertia obtained by (2-9a)

$$I = \int y^2 \, dA$$

is dependent upon the location of the x-axis as well as upon the size and shape of the area. To emphasize this, perhaps it would have been better to designate the moment of inertia as I_x. Thus

$$I_x = \int y^2 \, dA \qquad \text{(2-9b)}$$

In referring to this quantity, we should say the "*moment of inertia with respect to the x-axis*" because it often happens that the moment of inertia with respect to some other axis is calculated and it is necessary to distinguish between the two. In the bending stress problem, we need the value of the moment of inertia with respect to an axis that is coincident with the neutral axis and hence passes through the centroid of the area. This is the value of I that must be used in the stress equations (2-11), (2-12), and (2-13), but in future calculations it will sometimes be necessary to calculate I with respect to axes that do not pass through the centroid. To avoid confusion, when the axis does pass through the centroid we will place a bar over the symbol x so that $I_{\bar{x}}$ will indicate a *moment of inertia with respect to an axis through the centroid*.

In earlier courses in mechanics, when determining centroids, you became familiar with an integral that was similar to the integral in the moment of inertia equation except that the exponent of y was 1 instead of 2. In using the centroid equation, there was very little restriction on the choice of element, but in the moment of inertia equation, if the element is not chosen correctly, there is every likelihood that a serious error will be introduced. For example, if we are attempting to find I_x for the semicircular area of Fig. 2-7, any of the given elements might seem to be satisfactory. However, only two of these elements will give the correct answer. The reason for this is to be found in the derivation of the stress-moment equation. In (2-3) an incremental force was found by multiplying the incremental area by the stress acting on that area. The element of area had been chosen so that the stress was *constant* over the element; this was done by making the element *parallel* to the neutral axis. *Consequently, in determining the moment of inertia of an area with respect to any axis, the element of area must be chosen so that all points within the element are substantially the same distance from the axis.*

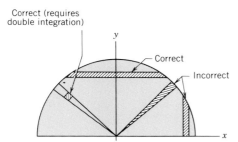

FIG. 2-7 Elements of area to determine I_x.

Example 2-2

Calculate the moment of inertia of the area given in Fig. 2-8a with respect to the x-axis.

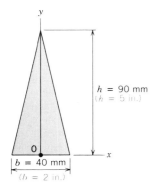

FIG. 2-8a

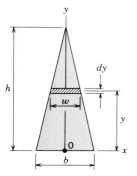

FIG. 2-8b

Solution.   Taking a typical element of area as in Fig. 2-8b, from Eq. (2-9b), we require

$$I_x = \int y^2 \, dA = \int_0^h y^2 w \, dy$$

where w is a variable and hence must be expressed in terms of y before integrating.

From similar triangles

$$\frac{w}{h-y} = \frac{b}{h}$$

$$w = \frac{b}{h}(h-y)$$

$$I_x = \int_0^h y^2 \frac{b}{h}(h-y) \, dy$$

$$= \frac{b}{h} \int_0^h (hy^2 - y^3) \, dy$$

$$= \frac{b}{h} \left[\frac{hy^3}{3} - \frac{y^4}{4} \right]_0^h$$

$$= \frac{b}{h} \left[\left(\frac{h^4}{3} - \frac{h^4}{4} \right) - (0-0) \right]$$

$$= bh^3 \left(\frac{1}{3} - \frac{1}{4} \right) = \frac{1}{12} bh^3 = \frac{1}{12} 40 \times 90^3 = \text{mm} \cdot \text{mm}^3 = \mathbf{2.43 \times 10^6 \ mm^4}$$

Example 2-2

Calculate the moment of inertia of the area given in Fig. 2-8a with respect to the x-axis.

Solution.   Taking a typical element of area as in Fig. 2-8b, from Eq. (2-9b), we require

$$I_x = \int y^2 \, dA = \int_0^h y^2 w \, dy$$

where w is a variable and hence must be expressed in terms of y before integrating.

From similar triangles

$$\frac{w}{h-y} = \frac{b}{h}$$

$$w = \frac{b}{h}(h-y)$$

$$I_x = \int_0^h y^2 \frac{b}{h}(h-y)\,dy$$

$$= \frac{b}{h}\int_0^h (hy^2 - y^3)\,dy$$

$$= \frac{b}{h}\left[\frac{hy^3}{3} - \frac{y^4}{4}\right]_0^h$$

$$= \frac{b}{h}\left[\left(\frac{h^4}{3} - \frac{h^4}{4}\right) - (0-0)\right]$$

$$= bh^3\left(\frac{1}{3} - \frac{1}{4}\right) = \frac{1}{12}bh^3 = \frac{1}{12}2 \times 5^3 = \textbf{20.83 in.}^4$$

Problems 2-9 to 2-15

It often happens that we have the value of the moment of inertia about an axis, say $\bar{x}$, through the centroid and find it necessary to calculate the moment of inertia about another axis that is parallel to $\bar{x}$. If we had a beam consisting of an S-shape alone, the neutral axis would coincide with the centroidal axis and the value of $I_{\bar{x}}$ to be used in (2-13) could be obtained from tables. However, if a cover plate is added as in Fig. 2-9a, the neutral axis for the new beam no longer coincides with the centroidal axis of the S-shape. The tabulated value of $I_{\bar{x}}$ for the S-shape is no longer of direct use in (2-13). To get the contribution to I that is made by the S-shape, we could work from first principles, but the answer can be more readily obtained by using the theorem that is derived next.

In Fig. 2-9b an area is given that has no restrictions on its shape. The $\bar{x}$-axis passes through its centroid; we will refer to the moment of inertia about the axis as $I_{\bar{x}}$ to emphasize that it is with respect to a centroidal axis. We will start as though we are going to find I_x about an axis *parallel* to the $\bar{x}$-axis from first principles.

The distance between the x-axis and the parallel axis through the centroid of the area will be dimensioned $\bar{Y}$ as in Fig. 2-9b. The element of area, dA, meets the requirements for use in (2-9b) and hence

$$I_x = \int (y + \bar{Y})^2\,dA = \int (y^2 + 2y\bar{Y} + \bar{Y}^2)\,dA$$

$$= \int y^2\,dA + \int 2y\bar{Y}\,dA + \int \bar{Y}^2\,dA$$

$$= \int y^2\,dA + 2\bar{Y}\int y\,dA + \bar{Y}^2\int dA$$

This equation can be greatly simplified by an inspection of its terms. The first integral is the moment of inertia of the area with respect to the $\bar{x}$-axis and hence may be written $I_{\bar{x}}$. The integral in the second term is the moment of the area about the $\bar{x}$-axis that passes through the centroid.

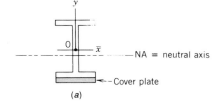

(a)

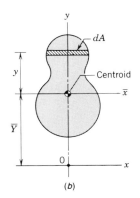

(b)

FIG. 2-9 (a) S-shape with cover plate.
(b) Parallel axis.

Hence the integral is zero and the second term disappears. The third integral will be the area; therefore, the equation reduces to

$$I_x = I_{\bar{x}} + \text{area} \times \bar{Y}^2 \qquad \text{(2-14)}$$

This is the mathematical expression for the *parallel axis theorem*, which may be stated:

> *The moment of inertia of an area with respect to any axis in the plane of the area is equal to the moment of inertia with respect to a parallel axis through the centroid of the area plus the area multiplied by the distance between the axes squared.*

The parallel axis theorem is a very simple and useful theorem, but it can be used incorrectly with perhaps disastrous results. It is essential to keep in mind that *one axis must be through the centroid of the area*; otherwise the term containing $\int y \, dA$ is not zero and must be accounted for.

Note also from (2-14) that the moment of inertia with respect to any axis other than through the centroid is always greater than $I_{\bar{x}}$; hence, in a set of I values calculated for various parallel axes, $I_{\bar{x}}$ is the minimum value.

Example 2-3

Determine the maximum bending stresses in tension and compression for the beam illustrated in Fig. 2-10a.

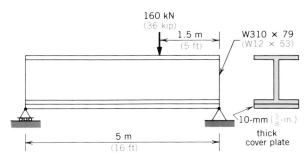

FIG. 2-10a

Solution. We must first locate the centroid of the built-up section before determining the moment of inertia. Subscripts 1 and 2 will be used to designate the properties of the S-shape and the plate, respectively.

From Table B-1

$$A_1 = 10\ 100\ \text{mm}^2$$

$$I_{\bar{x}_1} = 177 \times 10^6\ \text{mm}^4$$

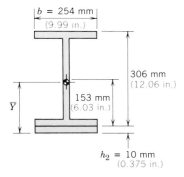

FIG. 2-10b

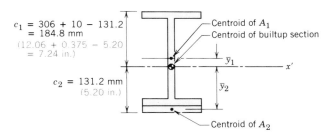

FIG. 2-10c

Also

$$A_2 = bh_2 = 254 \times 10 \text{ mm} \cdot \text{mm} = 2540 \text{ mm}^2$$

$$I_{\bar{x}_2} = \tfrac{1}{12} bh_2^3 \quad \text{(Case 4, Table H)}$$

$$= \tfrac{1}{12} \times 254 \times 10^3 \text{ mm} \cdot \text{mm}^3 = 0.02 \times 10^6 \text{ mm}^4$$

Referring to Fig. 2-10b, to find the location of the centroid of the built-up section with respect to the bottom of the section, we equate moments of area. Thus

$$A_1 y_1 + A_2 y_2 = (A_1 + A_2)\, \bar{Y}$$

$$10\ 100 \times 163 + 2540 \times 5 = 12\ 640\, \bar{Y}$$

$$\bar{Y} = 131.2 \text{ mm}$$

If, as shown in Fig. 2-10c, we designate as $\bar{y}_1$ and $\bar{y}_2$ the distance between the centroid of each area and the centroid of the built-up section, we obtain

$$\bar{y}_1 = 163.0 - 131.2 = 31.8 \text{ mm}$$

$$\bar{y}_2 = 131.2 - 5 = 126.2 \text{ mm}$$

The moment of inertia of the built-up section, $I_{\bar{x}}$, can now be calculated by using the parallel axis Theorem

$$I_{\bar{x}} = I_{\bar{x}_1} + A_1(\bar{y}_1)^2 + I_{\bar{x}_2} + A_2(\bar{y}_2)^2$$

$$= 177 \times 10^6 + 10\ 100 \times 31.8^2 + 0.02 \times 10^6 + 2540 \times 126.2^2$$

$$= (177 + 10.21 + 0.02 + 40.45)\, 10^6 \text{ mm}^4$$

$$= 227.7 \times 10^6 \text{ mm}^4$$

The maximum bending moment, which occurs at the section where the load is applied, is equal to

$$M = \frac{1.5 \times 160 \times 3.5}{5} \frac{\text{m} \cdot \text{kN} \cdot \text{m}}{\text{m}} = 168 \text{ kN} \cdot \text{m}$$

$$= 168 \times 10^6 \text{ N} \cdot \text{mm}$$

and hence the maximum bending stresses at that section are obtained as

$$\sigma = \frac{Mc_1}{I_{\bar{x}}} = \frac{168 \times 10^6 \times 184.8}{227.7 \times 10^6} \frac{\text{N} \cdot \text{mm} \cdot \text{mm}}{\text{mm}^4}$$

$$= 136.3 \frac{\text{N}}{\text{mm}^2} = 136.3 \text{ MPa} \qquad \text{(compression at the top)}$$

and

$$\sigma = \frac{Mc_2}{I_{\bar{x}}} = \frac{168 \times 10^6 \times 131.2}{227.7 \times 10^6} \frac{\text{N} \cdot \text{mm} \cdot \text{mm}}{\text{mm}^4}$$

$$= \mathbf{96.8 \text{ MPa}} \qquad \text{(tension at the bottom)}$$

Example 2-3

Determine the maximum bending stresses in tension and compression for the beam illustrated in Fig. 2-10a.

Solution. We must first locate the centroid of the built-up section before determining the moment of inertia. Subscripts 1 and 2 will be used to designate the properties of the S-shape and the plate, respectively.

From Table B-1

$$A_1 = 15.6 \text{ in.}^2$$

$$I_{\bar{x}_1} = 425 \text{ in.}^4$$

Also

$$A_2 = bh_2 = 9.99 \times 0.375 \text{ in.} \cdot \text{in.} = 3.75 \text{ in.}^2$$

$$I_{\bar{x}_2} = \tfrac{1}{12} bh_2^3 \qquad \text{(Case 4, Table H)}$$

$$= \tfrac{1}{12} \times 9.99 \times (0.375)^3 \text{ in.} \cdot \text{in.}^3 = 0.044 \text{ in.}^4$$

Referring to Fig. 2-10b, to find the location of the centroid of the built-up section with respect to the bottom of the section, we equate moments of area. Thus

$$A_1 y_1 + A_2 y_2 = (A_1 + A_2) \bar{Y}$$

$$15.6 \times 6.41 + 3.75 \times 0.1875 = 19.35 \bar{Y}$$

$$\bar{Y} = 5.20 \text{ in.}$$

If, as shown in Fig. 2-10c, we designate as $\bar{y}_1$ and $\bar{y}_2$ the distance between the centroid of each area and the centroid of the built-up section, we obtain

$$\bar{y}_1 = 6.41 - 5.20 = 1.21 \text{ in.}$$

$$\bar{y}_2 = 5.20 - \frac{0.375}{2} = 5.01 \text{ in.}$$

The moment of inertia of the built-up section, $I_{\bar{x}}$, can now be calculated by using the parallel axis Theorem

$$I_{\bar{x}} = I_{\bar{x}_1} + A_1(\bar{y}_1)^2 + I_{\bar{x}_2} + A_2(\bar{y}_2)^2$$

$$= 425 + 15.6 \times (1.21)^2 + 0.044 + 3.75 \times (5.01)^2$$

$$= (425 + 22.84 + 0.044 + 94.125) \text{ in.}^4$$

$$= 542.01 \text{ in.}^4$$

The maximum bending moment, which occurs at the section where the load is applied, is equal to

$$M = \frac{5 \times 36 \times 11}{16} \frac{\text{ft} \cdot \text{kip} \cdot \text{ft}}{\text{ft}}$$

$$= 123.75 \text{ kip} \cdot \text{ft}$$

and hence the maximum bending stresses at that section are obtained as

$$\sigma = \frac{Mc_1}{I_{\bar{x}}} = \frac{123.75 \times 12 \times 7.24}{542.01} \frac{\text{kip} \cdot \text{ft} \cdot \text{in./ft} \cdot \text{in.}}{\text{in.}^4}$$

$$= \textbf{19.84 kip/in.}^2 \qquad \text{(compression at the top)}$$

and

$$\sigma = \frac{Mc_2}{I_{\bar{x}}} = \frac{123.75 \times 12 \times 5.20}{542.01} \frac{\text{kip} \cdot \text{ft} \cdot \text{in./ft} \cdot \text{in.}}{\text{in.}^4}$$

$$= \textbf{14.25 kip/in.}^2 \qquad \text{(tension at the bottom)}$$

Problems 2-16 to 2-24

2-4 DESIGN OF BEAMS

Up to this point we have looked at the beam problem as a problem in *analysis*; that is, for a given set of loads, span, and cross section we have been calculating the stress. The more commonly encountered problem is to select a standard section, or *design* a member, for a given span and loads without exceeding a certain allowable stress. Under some conditions the allowable stress may be dependent upon the dimensions and shape of the cross section, in which case the selection of the member becomes more difficult. For the present we will take the allowable stress as though it depends only on the strength of the material and the safety factor.

A trial member will be acceptable when the stress obtained from (2-13) is equal to, or less than, the allowable stress, that is, if

$$\frac{Mc}{I} \leqslant \sigma_a \qquad \qquad \textbf{(2-15)}$$

For design purposes this inequality is more useful in the form

$$\frac{I}{c} \geqslant \frac{M}{\sigma_a} \qquad \qquad \textbf{(2-16)}$$

In the usual design process the maximum bending moment is taken from the bending moment diagram and the allowable stress determined (quite frequently in accordance with the rules of some legally constituted code) from standard strength tests in combination with a safety factor. The right-hand side of (2-16) is then known, and it remains to select or design a member that will satisfy the inequality. When a standard section is to be used, the tables could be searched until a member is found such that the combination of I and c satisfies (2-16). This takes more time than is really necessary, since the tables also provide the value of I/c for each member under the heading S, the *section modulus*. That is, the section modulus is defined as

$$S = \frac{I}{c} \qquad \textbf{(2-17)}$$

With tabulated values of S available it is much more convenient to use (2-16) in the form

$$S \geqslant \frac{M}{\sigma_a} \qquad \textbf{(2-18)}$$

To select a member, the S column is consulted and any member that satisfies (2-18) could be used. The members with very high values of S will obviously be understressed and wasteful of material. The best design, if there are no other constraints, will be that which satisfies (2-18) with the minimum amount of material. The smallest acceptable S does not necessarily coincide with the most economical member. To select the lightest and most economical standard section, the listed values of mass should be examined to find the lightest member with an acceptable S. The problem becomes much more complex if a built-up member is being designed because its cost will depend upon the combined costs of web plate, angles and cover plates as well as fabrication costs so that the lightest member is not necessarily the most economical.

Problems 2-25 to 2-27

2-5 ECCENTRIC AXIAL LOADS

Tension and compression members in Chapter 1 were arranged so that the load was applied at the centroid of the cross-sectional area. This ensured that the stress was uniformly distributed over the cross section and a very simple stress equation resulted. We will now determine the stress when an axial load is applied, not at the centroid, but displaced from the centroid by a distance e, the *eccentricity*. The system is then as shown in Fig. 2-11a, where the load location would suggest that there is a greater than average tensile stress in the bottom fibers and a smaller stress at the top fibers. The amount of the variation cannot be found with the given load arrangement, so we will alter it to that shown in Fig. 2-11b, where the *couple* and the *axial load* together are statically equivalent to

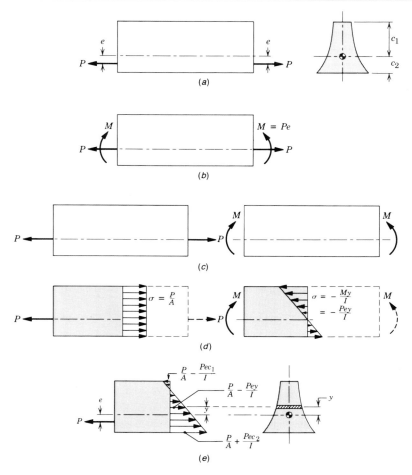

FIG. 2-11 (*a*) **Eccentrically loaded member.** (*b*) **Statically equivalent loads.** (*c*) **Problem decomposed into elementary problems.** (*d*) **Elementary problems solved for stress.** (*e*) **Elementary solutions superimposed.**

the original load. However, we still have no equations for getting stress from this system and must resort to the method of superposition. The procedure in the *method of superposition* is

1. Break an insoluble, complex problem into two or more soluble, elementary problems.
2. Solve the elementary problems.
3. Superimpose the solutions to get the answer to the original problem.

The two elementary problems are shown in Fig. 2-11*c* and the solutions, based on (1-1a) and (2-11), are shown in Fig. 2-11*d*. Superimposing the loads and stresses in Fig. 2-11*d* gives the original load system and the resulting stress, as shown in Fig. 2-11*e*. The general stress formula is

$$\sigma = \frac{P}{A} - \frac{Pey}{I}$$ (2-19)

The fact that *e* was considered to be measured downward and *y* upward must be taken into account if (2-19) is to give the correct signs on the components of stress. Usually, in practice, the signs are determined by inspection and the stresses obtained for the extreme fibers using

$$\sigma = \frac{P}{A} \pm \frac{Pec}{I}$$ (2-20)

Example 2-4

A beam of rectangular cross section carries a compressive load.

    a. Compare the stresses when the load is applied at the extreme fibers with the stress when the load is concentric.

    b. How much eccentricity is permitted if the material cannot carry a tensile stress?

Solution

a. When a concentric compressive load is applied to a rectangular cross section of width *b* and depth *h*, the resultant compressive stress is

$$\sigma_c = -\frac{P}{A} = -\frac{P}{bh}$$

For a compressive load applied as shown in Fig. 2-12*a*, the stresses in the extreme fibers are

$$\sigma_1 = -\frac{P}{bh} + \frac{Pey}{bh^3/12} = -\frac{P}{bh} + \frac{P(h/2)(h/2)}{bh^3/12}$$

$$= -\frac{P}{bh} + \frac{3P}{bh} = +\frac{2P}{bh}$$

$$\sigma_2 = -\frac{P}{bh} - \frac{Pey}{I} = -\frac{P}{bh} - \frac{3P}{bh} = -\frac{4P}{bh}$$

The eccentricity of load, therefore, produces stresses that at the top are *twice* as high, and at the bottom are *four* times as high, as the constant stress due to concentric loading.

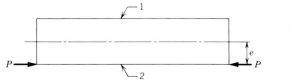

FIG. 2-12*a* Beam elevation and cross section.

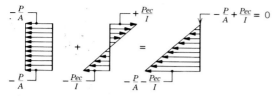

FIG. 2-12b Stress distributions.

b. To obtain the maximum allowable eccentricity when a tensile stress is not permitted anywhere on the cross section, consider the combined effects of the compressive stress and the bending stress illustrated in Fig. 2-12b. We see that in this case the top fiber governs the maximum eccentricity when the net stress due to combined compression and bending is zero. Then

$$\sigma = 0 = -\frac{P}{A} + \frac{Pec}{I}$$

and substituting $A = bh$, $c = h/2$ and $I = bh^3/12$, we get

$$e = \frac{h}{6}$$

For any rectangular cross section carrying a compressive load, all the fibers will be in compression when the load is concentric. There will be no tension under eccentric load provided that the eccentricity is no more than $\pm\frac{1}{6}h$, which is the case if the load falls within the middle one-third of the section. When a structure is made of a material that is incapable of withstanding tension, such as masonry or concrete (unreinforced), the designer must ensure that the load always falls within the middle one-third of the section. This constraint plays an important part in the design of certain retaining walls and the design of gravity dams. In the latter case the constraint applied to the dam when the head-pond is full and again when it is empty dictates the profile of the downstream face and the upstream face of the dam.

The bending moment on all sections of the eccentrically loaded beam was taken as *Pe*, which is correct for a straight member. However, a member that is straight before loading will curve when the load is applied. Under a *tensile load* the curvature will be in the direction shown in Fig. 2-13a. This deflection moves the centroid in the midregion of the member nearer to the line of action of the force and reduces the eccentricity to *e'*. The actual bending moment in the midregion is then less than that which was used to develop (2-19) and (2-20), and consequently the second term in these equations is larger than it should be. Although the equations are, therefore, in error, the error is slight because the amount of curvature is usually small. This means that *e'* is only slightly smaller than *e;* at any rate the error is on the side of safety, since the actual stresses on the cross section are slightly reduced by the lateral deflection.

When a *compressive load* is applied, the stress equations still apply, but with the sense of all stresses opposite to those of the tensile member. The deflection of the member is reversed as shown in Fig. 2-13b so that

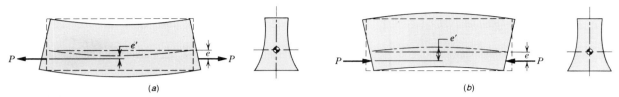

FIG. 2-13 (*a*) Deformation of eccentrically loaded tension member. (*b*) Deformation of eccentrically loaded compression member.

the eccentricity in the midregion, e', is larger than the eccentricity at the ends. This means that the stress equations are again in error, but this time the error is on the unsafe side, since the actual maximum stress is larger than the value given by the equations.

We cannot at this stage calculate e', but we can profit from an examination of how this might be done. The deflected curve could be found by an iterative process that starts, as a first approximation, with the constant bending moment that would exist in a straight member. That bending moment can be used to obtain the curvature and hence a displacement from the straight line. The original eccentricity together with the displacement give a second approximation to the deflected curve. The next iteration would use the deflected curve to determine a new larger bending moment and a new curve of increased deflections. With each iteration the change in the moments and deflections would be expected to become less and the iterations would be continued until the changes became insignificant. Under certain conditions this process converges to a solution in a few iterations. However, under some conditions each iteration gives a larger deflection and no solution can finally be obtained. This indicates that the member is overloaded and is in the process of *buckling*. In practice, buckling loads are not determined in this way, but the above reasoning will provide a useful background to the rigorous analysis of buckling that will be given in Chapter 4.

Problems 2-28 to 2-30

2-6 SHEARING STRESS IN BEAMS

When we consider the distribution of bending stress on the cross section of a beam, it is evident that material near the neutral axis contributes very little toward its moment-carrying capacity. If we had a beam such as that shown in Fig. 2-14*a*, a large portion of the material is near the neutral axis and is not effective in carrying moment. Relocating some of this material to give the section in Fig. 2-14*b* will improve its ability to carry moment. A gain is made in two ways; some material has been moved to a location where it will be more highly stressed and, in addition, the force on this material acts at a greater distance from the neutral axis. Both effects enhance the moment-carrying capacity.

Without changing the total amount of material, the rearrangement as we progress from the shape in Fig. 2-14*a* through *b* and *c* to *d*

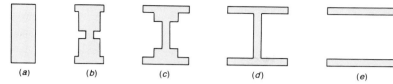

FIG. 2-14 Variations in cross-sectional shape.

will evidently make further improvements. If we calculated the moment of inertia of each of these sections, we would find it increasing from Fig. 2-14*a* to *d* with a corresponding increase in the allowable bending moment.

Carrying this progression to the extreme would lead us to Fig. 2-14*e*, where we have gone beyond the form of the commonly used beam cross sections by completely removing the web. This last step will not in fact give the expected increase but rather will produce a member that is very weak in bending. Removing the web permits each flange to act independently, resulting in a strength that is merely the sum of the strengths of the two flanges, each acting as a separate beam. Each of these flat plates has a very low moment of inertia about its own neutral axis and consequently a low bending strength.

The reason that we were on the verge of taking the dangerous step from Fig. 2-14*d* to *e* is that in focusing our attention on the *bending load*, we have forgotten that there is a *shear load* to be carried by the beam. In Chapter 1 we observed that the web seems to carry the shear load; yet in the cross section shown in Fig. 2-14*e* there is no web to carry this load. We can look upon the web as being necessary for carrying shear or for ensuring that all components of the member act as a single unit; in either case it is an essential part of the member and must not be neglected in design.

The approximate, intuitive method for finding the shearing stress in the web, introduced in Chapter 1, gives quite good results and is a useful method under some conditions. However, a more thorough analysis of shear is necessary for a complete understanding of beam action. We will find that shear is a more complex phenomenon than we might anticipate at this point. Some idea of the complexity of the design of the shear carrying part of a beam can be obtained by considering the modes of failure illustrated in Fig. 2-15.

For the beam loaded as shown in Fig. 2-15*a*, we will assume that the dimensions and material properties are such that there is *no failure by bending*. The load will be taken as being increased progressively *until failure occurs due to shear*. The largest shear load exists in the end regions so that we will focus our attention there and, in a hypothetical experiment, watch this region for signs of failure.

Imagine that we apply a progressively increasing set of loads to a *timber beam* as shown in Fig. 2-15*b*. From our experience with bolt design, and the shear mode of failure of bolts, we might expect the beam to fail along a plane such as *A-B*. We might predict the shear load at failure to be the strength of the wood in shear multiplied by the area of the cross section at *A-B* and expect the failure to take place by one side of the plane *A-B* sliding over the other side. A test would reveal that our

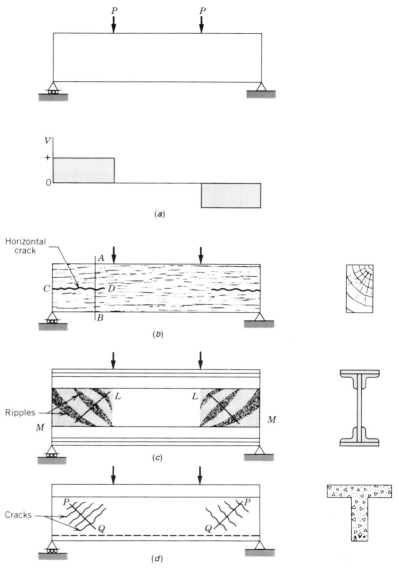

FIG. 2-15 Various modes of shear failure. (*a*) Beam and shear diagram. (*b*) Timber beam. (*c*) Plate girder. (*d*) Reinforced concrete beam.

prediction is quite wrong. First, in an actual beam we would find failure starting at a much lower load than that predicted by our calculation. Furthermore, we would find cracks developing not on plane *A-B* or a parallel plane but rather on a horizontal plane such as *C-D*. This is remarkably different from the mode of failure that we had good reason to expect.

If the member had been a *plate girder* with a web that was not strong enough to carry the shear load, the web would fail in another unexpected mode. Ripples would develop as shown in Fig. 2-15c with axes at 45° to the horizontal.

The *reinforced concrete member* in Fig. 2-15d would fail by cracks opening as shown; the cracks would occur again at about 45°, but in the direction opposite to the ripples that occurred in the plate girder.

The modes of failure discussed for the timber beam, steel plate girder, and reinforced concrete beam obviously point toward much more complex relationships between the shearing load and stress than would be expected from the treatment given in Chapter 1. We will now explore those relationships.

Returning to the timber beam, we will first account for the failure on the *horizontal* plane rather than on the *vertical* plane. From a beam of any cross-sectional shape, an element such as that shown in Fig. 2-16a is isolated as a free body depicted in Fig. 2-16b. The stresses that we would expect to be acting on the surfaces of the free body are shown. We will

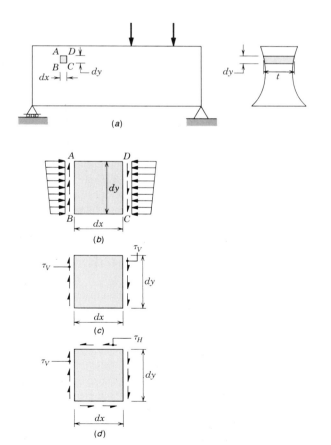

FIG. 2-16 Presence of vertical and horizontal shear stresses acting on a typical element in a beam.

find it profitable to use the moment equilibrium equation. If the normal stresses are included, they will add much to the complexity of the equation, but in the end all terms resulting from the normal stress will either cancel out or be removed as terms of higher differential order. Consequently, we will omit these normal stresses and take the element as in Fig. 2-16c. It is evident that rotational equilibrium cannot exist with the given stress; hence an additional horizontal stress, as seen in Fig. 2-16d, is required. For moments about any point the moment equilibrium equation becomes

$$\tau_H \, dx \, t \, dy - \tau_V \, dy \, t \, dx = 0$$

which simplifies to

$$\boxed{\tau_H = \tau_V} \tag{2-21}$$

This very simple, but useful relationship comes upon us quite unexpectedly. It indicates that if there is a shearing stress on one plane, there is also an equal shearing stress on a second plane at right angles to the first plane. Thus, when a shear load is imposed on a vertical section of a beam, the expected *vertical* shear stress will exist but there will also be an unexpected shear stress of equal magnitude on a *horizontal* plane. This is not just an academically interesting or a hypothetical stress, it is very real and is, in fact, the stress that causes timber to fail as seen in Fig. 2-15b. To explain this failure, we must remember that wood has different strengths in different directions, a fact well known to anyone who has used an axe to split firewood: The axe is used to split the wood in the longitudinal and not in the transverse direction because the grain runs longitudinally and wood has a much lower splitting strength in that direction. With respect to the timber beam, the grain running in the direction of the length of the beam makes the beam quite strong to resist the vertical shear stress, but the horizontal shear stress (equal in magnitude to the vertical) acts on a plane that is weak in shear. Consequently, the load that puts vertical shear on the timber beam causes a horizontal shear failure and produces the crack that is seen in Fig. 2-15b.

The next step must be to develop an equation for finding the magnitude of the shearing stress on a cross section when the shearing force and the properties of the section are known. A first guess would be to divide the vertical force by the area, but in many cases this would give a stress that is very much in error. To solve rigorously, consider two neighboring cross sections A-B and C-D in the region of the beam in Fig. 2-17a, where there is shear. Since there is shear present, there will be a varying bending moment and the element between the cross sections, when isolated, has a free-body diagram as shown in Fig. 2-17b; the stresses on the surface of the free body are shown in Fig. 2-17c. The distribution of normal stresses is known to be linear while the shear stress distribution has yet to be established. We will further reduce the free body by passing a cutting plane at a distance Y above the $\bar{x}$-axis and using the part above this plane as the free body. The horizontal shear on this section is provided for in the new free-body diagram in Fig. 2-17d.

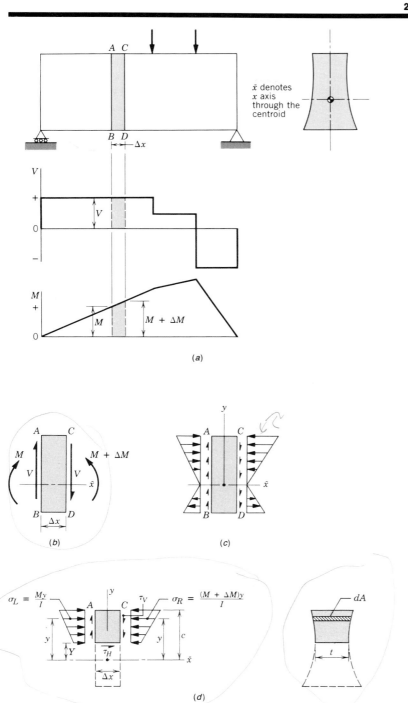

$$\sigma_L = \frac{My}{I} \qquad \tau_V \qquad \sigma_R = \frac{(M + \Delta M)y}{I}$$

(d)

FIG. 2-17 Derivation of shearing stress in a beam. (a) Beam, shear, and bending moment diagrams. (b) Forces on the element. (c) Bending stresses. (d) Horizontal shear stress τ_H acts on the element.

Because the moment is larger on the right side of the element than on the left, the normal stresses are larger and it is evident that in order to maintain equilibrium, there must exist a horizontal force which can only come from horizontal shear on the bottom surface of the element. Neglecting shear, the unbalanced longitudinal force on the element, acting toward the left, is given by

$$F_L = \int_Y^c \sigma_R \, dA - \int_Y^c \sigma_L \, dA$$

Using (2-11) to substitute moments for normal stresses, we obtain

$$F_L = \int_Y^c \frac{(M + \Delta M)}{I} y \, dA - \int_Y^c \frac{M}{I} y \, dA$$

$$= \int_Y^c \frac{My}{I} \, dA + \int_Y^c \frac{\Delta My}{I} \, dA - \int_Y^c \frac{My}{I} \, dA$$

$$= \int_Y^c \frac{\Delta My}{I} \, dA = \frac{\Delta M}{I} \int_Y^c y \, dA$$

Substituting for ΔM from (1-6a) gives

$$F_L = \frac{V \Delta x}{I} \int_Y^c y \, dA$$

Since the integral will probably be solved separately, it is convenient to let

$$Q = \int_Y^c y \, dA \qquad (2\text{-}22)$$

Then

$$F_L = \frac{V \Delta x \, Q}{I} \qquad (2\text{-}23)$$

This equation will be useful later in the design of rivets or welds to connect angles to plates in plate girders. It is immediately useful to determine horizontal shearing stress by substituting into $\sum F_x = 0$, which is

$$-F_L + \tau_H \Delta x \, t = 0$$

to get

$$\frac{-V \Delta x \, Q}{I} + \tau_H \Delta x \, t = 0$$

Rearranging

$$\tau_H = \frac{V \Delta x\, Q}{I \Delta x\, t}$$

$$\tau_H = \frac{VQ}{It} \qquad\qquad \textbf{(2-24a)}$$

Or using (2-21)

$$\tau_V = \frac{VQ}{It} \qquad\qquad \textbf{(2-24b)}$$

Designers who are thoroughly familiar with these equations do not need to be reminded that the shear is horizontal or vertical and prefer to use

$$\tau = \frac{VQ}{It} \qquad\qquad \textbf{(2-24c)}$$

This equation gives the *shearing stress*, τ, at any location chosen on a cross section and is the shear counterpart of the bending equation (2-13). It is, however, not nearly as simple an equation and shows that the shearing stress is not constant over the surface, nor even linearly varying, but varies in such a manner that we cannot name the function except in a few simple cases. The equation presents many opportunities to make errors and must be used with great care. It is best to keep the derivation in mind in order to minimize the chance of error. Because some of the important characteristics of the equation may have been obscured by the details in the derivation, it is worthwhile reviewing the meaning and limitation of the terms in the equation.

τ The equation is usually used to calculate the shearing stress, *either horizontal or vertical*, at a point in a beam such as P in Fig. 2-18. Although we may say that the stress at P is being calculated, the stress is the same at all points on P-P' of the cross section.

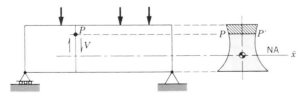

FIG. 2-18

V The stress is the result of a shearing force V, acting on the transverse cross section through P.

t The length of the line P-P' is the value of t.

I   The *I* came from the stress-moment equation and is the moment of inertia of the entire cross-sectional area *with respect to the neutral axis* of the beam, that is, $I_{\bar{x}}$ in Figs. 2-17 and 2-18.

Q   The quantity *Q* is defined by (2-22) and will be recognized as the *moment of an area*. The area is that portion of the cross section between *P-P'* and the boundary of the cross section. The boundary may be taken on *either* side of *P-P'* *but not on both*. In practice it is seldom necessary to perform an integration, the moment usually being calculated by summing the products of rectangular areas by their respective centroidal distances from the neutral axis.

Example 2-5

For the beam given in Fig. 2-19*a* calculate the vertical (and consequently the horizontal) shearing stress at *A*, *B*, *C*, *D*, and *E*.

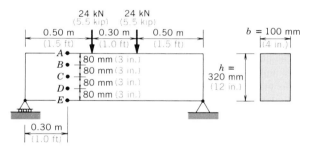

FIG. 2-19*a*

Solution.   To solve for the shear stress at the required levels of the beam, we will repeatedly use (2-24c),

$$\tau = \frac{VQ}{IT}$$

Certain quantities are constant, namely:

$$V = 24 \text{ kN} = 24 \times 10^3 \text{ N}$$

$$I = \frac{bh^3}{12} = \frac{100 \times 320^3}{12} \text{ mm} \cdot \text{mm}^3 = 273 \times 10^6 \text{ mm}^4$$

$$t = b = 100 \text{ mm}$$

The only variable is *Q*; this will be evaluated separately for each level *A* to *E*. From symmetry of the levels we recognize that $Q_A = Q_E$ and $Q_B = Q_D$; hence only three *Q* calculations will be required.

Levels B *and* D. Q_B can be obtained by considering the moment of the cross-hatched area between level B and the boundary of the cross section as shown in Fig. 2-19b.

$$Q_B = \underbrace{(100 \times 80)}_{\text{area}} \underset{\underset{\text{to centroid of area being}}{\underset{\text{considered}}{}}}{120} = 960 \times 10^3 \text{ mm}^3$$

distance from neutral axis to centroid of area being considered

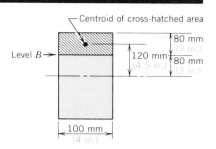

Level B → 120 mm (4.5 in.) 80 mm (3 in.) 80 mm (3 in.)

Centroid of cross-hatched area

100 mm (4 in.)

FIG. 2-19b

then

$$\tau_B = \frac{VQ_B}{It} = \frac{24 \times 10^3 \times 960 \times 10^3}{273 \times 10^6 \times 100} \frac{\text{N} \cdot \text{mm}^3}{\text{mm}^4 \cdot \text{mm}}$$

$$= 0.84 \frac{\text{N}}{\text{mm}^2} = \textbf{0.84 MPa}$$

Since $Q_D = Q_B$, $\tau_D = \tau_B$

$$\tau_D = \textbf{0.84 MPa}$$

Level C. For Q_c we consider the moment of the total area above the neutral axis as illustrated in Fig. 2-19c.

$$Q_c = (100 \times 160) 80 = 1280 \times 10^3 \text{ mm}^3$$

and

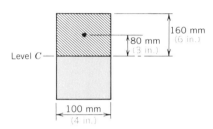

Level C — 80 mm (3 in.) 160 mm (6 in.)

100 mm (4 in.)

FIG. 2-19c

$$\tau_c = \frac{VQ_c}{It} = \frac{24 \times 10^3 \times 1280 \times 10^3}{273 \times 10^6 \times 100} \frac{\text{N} \cdot \text{mm}^3}{\text{mm}^4 \cdot \text{mm}}$$

$$= 1.12 \frac{\text{N}}{\text{mm}^2} = \textbf{1.12 MPa}$$

Levels A *and* E. Since no area exists between these levels and the boundary of the cross section,

$$Q_A = Q_E = 0 \quad \text{and} \quad \tau_A = \tau_E = \textbf{0}$$

Summarizing, we can see that the shear stress at the section under consideration varies between 0 at the outer fibers at levels A and E and a maximum stress at the neutral axis at level C. Plotting the results, we get the *parabolic shear stress distribution* of Fig. 2-19d, which is characteristic of all rectangular cross sections.

Example 2-5

For the beam given in Fig. 2-19a calculate the vertical (and consequently the horizontal) shearing stress at A, B, C, D, and E.

Solution. To solve for the shear stress at the required levels of the beam, we will repeatedly use (2-24c),

$$\tau = \frac{VQ}{IT}$$

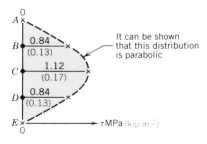

0
A
B 0.84 (0.13)
C 1.12 (0.17)
D 0.84 (0.13)
E
0

It can be shown that this distribution is parabolic

τ MPa (kip/in.²)

FIG. 2-19d

Certain quantities are constant, namely:

$$V = 5500 \text{ lb}$$

$$I = \frac{bh^3}{12} = \frac{4 \times (12)^3}{12} \text{ in.} \cdot \text{in.}^3 = 576 \text{ in.}^4$$

$$t = b = 4 \text{ in.}$$

The only variable is Q; this will be evaluated separately for each level A to E. From symmetry of the levels we recognize that $Q_A = Q_E$ and $Q_B = Q_D$, hence only three Q calculations will be required.

Levels B *and* D. Q_B can be obtained by considering the moment of the cross-hatched area between level B and the boundary of the cross section as shown in Fig.-2-19b.

$$Q_B = \underbrace{(4 \times 3)}_{\text{area}}\,4.5 = 54 \text{ in.}^3$$

$\llcorner$ distance from neutral axis to centroid of area being considered

then

$$\tau_B = \frac{VQ_B}{It} = \frac{5.5 \times 54}{576 \times 4} \frac{\text{kip} \cdot \text{in.}^3}{\text{in.}^4 \cdot \text{in.}}$$

$$= \textbf{0.13 kip/in.}^2$$

Since $Q_D = Q_B$, $\tau_D = \tau_B$

$$\tau_D = \textbf{0.13 kip/in.}^2$$

Level C.   For Q_c we consider the moment of the total area above the neutral axis as illustrated in Fig. 2-19c.

$$Q_c = (4 \times 6)\,3 = 72 \text{ in.}^3$$

and

$$\tau_c = \frac{VQ_c}{It} = \frac{5.5 \times 72}{576 \times 4} \frac{\text{kip} \cdot \text{in.}^3}{\text{in.}^4 \cdot \text{in.}}$$

$$= \textbf{0.17 kip/in.}^2$$

Levels A *and* E.   Since no area exists between these levels and the boundary of the cross section,

$$Q_A = Q_E = 0 \quad \text{and} \quad \tau_A = \tau_E = 0$$

Summarizing, we can see that the shear stress at the section under consideration varies between 0 at the outer fibers at levels A and E and a maximum stress at the neutral axis at level C. Plotting the results, we get the *parabolic shear stress distribution* of Fig. 2-19d, which is characteristic of all rectangular cross sections.

Example 2-6

Two 30-mm × 374-mm cover plates are attached to a W360 × 196 by means of 20-mm rivets. If the allowable stress in shear is 70 MPa, calculate the maximum allowable spacing of the rivets in Fig. 2-20a.

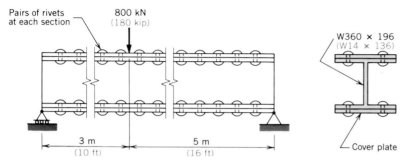

FIG. 2-20a

Solution.   To arrive at the maximum shear stress, we must use the *maximum* shear force. This occurs between the left support and the point of load application, and is equal to

$$V = \tfrac{5}{8}(800) = 500 \text{ kN} = 500 \times 10^3 \text{ N}$$

For the W-shape we obtain from Table B-1 the following quantities:

$$I = 636 \times 10^6 \text{ mm}^4$$

$$h = 372 \text{ mm}$$

$$b = 374 \text{ mm}$$

Where h and b are the dimensions defined in Fig. 2-20b. The moment of inertia of the cover plates, I_{CP}, is determined as

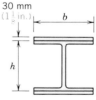

FIG. 2-20b

$$I_{CP} = 2\left[\frac{1}{12} 374 \times 30^3 + \left(\frac{372 + 30}{2} \right)^2 30 \times 374 \right]$$

$$= 2[0.8 \times 10^6 + 453.3 \times 10^6]$$

$$= 2 \times 454.1 \times 10^6$$

$$= 908 \times 10^6 \text{ mm}^4$$

The total I of the built-up section is then equal to

$$I_{\text{tot}} = (636 + 908)10^6 = 1544 \times 10^6 \text{ mm}^4$$

Next we find Q at the level between the W-shape and the cover plates as

$$Q = 30 \times 374 \left(\frac{372 + 30}{2} \right) = 2.255 \times 10^6 \text{ mm}^3$$

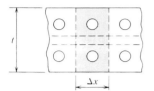

FIG. 2-20c

Since the allowable shear stress is 70 MPa, each rivet can resist a shear force of

$$70 \times A = 70 \frac{\pi \times 20^2}{4} = 21\,991\ N = 22 \times 10^3\ N$$

If pairs of rivets are spaced at a distance Δx as shown in Fig. 2-20c, each pair carries the unbalanced longitudinal force between sections Δx mm apart. Then from (2-23)

$$F_L = \frac{V \Delta x\, Q}{I}$$

Equating this load to the load capacity of the pair of rivets, we find that

$$\frac{V \Delta x\, Q}{I} = 2 \times 22 \times 10^3\ N$$

$$\Delta x = \frac{I}{VQ} \times 44 \times 10^3$$

$$= \frac{1544 \times 10^6 \times 44 \times 10^3}{500 \times 10^3 \times 2.255 \times 10^6} \frac{mm^4 \cdot N}{N \cdot mm^3} = 60.2\ mm$$

The maximum spacing of pairs of rivets is, therefore, **60 mm.**

Example 2-6

Two $1\frac{1}{8}$-in. $\times$ 14.73-in. cover plates are attached to a W14 $\times$ 136 by means of $\frac{3}{4}$-in. rivets. If the allowable stress in shear is 10 ksi, calculate the maximum allowable spacing of the rivets in Fig. 2-20a.

Solution. To arrive at the maximum shear stress, we must use the *maximum* shear force. This occurs between the left support and the point of load application, and is equal to

$$V = \tfrac{16}{26}(180) = 110.8\ kip$$

For the W-shape we obtain from Table B-1 the following quantities:

$$I = 1530\ in.^4$$

$$h = 14.66\ in.$$

$$b = 14.73\ in.$$

where h and b are the dimensions defined in Fig. 2-20b. The moment of inertia of the cover plates, I_{CP}, is determined as

$$I_{CP} = 2\left[\frac{1}{12} 14.73 \times 1.125^3 + \left(\frac{14.66 + 1.125}{2} \right)^2 1.125 \times 14.73 \right]$$

$$= 2[1.748 + 1.032 \times 10^3]$$

$$= 2 \times 1.034 \times 10^3$$

$$= 2068\ in.^4$$

The total I of the built-up section is then equal to

$$I_{tot} = 1530 + 2068 = 3598 \text{ in.}^4$$

Next we find Q at the level between the W-shape and the cover plates as

$$Q = 1.125 \times 14.73 \left(\frac{14.66 + 1.125}{2} \right) = 130.8 \text{ in.}^3$$

Since the allowable shear stress is 10 ksi, each rivet can resist a shear force of

$$10 \times A = 10 \frac{\pi \times 0.75^2}{4} = 4.42 \text{ kip}$$

If pairs of rivets are spaced at a distance Δx as shown in Fig. 2-20c, each pair carries the unbalanced longitudinal force between sections Δx in. apart. Then from (2-23)

$$F_L = \frac{V \Delta x Q}{I}$$

Equating this load to the load capacity of the pair of rivets, we find that

$$\frac{V \Delta x Q}{I} = 2 \times 4.42 \text{ kip}$$

$$\Delta x = \frac{I}{VQ} \times 8.84$$

$$= \frac{3598 \times 8.84}{110.8 \times 130.8} \frac{\text{in.}^4 \cdot \text{kip}}{\text{kip} \cdot \text{in.}^3} = 2.20 \text{ in.}$$

The maximum spacing of pairs of rivets is, therefore, **$2\frac{3}{16}$ In.**

Problems 2-31 to 2-43

2-7 CURVED BEAMS

In many machine elements such as those shown in Fig. 2-21, the stress that controls the design results from bending in a member that is not straight. The relationships between rotation and moment (2-10) and between stress and moment (2-11), which were derived for straight beams, must be reexamined to see wherein they are valid for *curved beams*. To do this, we will start from first principles and derive the *rotation* and *stress* equations for curved beams.

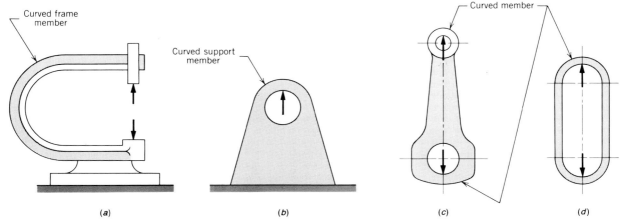

FIG. 2-21 (*a*) Punch-press frame. (*b*) Journal-bearing support. (*c*) Connecting rod. (*d*) Chain link.

Consider a curved member with a radius of curvature of the centroidal axis, R_c, and a symmetrical cross section of general form, as in Fig. 2-22*a*. The element defined by the two planes *A-B* and *C-D*, which pass through the center of curvature and are normal to the centroidal axis, will be examined for strain and stress. When a moment *M* is applied, as in Fig. 2-22*b*, the planes rotate relative to one another through an angle α. We will, as before, assume that plane sections remain plane. We will not, however, assume that the neutral axis passes through the centroid, but that it is located at an undetermined distance *Y* from the centroid. The fibers at a typical radius *R*, or at *y* from the neutral axis, are initially of length

$$G\text{-}H = R\,\Delta\theta$$

When the rotation α occurs, these fibers change length by

$$H\text{-}H' = \alpha y$$

and the compressive strain is

$$\varepsilon = \frac{H\text{-}H'}{G\text{-}H} = \frac{\alpha y}{R\,\Delta\theta}$$

The stress in the fibers will be

$$\sigma = E\varepsilon = \frac{E\alpha y}{R\,\Delta\theta} \qquad \textbf{(2-25)}$$

and the force on an element of cross-sectional area, *dA*, will be

$$dF = \sigma\,dA = \frac{E\alpha y}{R\,\Delta\theta}\,dA \qquad \textbf{(2-26)}$$

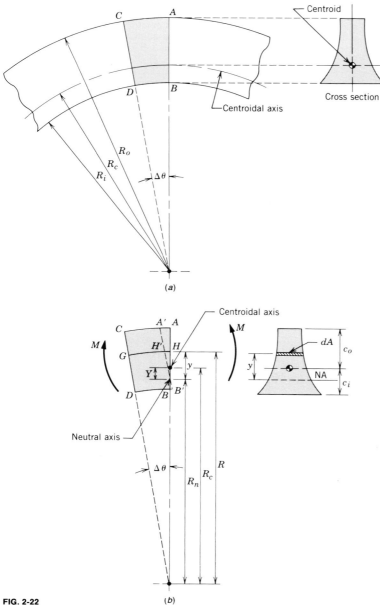

FIG. 2-22

(a)

(b)

Since the total normal force on the cross section must be zero,

$$\int_{\text{area}} dF = 0$$

$$\int_{\text{area}} \frac{E\alpha y}{R\,\Delta\theta}\, dA = 0$$

Substituting $R\text{-}R_n$ for y gives

$$\frac{E\alpha}{\Delta\theta}\int_{area}\frac{R\text{-}R_n}{R}\,dA=0$$

$$\int_{R_i}^{R_0}\frac{R\text{-}R_n}{R}\,dA=0 \qquad\qquad \textbf{(2-27)}$$

This equation determines the value of R_n that locates the neutral axis. It can be rearranged into a more convenient form by the following manipulations:

$$\int_{R_i}^{R_0}\frac{R}{R}\,dA-\int_{R_i}^{R_0}\frac{R_n}{R}\,dA=0$$

$$-R_n\int_{R_i}^{R_0}\frac{dA}{R}=-\int_{R_i}^{R_0}dA=A$$

$$R_n=\frac{A}{\displaystyle\int_{R_i}^{R_0}\frac{dA}{R}} \qquad\qquad \textbf{(2-28)}$$

The position of the neutral axis, measured inwardly from the centroidal axis, is given by

$$Y=R_c-R_n=R_c-\frac{A}{\displaystyle\int_{R_i}^{R_0}\frac{dA}{R}} \qquad\qquad \textbf{(2-29)}$$

Returning to (2-26), the force dF at a distance y from the neutral axis is equivalent to an increment of moment

$$dM=y\,dF$$

$$=\frac{E\alpha y^2}{\Delta\theta\,R}\,dA$$

The total bending moment will be

$$M=\int_{area}\frac{E\alpha}{\Delta\theta}\frac{y^2}{R}\,dA$$

$$=\frac{E\alpha}{\Delta\theta}\int_{R_i}^{R_0}\frac{(R\text{-}R_n)^2}{R}\,dA$$

$$= \frac{E\alpha}{\Delta\theta} \int_{R_i}^{R_0} (R\text{-}R_n) \frac{(R\text{-}R_n)}{R} \, dA$$

$$= \frac{E\alpha}{\Delta\theta} \left[\int_{R_i}^{R_0} (R\text{-}R_n) \, dA - R_n \int_{R_i}^{R_0} \frac{R\text{-}R_n}{R} \, dA \right]$$

The first integral is the moment of the cross-sectional area with respect to the neutral axis and can be written

$$\int_{R_i}^{R_0} (R\text{-}R_n) \, dA = YA$$

By (2-27) the second integral is seen to be zero and the moment equation reduces to

$$M = \frac{E\alpha YA}{\Delta\theta}$$

The angle of rotation for a given moment can then be determined by

$$\alpha = \frac{M \Delta\theta}{EYA} \tag{2-30}$$

Substituting into (2-25) gives stress

$$\sigma = \frac{Ey}{R \Delta\theta} \frac{M \Delta\theta}{EYA} = \frac{My}{YAR}$$

The stress in this development was taken as compression. To restore the equation to the standard sign convention, the sign needs to be changed, giving

$$\sigma = - \frac{My}{YAR} \tag{2-31}$$

At the outer surface the stress is

$$\sigma_0 = - \frac{M(c_0 + Y)}{YAR_0} \tag{2-32}$$

At the inner surface the stress is

$$\sigma_i = - \frac{M(-c_i + Y)}{YAR_i} = \frac{M(c_i - Y)}{YAR_i} \tag{2-33}$$

If we had not recognized that the curved beam differs from the straight beam, we would have determined stresses by (2-13) and have found

$$\boxed{\sigma'_0 = - \frac{Mc_0}{I}} \tag{2-34}$$

and

$$\sigma_i' = \frac{Mc_i}{I}$$

(2-35)

The ratio of the extreme stresses to the straight beam stresses are given by

$$\bar{\sigma}_0 = \frac{\sigma_0}{\sigma_0'} = \frac{M(c_0 + Y)}{YAR_0} \frac{I}{Mc_0}$$

or

$$\bar{\sigma}_0 = \left(1 - \frac{Y}{c_0}\right) \times \frac{I}{YAR_0}$$

(2-36)

and

$$\bar{\sigma}_i = \frac{\sigma_i}{\sigma_i'} = \frac{M(c_i + Y)}{YAR_i} \frac{I}{Mc_i}$$

or

$$\bar{\sigma}_i = \left(1 - \frac{Y}{c_i}\right) \times \frac{I}{YAR_i}$$

(2-37)

Because of the nature of the relationships, the stress equations cannot be further simplified. To determine the maximum stress in a given beam, stresses would first be calculated by (2-34) and (2--35). Then, using Y from (2-29) and the other known dimensions, $\bar{\sigma}_0$ and $\bar{\sigma}_i$ would be calculated, thus giving correction factors to be applied to the straight beam stresses in order to take into account the effect of curvature. This is done by

$$\sigma_0 = \bar{\sigma}_0 \sigma_0'$$

(2-38)

and

$$\sigma_i = \bar{\sigma}_i \sigma_i'$$

(2-39)

Although this procedure may be slightly more lengthy than is necessary, it gives more insight into the nature of curved beams and frequently leads to much simplified solutions. This happens when the proportions are such that $\bar{\sigma}_0$ and $\bar{\sigma}_i$ are so close to unity that they can be ignored and the simpler straight beam equations taken as applying to the curved member.

A measure of the *flexibility* of the curved beam in bending is given by (2-30). In this equation the length of the element of beam is given indirectly by $\Delta\theta$ although a more direct dimension would be l, the length measured along the centroidal axis. These are related by

$$l = R_c \Delta\theta$$

or

$$\Delta\theta = \frac{l}{R_c}$$

Substituting into (2-30) gives the flexibility equation

$$\boxed{\alpha = \frac{Ml}{EYAR_c}} \qquad \textbf{(2-40)}$$

Using (2-10) and (2-40), the flexibility of a curved beam relative to an equal length of straight beam is given by

$$\bar{\alpha} = \frac{\dfrac{Ml}{EYAR_c}}{\dfrac{Ml}{EI}}$$

or

$$\boxed{\bar{\alpha} = \frac{I}{YAR_c}} \qquad \textbf{(2-41)}$$

The quantity obtained from (2-41) is useful in determining the deflection of curved beams and also in indicating when the effect of curvature on flexibility is negligible.

Example 2-7

For the curved member of Fig. 2-23a, (a) determine the maximum tensile and compressive bending stress, and (b) compare the flexibility of the ring with that of a straight member.

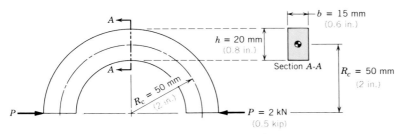

FIG. 2-23a

Solution.   The maximum bending moment occurs on section *A-A* and is

$$M_{max} = P \times R_c$$

The bending stress is the same at both surfaces of the straight beam and by (2-34) and (2-35) is

$$\sigma'_0 = \sigma'_i = \frac{M(h/2)}{\frac{1}{12}bh^3} = \frac{6M}{bh^2} = \frac{6PR_c}{bh^2}$$

$$= \frac{6 \times 2000 \times 50}{15 \times 20^2} \quad \frac{\text{N} \cdot \text{mm}}{\text{mm} \cdot \text{mm}^2}$$

$$= 100.0 \frac{\text{N}}{\text{mm}^2} = 100.0 \text{ MPa}$$

To evaluate (2-29), we will refer to Fig. 2-23*b* and do the integration separately. Thus

$$\int_{R_i}^{R_0} \frac{dA}{R} = \int_{R_i}^{R_0} \frac{b \, dR}{R} = \left[b \ln R \right]_{R_i}^{R_0}$$

$$= b \ln \frac{R_0}{R_i} = 15 \ln \left(\frac{50 + 10}{50 - 10} \right)$$

$$= 6.082 \text{ mm}$$

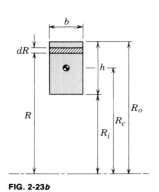

FIG. 2-23*b*

Substituting into (2-29) gives

$$Y = R_c - \frac{bh}{6.082} = 50 - \frac{15 \times 20}{6.082} = 0.674 \text{ mm}$$

$$A = bh = 15 \times 20 = 300 \text{ mm}^2$$

$$I = \frac{1}{12}bh^3 = \frac{1}{12} \times 15 \times 20^3 = 10\,000 \text{ mm}^4$$

Substituting into (2-36) and (2-37), we obtain

$$\bar{\sigma}_0 = \left(1 + \frac{0.674}{10} \right) \frac{10\,000}{0.674 \times 300 \times 60} \frac{\text{mm}^4}{\text{mm}^4} = 0.880$$

$$\bar{\sigma}_i = \left(1 - \frac{0.674}{10} \right) \frac{10\,000}{0.674 \times 300 \times 40} \frac{\text{mm}^4}{\text{mm}^4} = 1.153$$

$$\sigma_0 = \bar{\sigma}_0 \sigma'_0 = 0.880 \times 100 = \textbf{88.0 MPa}$$

$$\sigma_i = \bar{\sigma}_i \sigma'_i = 1.153 \times 100 = \textbf{115.3 MPa}$$

The maximum bending stresses are 88.0 and 115.3 MPa. By inspection σ_0 is tensile and σ_i compressive. Substituting into (2-41) gives the relative flexibility,

$$\bar{\alpha} = \frac{I}{YAR_c} = \frac{10\,000}{0.674 \times 300 \times 50} \frac{\text{mm}^4}{\text{mm}^4} = 0.989$$

Example 2-7

For the curved member of Fig. 2-23a, (a) determine the maximum tensile and compressive bending stress, and (b) compare the flexibility of the ring with that of a straight member.

Solution. The maximum bending moment occurs on section A-A and is

$$M_{max} = P \times R_c$$

The bending stress is the same at both surfaces of the straight beam and by (2-34) and (2-35) is

$$\sigma_0' = \sigma_i' = \frac{M(h/2)}{\frac{1}{12}bh^3} = \frac{6M}{bh^2} = \frac{6PR_c}{bh^2}$$

$$= \frac{6 \times 0.5 \times 2}{0.6 \times 0.8^2} \frac{\text{kip} \cdot \text{in.}}{\text{in.} \cdot \text{in.}^2}$$

$$= 15.63 \text{ kip/in.}^2$$

To evaluate (2-29), we will refer to Fig. 2-23b and do the integration separately. Thus

$$\int_{R_i}^{R_0} \frac{dA}{R} = \int_{R_i}^{R_0} \frac{b\,dR}{R} = \left[b \ln R \right]_{R_i}^{R_0}$$

$$= b \ln \frac{R_0}{R_i} = 0.6 \ln \left(\frac{2 + 0.4}{2 - 0.4} \right)$$

$$= 0.2433 \text{ in.}$$

Substituting into (2-29) gives

$$Y = R_c - \frac{bh}{0.2433} = 2 - \frac{0.6 \times 0.8}{0.2433} = 0.027 \text{ in.}$$

$$A = bh = 0.6 \times 0.8 = 0.48 \text{ in.}$$

$$I = \frac{1}{12}bh^3 = \frac{1}{12} \times 0.6 \times 0.8^3 = 0.0256 \text{ in.}^4$$

Substituting into (2-36) and (2-37), we obtain

$$\bar{\sigma}_0 = \left(1 + \frac{0.027}{0.4} \right) \frac{0.0256 \text{ in.}^4}{0.027 \times 0.48 \times 2.4 \text{ in.}^4} = 0.879$$

$$\bar{\sigma}_i = \left(1 - \frac{0.027}{0.4} \right) \frac{0.0256 \text{ in.}^4}{0.027 \times 0.48 \times 1.6 \text{ in.}^4} = 1.151$$

$$\sigma_0 = \bar{\sigma}_0 \sigma_0' = 0.879 \times 15.63 = \textbf{13.74 kip/in.}^2$$

$$\sigma_i = \bar{\sigma}_i \sigma_i' = 1.151 \times 15.63 = \textbf{17.99 kip/in.}^2$$

The maximum bending stresses are 13.74 and 17.99 kip/in.^{2}. By inspection σ_0 is tensile and σ_i compressive. Substituting into (2-41) gives the relative flexibility,

$$\bar{\alpha} = \frac{I}{YAR_c} = \frac{0.0256 \text{ in.}^4}{0.027 \times 0.48 \times 2 \text{ in.}^4} = 0.988$$

Example 2-7 shows that for the proportions of that particular beam, the maximum stress obtained from the straight beam equation, if not corrected, would be 15% in error while the flexibility would be only 1.1% in error. The straight beam equation for stress is evidently not accurate enough for use in this case, but the error is less for more slender beams. This can be seen from Fig. 2-24, where the effect of R_c/h on $\bar{\sigma}_i$, $\bar{\sigma}_0$, and $\bar{\alpha}$ are presented for a *rectangular cross section*. The curves indicate that for small ratios of R_c/h, that is, relatively sharply curved beams, $\bar{\sigma}_0$, $\bar{\alpha}$, and particularly $\bar{\sigma}_i$ depart significantly from the value of unity associated with straight beams. Note, however, when R_c/h is greater or equal to 6, the

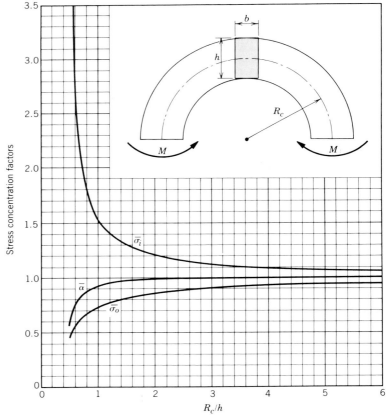

FIG. 2-24 Stress concentration factors for a curved beam of rectangular cross section.

straight beam equation will give stresses that are always less than 6% in error. Because $\bar{\sigma}_i$ and $\bar{\sigma}_0$ are multiplying factors that account for the deviation of curved beam stresses from straight beam stresses, the two factors can be viewed as *stress concentration factors*.

Curved beams, of course, can have other than rectangular cross sections. For an *elliptical cross section*, Fig. 2-25 shows similar trends for the effect of R_c/h on $\bar{\sigma}_i$, $\bar{\sigma}_0$, and $\bar{x}$ as for the rectangular section.

Section *A-A* in Example 2-7 is subjected to a direct *normal load* as well as the *bending load*. The normal load gives rise to additional compressive stress that should be combined with σ_i due to bending to give the maximum stress. The normal load divided by the cross-sectional area gives the average direct stress. A stress concentration factor is required for the evaluation of the maximum direct stress. Unfortunately, this factor is not the same as the factor for bending. However, in practice the direct stress is usually small relative to the bending stress; consequently, its determination is not critical and an estimated stress concentration factor is used. Careful designers make an allowance for the direct stress by initially ensuring that the bending stress is lower than

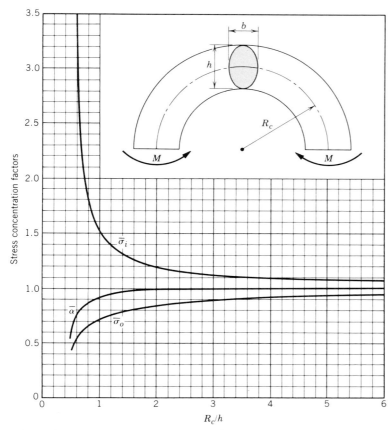

FIG. 2-25 Stress concentration factors for a curved beam of elliptical cross section.

the allowable stress by a small margin and then checking to see that the direct stress does not exceed that margin.

It may seem that we have carried out an analysis that has only *one basic assumption*, namely, that *plane sections remain plane* when a moment is applied. As with the straight beam analysis, that assumption is valid for pure bending load. Without recognizing it, we made *another assumption*; when the element *G-H*, in Fig. 2-22, changed its length to *G-H'* we assumed that the element did not move radially. If *G-H* is in compression and has no lateral support, it will move outwardly because of its curvature. Of course, it is not entirely free to move in that way, but the tendency will cause some radial stress to develop and some small radial motion to take place. Although the exact solution can only be obtained by resorting to advanced levels of the theory of elasticity, in most cases the exact solution is very close to that given here. For example, the value of $\bar{\sigma}_i$ for a rectangular section with $R_c/h = 1$, when calculated by (2-37), is 1.523 whereas the exact value is 1.527. For curvatures likely to be encountered in practice, the equations that we have developed are accurate enough for design purposes.

When the cross section is a hollow circle, as in a pipe elbow, or when it has the form of an S-shape or W-shape, the effect of radial motion is not negligible. Stresses and flexibility may then be very different from those indicated by the above equations. Therefore, such sections must not be analyzed by the methods of this book.

Problems 2-44 to 2-48

2-8 SUMMARY

In engineering practice a large number of beams are standard rolled steel sections chosen to satisfy (2-18), that is,

$$S \geqslant \frac{M}{\sigma_a}$$

The bending moment is the maximum that occurs in the beam when subjected to known loads. The determination of the loads themselves may require considerable effort and skill and be somewhat in error because the forces that will act on the structure must be estimated. Often the law will require that the rules of a certain code of practice be satisfied, and the loads may be specified in the code. For example, the applicable code may specify that the floor of an office building must be capable of carrying 4 kN/m^2 (80 lb/ft^2). As long as there is nothing unusual about the building under design, the engineer can safely assume that the code will be conservative.

The allowable stress will, in most cases, be also established by a code. The allowable stress in tension may be simply the strength of the material divided by the required safety factor. However, a flange in compression may, under certain conditions, fail by buckling in a manner somewhat similar to the compression members of a truss. When buckling is a possible mode of failure, this must be taken into account in determining the allowable stress for compression. Buckling of a compression flange is too complex a phenomenon to solve by analytical methods; hence empirical equations, as specified in the applicable code, must be used for the compressive design stress (discussed further in Chapter 4).

The proportions of standard rolled sections are such that shear in the web is seldom a controlling factor in the design. Unless the member is very short, the shearing stress will not govern the design, but nevertheless must be checked by the basic shear stress equation (2-24), that is,

$$\tau = \frac{VQ}{It}$$

In plate girders and reinforced concrete members, shear is more likely to have an influence on the final design. As shown in Fig. 2-15 the failures are not as simple as in the wooden member. The failure of the steel web, as seen in Fig. 2-15c, appears to be a compression failure with the material failing by buckling along a line at 45° to the horizontal. It is interesting to note that if the structure were a truss with diagonals in the L-M direction, these diagonals would be in compression and, if not strong enough, would fail in buckling.

The reinforced concrete member fails by cracks opening as though there were a tensile stress in the direction shown in Fig. 2-15d. If the structure had been a truss with some members in the P-Q direction, these diagonals would have been in tension. Since concrete is very weak in tension, the beam behaves as though it had diagonals in the P-Q direction and fails by cracking normal to the tensile load.

The peculiar modes of failure found in the web of these members suggest that we need to know more about what happens on inclined planes in stressed material. This is the subject of Chapter 5.

PROBLEMS

2-1 When subjected to constant bending moment, the elongation in a length of 20 mm (0.80 in.) on the outer fiber of a steel beam is found to be 0.015 mm (0.0006 in.). If the outer fiber of the beam is 250 mm (10 in.) from the neutral axis, determine the strain and stress at points 50, 100, 150, 200, and 250 mm (2, 4, 6, 8, 10 in.) from the neutral axis.

Two transverse planes are 100 mm (4 in.) apart. Determine the rotation of one plane relative to the other when the bending moment is applied.

2-2 For the beam shown in the illustration, the stress at A is known to be 135 MPa (20 ksi) compression and, at B, 135 MPa (20 ksi) tension. Calculate:

a. The bending moment on the section A-B.
b. The value of P.

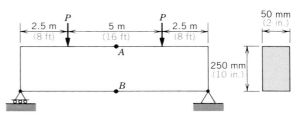

PROBLEM 2-2

2-3 The strain at point *A* in the steel beam is measured as 450×10^{-6} when the beam is under the loading system shown. What value of *P* will cause this strain?

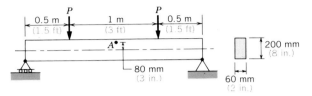

PROBLEM 2-3

2-4 Calculate the maximum bending stress in the given beam.

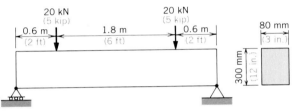

PROBLEM 2-4

2-5 Gages on the rectangular steel member indicate longitudinal strains of -0.003 at *A* and $+0.003$ at *B*. If the stress–strain curve for the material can be taken as given below, determine the load *P*.

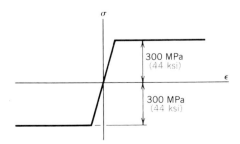

PROBLEM 2-5

2-6 Calculate the bending stress at *P* and at *Q* for the given steel beam.

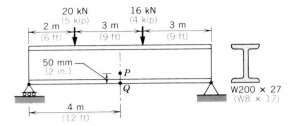

PROBLEM 2-6

2-7 If the bending stress is limited to 105 MPa (15 ksi), determine the maximum allowable load *F*.

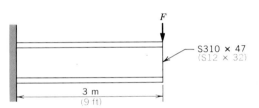

PROBLEM 2-7

2-8 For the given beams, calculate the maximum bending stress.

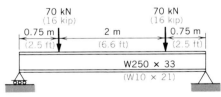

PROBLEM 2-8a

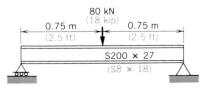

PROBLEM 2-8b

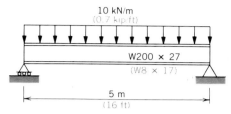

PROBLEM 2-8c

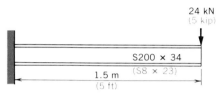

PROBLEM 2-8d

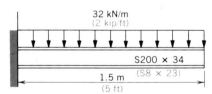

PROBLEM 2-8e

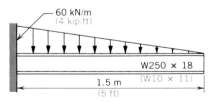

PROBLEM 2-8f

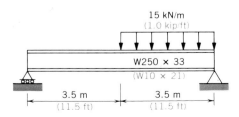

PROBLEM 2-8g

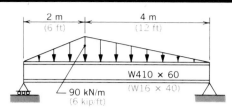

PROBLEM 2-8h

2-9 Calculate I_x for the given hollow section.

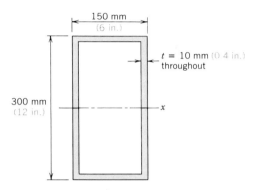

PROBLEM 2-9

2-10 Calculate the value of I_x for the following sections, ignoring the radii, and compare the values found with those listed in the tables.

a. S510 × 143 (S20 × 96).
b. W610 × 195 (W24 × 130).
c. W360 × 509 (W14 × 342).
d. W200 × 27 (W8 × 17).

2-11 Calculate from first principles the moment of inertia of the area with respect to the centroidal axis.

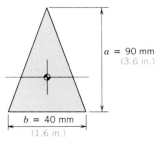

PROBLEM 2-11

2-12 Calculate from first principles the moment of inertia of the circular area with respect to a diameter.

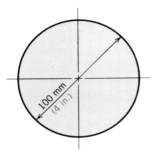

PROBLEM 2-12

2-13 Calculate the moment of inertia (a) with respect to the y-axis, and (b) with respect to the x-axis.

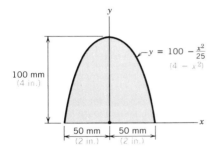

$$y = 100 - \frac{x^2}{25}$$
$$(4 - x^2)$$

PROBLEM 2-13

2-14 Calculate the moment of inertia, with respect to the x-axis, of the given areas.

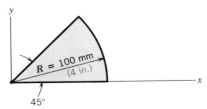

PROBLEM 2-14a

PROBLEM 2-14b

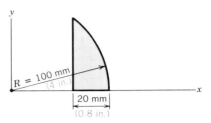

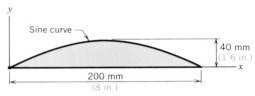

PROBLEM 2-14c

2-15 Sheet metal can be pressed to form corrugated panels that serve many useful purposes. The bending load that can be carried by a corrugated panel is many times that of a flat sheet of equal thickness. In order to determine the strength of a corrugated panel, it is necessary to calculate the moment of inertia of the corrugated cross section.

  Determine the moment of inertia, with respect to the x-axis, of the given typical portion of the cross section of a corrugated panel. Take advantage of the fact that the thickness is small, relative to the other dimensions.

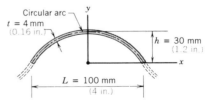

PROBLEM 2-15

2-16 Calculate the value of I_x for the following sections, ignoring the fillets and rounds. Compare the values found with those listed in the tables.

  a. L125 × 125 × 13 (L5 × 5 × $\frac{1}{2}$).
  b. L65 × 65 × 10 (L2$\frac{1}{2}$ × 2$\frac{1}{2}$ × $\frac{3}{8}$).
  c. L200 × 100 × 20 (L8 × 4 × $\frac{3}{4}$).
  d. L125 × 75 × 13 (L5 × 3 × $\frac{1}{2}$).

2-17 Calculate I with respect to the neutral axis for the given built-up sections.

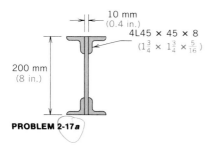

PROBLEM 2-17a

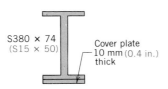

PROBLEM 2-17b

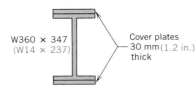

PROBLEM 2-17c

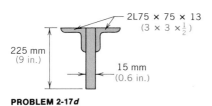

PROBLEM 2-17d

2-18 a. Calculate the maximum bending stress in the beam.

b. Determine the percentage error in the maximum stress as calculated by the approximate method of Section 1-6.

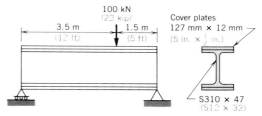

PROBLEM 2-18

2-19 to 2-21 Calculate the maximum bending stress (a) in tension, and (b) in compression.

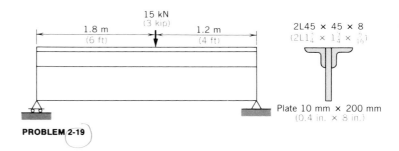

PROBLEM 2-19

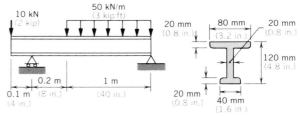

PROBLEM 2-20

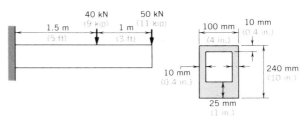

PROBLEM 2-21

2-22 In constructing concrete floor slabs, it is common practice to use a steel panel as the formwork under the slab. The panel performs a second useful function after the concrete has set by acting as reinforcement for the concrete.

Before the concrete sets, the panel must be strong enough to support the weight of the fluid concrete. To give added strength to the panel, it is usually corrugated. A typical cross section is shown.

A panel having the given cross section has a length of 2 m (6 ft) normal to the plane of the figure and is simply supported at the ends.

Calculate the maximum bending stress in the steel due to the weight of the slab before the concrete has set.

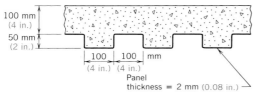

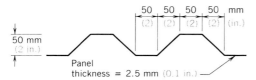

PROBLEM 2-22

2-23 Repeat Prob. 2-22 for a panel having the given shape. Note that the amount of steel is substantially the same in both cases.

PROBLEM 2-23

2-24 A cofferdam is to be built as shown with a watertight wall consisting of 300 mm × 300 mm (12 in. × 12 in.) Douglas fir piles driven side by side.

The piles may be taken as being simply supported by the wale. The lake bottom prevents lateral motion of the piles and also offers resistance to rotation, which is somewhere between simply supported and built-in.

Calculate the maximum bending stress assuming that the piles are simply supported at the level of the lake bottom.

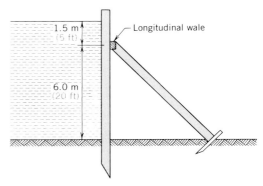

PROBLEM 2-24

2-25 If the allowable stress is 200 MPa (29 ksi) in tension and compression, select the lightest member to safely carry the given loads from (a) the S-shapes, and (b) the W-shapes.

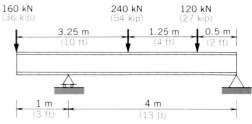

PROBLEM 2-25

2-26 Select a standard Douglas fir member from the sizes given in Table F. Use the allowable stress given in Table A.

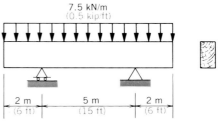

PROBLEM 2-26

2-27 For an allowable stress of 180 MPa (26 ksi) select the most economical S-shape to safely carry the specified loads.

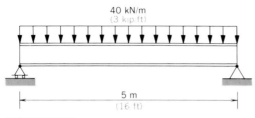

40 kN/m
(3 kip/ft)

5 m
(16 ft)

PROBLEM 2-27*a*

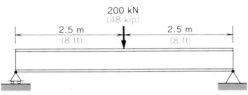

200 kN
(48 kip)

2.5 m   2.5 m
(8 ft)   (8 ft)

PROBLEM 2-27*b*

200 kN
(48 kip)

1.25 m   3.75 m
(4 ft)   (12 ft)

PROBLEM 2-27*c*

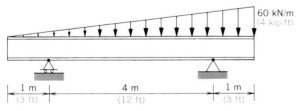

60 kN/m
(4 kip/ft)

1 m   4 m   1 m
(3 ft)   (12 ft)   (3 ft)

PROBLEM 2-27*d*

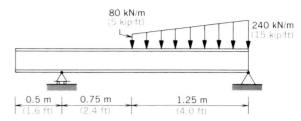

80 kN/m
(5 kip/ft)

240 kN/m
(15 kip/ft)

0.5 m   0.75 m   1.25 m
(1.6 ft)   (2.4 ft)   (4.0 ft)

PROBLEM 2-27*e*

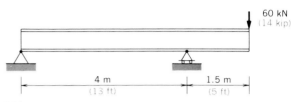

60 kN
(14 kip)

4 m   1.5 m
(13 ft)   (5 ft)

PROBLEM 2-27*f*

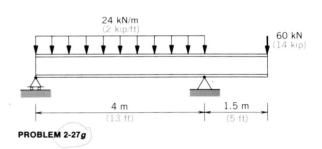

24 kN/m
(2 kip/ft)

60 kN
(14 kip)

4 m   1.5 m
(13 ft)   (5 ft)

PROBLEM 2-27*g*

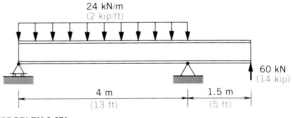

24 kN/m
(2 kip/ft)

60 kN
(14 kip)

4 m   1.5 m
(13 ft)   (5 ft)

PROBLEM 2-27*h*

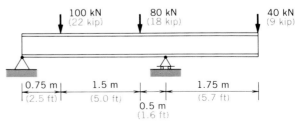

100 kN
(22 kip)

80 kN
(18 kip)

40 kN
(9 kip)

0.75 m   1.5 m   1.75 m
(2.5 ft)   (5.0 ft)   (5.7 ft)

0.5 m
(1.6 ft)

PROBLEM 2-27*i*

2-28 a. Calculate the stress at the upper surface of the S380 × 74 (S15 × 50) and at the lower surface when e is 50 mm (2.0 in.).

b. What is the upper limit on e if no compressive stress is permitted?

c. If the allowable stress is 150 MPa (22 ksi), by how much must the load be reduced, for an e of 80 mm (3.1 in.), from the safe load without eccentricity?

PROBLEM 2-28

2-29 The given T-bar is made of gray cast iron having properties given in Table A. The required safety factor is 3, based on the ultimate.

Calculate the stress at A and the stress at B and state whether the member is safe or not for the following loads.

a. $P_1 = 40$ kN (9.0 kip) and $P_2 = 0$.
b. $P_1 = 0$ and $P_2 = 40$ kN (9.0 kip).
c. $P_1 = 40$ kN (9.0 kip) and $P_2 = 40$ kN (9.0 kip).

Determine the maximum allowable load

d. P_1 when $P_2 = 0$.
e. P_2 when $P_1 = 0$.
f. P_1 and P_2 when the loads are equal.

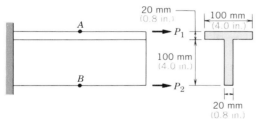

PROBLEM 2-29

2-30 Strain gages indicate that the strain at A is 0.0008 and at B is 0.0012. Determine the load P and its eccentricity e.

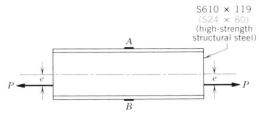

PROBLEM 2-30

2-31 a. Calculate the maximum horizontal shearing stress in the timber beam.

b. If the beam is Douglas fir, is it safe?

c. Determine the minimum thickness of the beam to ensure that the shear stress is not excessive.

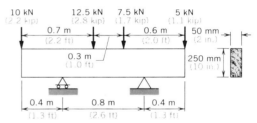

PROBLEM 2-31

2-32 a. Calculate the vertical shearing stress at P.

b. Calculate the maximum vertical shearing stress in the beam.

c. Calculate the percentage error if the maximum stress is determined by the approximate method of Section 1-6.

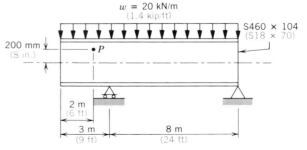

PROBLEM 2-32

2-33 a. Calculate the required strength of glue in the joints $A, B, C, \ldots, G$ of the given laminated beam.

b. If the joints in the beam are to be secured by nails instead of glue, and each nail can safely carry a shear load of 1.5 kN (0.34 kip), determine the nail spacing, L, required for joints $A, B, \ldots, G$.

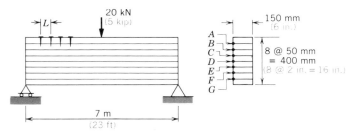

PROBLEM 2-33

2-34 Determine the shear stress in the glued joints of the given laminated beam.

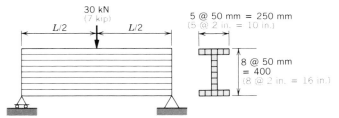

PROBLEM 2-34

2-35 A large beam is made by cutting a W360 × 196 (W14 × 136) into two parts and welding a plate between the parts as shown.

a. It is welded from both sides and each weld penetrates 6 mm (0.2 in.). Calculate the maximum shearing stress in the weld metal.

b. If the welds have full penetration and are intermittent with a spacing of 200 mm (8 in.), how long must each segment of weld be in the region to the left of the load for an allowable shear stress of 90 MPa (13 ksi) in the weld?

c. Under the same conditions as in part (b), how long must each segment be to the right of the load?

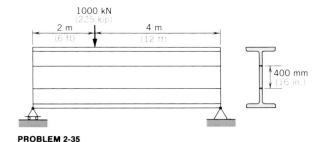

PROBLEM 2-35

2-36 Calculate the shearing stress in the welds that join the angles to the web plate.

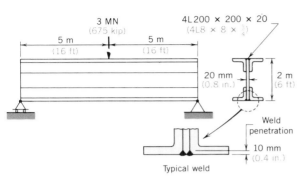

PROBLEM 2-36

2-37 A beam of circular cross section is subjected to a transverse shear force V. Show that the maximum horizontal shear stress has a value of $\tau_{max} = \frac{4}{3}(V/A)$, where A is the cross-sectional area.

2-38 Find the location and the value of the maximum horizontal shear stress in the given beams.

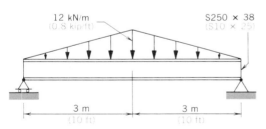

PROBLEM 2-38a

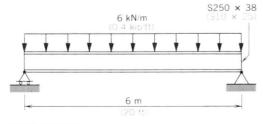

PROBLEM 2-38b

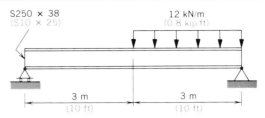

PROBLEM 2-38c

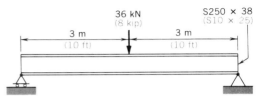

PROBLEM 2-38d

2-39 A simply supported beam is constructed from timber and has the cross-sectional dimensions shown. For the given loading find the maximum horizontal shear stress and its location.

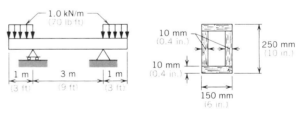

PROBLEM 2-39

2-40 Find the maximum horizontal shear stress and its location, for the given beam.

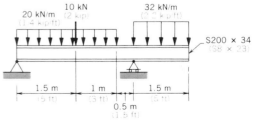

PROBLEM 2-40

2-41 For the simply supported timber beam find the maximum uniformly distributed load, *w*, that the beam can support if the allowable horizontal shear stress is 0.7 MPa (0.1 ksi).

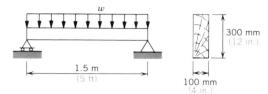

PROBLEM 2-41

2-42 If the maximum horizontal shear stress cannot exceed 0.7 MPa (0.1 ksi) for the timber beam, find the maximum length *L* for a loading of 8 kN · m (0.5 kip/ft).

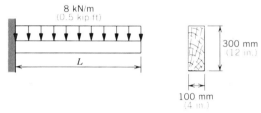

PROBLEM 2-42

2-43 The angles of the plate girder are attached by 30-mm (1.2 in.) rivets spaced at 120 mm (4.8 in.). Calculate the shearing stress in the rivets.

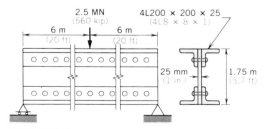

PROBLEM 2-43

2-44 Determine the maximum stress at section *A-A* in the crane hook.

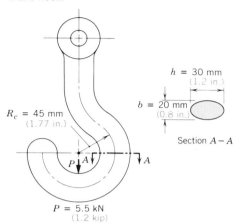

PROBLEM 2-44

2-45 The punch-press frame is made of gray cast iron having properties given in Table A. Determine the safe load for a safety factor of 3 based on the ultimate.

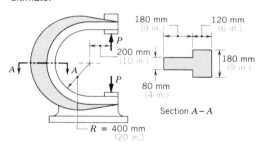

PROBLEM 2-45

2-46 A C-clamp is to be constructed from AISI 1090 bar stock of circular cross section. What diameter of rod must be used if the loading arrangement is as shown and the required safety factor is 1.25 on yield.

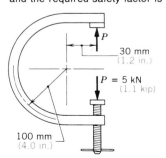

PROBLEM 2-46

2-47 For the C-clamp of Prob. 2-46, select a more economical cross section and calculate the resulting percent saving in weight.

2-48 A coping saw is constructed with a U-frame as shown having a rectangular cross section. If the tension in the blade is 300 N (67 lb), calculate the maximum stress at *A-A* and compare this to the stress at section *B-B*.

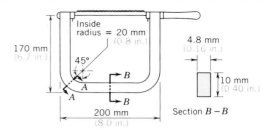

PROBLEM 2-48

Advanced Problems

2-A Is the given section acceptable if the allowable stress in bending is 140 MPa (20 ksi)? If not, design cover plates to reinforce the member. Do not make the plates any longer than necessary.

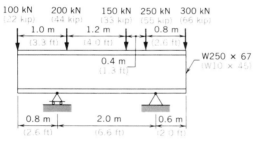

PROBLEM 2-A

2-B For the given plate girder and an allowable bending stress of 150 MPa (22 ksi) select equal leg angles.

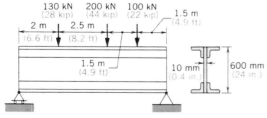

PROBLEM 2-B

2-C A W250 × 33 (W10 × 21) beam is located in an exterior wall of a building in such a way that, in winter, the flanges are cooled to −20°C (−4°F) while the remainder of the steel is at 15°C (59°F). Calculate the thermal stresses in the member if the member is not constrained.

2-D If the W-shape given in Prob. 2-C has only one flange cooled, determine the thermal stresses.

2-E A strain gauge is placed at location *A* of the crane shown. What value of strain will be recorded for the loads indicated? The member is a steel pipe with a 50 mm (2.00 in.) outside diameter and 46 mm (1.84 in.) inside diameter.

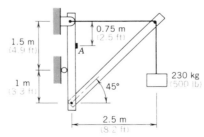

PROBLEM 2-E

2-F The stress found in Prob. 2-24 exceeds the allowable stress as given in Table A. Relocating the wale at a lower elevation, if not carried too far, will reduce the stress.

a. Find, if possible, a location that will reduce the stress to the allowable or less.

b. What is the location for minimum stress?

2-G A semicircular arch has a rectangular cross section 400 mm (16 in.) deep and 100 mm (4 in.) wide. It is made of laminated and glued Western hemlock. Calculate the maximum bending stress due to the weight of the arch only.

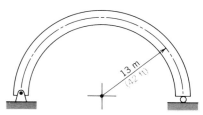

PROBLEM 2-G

2-H An element of a leaf spring is to be made from a bar of AISI 1020 that is 25 mm (1 in.) wide, 10 mm (0.4 in.) thick, and 400 mm (16 in.) long.

Initially the bar is curved to a radius of 2000 mm (80 in.) and in this state is stress free.

Before being installed in the spring, the leaf is placed on the flat platen of a press and the head (also flat) lowered until the leaf is pressed out straight. It can be assumed that there is no friction between the spring and the press surfaces.

a. Draw a profile of stress in the leaf while it is in the press.
b. Determine the bending moment in the leaf while it is in the press.

The head is raised and the leaf removed from the press.

c. Determine the radius of curvature.
d. Draw the stress profile.

DEFLECTIONS DUE TO BENDING

3-1 INTRODUCTION

We have seen that the elongation of a single truss member, when load is applied, is usually quite insignificant. However, when the accumulative effect of elongation in all members of a structure is taken into account, the displacements of some parts of the structure are found to be appreciable. When a beam is subjected to lateral loading, the deflection is usually much larger than for an axially loaded member. We have an interest in beam deflection because the design of floor beams is frequently controlled by the limit on deflection imposed by a code rather than by the allowable stress. We must also understand beam deflection in order to solve certain statically indeterminate problems that occur in beam design. Figure 3-3 shows one of many possible types of indeterminate beams.

3-2 BEAM FLEXURE EQUATION

Our objective is to write the equation that will give the deflection of a beam as a function of location along the beam when the loads and physical properties of the beam are known. This is illustrated in Fig. 3-1a, where a beam, which was initially straight, has loads imposed on it causing it to deflect. We would like an equation relating the deflection, y, to location along the beam, x, and to the loads and properties of the beam. The x-axis is placed, for convenience, to coincide with the centerline of the beam. The equation that we are attempting to establish will give the vertical displacement of the particles that were located on the x-axis before the load was applied. We could equally well work with the displacements of particles on the top surface or on the bottom surface if there were reason to do so. In all practical cases the axial displacement can be ignored without introducing a significant error.

A typical element of the beam is shown in Fig. 3-1b, where two neighboring planes, which were initially normal to the undeflected centerline, are shown to be rotated, through an angle α, relative to one another. The bending moments acting on the planes are not identical for the given loading, but if the planes are separated by a distance dL, the moment difference is incremental and may be ignored. The element is similar to that of Figs. 2-3, 2-4 and 2-5 except that the length is designated as dL instead of l. With this change in dimension (2-10) becomes

$$\alpha = \frac{M\,dL}{EI} \qquad \text{(3-1)}$$

The deflection of beams in real structures is always small in comparison with that shown in the diagram; consequently, the slope is always small and the length of the element, dL, is substantially equal to the horizontal projection of that length, designated as dx. Substituting dx for dL in (3-1) gives

$$\alpha = \frac{M\,dx}{EI} \qquad \text{(3-2a)}$$

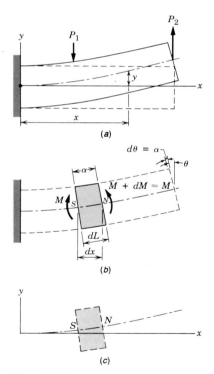

FIG. 3-1 Deflection of a cantilever beam. (a) Deflection under load. (b) Deflected element. (c) Elastic deflection curve.

Note that the relative rotation between the two planes contributes an increment, $d\theta$, to the total rotation, θ, at the end of the beam. Since $d\theta = \alpha$,

$$d\theta = \frac{M\,dx}{EI} \qquad\qquad \text{(3-2b)}$$

If we want the total rotation at the free end, we can integrate (3-2b) over the length of the beam to arrive at

$$\theta = \int d\theta = \int \frac{M\,dx}{EI} \qquad\qquad \text{(3-2c)}$$

When E and I are constant along the length of the beam, which is the case in most beam problems,

$$\theta = \frac{1}{EI} \int M\,dx \qquad\qquad \text{(3-2d)}$$

Although (3-2b) to (3-2d) are not required at this time, they will be useful in later sections of this chapter.

The cross-sectional planes of Fig. 3-1 were initially normal to the centerline of the beam and will be assumed to remain normal. Hence the slope of the deflection curve at N is equal to the rotation of the normal plane at N and the slope of the curve at S is equal to the rotation of the normal plane at S. This can be written

$$\text{slope at } N = \text{rotation at } N \qquad\qquad \text{(3-3)}$$

$$\text{slope at } S = \text{rotation at } S \qquad\qquad \text{(3-4)}$$

Subtracting (3-4) from (3-3) gives

$$\text{slope at } N - \text{slope at } S = \text{rotation at } N - \text{rotation at } S \qquad \text{(3-5)}$$

Expressing slope as a differential and recalling that α is the relative rotation, we obtain

$$\left.\frac{dy}{dx}\right]_N - \left.\frac{dy}{dx}\right]_S = \alpha \qquad\qquad \text{(3-6)}$$

Substituting for α from (3-2a) yields

$$\left.\frac{dy}{dx}\right]_N - \left.\frac{dy}{dx}\right]_S = \frac{M\,dx}{EI}$$

$$\left\{\left.\frac{dy}{dx}\right]_N - \left.\frac{dy}{dx}\right]_S\right\} \bigg/ dx = \frac{M}{EI} \qquad\qquad \text{(3-7)}$$

Upon recalling the definition of differentiation, the left side of the equation will be recognized as

$$\frac{d}{dx}\left(\frac{dy}{dx}\right) \qquad \text{or} \qquad \frac{d^2 y}{dx^2}$$

and then (3-7) can be written as

$$\frac{d^2y}{dx^2} = \frac{M}{EI}$$

(3-8)

This is the *elastic flexure equation for beams in bending.* We may have expected a more explicit equation, but we cannot proceed beyond this differential equation until we have an actual case for which we can integrate (3-8) to get an equation for that particular case. To understand (3-8), it is necessary to work many problems such as Example 3-1.

Example 3-1

Determine the deflection at the end of the cantilever shown in Fig. 3-2a.

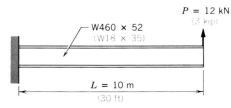

FIG. 3-2*a*

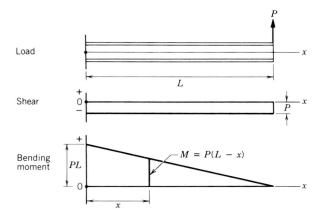

FIG. 3-2*b*

Solution. Since the bending moment, *M*, is a variable, we must first obtain an equation for the moment in terms of *x*. If we refer to Fig. 3-2b, the moment equation is obtained as

$$M = P(L - x)$$

Substituting for M into (3-8) gives

$$\frac{d^2y}{dx^2} = \frac{P}{EI}(L - x) = \frac{P}{EI}(-x + L)$$

Integrating gives either

$$\frac{dy}{dx} = \frac{P}{EI}\left(-\frac{1}{2}x^2 + Lx\right) + C_0$$

or

$$\frac{dy}{dx} = \frac{P}{EI}\left(-\frac{1}{2}x^2 + Lx + C_1\right)$$

If the problem is solved using each of these equations in turn, the final answers will be identical. However, the *second form will usually lead to simpler expressions* throughout the solution; consequently, this form will be used here and it is recommended that in all beam deflection problems the corresponding form be used. Thus using

$$\frac{dy}{dx} = \frac{P}{EI}\left(-\frac{1}{2}x^2 + Lx + C_1\right)$$

we could integrate again, but it is best to evaluate constants of integration as early as possible in any problem. In this case C_1 can be evaluated because we know one value of dy/dx. At the left support neither the vertical section nor the centerline of the member can rotate; hence the elastic curve must remain horizontal or

$$\frac{dy}{dx}\bigg]_{x=0} = 0$$

Then

$$\frac{P}{EI}\left(-\frac{1}{2}0^2 + L0 + C_1\right) = 0$$

Therefore,

$$C_1 = 0$$

Substituting for C_1 gives

$$\frac{dy}{dx} = \frac{P}{EI}\left(-\frac{1}{2}x^2 + Lx\right)$$

Integrating again, we find that

$$y = \frac{P}{EI}\left(-\frac{1}{6}x^3 + \frac{1}{2}Lx^2 + C_2\right)$$

Since there can be no deflection at the left support, the equation must be such that

$$y]_{x=0} = 0$$

or

$$\frac{P}{EI}\left(-\frac{1}{6}0^3 + \frac{1}{2}L0^2 + C_2\right) = 0$$

Therefore,

$$C_2 = 0$$

and

$$y = \frac{P}{EI}\left(-\frac{1}{6}x^3 + \frac{1}{2}Lx^2\right)$$

This equation gives the deflection at all points along the length of the member. The deflection at the right end is obtained by substituting L for x to give

$$\delta = y]_{x=L} = \frac{P}{EI}\left(-\frac{1}{6}L^3 + \frac{1}{2}LL^2\right) = \frac{1}{3}\frac{PL^3}{EI}$$

This is the general deflection equation for a cantilever beam with a concentrated load at the end. For the particular case given

$$P = 12 \text{ kN} = 12 \times 10^3 \text{ N}$$

$$L = 10 \text{ m} = 10 \times 10^3 \text{ mm}$$

$$E = 200\,000 \text{ MPa} \qquad \text{(Table A)}$$

$$= 0.2 \times 10^6 \frac{\text{N}}{\text{mm}^2}$$

$$I = 212 \times 10^6 \text{ mm}^4 \qquad \text{(Table B-1)}$$

$$\delta = \frac{1}{3}\frac{12 \times 10^3 \times (10 \times 10^3)^3}{0.2 \times 10^6 \times 212 \times 10^6}\frac{\text{N} \cdot \text{mm}^3}{\text{N/mm}^2 \cdot \text{mm}^4}$$

$$= \textbf{94.3 mm}$$

In this problem the constants of integration were found to be zero. If they had been overlooked in the integration, the results would not have been altered. This is not normally the case and you should not conclude that constants of integration can be ignored; in most cases this would produce erroneous results.

Example 3-1

Determine the deflection at the end of the cantilever shown in Fig. 3-2a.

Solution. Since the bending moment, M, is a variable, we must first obtain an equation for the moment in terms of x. If we refer to Fig. 3-2b, the moment equation is obtained as

$$M = P(L - x)$$

Substituting for M into (3-8) gives

$$\frac{d^2y}{dx^2} = \frac{P}{EI}(L - x) = \frac{P}{EI}(-x + L)$$

Integrating gives either

$$\frac{dy}{dx} = \frac{P}{EI}\left(-\frac{1}{2}x^2 + Lx\right) + C_0$$

or

$$\frac{dy}{dx} = \frac{P}{EI}\left(-\frac{1}{2}x^2 + Lx + C_1\right)$$

If the problem is solved using each of these equations in turn, the final answers will be identical. However, the *second form will usually lead to simpler expressions* throughout the solution; consequently; this form will be used here and it is recommended that in all beam deflection problems the corresponding form be used. Thus using

$$\frac{dy}{dx} = \frac{P}{EI}\left(-\frac{1}{2}x^2 + Lx + C_1\right)$$

we could integrate again, but it is best to evaluate constants of integration as early as possible in any problem. In this case C_1 can be evaluated because we know one value of dy/dx. At the left support neither the vertical section nor the centerline of the member can rotate; hence the elastic curve must remain horizontal or

$$\frac{dy}{dx}\Bigg]_{x=0} = 0$$

Then

$$\frac{P}{EI}\left(-\frac{1}{2}0^2 + L0 + C_1\right) = 0$$

Therefore,

$$C_1 = 0$$

Substituting for C_1 gives

$$\frac{dy}{dx} = \frac{P}{EI}\left(-\frac{1}{2}x^2 + Lx\right)$$

Integrating again, we find that

$$y = \frac{P}{EI}\left(-\frac{1}{6}x^3 + \frac{1}{2}Lx^2 + C_2\right)$$

Since there can be no deflection at the left support, the equation must be such that

$$y]_{x=0} = 0$$

or

$$\frac{P}{EI}\left(-\frac{1}{6}0^3 + \frac{1}{2}L0^2 + C_2\right) = 0$$

Therefore,

$$C_2 = 0$$

and

$$y = \frac{P}{EI}\left(-\frac{1}{6}x^3 + \frac{1}{2}Lx^2\right)$$

This equation gives the deflection at all points along the length of the member. The deflection at the right end is obtained by substituting L for x to give

$$\delta = y]_{x=L} = \frac{P}{EI}\left(-\frac{1}{6}L^3 + \frac{1}{2}LL^2\right) = \frac{1}{3}\frac{PL^3}{EI}$$

This is the general deflection equation for a cantilever beam with a concentrated load at the end. For the particular case given

$$P = 3 \text{ kip}$$

$$L = 30 \text{ ft} = 360 \text{ in.}$$

$$E = 29 \times 10^3 \text{ kip/in.}^2 \qquad \text{(Table A)}$$

$$I = 510 \text{ in.}^4 \qquad \text{(Table B-1)}$$

$$\delta = \frac{1}{3}\frac{3 \times 360^3}{29 \times 10^3 \times 510}\frac{\text{kip} \cdot \text{in.}^3}{\text{kip/in.}^2 \cdot \text{in.}^4}$$

$$= \textbf{3.15 in.}$$

In this problem the constants of integration were found to be zero. If they had been overlooked in the integration, the results would not have been altered. This is not normally the case and you should not conclude that constants of integration can be ignored; in most cases this would produce erroneous results.

Problems 3-1 to 3-7

3-3 STATICALLY INDETERMINATE BEAMS

The cantilever shown in Fig. 3-3 is statically indeterminate and consequently the expression for M in terms of x, which is required for the integration of (3-8), cannot be written explicitly. A solution can be obtained by assigning a symbol, say R, to the right reaction. Then the bending moment can be expressed in terms of R and known quantities. By proceeding to integrate in the usual manner, an expression for displacement will evolve that contains three unknowns: two constants of integration and the unknown reaction R. However, three conditions on the deflection curve are known:

1. The deflection at the left end is zero.
2. The deflection at the right end is zero.
3. The slope of the deflection curve at the left is zero.

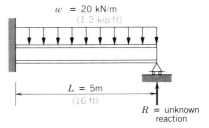

FIG. 3-3 Propped cantilever.

Consequently, three equations in three unknowns can be written and can be solved to give the value of R as well as the constants of integration. Many other statically indeterminate problems can be solved by similar techniques; however, an equation can be derived that has advantages over (3-8) for this type of problem.

To derive the new flexure equation, it will be necessary to establish a relationship between shear and distributed load. This can be done by considering a portion of the beam in Fig. 3-4a that is subjected to a distributed upward load. A free body taken from the beam is shown in Fig. 3-4b. Equating the sum of all vertical forces to zero gives

$$V - (V + dV) + q\, dx = 0$$

$$-dV = -q\, dx$$

$$\frac{dV}{dx} = q \tag{3-9}$$

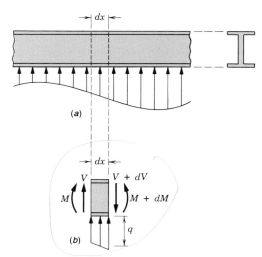

FIG. 3-4 (a) Distributed load. (b) Typical free body.

This equation will be set aside temporarily.

Differentiating both sides of (3-8) with respect to x gives

$$\frac{d^3y}{dx^3} = \frac{d}{dx}\left(\frac{M}{EI}\right)$$

If both E and I are constants

$$\frac{d^3y}{dx^3} = \frac{1}{EI}\frac{dM}{dx}$$

From (1-6b)

$$\frac{dM}{dx} = V$$

Hence

$$\frac{d^3y}{dx^3} = \frac{1}{EI} V \qquad \textbf{(3-10)}$$

Taking the derivative with respect to x gives

$$\frac{d^4y}{dx^4} = \frac{1}{EI}\frac{dV}{dx}$$

Substituting from (3-9), we obtain

$$\boxed{\frac{d^4y}{dx^4} = \frac{1}{EI} q} \qquad \textbf{(3-11)}$$

This is a *second flexure equation* that is often more convenient than (3-8), especially when moment cannot be written as an explicit function of x.

    The use of this equation will be illustrated by solving for the maximum stress in the propped cantilever shown in Fig. 3-3.

Example 3-2

Calculate the maximum bending stress for the beam and loading given in Fig. 3-5a.

    Solution. Displacement is not required, but, as in most statically indeterminate cases, the solution can be found through a consideration of displacement. From (3-11) we obtain

$$\frac{d^4y}{dx^4} = \frac{1}{EI} q = \frac{1}{EI}(-w)$$

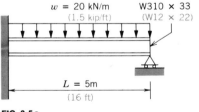

FIG. 3-5a

Integrating gives

$$\frac{d^3y}{dx^3} = \frac{1}{EI}(-wx + C_1)$$

We have no information that will enable us to evaluate C_1, so we proceed by integrating again

$$\frac{d^2y}{dx^2} = \frac{1}{EI}\left(-\frac{wx^2}{2} + C_1 x + C_2\right) \qquad \textbf{(1)}$$

From (3-8) the bending moment,

$$M = EI\frac{d^2y}{dx^2}$$

$$= -\frac{wx^2}{2} + C_1 x + C_2 \qquad \textbf{(2)}$$

At the right end the support is such that the bending moment must be zero or:

$$M]_{x=L} = 0$$

Therefore,

$$-\frac{wL^2}{2} + C_1 L + C_2 = 0 \tag{3}$$

No immediate use can be made of (3), since there are two unknowns and only one equation.

Integrating (1) gives

$$\frac{dy}{dx} = \frac{1}{EI}\left(-\frac{wx^3}{6} + \frac{C_1 x^2}{2} + C_2 x + C_3\right) \tag{4}$$

At the left support the slope of the curve is zero or

$$\frac{dy}{dx}\bigg]_{x=0} = 0$$

$$\frac{1}{EI}\left(-\frac{w0^3}{6} + \frac{C_1 0^3}{2} + C_2 0 + C_3\right) = 0$$

Therefore,

$$C_3 = 0$$

Substituting into (4) yields

$$\frac{dy}{dx} = \frac{1}{EI}\left(-\frac{wx^3}{6} + \frac{C_1 x^2}{2} + C_2 x\right)$$

$$y = \frac{1}{EI}\left(-\frac{wx^4}{24} + \frac{C_1 x^3}{6} + \frac{C_2 x^2}{2} + C_4\right)$$

From the left support condition

$$y]_{x=0} = 0$$

we obtain

$$\frac{1}{EI}\left(-\frac{w0^4}{24} + \frac{C_1 0^4}{6} + \frac{C_2 0^2}{2} + C_4\right) = 0$$

Therefore,

$$C_4 = 0$$

$$y = \frac{1}{EI}\left(-\frac{wx^4}{24} + \frac{C_1 x^3}{6} + \frac{C_2 x^2}{2}\right)$$

From the right support condition

$$y]_{x=L} = 0$$

we obtain

$$\frac{1}{EI}\left(-\frac{wL^4}{24} + \frac{C_1 L^3}{6} + \frac{C_2 L^2}{2} \right) = 0 \tag{5}$$

There are two unknowns, C_1 and C_2, in two equations, (3) and (5). From (3)

$$LC_1 + C_2 = \frac{wL^2}{2} \tag{6}$$

From (5)

$$\frac{L}{6} C_1 + \frac{1}{2} C_2 = \frac{wL^2}{24} \tag{7}$$

(6) $-2 \times$ (7) gives

$$\tfrac{2}{3} LC_1 + 0C_2 = \tfrac{5}{12} wL^2$$

$$C_1 = \tfrac{5}{8} wL$$

(6) $-6 \times$ (7) gives

$$0C_1 - 2C_2 = \tfrac{1}{4} wL^2$$

$$C_2 = -\tfrac{1}{8} wL^2$$

Substituting into (2), we obtain

$$M = -\frac{wx^2}{2} + \frac{5}{8} wLx - \frac{1}{8} wL^2$$

$$= \frac{wL^2}{8}\left[-4\left(\frac{x}{L}\right)^2 + 5\left(\frac{x}{L}\right) - 1 \right]$$

Cautionary note: If the initial assignment had been to select a safe section, we would want to know the maximum bending moments M_{max}. We know that differentiation is useful in finding maximum values and consequently we might equate dM/dx to zero to find the location of M_{max} and thence its value. A design based on the bending moment thus obtained will be very much in error. The error comes from treating a peak value, or stationary value, as though it is the greatest value a function can have. Within the length of the beam, which is the valid region of the equation for M, *the greatest moment is not always the peak moment.* An examination of the bending moment diagram of Fig. 3-5b will make this point clear.

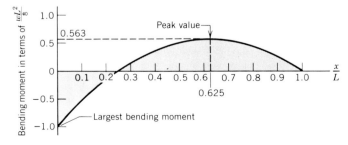

FIG. 3-5b **Bending moment in propped cantilever.**

Bending moment for largest stress $= 1 \times \dfrac{wL^2}{8}$

$$= \dfrac{20 \times 5^2}{8} \text{kN/m} \cdot \text{m}^2 = 62.5 \text{ kN} \cdot \text{m}$$

$$= 62.5 \times 10^6 \text{ N} \cdot \text{mm}$$

and maximum bending stress

$$\sigma = \dfrac{M}{S} = \dfrac{62.5 \times 10^6}{415 \times 10^3} \dfrac{\text{N} \cdot \text{mm}}{\text{mm}^3}$$

$$= 151 \dfrac{\text{N}}{\text{mm}^2}$$

$$= \mathbf{151 \; MPa}$$

Example 3-2

Calculate the maximum bending stress for the beam and loading given in Fig. 3-5a.

Solution. Displacement is not required, but, as in most statically indeterminate cases, the solution can be found through a consideration of displacement. From (3-11) we obtain

$$\dfrac{d^4y}{dx^4} = \dfrac{1}{EI}q = \dfrac{1}{EI}(-w)$$

Integrating gives

$$\dfrac{d^3y}{dx^3} = \dfrac{1}{EI}(-wx + C_1)$$

We have no information that will enable us to evaluate C_1, so we proceed by integrating again

$$\dfrac{d^2y}{dx^2} = \dfrac{1}{EI}\left(-\dfrac{wx^2}{2} + C_1 x + C_2\right) \tag{1}$$

From (3-8) the bending moment,

$$M = EI\dfrac{d^2y}{dx^2}$$

$$= -\dfrac{wx^2}{2} + C_1 x + C_2 \tag{2}$$

At the right end the support is such that the bending moment must be zero, or

$$M]_{x=L} = 0$$

Therefore,

$$-\dfrac{wL^2}{2} + C_1 L + C_2 = 0 \tag{3}$$

No immediate use can be made of (3) since there are two unknowns and only one equation.

Integrating (1) gives

$$\frac{dy}{dx} = \frac{1}{EI}\left(-\frac{wx^3}{6} + \frac{C_1 x^2}{2} + C_2 x + C_3\right) \qquad (4)$$

At the left support the slope of the curve is zero or

$$\frac{dy}{dx}\bigg]_{x=0} = 0$$

$$\frac{1}{EI}\left(-\frac{w0^3}{6} + \frac{C_1 0^3}{2} + C_2 0 + C_3\right) = 0$$

Therefore,

$$C_3 = 0$$

Substituting into (4) yields

$$\frac{dy}{dx} = \frac{1}{EI}\left(-\frac{wx^3}{6} + \frac{C_1 x^2}{2} + C_2 x\right)$$

$$y = \frac{1}{EI}\left(-\frac{wx^4}{24} + \frac{C_1 x^3}{6} + \frac{C_2 x^2}{2} + C_4\right)$$

From the left support condition

$$y]_{x=0} = 0$$

we obtain

$$\frac{1}{EI}\left(-\frac{w0^4}{24} + \frac{C_1 0^4}{6} + \frac{C_2 0^2}{2} + C_4\right) = 0$$

Therefore,

$$C_4 = 0$$

$$y = \frac{1}{EI}\left(-\frac{wx^4}{24} + \frac{C_1 x^3}{6} + \frac{C_2 x^2}{2}\right)$$

From the right support condition

$$y]_{x=L} = 0$$

we obtain

$$\frac{1}{EI}\left(-\frac{wL^4}{24} + \frac{C_1 L^3}{6} + \frac{C_2 L^2}{2}\right) = 0 \qquad (5)$$

There are two unknowns, C_1 and C_2, in two equations, (3) and (5). From (3)

$$LC_1 + C_2 = \frac{wL^2}{2} \qquad (6)$$

From (5)

$$\frac{L}{6} C_1 + \frac{1}{2} C_2 = \frac{wL^2}{24} \qquad (7)$$

(6) $-2 \times$ (7) gives

$$\tfrac{2}{3} L C_1 + 0 C_2 = \tfrac{5}{12} w L^2$$

$$C_1 = \tfrac{5}{8} w L$$

(6) $-6 \times$ (7) gives

$$0 C_1 - 2 C_2 = \tfrac{1}{4} w L^2$$

$$C_2 = -\tfrac{1}{8} w L^2$$

Substituting into (2), we obtain

$$M = -\frac{wx^2}{2} + \frac{5}{8} wLx - \frac{1}{8} wL^2$$

$$= \frac{wL^2}{8}\left[-4\left(\frac{x}{L}\right)^2 + 5\left(\frac{x}{L}\right) - 1 \right]$$

Cautionary note: If the initial assignment had been to select a safe section, we would want to know the maximum bending moments M_{max}. We know that differentiation is useful in finding maximum values and consequently we might equate dM/dx to zero to find the location of M_{max} and thence its value. A design based on the bending moment thus obtained will be very much in error. The error comes from treating a peak value, or stationary value, as though it is the greatest value a function can have. Within the length of the beam, which is the valid region of the equation for M, *the greatest moment is not always the peak moment.* An examination of the bending moment diagram of Fig. 3-5*b* will make this point clear.

Bending moment for largest stress $= 1 \times \dfrac{wL^2}{8}$

$$= \frac{1.5 \times 16^2}{8} \frac{\text{kip/ft} \cdot \text{ft}^2}{\text{ft}}$$

$$= 48.0 \text{ kip} \cdot \text{ft}$$

and maximum bending stress

$$\sigma = \frac{M}{S} = \frac{48 \times 12}{25.3} \frac{\text{kip} \cdot \text{ft} \cdot \text{in./ft}}{\text{in.}^3}$$

$$= \textbf{22.8 kip/in.}^2$$

Problems 3-8 to 3-14

In all preceding problems it was possible to integrate the appropriate flexural equation, evaluate the 2 or 4 constants of integration, and obtain a solution. In a great many practical problems the loads are such that the method is much more difficult to apply. To illustrate, the beam in Fig. 3-6*a* is easy to solve due to symmetry while the solution to the beam in Fig. 3-6*b* is extremely lengthy. For the latter case we do not yet know how to write a *single* equation for q and hence must write Eq. (3-11) *twice*, once for each region of the beam. When the two equations are integrated, 8 constants appear. The end conditions provide only four equations in these 8 unknowns. The other four equations can be obtained by considering conditions at midspan, where the displacements, rotations, bending moments, and shears must be continuous across the midspan interface. This will lead to a solution, *but it will be necessary to solve 8 equations in 8 unknowns.* Although some equations will be quite simple, solving the equations is still a long procedure with numerous opportunities to make errors. The beam in Fig. 3-6c would also involve 8 unknowns while those in Figs. 3-6d and 3-6e would have 12 constants of integration. The difficulty with the more complicated problems arises from our inability to write a single equation for either load intensity, or for bending moment, that is applicable to the whole beam. Fortunately, such an equation can be written with the aid of *singularity functions*.

3-4 SINGULARITY FUNCTIONS

Suppose that we are required to determine the deflection equation for the beam shown in Fig. 3-7. The bending moment requires three separate equations, each applicable to one portion of the beam:

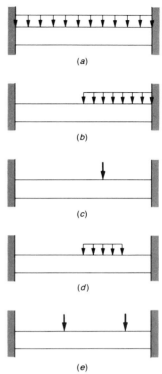

(a)

(b)

(c)

(d)

(e)

FIG. 3-6

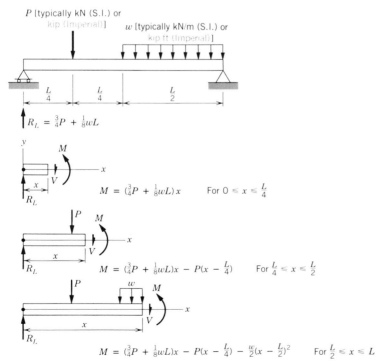

P [typically kN (S.I.) or kip (Imperial)]

w [typically kN/m (S.I.) or kip ft (Imperial)]

$\frac{L}{4}$ $\frac{L}{4}$ $\frac{L}{2}$

$R_L = \frac{3}{4}P + \frac{1}{8}wL$

M

$M = (\frac{3}{4}P + \frac{1}{8}wL)x$ For $0 \leq x \leq \frac{L}{4}$

$M = (\frac{3}{4}P + \frac{1}{8}wL)x - P(x - \frac{L}{4})$ For $\frac{L}{4} \leq x \leq \frac{L}{2}$

$M = (\frac{3}{4}P + \frac{1}{8}wL)x - P(x - \frac{L}{4}) - \frac{w}{2}(x - \frac{L}{2})^2$ For $\frac{L}{2} \leq x \leq L$

FIG. 3-7

$$M = \left(\frac{3}{4}P + \frac{1}{8}wL\right)x \qquad \textbf{(1)} \qquad \text{for } 0 \leqslant x \leqslant \frac{L}{4}$$

$$M = \left(\frac{3}{4}P + \frac{1}{8}wL\right)x - P\left(x - \frac{L}{4}\right) \qquad \textbf{(2)} \qquad \text{for } \frac{L}{4} \leqslant x \leqslant \frac{L}{2}$$

$$M = \left(\frac{3}{4}P + \frac{1}{8}wL\right)x - P\left(x - \frac{L}{4}\right) - \frac{w}{2}\left(x - \frac{L}{2}\right)^2 \quad \textbf{(3)} \qquad \text{for } \frac{L}{2} \leqslant x \leqslant L$$

The third equation contains the functions that appear in the previous two equations and would suffice to serve for the whole beam provided that there was some way to drop the inappropriate expressions according to the value of x. Notice that the variable part of the last term in (3),

$$\left(x - \frac{L}{2}\right)$$

becomes negative for x less than $L/2$. That is, the last term does not appear in any moment equation when

$$\left(x - \frac{L}{2}\right)$$

is negative. Then let us rewrite the last term as

$$-\frac{w}{2}\left\langle x - \frac{L}{2}\right\rangle^2$$

and *take the pointed brackets to mean that when the quantity inside the brackets is positive the brackets have the same effect as ordinary brackets, but when that quantity is negative the value inside will be taken as zero.*

The second term in moment equations (2) and (3) does not appear in (1), where the variable part

$$\left(x - \frac{L}{4}\right)$$

would be negative. In (2) and (3) this variable is always positive. Consequently, pointed brackets, with the meaning given above, when placed on the second term, thus

$$-P\left\langle x - \frac{L}{4}\right\rangle$$

will cause the term to be ineffective when x is less than $L/4$. We can then write

$$M = \left(\frac{3}{4}P + \frac{1}{8}wL\right)x - P\left\langle x - \frac{L}{4}\right\rangle - \frac{w}{2}\left\langle x - \frac{L}{2}\right\rangle^2 \qquad \textbf{(4)}$$

and have in a single expression all the information previously contained in (1), (2), and (3). The functions expressed by the pointed brackets are referred to as *singularity functions*. The general form of the singularity **function** is

$$y = \langle x - a \rangle^n$$

When $n = 0$ the function varies as in Fig. 3-8a and could be described as a step function. Since this is a frequently used form of the singularity function, the whole class is sometimes referred to, erroneously, as *step functions*.

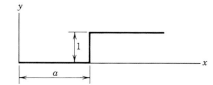

When $n = 1$ the function becomes a ramp as indicated in Fig. 3-8b.

It will be necessary to perform the operations of calculus on the singularity functions. In this respect they do not differ from the same variables with ordinary brackets. Thus

FIG. 3-8a

$$\frac{d}{dx} \langle x - a \rangle^n = n \langle x - a \rangle^{n-1}$$

and

$$\int \langle x - a \rangle^n dx = \frac{\langle x - a \rangle^{n+1}}{n+1} + c \qquad \text{for } n \neq -1$$

FIG. 3-8b

The following examples demonstrate the use and power of these functions, which were originally introduced by Clebsch in 1862 and used by Macaulay in 1919 to solve beam deflection problems.

Example 3-3

Determine the deflection of the beam given in Fig. 3-9a at A, B, and C in terms of w, L, E, and I.

Solution. We can determine moment equations for the two halves of the beam by taking free bodies as shown in Fig. 3-9b. The two moment equations can be combined, using a singularity function, into

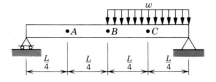

FIG. 3-9a

$$M = \frac{1}{8} wLx - \frac{w}{2} \left\langle x - \frac{L}{2} \right\rangle^2 \qquad \text{for } 0 \leq x \leq L$$

From (3-8)

$$\frac{d^2y}{dx^2} = \frac{M}{EI} = \frac{w}{2EI} \left[\frac{L}{4} x - \left\langle x - \frac{L}{2} \right\rangle^2 \right]$$

Integrating gives

$$\frac{dy}{dx} = \frac{w}{2EI} \left[\frac{L}{8} x^2 - \frac{1}{3} \left\langle x - \frac{L}{2} \right\rangle^3 + C_1 \right]$$

Integrating again, we obtain

$$y = \frac{w}{2EI} \left[\frac{L}{24} x^3 - \frac{1}{12} \left\langle x - \frac{L}{2} \right\rangle^4 + C_1 x + C_2 \right]$$

From the left end support condition

$$y]_{x=0} = 0$$

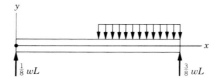

$$M = \tfrac{1}{8}\,wLx \qquad \text{for } 0 \le x \le \tfrac{L}{2}$$

$$M = \tfrac{1}{8}\,wLx - \tfrac{w}{2}\left(x - \tfrac{L}{2}\right)^2 \qquad \text{for } \tfrac{L}{2} \le x \le L$$

FIG. 3-9b

Therefore,

$$C_2 = 0$$

From the right end support condition

$$y]_{x=L} = 0$$

$$\frac{w}{2EI}\left[\frac{L}{24}L^3 - \frac{1}{12}\left(L - \frac{L}{2}\right)^4 + C_1 L\right] = 0$$

Therefore,

$$C_1 = -\frac{7}{192}L^3$$

The deflection equation then becomes

$$y = \frac{w}{2EI}\left[\frac{L}{24}x^3 - \frac{1}{12}\left\langle x - \frac{L}{2}\right\rangle^4 - \frac{7}{192}L^3 x\right]$$

$$= \frac{wL^4}{2EI}\left[\frac{1}{24}\left(\frac{x}{L}\right)^3 - \frac{1}{12}\left\langle \frac{x}{L} - \frac{1}{2}\right\rangle^4 - \frac{7}{192}\left(\frac{x}{L}\right)\right]$$

The downward deflections are then given by

$$\delta_A = -y]_{\frac{x}{L} = \frac{1}{4}} = \frac{-wL^4}{2EI}\left[\frac{1}{24}\left(\frac{1}{4}\right)^3 - \frac{1}{12}0 - \frac{7}{192}\left(\frac{1}{4}\right)\right]$$

$$= 0.00423 \frac{wL^4}{EI}$$

$$\delta_B = -y]_{\frac{x}{L}=\frac{1}{2}} = \frac{-wL^4}{2EI}\left[\frac{1}{24}\left(\frac{1}{2}\right)^3 - \frac{1}{12}0^4 - \frac{7}{192}\left(\frac{1}{2}\right)\right]$$

$$= 0.00651 \frac{wL^4}{EI}$$

$$\delta_C = -y]_{\frac{x}{L}=\frac{3}{4}} = \frac{-wL^4}{2EI}\left[\frac{1}{24}\left(\frac{3}{4}\right)^3 - \frac{1}{12}\left(\frac{3}{4}-\frac{1}{2}\right)^4 - \frac{7}{192}\left(\frac{3}{4}\right)\right]$$

$$= 0.00505 \frac{wL^4}{EI}$$

Example 3-4

For the beam of Fig. 3-10a, determine the magnitude of the left reaction in terms of w and L.

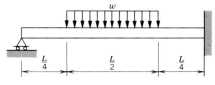

FIG. 3-10a

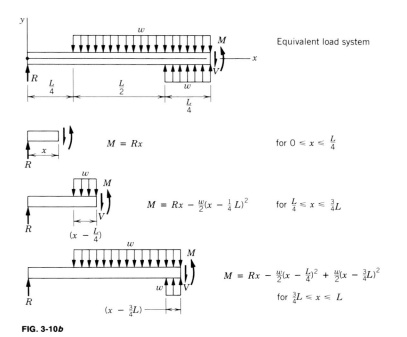

Equivalent load system

$$M = Rx \qquad \text{for } 0 \le x \le \frac{L}{4}$$

$$M = Rx - \frac{w}{2}(x - \tfrac{1}{4}L)^2 \qquad \text{for } \frac{L}{4} \le x \le \frac{3}{4}L$$

$$M = Rx - \frac{w}{2}(x - \tfrac{L}{4})^2 + \frac{w}{2}(x - \tfrac{3}{4}L)^2 \qquad \text{for } \frac{3}{4}L \le x \le L$$

FIG. 3-10b

Solution. We again start our solution by writing the moment equations for the beam regions as indicated in Fig. 3-10b. Note the use of the *equivalent load system* for the third region. The combined moment equation is

$$M = Rx - \frac{w}{2}\left\langle x - \frac{1}{4}L \right\rangle^2 + \frac{w}{2}\left\langle x - \frac{3}{4}L \right\rangle^2 \qquad \text{for } 0 \le x \le L$$

From (3-8)

$$\frac{d^2y}{dx^2} = \frac{M}{EI} = \frac{w}{2EI}\left[\frac{2R}{w}x - \left\langle x - \frac{1}{4}L \right\rangle^2 + \left\langle x - \frac{3}{4}L \right\rangle^2\right]$$

Integrating gives

$$\frac{dy}{dx} = \frac{w}{2EI}\left[\frac{R}{w}x^2 - \frac{1}{3}\left\langle x - \frac{1}{4}L \right\rangle^3 + \frac{1}{3}\left\langle x - \frac{3}{4}L \right\rangle^3 + C_1\right] \tag{1}$$

From the right support condition

$$\frac{dy}{dx}\bigg]_{x=L} = 0$$

$$\frac{R}{w}L^2 - \frac{1}{3}\left(\frac{3}{4}L\right)^3 + \frac{1}{3}\left(\frac{1}{4}L\right)^3 + C_1 = 0$$

$$C_1 = -\frac{R}{w}L^2 + \frac{13}{96}L^3 \tag{2}$$

Substituting (2) into (1), we obtain

$$\frac{dy}{dx} = \frac{w}{2EI}\left[\frac{R}{w}(x^2 - L^2) - \frac{1}{3}\left\langle x - \frac{1}{4}L \right\rangle^3 + \frac{1}{3}\left\langle x - \frac{3}{4}L \right\rangle^3 + \frac{13}{96}L^3\right]$$

Integrating again

$$y = \frac{w}{2EI}$$

$$\times \left[\frac{R}{w}\left(\frac{1}{3}x^3 - L^2x\right) - \frac{1}{12}\left\langle x - \frac{1}{4}L \right\rangle^4 + \frac{1}{12}\left\langle x - \frac{3}{4}L \right\rangle^4 + \frac{13}{96}L^3x + C_2\right]$$

From the left support condition

$$y]_{x=0} = 0$$

Therefore,

$$C_2 = 0$$

At the right support

$$y]_{x=0} = 0$$

$$\frac{R}{w}\left(\frac{1}{3}L^3 - L^3\right) - \frac{1}{12}\left(\frac{3}{4}L\right)^4 + \frac{1}{12}\left(\frac{1}{4}L\right)^4 + \frac{13}{96}L^4 = 0$$

$$\frac{R}{w}\left(-\frac{2}{3}\right)L^3 + \frac{7}{64}L^4 = 0$$

$$R = \frac{3}{2}\frac{7}{64}wL = \mathbf{0.1641\ wL}$$

Problems 3-15 to 3-22

When load, shear, bending moment, rotation, and displacement are examined together an interesting relationship is found to exist. These quantities are shown together in Fig. 3-11. The curves, when placed in the order shown, display the relationships in a form that is very useful and easy to remember.

From the equations listed on the left side of Fig. 3-11 it can be seen that when the load intensity is given, one integration gives the shear and another integration gives the bending moment. Integrating bending moment and applying the factor $1/EI$ gives rotation, and another integration gives displacement. Of course, a constant of integration appears at each integration and the evaluation of these constants, on the basis of the beam support conditions, becomes the principal task in determining the deflection equation.

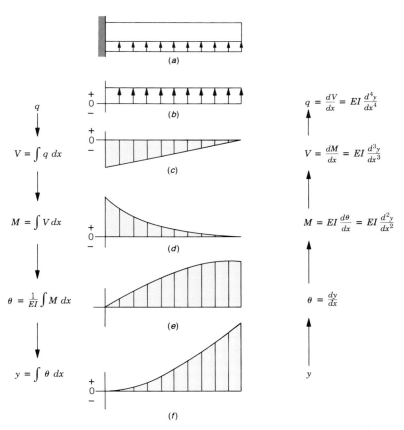

$$q$$

$$V = \int q\,dx$$

$$M = \int V\,dx$$

$$\theta = \frac{1}{EI}\int M\,dx$$

$$y = \int \theta\,dx$$

$$q = \frac{dV}{dx} = EI\frac{d^4y}{dx^4}$$

$$V = \frac{dM}{dx} = EI\frac{d^3y}{dx^3}$$

$$M = EI\frac{d\theta}{dx} = EI\frac{d^2y}{dx^2}$$

$$\theta = \frac{dy}{dx}$$

$$y$$

FIG. 3-11 Load-deflection relationships. (*a*) **Loaded member.** (*b*) **Load intensity.** (*c*) **Shear.** (*d*) **Bending moment.** (*e*) **Rotation.** (*f*) **Displacement.**

The reverse process, where we start with deflection and proceed to load intensity, merely requires a differentiation at each step with constants dropping out rather than appearing. The relationships, in differential form, are listed on the right side of Fig. 3-11. This column of equations should be read upward as indicated by the arrows.

Few practical problems start with a known displacement; hence the differential relationships may seem to be unimportant. However, they are often very convenient for checking. Usually, the displacement is obtained by integration from the known load intensity or the known bending moment and the differential relationships provide a check on the validity of the solutions.

3-5 SUPERPOSITION

Because certain beam problems occur repeatedly, general solutions have been worked out and made available in tables such as Table G. Similar tables are used by practicing engineers in order to save time and increase reliability. However, it would be impractical to provide solutions to all possible combinations of loads and end conditions. Many problems, which differ from those that have been tabulated, can be solved by combining known solutions. To illustrate this method, known as the *method of superposition*, let us work the following example.

Example 3-5

Determine the deflection at the center of the simply supported beam of Fig. 3-12*a* carrying a concentrated load at midspan as well as a uniformly distributed load.

Solution. A beam loaded in this manner does not appear in Table G; however, *we can resolve this complex problem into the two basic problems* illustrated in Figs. 3-12*b* and *c*. Standard solutions for the two basic problems can be obtained from Table G as

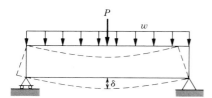

FIG. 3-12*a*

$$\delta_1 = \frac{PL^3}{48EI} \quad and \quad \delta_2 = \frac{5}{384}\frac{wL^4}{EI}$$

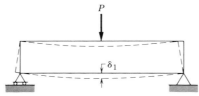

FIG. 3-12*b*

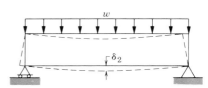

FIG. 3-12*c*

The required deflection for the original problem of Fig. 3-12a is then obtained by superimposing the two solutions, that is, by adding the deflections of the two basic problems. Thus

$$\delta = \delta_1 + \delta_2$$

$$= \frac{PL^3}{48EI} + \frac{5}{384}\frac{wL^4}{EI}$$

The method of superposition is most useful in solving statically indeterminate problems such as the propped cantilever beam shown in Fig. 3-3. We will demonstrate this method in Example 3-6.

Example 3-6

Determine the reaction at the roller under the end of the propped cantilever in Fig. 3-3.

Solution. By breaking the problem into the two cases shown in Figs. 3-13a and b, we have replaced the indeterminate problem by two simple problems that have known solutions. For the uniformly distributed load the deflection at the end is

$$\delta_1 = \frac{wL^4}{8EI} \qquad \text{from Table G}$$

and for the concentrated load

$$\delta_2 = \frac{PL^3}{3EI} \qquad \text{from Table G}$$

Superimposing the solutions will give, for the combined loads, deflection

$$\delta = \delta_1 + \delta_2 = \frac{wL^4}{8EI} + \frac{PL^3}{3EI}$$

in which all quantities are known except δ and P. But, from the original end conditions we know that the deflection at the right end is zero. Note that we are again making use of a displacement condition to solve a statically indeterminate problem. Equating the deflection to zero gives

$$\frac{wL^4}{8EI} + \frac{PL^3}{3EI} = 0$$

Solving for the unknown P will give the concentrated load, which is the reaction at the roller required to prevent any deflection at the end. This gives

$$P = -\frac{3}{8}wL$$

and substituting the quantities from Fig. 3-3, we get

$$P = -\frac{3}{8} \times 20 \times 5\frac{kN}{m} \cdot m = -\textbf{37.5 kN}$$

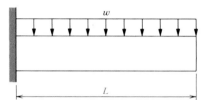

FIG. 3-13a

FIG. 3-13b

The negative sign tells us that the reaction is opposite to the direction assumed for P; that is, it indicates that the reaction is an upward force.

Example 3-6

For the propped cantilever of Fig. 3-3, solve for the reaction at the roller.

Solution. By breaking the problem into the two cases shown in Figs. 3-13a and b, we have replaced the indeterminate problem by two simple problems that have known solutions. For the uniformly distributed load the deflection at the end is

$$\delta_1 = \frac{wL^4}{8EI} \qquad \text{from Table G}$$

and for the concentrated load

$$\delta_2 = \frac{PL^3}{3EI} \qquad \text{from Table G}$$

Superimposing the solutions will give, for the combined loads, deflection

$$\delta = \delta_1 + \delta_2 = \frac{wL^4}{8EI} + \frac{PL^3}{3EI}$$

in which all quantities are known except δ and P. But, from the original end conditions we know that the deflection at the right end is zero. Note that we are again making use of a displacement condition to solve a statically indeterminate problem. Equating the deflection to zero gives

$$\frac{wL^4}{8EI} + \frac{PL^3}{3EI} = 0$$

Solving for the unknown P will give the concentrated load, which is the reaction at the roller required to prevent any deflection at the end. This gives

$$P = -\frac{3}{8} wL$$

and substituting the quantities from Fig. 3-3, we get

$$P = -\frac{3}{8} \times 1.5 \times 16 \frac{\text{kip} \cdot \text{ft}}{\text{ft}} = -\textbf{24.0 kip}$$

The negative sign tells us that the reaction is opposite to the direction assumed for P; that is, it indicates that the reaction is an upward force.

The principle of superposition that has been illustrated here has widespread application in engineering. There is one restriction on this method that has not been mentioned: The *relationships must be linear*. In the cases of beam deflection, the deflection varies linearly with the load, which makes the addition of two force-load systems an acceptable process. There are many conditions causing nonlinearity such as material being strained beyond the proportional limit. In all nonlinear cases the method of superposition cannot be used.

Problems 3-23 to 3-30

3-6 MOMENT-AREA METHOD

When the deflection of a beam is required, the integrations summarized in Fig. 3-11 provide a convenient method. In combination with singularity functions they constitute a powerful method for determining the deflection equation of a beam and for solving certain indeterminate problems. Frequently, however, the complete deflection equation is not required and the equation is only used to calculate the displacement at one or two locations. Such cases can often be solved more readily by the theorems of the *moment-area method*, which will now be developed.

Consider the case of a beam where the bending moment equation is known, such as in the Fig. 3-14a. We could obtain information about the rotation of the beam cross sections by integrating M in

$$\theta = \frac{1}{EI} \int M \, dx \qquad \text{from (3-2d)}$$

or we could choose to integrate M/EI in

$$\theta = \int \frac{M}{EI} dx \qquad \text{from (3-2c)}$$

Although the first form seems more convenient, the second will be used in this development. Since the moment is known, M/EI can be plotted as in Fig. 3-14b. Integrating (3-2c) between the limits x_A and x_B will give the rotation of the section at B relative to the section at A. The integration would be the same if the problem had been to find the area of the M/EI diagram between A and B. This gives us the *first theorem of the moment-area method*:

When considering two points on a beam, the rotation at the point on the right, relative to the one on the left, is equal to the area of the M/EI diagram between the points.

This will prove to be useful, but we are more likely to be interested in *displacement*, so let us again consider the same points and the contribution to displacement due to the flexure of an incremental length of beam at a typical location between the points. This is shown in Fig. 3-14c, where only the bending in the incremental length, dx, is taken as contributing to deflection. The rotation of the right end of the element will be, by the first theorem, the shaded area, dA, of the M/EI diagram. The point B will move a distance equal to the rotation, that is, the area dA, multiplied by the distance a. Or

$$d\delta_B = a \, dA$$

Note that the right side of the equation is the moment of the area dA about the point B.

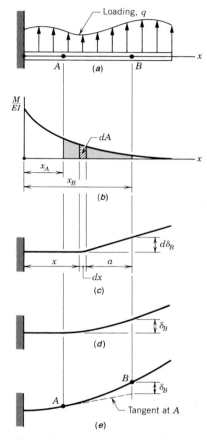

FIG. 3-14 Moment-area concepts. (*a*) Beam under loading, *q*. (*b*) *M/EI* diagram. (*c*) Bending in the incremental length *dx* only. (*d*) Bending in the length *AB*. (*e*) Deflected shape of the entire beam.

Now consider the effect of flexure in all parts of the beam between A and B. This will give a deflection curve as in Fig. 3-14d, and the deflection at B can be obtained by

$$\delta_B = \int_{x_a}^{x_b} a \, dA$$

The quantity being integrated has been observed to be the moment of the incremental area so that the integral is equal to the moment of the area of the M/EI diagram, between points A and B, about point B. This may be considered the *second theorem of the moment-area method* and stated:

For any two points on a beam the displacement *of the point on the right, measured from the tangent to the deflection curve at the point on the left, is equal to the moment, with respect to the point on the right, of the area of the* M/EI *diagram between the points.*

When the flexure of the remaining part of the beam is taken into account, the deflection curve will be as shown in Fig. 3-14e. This change influences the displacements at A and B but does not alter the relative displacement of B when measured from the tangent at A and does not invalidate the theorem.

The following examples will illustrate the use of the moment-area theorems and are worthy of careful study. Solutions to the first two example problems could be obtained by other methods with little additional effort. The third problem shows the ease with which certain statically indeterminate problems can be solved by the moment-area method. You will fully appreciate the advantage if you solve the third problem by other methods.

Example 3-7

For the cantilever of Fig. 3-15, determine the rotation and vertical displacement of the free end in terms of E, I, P, and L.

Solution. The required dimensions are indicated on the M/EI diagram of Fig. 3-16. The rotation at end B can be determined by application of the first theorem

$$\theta_B - \theta_A = \text{area of diagram between } A \text{ and } B$$

$$= \frac{1}{2} \times \frac{3}{4}\frac{PL}{EI} \times \frac{3}{4}L = \frac{9}{32}\frac{PL^2}{EI}$$

From the left support condition

$$\theta_A = 0$$

Hence

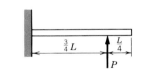

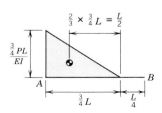

FIG. 3-15

FIG. 3-16

$$\theta_B = \frac{9}{32} \frac{PL^2}{EI}$$

Utilizing the second theorem, the displacement of point B from the tangent at A, which equals the moment of area of diagram about B, is

$$\left(\frac{1}{2} \times \frac{3}{4} \frac{PL}{EI} \times \frac{3}{4} L\right)\left(\frac{1}{2} L + \frac{1}{4} L\right) = \frac{27}{128} \frac{PL^3}{EI}$$

Since the tangent at A remains fixed, the displacement at B is

$$\delta_B = \frac{27}{128} \frac{PL^3}{EI}$$

Example 3-8

Determine the displacement of the beam given in Fig. 3-17a at the point of load application.

Solution. The end reactions, M/EI diagram, and deflected shape for the beam are indicated in Fig. 3-17b.

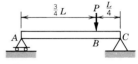

FIG. 3-17a

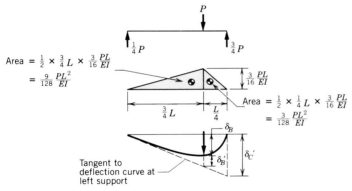

Area $= \frac{1}{2} \times \frac{3}{4} L \times \frac{3}{16} \frac{PL}{EI}$
$= \frac{9}{128} \frac{PL^2}{EI}$

Area $= \frac{1}{2} \times \frac{1}{4} L \times \frac{3}{16} \frac{PL}{EI}$
$= \frac{3}{128} \frac{PL^2}{EI}$

FIG. 3-17b

From the second theorem

$$\delta'_B = \frac{9}{128} \frac{PL^2}{EI} \times \frac{1}{3} \times \frac{3}{4} L = \frac{9}{512} \frac{PL^3}{EI}$$

and

$$\delta'_C = \frac{9}{128} \frac{PL^2}{EI} \times \left(\frac{1}{3} \times \frac{3}{4} L + \frac{1}{4} L\right) + \frac{3}{128} \frac{PL^2}{EI} \times \frac{2}{3} \times \frac{1}{4} L$$

$$= \frac{9}{256} \frac{PL^3}{EI} + \frac{1}{256} \frac{PL^3}{EI} = \frac{10}{256} \frac{PL^3}{EI}$$

By similar triangles

$$\frac{\delta_B + \delta'_B}{\frac{3}{4}L} = \frac{\delta'_C}{L}$$

Hence

$$\delta_B = \frac{3}{4}\delta'_C - \delta'_B = \left(\frac{3}{4}\frac{10}{256} - \frac{9}{512}\right)\frac{PL^3}{EI} = \frac{6}{512}\frac{PL^3}{EI}$$

Example 3-9

Determine the reaction at the roller under the right end of the statically indeterminate beam of Fig. 3-18a.

Solution. The deflected shape, greatly exaggerated, together with the unknown roller reaction, R, are indicated in Fig. 3-18b. We will find the moment-area method easier to use if we resolve the moment into two parts that when superimposed will be equivalent to the indeterminate cantilever problem. The two parts are shown in Fig. 3-18c, where the first part is the moment diagram due to P and the second part is the moment diagram due to the unknown reaction R.

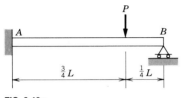

FIG. 3-18a

FIG. 3-18b

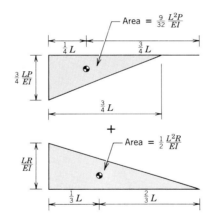

FIG. 3-18c

Since the tangent at A remains stationary when the load is applied, any displacement relative to that tangent is an absolute displacement. Using the second theorem to state the displacement at point B and equating to zero because the support does not allow any displacement at B gives an equation in the unknown R.

$$\delta_B = -\frac{9}{32}\frac{L^2P}{EI}\frac{3}{4}L + \frac{1}{2}\frac{L^2R}{EI}\frac{2}{3}L = 0$$

$$\frac{1}{3}\frac{L^3}{EI}R = \frac{27}{128}\frac{L^3}{EI}P$$

$$R = \frac{81}{128}P$$

Problems 3-31 to 3-36

3-7 DEFLECTION OF LONG-RADIUS CURVED BEAMS

We will now consider members having a radius of curvature that is large enough to permit us to use, without significant error, the angle-moment and stress-moment equations, (2-10) and (2-11), which were developed for straight beams. The methods of Section 2-7 can be used to establish the lower limit on the radius for a given cross section and a given acceptable level of error. When the radius is smaller than this lower limit, accurate results can still be obtained by combining the methods illustrated in this section with those of Section 2-7.

No attempt will be made to apply the theorems of the moment-area method or to modify them for use in curved members. Instead we will use the principles upon which the moment-area method was based and will find that some otherwise difficult problems are quite easy to solve by this approach. Since we have all the necessary equations at our disposal and there are no theorems to prove, we will use examples to present the methods for solving curved beam problems.

Example 3-10

Determine the increase in the dimension D, in terms of P, R, E, and I, when the load P is applied to the open ring of Fig. 3-19a.

Solution. The portions of the ring to the left of a vertical plane through the center can be disregarded, since they do not contribute to the required deflection. The remaining C is symmetrical about a horizontal plane, so we will solve for only one-half of the C, thus reducing the problem to finding the displacement at the point of load application in Fig. 3-19b.

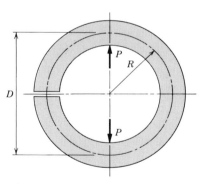

FIG. 3-19a

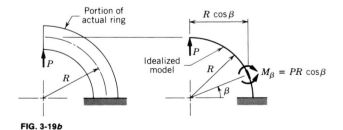

FIG. 3-19b

Consider an incremental length of the ring subtending an angle $d\beta$ as indicated in Fig. 3-19c. The moment will cause one end of the element to rotate relative to the other through an angle

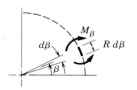

FIG. 3-19c

$$\alpha = \frac{M_\beta R \, d\beta}{EI} \qquad \text{from (3-1)}$$

$$= \frac{PR^2}{EI} \cos \beta \, d\beta \qquad \text{(3-12)}$$

If only this element is considered to be flexible, the part will displace as in Fig. 3-19d. The end will move vertically.

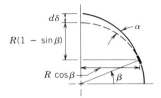

FIG. 3-19d

$$d\delta = R \cos \beta \, \alpha$$

$$= \frac{PR^3}{EI} \cos^2 \beta \, d\beta \qquad \textbf{(3-13)}$$

Since we want the total motion at the end, we must sum the contributions made by all the elements in the quadrant. This is done by integrating (3-13) to get the vertical displacement. Then

$$\delta = \frac{PR^3}{EI} \int_0^{\pi/2} \cos^2 \beta \, d\beta = \frac{PR^3}{EI} \left[\frac{1}{2} \beta + \frac{1}{4} \sin 2\beta \right]_0^{\pi/2} = \frac{PR^3}{EI} \frac{\pi}{4}$$

Since the lower portion of the ring will deflect an equal amount, the change in D is

$$\Delta D = 2\delta = \frac{\pi}{2} \frac{PR^3}{EI}$$

Example 3-11

The ends of the open ring in Example 3-10 are welded together to form the closed ring of Fig. 3-20a. Calculate the maximum bending moment in the closed ring and compare it with the maximum moment in the open ring.

Solution. This is a statically indeterminate problem, since the moment is not known at any section in the ring. The moment at point B will be taken as M_0.

As in the previous example, only one quadrant of the ring needs to be considered because of the two planes of symmetry. We can then model the problem as shown in Fig. 3-20b. Note for point A that the roller arrangement permits horizontal displacements without allowing either rotation or vertical motion. Visualize the deformed shape of the closed ring to satisfy yourself that the modeling at A is correct.

Consider now an incremental length of ring as in Fig. 3-20c and the contribution that it will make to rotation at point B.

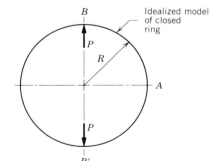

FIG. 3-20a

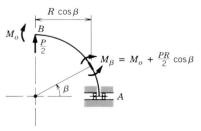

FIG. 3-20b

$$d\theta = \alpha = \frac{M_\beta R \, d\beta}{EI} = \frac{1}{EI} \left(M_0 + \frac{PR}{2} \cos \beta \right) R \, d\beta$$

The rotation of the section at B relative to A will be obtained by integrating over the quadrant. But from symmetry there is no rotation at either A or B; therefore,

$$\int_0^{\pi/2} d\theta = 0$$

$$\frac{1}{EI} \int_0^{\pi/2} \left(M_0 + \frac{PR}{2} \cos \beta \right) R \, d\beta = 0$$

$$M_0\beta + \frac{PR}{2}\sin\beta \bigg]_0^{\pi/2} = 0$$

$$M_0\frac{\pi}{2} + \frac{PR}{2} = 0$$

$$M_0 = -\frac{1}{\pi}PR$$

and

$$M_\beta = -\frac{1}{\pi}PR + \frac{1}{2}PR\cos\beta$$

$$= PR\left(-\frac{1}{\pi} + \frac{1}{2}\cos\beta\right)$$

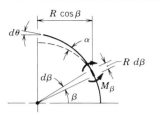

FIG. 3-20c

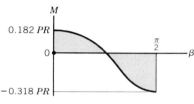

FIG. 3-20d

Plotting the variation of moment with β in Fig. 3-20d, we see that large bending moments occur at A and B. The greatest bending moment occurs at points B and B' of Fig. 3-20a, and is

$$M_{\max} = M_0 = \textbf{0.318 } \textbf{\textit{PR}}$$

Recall that for the open ring

$$M_{\max} = PR$$

Therefore, welding the ends of the ring together reduces the bending moment to 32% of its former value and thus more than triples the load-carrying capacity of the ring. This is probably a good return for the extra cost of analysis and of welding the ring.

Example 3-12

The end frame of a machine as shown in Fig. 3-21a is intended to support a horizontal shaft. Under normal conditions the shaft pushes downward on the support, but under some conditions the shaft may tend to lift, exerting an upward force of 60 kN at the middle of the arch over the shaft. Calculate the maximum stress in the frame due to the upward force.

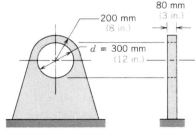

FIG. 3-21a

Solution. This example represents a very practical engineering problem and is typical of such problems in that it does not conform exactly to theory. The arched top resembles a curved beam, but there is no clearly defined location for the end support where we are able to say that certain end conditions exist. To find an accurate answer would require methods that are used by specialists in stress analysis. Under these conditions the designer uses some reasonably accurate method to find an approximate stress and then, if the safety of the part seems to be in doubt, either strengthens the part or calls on a stress analyst to determine the stresses more precisely. The following solution illustrates the calculations that might be carried out as a first approximation.

Let us start by replacing the actual arch by an arch having a uniform cross section and terminating at fixed supports. The location of the supports is a matter of judgment on the part of the designer. We will assume the supports to be located as indicated in Fig. 3-21b.

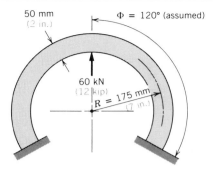

50 mm
(2 in.)

Φ = 120° (assumed)

60 kN
(12 kip)

R = 175 mm
(7 in.)

FIG. 3-21b

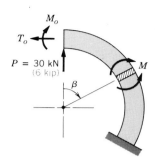

M_o

T_o

$P = 30$ kN
(6 kip)

M

β

FIG. 3-21c

From symmetry it is apparent that the cross section at the top of the arch does not rotate and does not move in the horizontal direction. These known displacement conditions will be used to determine the unknown values in the indeterminate structure. Again because of symmetry, it is sufficient to work with only half of the arch as shown in Fig. 3-21c.

The tensile force T_0 is real and cannot be neglected. The moment at a typical section can be obtained from the free body of Fig. 3-21d.

$$M = PR \sin \beta - T_0 R(1 - \cos \beta) + M_0 \tag{1}$$

The flexibility of an increment of arch will give motions as in Fig. 3-21e.

$$d\theta = \frac{MR\,d\beta}{EI} = \frac{[PR \sin \beta - T_0 R(1 - \cos \beta) + M_0]\,R\,d\beta}{EI}$$

$$d\delta = R(1 - \cos \beta)\,d\theta = \frac{R^2(1 - \cos \beta)[PR \sin \beta - T_0 R(1 - \cos \beta) + M_0]}{EI}\,d\beta$$

From the known displacements at the top of the arch

$$\int_0^\Phi d\theta = 0 \qquad \text{or} \qquad \frac{R}{EI}\int_0^\Phi [PR \sin \beta - T_0 R(1 - \cos \beta) + M_0]\,d\beta = 0 \tag{2}$$

and

$$\int_0^\Phi d\delta = 0 \qquad \text{or} \qquad \frac{R^2}{EI}\int_0^\Phi (1 - \cos \beta)[PR \sin \beta - T_0 R(1 - \cos \beta) + M_0]\,d\beta = 0 \tag{3}$$

By performing the integration in (2) and (3) and solving the simultaneous equations (operations that are lengthy and must be done with great care), T_0 and M_0 are found to be

$$M_0 = -0.412PR$$
$$T_0 = 0.519P$$

Substituting into the bending moment equation (1) gives

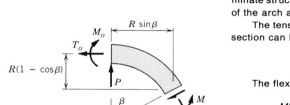

$R \sin\beta$

M_o

T_o

$R(1 - \cos\beta)$

P

β

M

FIG. 3-21d

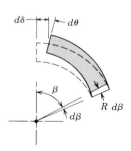

$d\delta$

$d\theta$

β

$R\,d\beta$

$d\beta$

FIG. 3-21e

$$M = PR \sin \beta - 0.519 PR(1 - \cos \beta) - 0.412 PR$$

$$= (\sin \beta + 0.519 \cos \beta - 0.931) PR$$

The maximum bending moment occurs at the top of the arch, where

$$M_{max} = M_0 = -0.412 PR$$

$$= -0.412 \times 30 \times 175 \text{ kN} \cdot \text{mm}$$

$$= -2163 \text{ kN} \cdot \text{mm} = -2\,1630\,000 \text{ N} \cdot \text{mm}$$

The maximum bending stress is

$$\sigma = \frac{Mc}{I} = \frac{2\,163\,000 \times 25}{\frac{1}{12} 80 \times 50^3} \frac{\text{N} \cdot \text{mm} \cdot \text{mm}}{\text{mm} \cdot \text{mm}^3}$$

$$= 64.9 \frac{\text{N}}{\text{mm}^2} = 64.9 \text{ MPa}$$

Figure 2-24 shows that if the curvature is taken into account, we must include a factor of 1.10 for stress concentration, which gives

$$\sigma = 1.10 \times 64.9 = \textbf{71.4 MPa}$$

Example 3-12

The end frame of a machine as shown in Fig. 3-21*a* is intended to support a horizontal shaft. Under normal conditions the shaft pushes downward on the support, but under some conditions the shaft may tend to lift, exerting an upward force of 12 kip at the middle of the arch over the shaft. Calculate the maximum stress in the frame due to the upward force.

Solution. This example represents a very practical engineering problem and is typical of such problems in that it does not conform exactly to theory. The arched top resembles a curved beam, but there is no clearly defined location for the end support where we are able to say that certain end conditions exist. To find an accurate answer would require methods that are used by specialists in stress analysis. Under these conditions the designer uses some reasonably accurate method to find an approximate stress and then, if the safety of the part seems to be in doubt, either strengthens the part or calls on a stress analyst to determine the stresses more precisely. The following solution illustrates the calculations that might be carried out as a first approximation.

Let us start by replacing the actual arch by an arch having a uniform cross section and terminating at fixed supports. The location of the supports is a matter of judgment on the part of the designer. We will assume the supports to be located as indicated in Fig. 3-21*b*.

From symmetry it is apparent that the cross section at the top of the arch does not rotate and does not move in the horizontal direction. These known displacement conditions will be used to determine the unknown values in the indeterminate structure. Again because of symmetry, it is sufficient to work with only half of the arch as shown in Fig. 3-21*c*.

The tensile force T_0 is real and cannot be neglected. The moment at a typical section can be obtained from the free body of Fig. 3-21d.

$$M = PR \sin \beta - T_0 R (1 - \cos \beta) + M_0$$

The flexibility of an increment of arch will give motions as in Fig. 3-21e.

$$d\theta = \frac{MR\,d\beta}{EI} = \frac{[PR \sin \beta - T_0 R(1 - \cos \beta) + M_0] R\,d\beta}{EI}$$

$$d\delta = R(1 - \cos \beta)\,d\theta = \frac{R^2(1 - \cos \beta)[PR \sin \beta - T_0 R(1 - \cos \beta) + M_0]}{EI}\,d\beta$$

From the known displacements at the top of the arch

$$\int_0^\Phi d\theta = 0 \qquad \text{or} \qquad \frac{R}{EI} \int_0^\Phi [PR \sin \beta - T_0 R(1 - \cos \beta) + M_0]\,d\beta = 0 \qquad \text{(2)}$$

and

$$\int_0^\Phi d\delta = 0 \qquad \text{or} \qquad \frac{R^2}{EI} \int_0^\Phi (1 - \cos \beta)[PR \sin \beta - T_0 R(1 - \cos \beta) + M_0]\,d\beta = 0 \qquad \text{(3)}$$

By performing the integration in (2) and (3) and solving the simultaneous equations (operations that are lengthy and must be done with great care) T_0 and M_0 are found to be

$$M_0 = -0.412PR$$

$$T_0 = 0.519P$$

Substituting into the bending moment equation (1) gives

$$M = PR \sin \beta - 0.519PR(1 - \cos \beta) - 0.412PR$$

$$= (\sin \beta + 0.519 \cos \beta - 0.931)PR$$

The maximum bending moment occurs at the top of the arch, where

$$M_{max} = M_0 = -0.412PR$$

$$= -0.412 \times 6 \times 7 \text{ kip} \cdot \text{in.}$$

$$= -17.3 \text{ kip} \cdot \text{in.}$$

The maximum bending stress is

$$\sigma = \frac{Mc}{I} = \frac{17.3 \times 1}{\frac{1}{12}3 \times 2^3} \frac{\text{kip} \cdot \text{in.} \cdot \text{in.}}{\text{in.} \cdot \text{in.}^2}$$

$$= 8.65 \text{ kip/in.}^2$$

Figure 2-24 shows that if the curvature is taken into account, we must include a factor of 1.10 for stress concentration, which gives

$$\sigma = 1.10 \times 8.65 = \textbf{9.52 kip/in.}^2$$

Problems 3-37 to 3-39

3-8 SUMMARY

The main purpose of this chapter was to develop the flexure formulas, (3-8) and (3-11), and to provide some experience in applying them. Statically indeterminate cases were encountered and some insight gained as to the difficulty and importance of this category of problems.

Superposition was presented as the preferred method for solving certain problems. However, becoming familiar with superposition was more important than finding solutions to the problems because superposition has application in many areas of stress analysis and will be used frequently in our future studies.

Moment-area was found to be a convenient method for solving various problems. It is a method that becomes quite complicated and requires further development when more advanced structures are encountered. At the present stage it is sufficient for you to be acquainted with the fundamentals of the method. Deflection of long-radius curved beams was introduced to illustrate the power of the principles underlying the moment-area method and so that you would appreciate the differences between straight and curved beams.

This chapter afforded an opportunity to become familiar with singularity functions, and you have seen that certain problems can be greatly simplified by their use. It must be appreciated that merely an introduction to the topic has been given; there is much more to be learned by those who have a special interest. To illustrate a serious limitation at our present stage, we can express distributed loads that are variable and are intermittent, but we cannot write a load function for concentrated loads. If we had taken the next step and dealt with the concentrated load, we would have encountered the source of the expression "singularity function," but having regard for the scope of this book we have stopped short of that step.

PROBLEMS

3-1 Calculate the deflection at the right end of the given cantilever beam.

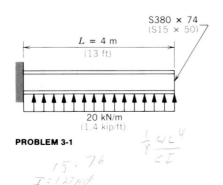

PROBLEM 3-1

3-2 Calculate the deflection at point A and point B for the simply supported beam.

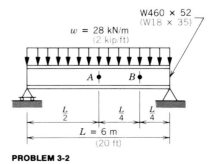

PROBLEM 3-2

3-3 a. Select a standard S-section for an allowable bending stress of 140 MPa (20 ksi).
   b. For the member chosen in part (a), determine the deflection at the end.

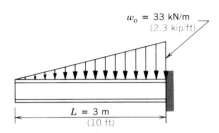

w_o = 33 kN/m
(2.3 kip/ft)

L = 3 m
(10 ft)

PROBLEM 3-3

3-4 Determine the rotation of the beam at the left support.

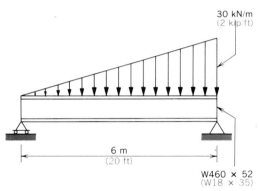

30 kN/m
(2 kip/ft)

6 m
(20 ft)

W460 × 52
(W18 × 35)

PROBLEM 3-4

3-5 Calculate the deflection at point A.

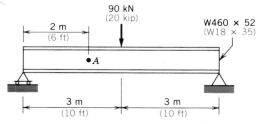

90 kN
(20 kip)

2 m
(6 ft)

W460 × 52
(W18 × 35)

•A

3 m
(10 ft)

3 m
(10 ft)

PROBLEM 3-5

3-6 Determine the maximum deflection in the given beam.

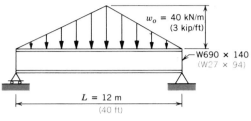

w_o = 40 kN/m
(3 kip/ft)

W690 × 140
(W27 × 94)

L = 12 m
(40 ft)

PROBLEM 3-6

3-7 Calculate the maximum deflection.

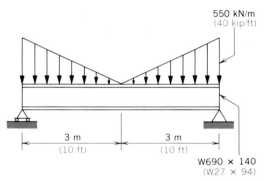

550 kN/m
(40 kip/ft)

3 m
(10 ft)

3 m
(10 ft)

W690 × 140
(W27 × 94)

PROBLEM 3-7

3-8 Draw bending moment diagrams and displacement diagrams for the two given beams. Compare the curves.

More engineering effort is required to solve the indeterminate case. In practice, is there any advantage in specifying that one end be built in thus making the beam statically indeterminate?

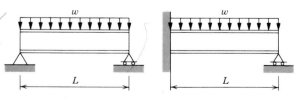

w

L

w

L

PROBLEM 3-8

3-9 a. Calculate the maximum bending stress.
   b. Is the beam safe?

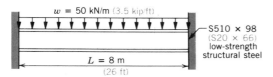

PROBLEM 3-9

3-10 If the beam in Prob. 3-9 has simply supported ends, select the lightest safe S-shape.

3-11 Calculate the reaction at the left support due to the given distributed load.

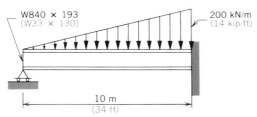

PROBLEM 3-11

3-12 Determine the maximum bending stress in the given beam due to the linearly varying distributed load.

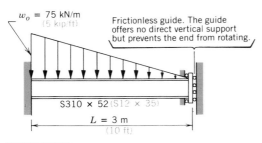

PROBLEM 3-12

3-13 Calculate the maximum bending stress.

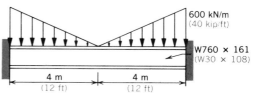

PROBLEM 3-13

3-14 Determine the right reaction.

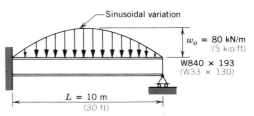

PROBLEM 3-14

3-15 Determine the deflection at midspan.

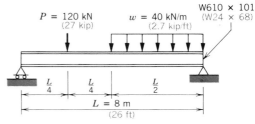

PROBLEM 3-15

3-16 Determine the deflection at midspan.

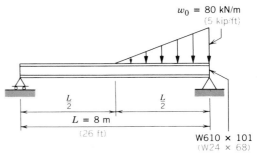

PROBLEM 3-16

3-17 Use singularity functions to solve Prob. 3-7.

3-18 Determine the location and magnitude of the maximum deflection.

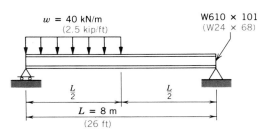

PROBLEM 3-18

3-19 Calculate the bending stress at the right support.

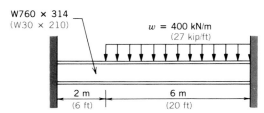

PROBLEM 3-19

3-20 Determine the deflection at midspan.

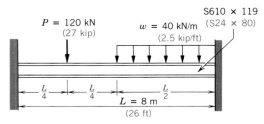

PROBLEM 3-20

3-21 A couple, C, is applied to the given cantilever beam. Draw the bending moment diagram.

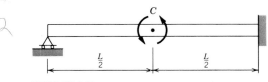

PROBLEM 3-21

3-22 Determine the value of the center reaction in terms of P, w, and L.

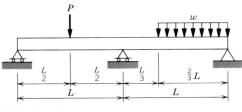

PROBLEM 3-22

3-23 Determine the deflection at location A by the method of superposition for the given beams. In all cases the member is an S250 × 38 (S10 × 25).

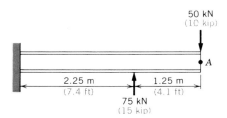

PROBLEM 3-23a

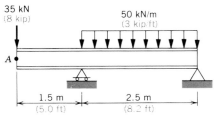

PROBLEM 3-23b

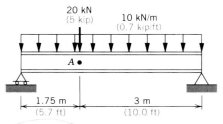

PROBLEM 3-23c

3-24 Before the load is applied to the given beam, there is a space of 10 mm (0.4 in.) between the beam and the roller at midspan. Using superposition, determine the reaction at the center support when the load is applied to the beam.

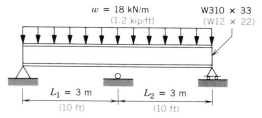

PROBLEM 3-24

3-25 Calculate the force on the support at midspan.

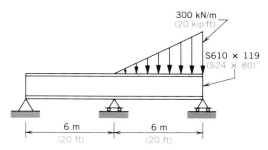

PROBLEM 3-25

3-26 Calculate the force on the support at midspan.

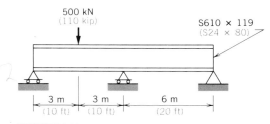

PROBLEM 3-26

3-27 Calculate the maximum bending stress.

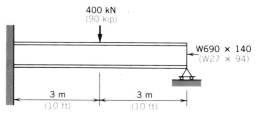

PROBLEM 3-27

3-28 Calculate the maximum bending stress.

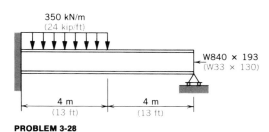

PROBLEM 3-28

3-29 Calculate the load on the right support.

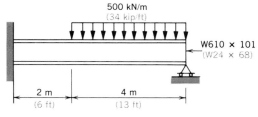

PROBLEM 3-29

3-30 Determine the maximum stress in the members due to a load P of 40 kN (9 kip).

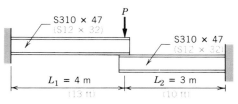

PROBLEM 3-30

3-31 Use the moment-area method to determine the deflection at A in terms of P, E, I, and L for the given beams and sketch the deflection curve for each beam. (Note that, although the total load is the same in all three cases, the displacements at A are far from equal.)

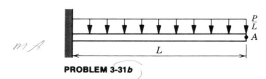

PROBLEM 3-31a

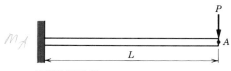

PROBLEM 3-31b

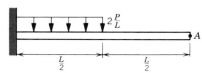

PROBLEM 3-31c

3-32 Use the moment-area method to determine the deflection at the end of the cantilever beam.

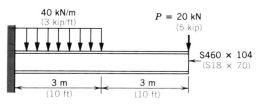

PROBLEM 3-32

3-33 Use the moment-area method to determine the deflection at the end of the beam due to:

a. Load P only.

b. Load P and the weight of the beam.

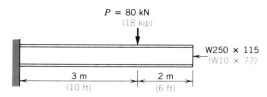

PROBLEM 3-33

3-34 Use the moment-area method to determine the deflection at the end of the reinforced cantilever.

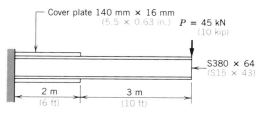

PROBLEM 3-34

3-35 Use the moment-area method to determine the deflection at A in terms of P, E, I, and L for the given beams and sketch the deflection curve for each beam.

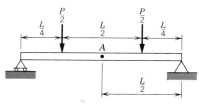

PROBLEM 3-35a

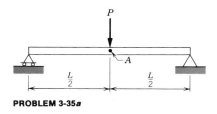

PROBLEM 3-35b

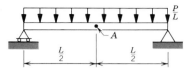

PROBLEM 3-35c

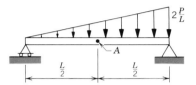

PROBLEM 3-35d

3-36 Use the moment-area method to find the maximum bending moment in each of the statically indeterminate beams.

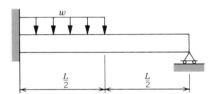

PROBLEM 3-36a

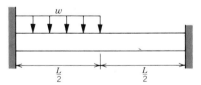

PROBLEM 3-36b

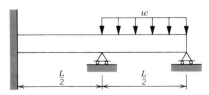

PROBLEM 3-36c

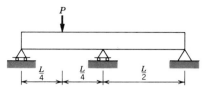

PROBLEM 3-36d

3-37 A steel ring is formed by forging a rod as shown. Welding the ends of the rod together would greatly strengthen the ring; however, this has not been done and the ends can be treated as being parallel planes that just touch when the ring is unloaded. When a 3.0 kN (0.7 kip) load, P, is applied, determine (a) the angle between the ends of the rod, and (b) the width of the gap that will appear.

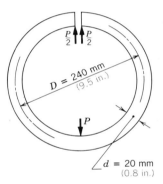

PROBLEM 3-37

3-38 Repeat Prob. 3-37 when the P forces act along the horizontal diameter.

3-39 A semicircular arch has a uniform cross section 400 mm (16 in.) deep and 100 mm (4 in.) wide. It is made of Western hemlock, laminated and glued. Determine, due to the weight of the arch only:

a. The amount of horizontal motion at the right support.
b. The horizontal component of the reaction at the pin if the right end is pin-connected.

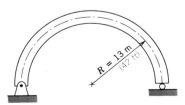

PROBLEM 3-39

Advanced Problems

3-A If the allowable steel stress is 140 MPa (20 ksi) for both the rod and the beam, what is the maximum allowable distributed load that can be carried by the beam?

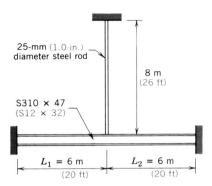

25-mm (1.0 in.) diameter steel rod

8 m (26 ft)

S310 × 47 (S12 × 32)

L_1 = 6 m (20 ft) L_2 = 6 m (20 ft)

PROBLEM 3-A

3-B A chain is composed of links that are formed by bending lengths of 30-mm (1.2-in.) round bar stock into circles of 120-mm (4.8-in.) mean diameter and welding the ends together. The material is AISI 4340. For a safety factor of 1.5 on yield, determine the load that the chain can carry safely.

3-C Bar stock having a diameter of 30 mm (1.2 in.) is used to form welded chain links. The links have semicircular ends and straight sides, as shown in the figure, and are made of AISI 4340. For a safety factor of 1.5 on yield, determine the safe load that can be carried by a chain composed of these links.

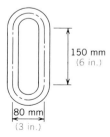

150 mm (6 in.)

80 mm (3 in.)

PROBLEM 3-C

3-D The link described in Prob. 3-C has a spreader bar added as shown. The bar prevents the midside points of the link from moving toward one another. Determine the safe load for a safety factor of 1.5 on yield.

PROBLEM 3-D

3-E In a power plant a 300-mm (12-in.) steam main must connect two fixed units that are 200 m (60 ft) apart. When constructed, the line will be stress-free at 20°C (68°F); in service it carries steam at 200°C (390°F). A straight pipe would be greatly overloaded by the thermal expansion. To reduce the forces, an expansion loop as shown is proposed.

Because of the effects mentioned near the end of Section 2-7, the bending flexibility of the curved pipe, for the dimensions in this case, can be taken as 2.9 times the flexibility of a straight length of pipe.

Determine the axial force in the straight portion of the pipe due to temperature change:

a. For a straight pipe running between the fixed ends.

b. For the given configuration.

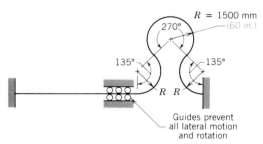

R = 1500 mm (60 in.)

270°

135°   135°

R R

Guides prevent all lateral motion and rotation

PROBLEM 3-E

3-F The flywheel of a punch press consists of a hub, an outer rim, and 16 radial spokes. The hub has a diameter of 300 mm (12 in.) and the inside diameter of the rim is 1500 mm (60 in.). The spokes consist of 12-mm (0.5-in.) rods 600 mm (24 in.) long and are securely welded to the hub and the rim.

During the punching operation a torque of 1.0 kN·m (740 lb·ft) is applied to the hub of the flywheel. The inertia of the hub and the spokes may be considered negligible relative to that of the rim.

When the torque is applied, determine:

a. The angular displacement of the hub relative to the rim.
b. The maximum bending stress in the spokes.

3-G A cantilever beam consists of a W610 × 195 (W24 × 130) 5.0 m (16 ft) long. The left end is built-in and the right end is supported by a vertical rod having a diameter of 50 mm (2 in.). The rod is 45 m (150 ft) long

and has its upper end attached to a fixed support that is directly above the right end of the beam.

The material is high-strength structural steel and is unstressed when the temperature is 30°C (86°F).

a. Calculate the stress in the rod and the bending stress in the beam when a uniformly distributed downward load of 200 kN/m (14 kip/ft) is applied to the beam.
b. Determine the maximum safe uniformly distributed load.
c. Calculate the stresses when the load in (a) is applied and the temperature is −20°C (−4°F).
d. Determine the maximum safe uniformly distributed load when the temperature is −20°C (−4°F).

4

COLUMNS I

4-1 INTRODUCTION

Truss members that are subjected to tensile load are relatively easy to design. If the load is applied at the centroid of the cross-sectional area, it is only necessary to ensure that the average stress on the net cross-sectional area does not exceed the allowable stress for the material. The length of the member and the shape of the cross section are not factors in the design. You have designed some compression members, that is, columns, on the basis of a given allowable compressive stress, but you were warned that this allowable compressive stress was not as readily determined as the allowable stress for tension. The material covered to this point provides sufficient background to enable us now to derive the strength equation for columns. We will find that the length of the member and the shape of the cross-sectional area have more influence on column strength than the yield stress or the ultimate stress. Column failure, described as *buckling*, is due to instability and occurs as a complete collapse, frequently without warning. Consequently, the greatest care must be taken in the design of columns and a larger safety factor is applied.

The member shown in Fig. 4-1 is in the process of failing by buckling. Note that under an axial load the member is deflecting *laterally* in a curve similar to that of a beam under a bending load. This is a characteristic of buckling—in reality it is a bending failure resulting from an axial compressive load. If the member is very short, or is laterally supported so that it cannot bend, there will be no buckling and the strength will be larger than when there is buckling. The column in Fig. 4-1 did not fail completely because the testing machine does not apply load but actually applies a displacement and measures the resistance that the member offers to the displacement. Consequently, when a column begins to buckle in a testing machine, the load adjusts itself to that required to hold the member in its deflected configuration.

Let us imagine that we are conducting load tests on members in compression by means of a simple testing machine similar to the one that we used in Chapter 1 but arranged to exert a compressive force as in Fig. 4-2b. The stress–strain curve for the material will be taken as shown in Fig. 4-2a, which is realistic for steel provided that the specimen used in the compressive test is either short enough or is laterally constrained to prevent buckling.

The simple testing machine in Fig. 4-2b applies a displacement to one end of the column while the torque required to advance the screw is a measure of the load on the column. Modern testing machines function in a similar manner except that load measurements are made by more accurate methods. From our knowledge of stress–strain relations, but lacking knowledge of column action, we might predict the curve relating force and axial displacement to be similar to curve A in Fig. 4-2c. We would expect the relationship to be linear until the stress reaches the proportional limit; in fact, we would find this to be the case *if we provided lateral support to the column to prevent it from buckling*. For the column with no lateral constraint, other than at the simply supported ends, test data is more likely to give curve B. Curve B coincides with the predicted curve for low loads but deviates abruptly and becomes horizontal. The

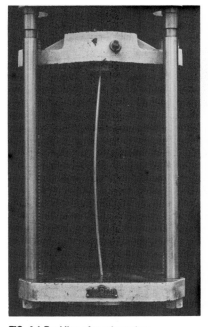

FIG. 4-1 Buckling of a column in a compression testing machine.

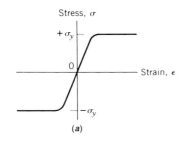

(a)

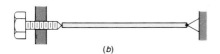

(b)

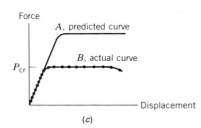

(c)

FIG. 4-2 (a) Stress-strain relations.
(b) Simple-compression test machine.
(c) Force-displacement relations.

force at which this occurs bears no relationship to the yield stress and would occur at a different level if we repeated the test with another member having a different length or a different cross section. We are, of course, interested in predicting this force, referred to as the *critical load*, P_{cr}. A great many tests would be required to establish the equation for the critical load in terms of the column dimensions. Euler developed an equation for the critical load more than 200 years ago,[*] but there was no experimental evidence to verify his conclusion so that the Euler buckling equation was not accepted for many years. Today we know from experimental evidence that the Euler equation is correct and it now forms the basis of all column design.

4-2 EULER'S BUCKLING LOAD

If we conducted a column load test and measured force when increments of displacement were applied, the data would give a curve such as that in Fig. 4-3b. The displacement in the initial straight-line portion of the curve would be the same as that calculated from $\sigma = P/A$, $\varepsilon = \sigma/E$ and $\Delta L = \varepsilon L$. If the load is applied at the centroid of the cross-sectional area and the member is initially perfectly straight, that is, if it is an *ideal column*, the curve will deviate quite abruptly from the initial straight line and the force will remain constant over a considerable displacement before beginning to drop. Since this constant force is the largest that the column is capable of supporting, it is of great interest to designers. Observations made on the lateral deflection would reveal that there is no appreciable lateral deflection during the initial straight-line action, but bending takes place when the critical load is applied and then there is substantial lateral deflection. In the process of buckling, as displacement is increased, the lateral deflection increases although the force necessary to cause the displacement remains constant at the critical load, P_{cr}. If we attempt to derive an equation for P_{cr} directly, we will have great difficulty, but if we set as our objective the determination of the elastic deflection curve for a column such as that of Fig. 4-3c and use the techniques of Chapter 3, we will discover that the deflection curve cannot be determined but that the equation for P_{cr} appears.

From experience gained in dealing with the deflection of beams, we know that the first step is to write an equation for the bending moment at any location along the beam. To obtain this expression for a typical location, x units along the beam, we take a section at x and the free body as shown in Fig. 4-3d. Since we have no interest in T and V, we can obtain a useful equation that does not contain them by writing the equilibrium equation for moments about N. This gives

$$P_{cr}y + M = 0 \qquad \text{or} \qquad M = -P_{cr}y$$

[*]Leonhard Euler was a Swiss mathematician. He derived the critical load equation in the mid-eighteenth century.

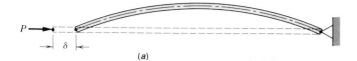

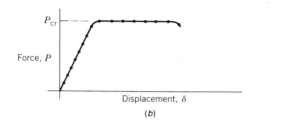

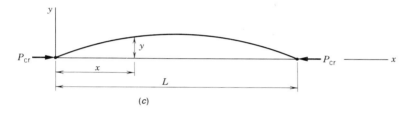

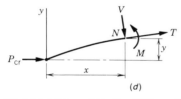

FIG. 4-3 (*a*) Column deflection. (*b*) Force displacement relation. (*c*) Buckling deflection curve. (*d*) Free-body diagram.

Substituting into the known relationship between bending moment and displacement (3-8) gives

$$\frac{d^2 y}{dx^2} = -\frac{P_{cr}}{EI} \, y \qquad \textbf{(4-1)}$$

The corresponding equation for beam problems had x as the only variable on the right-hand side and consequently could be integrated. In (4-1) y is a function of x, but the function is unknown at this stage so that it cannot be substituted into (4-1) and then integrated.

Equation (4-1) is a common type of differential equation and its solution can be found in many textbooks on mathematics.* It is sufficient for our purposes to know that the solution

---

*Refer, for instance, to R. E. Gaskell, *Engineering Mathematics*, Holt, Rinehart and Winston, New York, 1958.

$$y = C_1 \sin \sqrt{\frac{P_{cr}}{EI}} x + C_2 \cos \sqrt{\frac{P_{cr}}{EI}} x \qquad \textbf{(4-2)}$$

does in fact satisfy (4-1), the governing differential equation. Our experience with beams in bending prepared us for the appearance of two constants of integration but not in the form in which they occur in (4-2). For a given column, E and I can be taken as known quantities, which leaves three unknowns, C_1, C_2, and P_{cr} in (4-2), although there are only two known properties of the deflection curve. Hence we are forewarned that all three unknowns cannot be evaluated, but we will still attempt to solve for C_1 and C_2 in the usual manner.

For the origin at the left end of the member and the known zero deflection at that end

$$y]_{x=0} = 0$$

Then from (4-2)

$$C_1 \sin \sqrt{\frac{P_{cr}}{EI}} 0 + C_2 \cos \sqrt{\frac{P_{cr}}{EI}} 0 = 0$$

$$C_1 \sin 0 + C_2 \cos 0 = 0$$

$$C_1 \times 0 + C_2 \times 1 = 0$$

Therefore,

$$C_2 = 0$$

and (4-2) reduces to

$$y = C_1 \sin \sqrt{\frac{P_{cr}}{EI}} x \qquad \textbf{(4-3)}$$

Since the deflection is also known to be zero at the right end,

$$y]_{x=L} = 0$$

For (4-3) to satisfy this condition

$$C_1 \sin \sqrt{\frac{P_{cr}}{EI}} L = 0 \qquad \textbf{(4-4)}$$

Based on our experience we would expect that this should give us C_1; however, the only value of C_1 that satisfies (4-4) is zero. This would give, when substituted into (4-3)

$$y = 0 \sin \sqrt{\frac{P_{cr}}{EI}} x = 0$$

which is the equation of a member that does not bend. This is a perfectly valid solution, but it corresponds to the undeflected column, which was the state of the column under loads less than the critical load. However, it is not the solution to a column that is in the act of buckling, so we must search for other solutions to (4-4). If we look a little more carefully at the physical situation, we will see that evaluation of C_1 is impossible. The deflection curve has been established by (4-3) as being a sinusoidal curve with C_1 the amplitude of the curve. When various end displacements are applied, all of which correspond to the critical load, there will be many different amplitudes to the lateral deflection, and hence a variety of values of C_1 all of which are applicable to the column subjected to the critical load. Thus there is no way that C_1 can be evaluated and we can concentrate our effort on using the second end condition to evaluate the other unknown, P_{cr}.

Returning to (4-4), we see that for nonzero C_1, the equation can be satisfied only when

$$\sin \sqrt{\frac{P_{cr}}{EI}} L = 0 \tag{4-5}$$

But the sine is zero for arguments of 0, π, 2π, 3π, ..., etc., radians. Hence (4-5) is satisfied when

$$\sqrt{\frac{P_{cr}}{EI}} L = 0 \tag{4-6a}$$

or

$$\sqrt{\frac{P_{cr}}{EI}} L = \pi \tag{4-6b}$$

or

$$\sqrt{\frac{P_{cr}}{EI}} L = 2\pi \tag{4-6c}$$

or

$$\sqrt{\frac{P_{cr}}{EI}} L = 3\pi \tag{4-6d}$$

and so on. The critical load is then, from (4-6a),

$$P_{cr} = 0 \tag{4-7a}$$

or from (4-6b),

$$P_{cr} = \frac{\pi^2}{L^2} EI \tag{4-7b}$$

or from (4-6c),

$$P_{cr} = 4\frac{\pi^2}{L^2} EI \tag{4-7c}$$

or from (4-6d),

$$P_{cr} = 9 \frac{\pi^2}{L^2} EI \tag{4-7d}$$

and so on.

The solution in (4-7a) tells us that all conditions are satisfied for no load and, from (4-3), no deflection; this case is obviously correct but of no engineering interest.

The solution in (4-7b) is useful, and its validity has been confirmed over the years by numerous experimental studies. This is *Euler's buckling equation*; because of its importance it is repeated here with its own equation number as

$$\boxed{P_{cr} = \frac{\pi^2 EI}{L^2}} \tag{4-8}$$

For Euler buckling, the deflection curve in (4-3) becomes

$$y = C_1 \sin \pi \frac{x}{L} \tag{4-9}$$

in which, as we have seen, C_1 cannot be evaluated without additional information. A sketch of (4-9) will show that the form of the deflection curve is similar to that of Fig. 4-3c.

The Euler equation shows that the strength of a column in buckling is not dependent on the strength of the material but on another material property, E, the modulus of elasticity. The strength also depends on the moment of inertia of the cross section, which should not surprise us because buckling is obviously a type of bending failure. The strength is also seen to depend on the length squared, which means that as the column length increases, the strength diminishes very rapidly.

The strengths indicated by (4-7c), (4-7d), and the equations that follow are all greater than the Euler buckling load. They give the strengths of columns having lateral support at certain intermediate points along the length. These supported columns can be treated by other methods; hence solutions other than that given in (4-7b) are of little practical importance.

Example 4-1

Calculate the Euler buckling load for an aluminum column 2 m long with a 50-mm square cross section. Assume pin-ended supports.

Solution. To solve for the critical buckling load by means of (4-8), we require E and I. We obtain

$$E = 70\,000 \text{ MPa} \qquad \text{from Table A}$$

$$I = \tfrac{1}{12} \times 50 \times 50^3 \text{ mm}^4$$

$$= 521 \times 10^3 \text{ mm}^4$$

Substituting into Eq. (4-8) gives

$$P_{cr} = \frac{\pi^2 EI}{L^2} = \frac{\pi^2 70 \times 10^3 \times 521 \times 10^3}{(2 \times 10^3)^2} \frac{\text{N/mm}^2 \cdot \text{mm}^4}{\text{mm}^2}$$

$$= 89\ 960\ \text{N}$$

$$= \textbf{90.0 kN}$$

Example 4-1

Calculate the Euler buckling load for an aluminum column 6.5 ft long with a 2-in. square cross section. Assume pin-ended supports.

Solution. To solve for the critical buckling load by means of (4-8), we require E and I. We obtain

$$E = 10 \times 10^3\ \text{kip/in.}^2 \qquad \text{from Table A}$$

$$I = \tfrac{1}{12} \times 2 \times 2^3\ \text{in.}^4$$

$$= 1.333\ \text{in.}^4$$

Substituting into Eq. (4-8) gives

$$P_{cr} = \frac{\pi^2}{L^2} EI = \frac{\pi^2\ 10 \times 10^3 \times 1.333}{(6.5 \times 12)^2} \frac{\text{kip/in.}^2 \cdot \text{in.}^4}{\text{in.}^2}$$

$$= \textbf{21.63 kip}$$

Example 4-2

The material of the column of Example 4-1 is redistributed to form a hollow square section with each external side dimension equal to 100 mm while maintaining the same cross-sectional area. What is the permissible column length for the same buckling load capacity?

Solution. The problem statement requires that the total cross-sectional area, A, remain at 2500 mm^2 but be redistributed into a hollow square section. Referring to Fig. 4-4a, we find that

$$A = 2500 = 100^2 - (100 - 2t)^2$$

$$4t^2 - 400t + 2500 = 0$$

$$t = \frac{400 \pm \sqrt{400^2 - 4 \times 4 \times 2500}}{2 \times 4}$$

$$= 93.3 \qquad \text{or} \qquad 6.70\ \text{mm}$$

The 93.3-mm value has no real significance. Using 6.70 mm as the thickness, we calculate I from the dimensions of Fig. 4-4b as

$$I = \tfrac{1}{12}(100 \times 100^3 - 86.6 \times 86.6^3)\ \text{mm}^4$$

$$= 3.65 \times 10^6\ \text{mm}^4$$

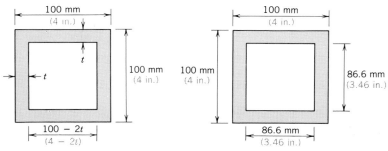

FIG. 4-4a   FIG. 4-4b

Solving for L^2 from (4-8) and substituting yields

$$L^2 = \frac{\pi^2 EI}{P_{cr}} = \frac{\pi^2 \times 70 \times 10^3 \times 3.65 \times 10^6}{90 \times 10^3} \frac{\text{N/mm}^2 \cdot \text{mm}^4}{\text{N}}$$

$$= 28.0 \times 10^6 \text{ mm}^2$$

$$L = 5.29 \times 10^3 \text{ mm} = \textbf{5.29 m}$$

Example 4-2

The material of the column of Example 4-1 is redistributed to form a hollow square section with each external side dimension equal to 4 in. while maintaining the same cross-sectional area. What is the permissible column length for the same buckling load capacity?

Solution. The problem statement requires that the total cross-sectional area, A, remain at 4 in.^{2} but be redistributed into a hollow square section. Referring to Fig. 4-4a, we find that

$$A = 4 = 4^2 - (4 - 2t)^2$$

$$4t^2 - 16t + 4 = 0$$

$$t^2 - 4t + 1 = 0$$

$$t = \frac{4 \pm \sqrt{(-4)^2 - 4 \times 1 \times 1}}{2 \times 1}$$

$$= 3.73 \quad \text{or} \quad 0.27 \text{ in.}$$

The 3.73-in. value has no real significance. Using 0.27 in. as the thickness, we calculate I from the dimensions of Fig. 4-4b as

$$I = \tfrac{1}{12}(4 \times 4^3 - 3.46 \times 3.46^3) \text{ in.}^4$$

$$= 9.39 \text{ in.}^4$$

Solving for L^2 from (4-8) and substituting yields

$$L^2 = \frac{\pi^2 EI}{P_{cr}} = \frac{\pi^2 \times 10 \times 10^3 \times 9.39}{21.63} \; \frac{\text{kip/in.}^2 \cdot \text{in.}^4}{\text{kip}}$$

$$= 42.85 \times 10^3 \text{ in.}^2$$

$$L = 207 \text{ in.} = \textbf{17.3 ft}$$

We see that the redistribution of material in Example 4-2 permits us to use a column more than twice as long as that of Example 4-1 without sacrificing any load-carrying capacity. This result is not really surprising: For the same volume of material, the hollow square section gives us an increased moment of inertia, and hence increased bending stiffness. Since buckling is an instability failure involving bending, we would then expect the column of Example 4-2 to have a greater permissible length than that of Example 4-1 for the same buckling loads. Conversely, if both columns had the same length, we would expect the hollow section to be able to resist a significantly higher buckling load.

Problems 4-1 to 4-11

4-3 ENERGY CONSIDERATIONS IN BUCKLING

The Euler buckling equation was derived for *ideal columns* that:

1. Are perfectly straight when unloaded.
2. Are subjected to axial loads that are applied at the centroid.
3. Have ends supported in such a manner that there is no resistance to end rotation. Pin-ended supports fulfill this condition.

It is reasonable to doubt, at this stage, that such a column would ever begin to bend under an axial load that produces no initial bending moment. In order to understand why columns do, in fact, bend, we must take energy into consideration. We will do this by discussing energy levels in general terms without reference to actual numbers.

Consider a straight column that is constrained by lateral supports, such as rows of rollers, so that it cannot bend and therefore cannot buckle. For applied axial displacements at one end, the required force can be readily calculated from elementary principles and the relationship represented by curve A in Fig. 4-5a obtained. Curve A will be straight until the proportional limit of the material is reached; then it will begin to curve. We are interested only in the straight portion and particularly at a displacement δ_1 where the force is P_1. In this state there is no reason to expect any lateral force against the constraining rollers, and the member is in an equilibrium state. All particles have been strained equally in compression, and there is a certain amount of strain energy stored in the particles that make up the column. If we now remove the rollers and, without moving the end supports, cause the column to curve slightly, there will be a change in energy. The arc length will be slightly larger

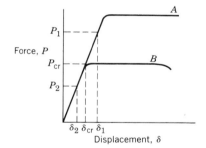

FIG. 4-5a

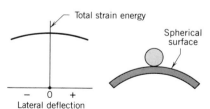

Lateral deflection
FIG. 4-5b   **FIG. 4-5c**

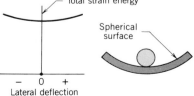

Lateral deflection
FIG. 4-5d   **FIG. 4-5e**

than the fixed distance between supports so that bending the column will release some of the axial strain, tending to reduce the strain energy. Since the column is now curved, there will now be some energy due to bending that was not stored in the straight column; this tends to increase the strain energy. These two effects oppose one another and which one will dominate depends upon certain conditions. For the conditions represented by δ_1 and P_1 in Fig. 4-5a, the total energy in the deflected state is less than when the column is straight, and if we plot the strain energy against the lateral displacement, the curve would have the shape shown in Fig. 4-5b. This is analogous to a ball resting on top of a sphere as depicted in Fig. 4-5c: The ball is in equilibrium but any small disturbance will cause it to roll off. In the case of the column, the ideal column is in equilibrium when it is straight, but any slight lateral deflection will cause a reduction in strain energy and, like the ball, is *unstable* and will not return to its initial shape but will collapse. In practice the state of unstable equilibrium cannot be reached because a member is never perfectly straight and the load cannot be applied exactly at the centroid of the cross-sectional area. Even if we could fulfill these conditions, any slight vibration or lateral load would still cause collapse. For the column compressed an amount δ_1 in Fig. 4-5a the load P_1 is unattainable and the column will buckle and will exert the smaller force, P_{cr}, against the end supports.

For another case the column may be in the state represented by δ_2 and P_2 in Fig. 4-5a. The strain energy calculations would give a curve shaped as in Fig. 4-5d. The same type of curve will be found for the potential energy of a ball inside a sphere as shown in Fig. 4-5e. If the ball is moved slightly, it will return to its original position. Similarly, the column is *stable* and will straighten after the removal of a temporary lateral force. *The transition between stable and unstable equilibrium for a straight column takes place at δ_{cr} when the force is P_{cr}. Consequently, an ideal column cannot carry a load greater than P_{cr} as determined by Euler's buckling equation, (4-8).*

In our analysis we have treated columns as though they were subjected to a compressive displacement and we then found the associated resisting force or the force required to impart the displacement. This is the manner in which testing machines function, as mentioned in Chapter 1, when tensile tests were being discussed. However, in actual structural applications the load is applied and we accept the displacement that follows. If a load less than the critical load, such as P_2 in Fig. 4-5a, is applied, the column shortens a small amount and carries the load safely in a stable equilibrium state. But a load greater than the critical load, such as P_1, cannot be carried by the column in a stable state, and the column will collapse suddenly and completely if such a load is applied. The dramatic nature of column collapse cannot be fully appreciated by observations made on buckling in a testing machine.

4-4 SLENDERNESS RATIO

The moment of inertia, *I*, of a cross section is an important parameter in beam and in column analysis. With respect to columns it is also convenient to use a related parameter, the *radius of gyration*, *r*. The expression "radius of gyration" has physical meaning in the dynamics of rotating bodies just as "moment of inertia" conveys real meaning in that field. In Mechanics of Materials we use the phrase "radius of gyration" because of the analogy to a similar quantity in dynamics, without any physical definition. The radius of gyration is defined by

$$r = \sqrt{\frac{I}{A}} \qquad\qquad (4\text{-}10)$$

where *A* is the cross-sectional area. The dimensions of *I* being mm^4 ($in.^4$) and *A* mm^2 ($in.^2$) give *r* as millimetres (inches), a linear dimension.

Using (4-10), we may change the Euler buckling equation (4-8) to

$$P_{cr} = \frac{\pi^2}{L^2} E r^2 A$$

$$P_{cr} = \frac{\pi^2 E}{(L/r)^2} A \qquad\qquad (4\text{-}11)$$

When the critical load is applied to a column, the average stress on a cross section is

$$\sigma_{cr} = \frac{P_{cr}}{A}$$

and substituting from (4-11)

$$\sigma_{cr} = \frac{\pi^2 E}{(L/r)^2} \qquad\qquad (4\text{-}12)$$

This is a convenient form of the Euler equation for the purpose of comparison of the load-carrying capacity of various members. It converts the critical load to a *critical stress*, σ_{cr}, which depends, as before, on *E* and also on the dimensionless quantity, *L/r*, called the *slenderness ratio*.

When a buckling column consists of a rod or a pipe, the direction in which it will bend is unpredictable and will probably be determined by a slight initial curvature or an eccentricity of the applied load. For circular cross sections, the value of *I* is the same for all diametral axes. The case is different when an S-shape is used for a column. For that shape and most others the *I* of the cross section will depend on the axis to which it is referred. Since the standard S-shapes have a smaller *I* with respect to the *y*-axis, a column made of an S-shape will fail by bending about the *y*-axis. Such a column will determine for itself which axis has the minimum *I*, or, in other terms, the minimum *r*, and will fail by bending about that axis. Consequently, unless there is good reason for doing otherwise, when (4-8) is used the *I* should be the *minimum* value for the section and in (4-12) the *minimum* radius of gyration should be substituted. Note that for S-shapes and W-shapes the axis of minimum *r* in Tables B-1 and B-2 is always the *y*-axis. A column consisting of a single angle will buckle about neither the x-axis nor the y-axis, but an inclined axis about which the I is a minimum. The direction of this axis and the value of the corresponding least radius of gyration are given in Tables B-4 and B-5.

Problems 4-12 to 4-18

4-5 COLUMN DESIGN

Critical stress as determined by (4-12) for steel is plotted against the slenderness ratio in Fig. 4-6. It can be seen that for large L/r values, that is, for very slender members, the buckling stress is so small that the members would be quite uneconomical. For this reason columns are seldom designed with a slenderness ratio greater than 200. For low values of L/r, columns appear to be remarkably strong. For example, a steel column with a slenderness ratio of 20 would appear to have, from (4-12), a critical stress of 5000 MPa (725 ksi). This is obviously not realistic for structural steel. We are evidently using the equation outside its applicable range. Since the column analysis was based on *elastic* behavior of the material, the analysis is not valid for stresses above the yield stress, σ_y, which for low-strength structural steel is about 230 MPa (33 ksi). Therefore, we should not use the curve, or (4-12), where stresses above the yield stress are indicated, that is, the curve above the dotted horizontal line in Fig. 4-6 is meaningless.

If we were to conduct a series of tests on low-strength steel columns over a range of slenderness ratios, we might get data as shown by the points in Fig. 4-7. We would find relatively good agreement with Euler's equation for L/r greater than 110, that is, for *long* columns, and good agreement with the yield stress for very *short* columns. However, the strength in the *intermediate* column range would be less than expected. This is due to failure by a combination of yielding and buckling for which

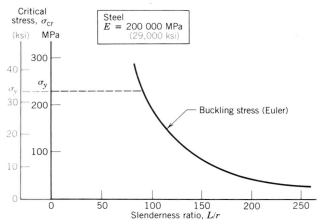

FIG. 4-6 Variation of column critical stress with slenderness ratio for steel.

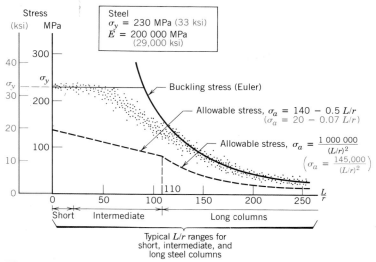

Steel
σ_y = 230 MPa (33 ksi)
E = 200 000 MPa
(29,000 ksi)

Buckling stress (Euler)

Allowable stress, $\sigma_a = 140 - 0.5\ L/r$
($\sigma_a = 20 - 0.07\ L/r$)

Allowable stress, $\sigma_a = \dfrac{1\ 000\ 000}{(L/r)^2}$

$\left(\sigma_a = \dfrac{145,000}{(L/r)^2}\right)$

Short   Intermediate   Long columns

Typical L/r ranges for
short, intermediate, and
long steel columns

FIG. 4-7

there is no satisfactory theory. Many empirical formulas have been devised to approximate the strength of columns with L/r less than 110. The formula to be used in any particular jurisdiction may be specified by the legally established code. Some codes treat columns in three categories: *short*, intermediate, and long. For the designs required in the exercises of this book we will use allowable stress equations for only two categories of length. These equations have a safety factor incorporated into them. The allowable stress for *low-strength structural steel* is given by:

$$\sigma_a = 140 - 0.5\left(\frac{L}{r}\right) \quad \text{for } 0 \leqslant \frac{L}{r} \leqslant 110 \qquad \textbf{(4-13a)}$$

$$\sigma_a = \frac{1\ 000\ 000}{\left(\dfrac{L}{r}\right)^2} \quad \text{for } \frac{L}{r} > 110 \qquad \textbf{(4-13b)}$$

In Imperial units the equations to give allowable stress in ksi are as follows.

For low-strength structural steel:

$$\sigma_a = 20 - 0.07\left(\frac{L}{r}\right) \quad \text{for } 0 \leqslant \frac{L}{r} \leqslant 110 \qquad \textbf{(4-13a)}$$

$$\sigma_a = \frac{145,000}{\left(\dfrac{L}{r}\right)^2} \quad \text{for } \frac{L}{r} > 110 \qquad \textbf{(4-13b)}$$

The allowable stresses determined by these equations are shown by the broken line in Fig. 4-7.

It is important to note that the allowable steel stress equations, (4-13a) and (4-13b), do not apply to other materials because of differences in the E value and the yield stress. For the design exercises of this book, equations similar to (4-13a) and (4-13b) will be used for wood and aluminum. They are for *wood (Douglas fir, western hemlock or eastern spruce)*:

$$\sigma_a = 5 - 0.01\left(\frac{L}{r}\right) \quad \text{for } 0 \leqslant \frac{L}{r} \leqslant 90 \tag{4-14a}$$

$$\sigma_a = \frac{33\,000}{\left(\dfrac{L}{r}\right)^2} \quad \text{for } \frac{L}{r} > 90 \tag{4-14b}$$

and for *aluminum* (6061-T6):

$$\sigma_a = 140 - 1.0\left(\frac{L}{r}\right) \quad \text{for } 0 \leqslant \frac{L}{r} \leqslant 70 \tag{4-15a}$$

$$\sigma_a = \frac{340\,000}{\left(\dfrac{L}{r}\right)^2} \quad \text{for } \frac{L}{r} > 70 \tag{4-15b}$$

For wood (Douglas fir, western hemlock or eastern spruce):

$$\sigma_a = 0.7 - 0.0014\left(\frac{L}{r}\right) \quad \text{for } 0 \leqslant \frac{L}{r} \leqslant 90 \tag{4-14a}$$

$$\sigma_a = \frac{4800}{\left(\dfrac{L}{r}\right)^2} \quad \text{for } \frac{L}{r} > 90 \tag{4-14b}$$

For aluminum (6061-T6):

$$\sigma_a = 20 - 0.14\left(\frac{L}{r}\right) \quad \text{for } 0 \leqslant \frac{L}{r} \leqslant 70 \tag{4-15a}$$

$$\sigma_a = \frac{50,000}{\left(\dfrac{L}{r}\right)^2} \quad \text{for } \frac{L}{r} > 70 \tag{4-15b}$$

In using equations (4-13a and b), (4-14a and b), and (4-15a and b) we must recognize that they are subject to the following important limitations:

1. The equations apply to *typical* strength levels and not to all possible strength levels. In actual design practice more complicated allowable stress equations take into account various strength levels for each material; however, for the design exercises of this book the three simple sets of equations will be adequate.
2. Since we have encountered the three classes of short, intermediate, and long compression members, we might expect that for each material three allowable stress equations are required to define strengths over the respective ranges. Although in actual design practice this is often done, two equations suffice for

purposes of this book. Note from Fig. 4-7 for steel and equations (4-13a), (4-14a), and (4-15a) that the first equation is always a straight line, which spans the L/r ranges for short and inter-mediate columns. The second equation for each material, that is, (4-13b), (4-14b), and (4-15b), refers to the long column range.

3. The allowable stress equations incorporate adequate safety factors for the typical strength levels to which the equations apply. This is in contrast to the buckling equations, (4-8) and (4-12), which refer to loads and stresses at failure and hence do not contain safety factors.

As you can see from the discussion concerning design equations, column design, even when it refers to the case of an ideal column with hinged ends, is not a simple matter. In practice, the designer deals with many codes for the design of columns in various materials and must therefore master a variety of design approaches.

Example 4-3

A W250 × 67 of low strength structural steel is to be used as a 7 m long column having hinged ends. What is the allowable axial load (a) if the column is free to fail in any direction, and (b) if the column is constrained from failure in the weak direction?

Solution

a. For the column section we obtain the following properties from Table B-1:

$$\text{area} = A = 8550 \text{ mm}^2$$

$$r_x = 110 \text{ mm}$$

$$r_y = 51 \text{ mm}$$

Since the column is free to fail in any direction, the lower bending stiffness about the y-axis will govern and lead to the largest possible L/r value. We obtain

$$\frac{L}{r_y} = \frac{7000}{51} = 137.3$$

which means that we are dealing with a column having $L/r > 110$ and Eq. (4-13b) applies. Then

$$\sigma_a = \frac{1\,000\,000}{(L/r)^2} = \frac{1\,000\,000}{137.3^2} = 53.1 \text{ MPa}$$

$$P_a = A\sigma_a = 8550 \times 53.1 \text{ mm}^2 \cdot \frac{\text{N}}{\text{mm}^2}$$

$$P_a = 454 \text{ kN}$$

b. Constraining the column in the weak direction means that the full column strength in the strong direction can be utilized and hence r_x governs. We obtain

$$\frac{L}{r_x} = \frac{7000}{110} = 63.6 < 110$$

and hence (4-13a) applies. Substituting into (4-13a) gives

$$\sigma_a = 140 - 0.5 \times 63.6 = 108.2 \text{ MPa}$$

$$P_a = A\sigma_a = 8550 \times 108.2 \text{ mm}^2 \cdot \frac{\text{N}}{\text{mm}^2}$$

$$P_a = \textbf{925 kN}$$

The example shows that the allowable load is more than doubled when failure can be prevented in the weak direction of the column section. Constraining a column in the weak direction is often possible in practice; you can gain an insight into this technique by examining bridges or steel frames of buildings under construction.

Example 4-3

A W10 × 45 of low strength structural steel is to be used as a 23 ft column having hinged ends. What is the allowable axial load (a) if the column is free to fail in any direction, and (b) if the column is constrained from failure in the weak direction?

Solution

a. For the column section we obtain the following properties from Table B-1:

$$\text{area} = A = 13.3 \text{ in.}^2$$

$$r_x = 4.32 \text{ in.}$$

$$r_y = 2.0 \text{ in.}$$

Since the column is free to fail in any direction, the lower bending stiffness about the y-axis will govern and lead to the largest possible L/r value. We obtain

$$\frac{L}{r_y} = \frac{23 \times 12}{2} = 138$$

which means that we are dealing with a column having $L/r > 110$ and Eq. (4-13b) applies. Then

$$\sigma_a = \frac{145,000}{(L/r)^2} = \frac{145,000}{138^2} = 7.61 \text{ kip/in.}^2$$

$$P_a = A\sigma_a = 13.3 \times 7.61 \text{ in.}^2 \cdot \frac{\text{kip}}{\text{in.}^2}$$

$$P_a = 101.2 \text{ kip}$$

b. Constraining the column in the weak direction means that the full column strength in the strong direction can be utilized and hence r_x governs. We obtain

$$\frac{L}{r_x} = \frac{23 \times 12}{4.32} = 63.9 < 110$$

and hence (4-13a) applies. Substituting into (4-13a) gives

$$\sigma_a = 20 - 0.07\left(\frac{L}{r}\right) = 20 - 0.07 \times 63.9 = 15.53 \text{ kip/in.}^2$$

$$P_a = A\sigma_a = 13.3 \times 15.53 \text{ in.}^2 \cdot \frac{\text{kip}}{\text{in.}^2}$$

$$P_a = \textbf{206.6 kip}$$

The example shows that the allowable load is more than doubled when failure can be prevented in the weak direction of the column section. Constraining a column in the weak direction is often possible in practice; you can gain an insight into this technique by examining bridges or steel frames of buildings under construction.

Problems 4-19 to 4-24

4-6 END CONDITIONS

The ends of all columns considered up to this point have been treated as though the end supports offered *no constraint* against rotation, as in Fig. 4-8a. This ideal condition could be attained by using a frictionless pin at each end or by placing spherical end caps on the column. In practice the end connections are frequently semirigid and offer some resistance to rotation. In such cases, the assumption that there is no constraint introduces an error on the side of safety.

When the ends are *completely constrained* against rotation, the shape of the deflected column will be as shown in Fig. 4-8b. To determine the strength of this column, we could repeat the analysis that led to Euler's equation for pin-ended columns, using the appropriate boundary conditions, and arrive at the buckling load. Instead we will use a more intuitive approach.

Inspection of the deflection curve shows that going from left to right, the deflected column is first curved upward, then downward, and then upward again. At the points where the curvature changes direction there will be a zero curvature, and hence no bending moment. These locations are known as *points of contraflexure*. The symmetry of the member suggests that we could assume the points of contraflexure to be located at the one-quarter and three-quarter points along the length. Since there is no bending moment at the points of contraflexure, the load-carrying capacity will not be altered if we cut the beam at these points and reconnect it with frictionless pins as in Fig. 4-8c. The action of the central region, shown as a free body in Fig. 4-8d, is the same as a pin-ended column and its critical load can be found by substituting L/2, referred to as the *effective length*, for the length in the Euler equation (4-8), which gives

$$P_{cr} = \frac{\pi^2 EI}{(L/2)^2} = 4\frac{\pi^2 EI}{L^2}$$

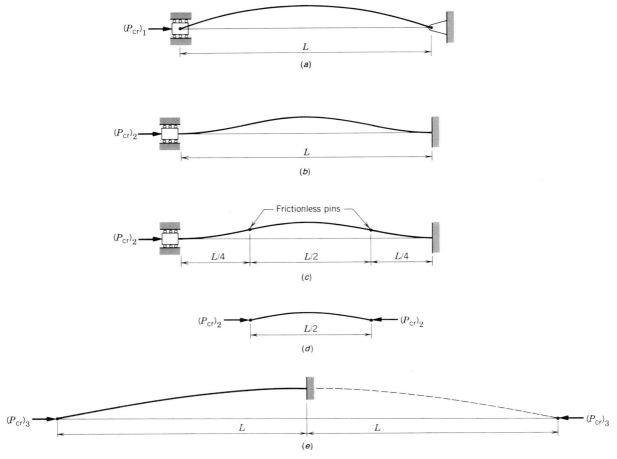

FIG. 4-8 Concept of effective lengths of columns for different end constraints. (*a*) Pin-ended column. (*b*) Column with built-in ends. (*c*) Modified column with built-in ends. (*d*) Central region of column with built-in ends. (*e*) Column with one end free and one end built-in.

This shows that for columns which are identical, except for the end constraints, *a built-in column is four times as strong as a pin-ended column*. Consequently, in designing real structures, there is a strong incentive for fixing the ends of columns against rotation. A rigorous analysis of the column with built-in ends will lead to the same conclusions as this solution based on intuition.

The column in Fig. 4-8*e* has one free and one built-in end. If we consider the column to be extended beyond the right support as a mirror image of the real part of the column, the deflection curve resembles a pin-ended column of length 2L. Using this *effective length* in the Euler equation (4-8), we obtain.

$$P_{cr} = \frac{\pi^2 EI}{(2L)^2} = \frac{1}{4}\frac{\pi^2 EI}{L^2}$$

It is seen that the column of Fig. 4-8e has a strength that is only *one-quarter* of the strength of a pin-ended column. Because of this very large reduction in strength when one end is able to move laterally, designers must exercise caution in cases where lateral motion may occur. This effect is present in the columns of buildings where the floors, above the first, are able to move horizontally; such motion is referred to as *sidesway*. Rigorous analysis would confirm the accuracy of this intuitive solution to the free-end column.

In order to prevent the influence of end constraints from being overlooked, we must always use the *effective length* in (4-13a and b), (4-14a and b), and (4-15a and b). To guard against such an error, we will rewrite the equations in terms of effective length.

For *low-strength structural steel*

$$\sigma_a = 140 - 0.5\left(\frac{L'}{r}\right) \qquad \text{for } 0 \leqslant \frac{L'}{r} \leqslant 110 \qquad \textbf{(4-16a)}$$

$$\sigma_a = \frac{1\,000\,000}{(L'/r)^2} \qquad \text{for } \frac{L'}{r} > 110 \qquad \textbf{(4-16b)}$$

for *wood* (Douglas fir, western hemlock and eastern spruce)

$$\sigma_a = 5 - 0.01\left(\frac{L'}{r}\right) \qquad \text{for } 0 \leqslant \frac{L'}{r} \leqslant 90 \qquad \textbf{(4-17a)}$$

$$\sigma_a = \frac{33\,000}{(L'/r)^2} \qquad \text{for } \frac{L'}{r} > 90 \qquad \textbf{(4-17b)}$$

for *aluminum* (6061-T6)

$$\sigma_a = 140 - 1.0\left(\frac{L'}{r}\right) \qquad \text{for } 0 \leqslant \frac{L'}{r} \leqslant 70 \qquad \textbf{(4-18a)}$$

$$\sigma_a = \frac{340\,000}{(L'/r)^2} \qquad \text{for } \frac{L'}{r} > 70 \qquad \textbf{(4-18b)}$$

In Imperial units the equations, in terms of effective length, become as follows.

For low-strength structural steel:

$$\sigma_a = 20 - 0.07\left(\frac{L'}{r}\right) \qquad \text{for } 0 \leqslant \frac{L'}{r} \leqslant 110 \qquad \textbf{(4-16a)}$$

$$\sigma_a = \frac{145{,}000}{(L'/r)^2} \qquad \text{for } \frac{L'}{r} > 110 \qquad \textbf{(4-16b)}$$

For wood (Douglas fir, western hemlock or eastern spruce):

$$\sigma_a = 0.7 - 0.0014\left(\frac{L'}{r}\right) \qquad \text{for } 0 \leqslant \frac{L'}{r} \leqslant 90 \qquad \textbf{(4-17a)}$$

$$\sigma_a = \frac{4800}{(L'/r)^2} \qquad \text{for } \frac{L'}{r} > 90 \qquad \textbf{(4-17b)}$$

For aluminum (6061-T6):

$$\sigma_a = 20 - 0.14 \left(\frac{L'}{r} \right) \qquad \text{for } 0 \leqslant \frac{L'}{r} \leqslant 70 \tag{4-18a}$$

$$\sigma_a = \frac{50,000}{(L'/r)^2} \qquad \text{for } \frac{L'}{r} > 70 \tag{4-18b}$$

Of course, the *effective length* L' must also be used in the buckling equations, (4-8) and (4-1?) therefore, it is useful to rewrite these equations as:

$$P_{cr} = \frac{\pi^2 EI}{(L')^2} \tag{4-19}$$

and

$$\sigma_{cr} = \frac{\pi^2 E}{(L'/r)^2} \tag{4-20}$$

The effective length, L', can be expressed as

$$L' = KL \tag{4-21}$$

where K is a factor that is determined by the end constraints. Figure 4-9 gives values of K for the most commonly encountered combinations of basic end conditions. In practice the constraints are not always clearly defined and the designer must then use judgment in arriving at the value for K. Although on some occasions values of K between those given above will be determined by engineering judgment, generally, when there is any doubt, the more conservative basic model will be used.

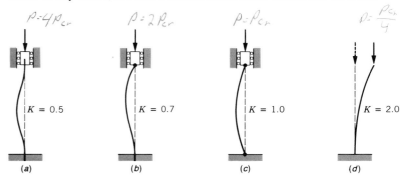

FIG. 4-9 Effect of end constraints on K.

Example 4-4

For the cases of Figs. 4-10*a* and *b*, determine the safe loads P_1 and P_2 if the vertical member in each case is a 7-m-long W360 × 147.

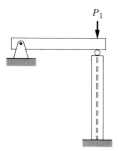

FIG. 4-10a

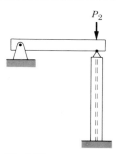

FIG. 4-10b

Solution. From Table B-1 we obtain for a W360 × 147 section:

$$r_{min} = r_y = 94 \text{ mm}$$

$$area = A = 18\ 800 \text{ mm}^2$$

Case 1. As shown in Fig. 4-10c, the roller at the top of the column permits lateral movement to take place; the buckling shape therefore corresponds to Fig. 4-9d. Hence

$$K = 2$$

and

$$L' = 2 \times 7000 \text{ mm} \qquad \text{from (4-21)}$$

$$= 14\ 000 \text{ mm}$$

$$\frac{L'}{r} = \frac{14\ 000}{94} \frac{\text{mm}}{\text{mm}} = 149 > 110$$

Therefore,

$$\sigma_a = \frac{1\ 000\ 000}{(L'/r)^2} \qquad \text{from (4-16b)}$$

$$= \frac{1\ 000\ 000}{149^2}$$

$$= 45.0 \text{ MPa} = 45.0 \frac{\text{N}}{\text{mm}^2}$$

$$\text{safe } P_1 = \sigma_a \times A$$

$$= 45.0 \times 18\ 800 \frac{\text{N}}{\text{mm}^2} \cdot \text{mm}^2$$

$$= 847\ 000 \text{ N}$$

$$= \textbf{847 kN}$$

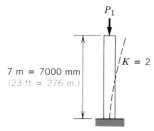

FIG. 4-10c

Case 2. For case 2 the hinge at the top constrains the column from moving laterally; hence the buckling shape shown in Fig. 4-10d corresponds to Fig. 4-9b.

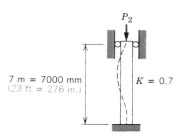

FIG. 4-10d

Then

$$K = 0.7$$

and

$$L' = 0.7 \times 7000 = 4900 \text{ mm} \qquad \text{from (4-21)}$$

$$\frac{L'}{r} = \frac{4900}{94} = 52.1 < 110$$

Therefore,

$$\sigma_a = 140 - 0.5 \frac{L'}{r} \qquad \text{from (4-16a)}$$

$$= 140 - 0.5 \times 52.1$$

$$= 114 \text{ MPa} = 114 \frac{\text{N}}{\text{mm}^2}$$

$$\text{safe } P_2 = \sigma_a \times A$$

$$= 114 \times 18\,800 \frac{\text{N}}{\text{mm}^2} \cdot \text{mm}^2$$

$$= 2\,142\,000 \text{ N}$$

$$= \mathbf{2142\ kN}$$

These cases clearly indicate the strong influence that end conditions have on the load-carrying capacity of compression members. If, in practice, the actual load was 2000 kN and the structure designed as in case 2, the disastrous results of building the structure as in case 1 can be imagined. This points out the importance of careful on-site inspection during construction.

Example 4-4

and b, determine the safe loads P_1 and P_2 if the vertical member in each case is a 23-ft-long W14 × 95.

Solution. From Table B-1 we obtain for a W14 × 95 section:

$$r_{\min} = r_y = 3.72 \text{ in.}$$

$$\text{area} = A = 29.1 \text{ in.}^2$$

Case 1. As shown in Fig. 4-10c, the roller at the top of the column permits lateral movement to take place; the buckling shape therefore corresponds to Fig. 4-9d. Hence

$$K = 2$$

and

$$L' = 2 \times 23 \times 12 \text{ in.} \qquad \text{from (4-21)}$$

$$= 552 \text{ in.}$$

$$\frac{L'}{r} = \frac{552}{3.72} \frac{\text{in.}}{\text{in.}} = 148 > 110$$

Therefore,

$$\sigma_a = \frac{145,000}{(L'/r)^2} \qquad \text{from (4-16b)}$$

$$= \frac{145,000}{148^2}$$

$$= 6.6 \text{ kip/in.}^2$$

$$\text{safe } P_1 = \sigma_a \times A$$

$$= 6.6 \times 29.1 \frac{\text{kip}}{\text{in.}^2} \cdot \text{in.}^2$$

$$= \textbf{192 kip}$$

Case 2.   For case 2 the hinge at the top constrains the column from moving laterally; hence the buckling shape shown in Fig. 4-10*d* corresponds to Fig. 4-9b. Then

$$K = 0.7$$

and

$$L' = 0.7 \times 23 \times 12 = 193.2 \text{ in.} \qquad \text{from (4-21)}$$

$$\frac{L'}{r} = \frac{193.2}{3.72} = 52 < 110$$

Therefore,

$$\sigma_a = 20 - 0.07 \frac{L'}{r} \qquad \text{from (4-16a)}$$

$$= 20 - 0.07 \times 52$$

$$= 16.36 \text{ kip/in.}^2$$

$$\text{safe } P_2 = \sigma_a \times A$$

$$= 16.36 \times 29.1 \frac{\text{kip}}{\text{in.}^2} \cdot \text{in.}^2$$

$$= \textbf{476 kip}$$

These cases clearly indicate the strong influence that end conditions have on the load carrying capacity of compression members. If, in practice, the actual load was 450 kip and the structure designed as in case 2, the disastrous results of building the structure as in case 1 can be imagined. This points out the importance of careful on-site inspection during construction.

PROBLEMS

4-1 A standard 63-mm ($2\frac{1}{2}$-in.) steel pipe is used as a pin-ended column 4 m (13 ft) long.

    a. Determine the Euler buckling load.
    b. What is the safe load, if the required safety factor is 2.00?
    c. Determine the stress when the safe load is applied.
    d. Calculate the apparent safety factor, based on yield, when the safe load is applied. The material is AISI 1020 steel.

4-2 A pin-ended column is 12 m (40 ft) long. Calculate the Euler buckling load if the shape is

    a. W360 × 147 (W14 × 95)
    b. S610 × 149 (S24 × 100)
    c. Which family of sections, the W-shapes or the S-shapes, appears to be better proportioned for use as columns?

4-3 A pin-ended column consists of a 40-mm (1.6-in.) rod, 1.5 m (5 ft) long. Calculate the Euler buckling load if the material is

    a. AISI 1020.
    b. AISI 8760.
    c. AISI 431.
    d. Gray cast iron.
    e. Aluminum.
    f. Magnesium.

4-4 Determine the Euler buckling load for a pin-ended column that is 8 m (26 ft) long and has a 125-mm × 300-mm (5-in. × 12-in.) cross section when the wood is

    a. Douglas fir.
    b. Western hemlock.
    c. Eastern spruce.

4-5 Determine the Euler buckling load for a pin-ended column 20 m (65 ft) long that is made of two S610 × 158 (S24 × 106) welded together as shown in the sketch.

PROBLEM 4-5

4-6 Determine the Euler buckling load if the welds specified for the column in Prob. 4-5 are inadvertently omitted.

4-7 Calculate the cross-sectional area of a solid square bar of AISI 1090 steel to safely carry an axial compressive load of 80 kN (18 kip). The factor of safety is 2 and the length is 3 m (10 ft) between pinned ends.

4-8 Calculate the critical buckling load for a 3.2-m (10-ft)-long steel column having the cross section shown in the illustration.

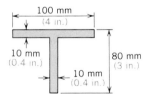

PROBLEM 4-8

4-9 A pin-ended column 13 m (43 ft) long is to be constructed by welding together four L150 × 150 × 20 (L6 × 6 × $\frac{3}{4}$). Determine the Euler buckling load if the angles are

    a. Welded together to form a hollow-square section.
    b. Welded together to form a star-shaped section.

4-10 A pin-ended column 8 m (26 ft) long is to be constructed by welding together two C250 × 45 (C10 × 30). Determine the Euler buckling load if the C-shapes are

a. Welded together back to back so that they form a section resembling an S-shape.

b. Welded together flange to flange so that they form a hollow rectangular section.

b. Placed side by side but not welded.

4-11 What are the fundamental assumptions that were made but not stated in the derivation of the Euler buckling equation?

4-12 Determine the slenderness ratio of a steel column that is 10 m (33 ft) long with each of the cross sections shown in the diagram. Note that the amount of steel is substantially the same in all cases, but its distribution is different.

200 mm (8 in.)
steel pipe

PROBLEM 4-12a

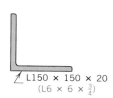

L150 × 150 × 20
(L6 × 6 × $\frac{3}{4}$)

PROBLEM 4-12b

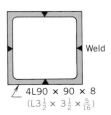

Weld

4L90 × 90 × 8
(L3$\frac{1}{2}$ × 3$\frac{1}{2}$ × $\frac{5}{16}$)

PROBLEM 4-12c

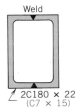

Weld

2C180 × 22
(C7 × 15)

PROBLEM 4-12d

4-13 A pin-ended column is 4.5 m (15 ft) long. Calculate the critical stress and the critical load if the member is an

a. L200 × 100 × 13 (L8 × 4 × $\frac{1}{2}$).
b. L150 × 150 × 13 (L6 × 6 × $\frac{1}{2}$).
c. L125 × 125 × 16 (L5 × 5 × $\frac{5}{8}$).

If the member is made of AISI 1020 steel, do the stress values seem realistic?

4-14 A pin-ended column is 4 m (13 ft) long. Calculate the critical stress and the critical load if the member is an

a. S200 × 27 (S8 × 18).
b. W200 × 27 (W8 × 17).
c. 250 mm pipe (10 in.).

If the member is made of AISI 1020 steel, do the stress values seem realistic?

4-15 A piece of 200-mm (8-in.) pipe is used as a pin-ended column. Calculate the critical stress and the critical load if the length of the column is

a. 12 m (39 ft).
b. 8 m (26 ft).
c. 4 m (13 ft).

If the member is made of AISI 1020 steel, do the stress values seem realistic?

4-16 For a particular job being done in your design office 88-mm (3$\frac{1}{2}$-in.) pipes made of AISI 1020 are frequently used for pin-ended compression members.

a. Make a diagram on which you plot curves of critical load and critical stress for lengths ranging from 1.0 to 5.0 m (3 to 15 ft).

b. If you were to give the curves to your assistant in order to save repetitious calculations, what warning would you issue?

4-17 Determine the length at which a pin-ended column made from an L75 × 50 × 8 (L3 × 2 × $\frac{5}{16}$) will buckle under a load of 70 kN (16 kip).

4-18 An aluminum bar is installed between fixed supports 0.8 m (2.6 ft) apart at a temperature of 20°C (68°F) such that it is precompressed with a force of 20 kN (4.5 kip). The temperature is then increased slowly. Determine the temperature at which you would expect buckling to occur.

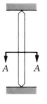

A   A

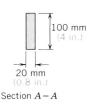

100 mm
(4 in.)

20 mm
(0.8 in.)

Section $A-A$

PROBLEM 4-18

4-19 A low-strength structural steel column is 6 m (20 ft) long, has pivoted ends, and carries a load of 500 kN (112 kip). Select the lightest section to meet the conditions of (4-13a) or (4-13b), whichever is applicable, from

a. The S-shapes.
b. The W-shapes.

4-20 A low-strength structural steel column is 4.80 m (16 ft) long, has pivoted ends, and carries a load of 500 kN (112 kip). Select the lightest section to meet the conditions of (4-13a) or (4-13b), whichever is applicable, from (a) the S-shapes, and (b) the W-shapes. Discuss which section is more economical in terms of the amount of material. What is the percentage saving in material?

4-21 A 5-m (16-ft) low-strength structural steel column with pin-connected ends is to have one of the four cross sections shown in Prob. 4-12.

a. For each of the cross sections calculate the load that could be safely carried by the column.
b. Compare the safe load in part (a) with the Euler critical load, P_{cr}.
c. Compare the safe load in part (a) with the safe tensile load if the allowable tensile stress is 140 MPa (20 ksi).

4-22 A 2-m (6.6-ft)-long column, made from an L75 × 50 × 8 (L3 × 2 × $\frac{5}{16}$), has pivoted ends. Based on the allowable stress equations, (4-13a) or (4-13b), whichever is applicable, determine the safe load.

4-23 A tubular compression member has pivoted ends and is 0.85 m (2.8 ft) long. The outside diameter of the tube is 30 mm (1.2 in.) and the wall thickness 2 mm (0.08 in.). Calculate the safe compressive load if the material is

a. Low-strength structural steel.
b. Aluminum alloy 6061-T6.

4-24 A tubular column has pivoted ends and is 0.85 m (2.8 ft) long. The mean diameter of the tube is 20 mm (0.8 in.). Determine the required wall thickness if the tube is to safely carry a compressive load of 4.5 kN (1 kip) and the material is

a. Low-strength structural steel.
b. Aluminum alloy 6061-T6.

4-25 A vertical Douglas fir pole is 100 mm (4 in.) in diameter. The lower end is embedded in concrete and the top is 8.0 m (26 ft) above ground level. A mass of 160 kg (350 lb) is placed on the top. Determine the safety factor.

4-26 Determine the safety factor in Prob. 4-25 if Eastern spruce is substituted for Douglas fir.

4-27 A vertical column 5 m (16 ft) long is built-in at the base and braced at the top by cables so that all sidesway is prevented. Determine the safe load if the member is a

a. W200 × 27 (W8 × 17).
b. S200 × 27 (S8 × 18).
c. 150-mm pipe (6-in. pipe).

Note that the amount of material is substantially the same in each column.

4-28 Determine the safe loads in Prob. 4-27 if the bracing can prevent sidesway in one vertical plane only. Consider the bracing to be in the plane that will be most effective in increasing the load-carrying capacity.

4-29 a. An aluminum (6061-T6) mast must safely support, at its upper end, a load with a total mass of 1000 kg (2200 lb). The free-standing mast is 3.0 m (10 ft) long and is made of tubing with an external diameter of 100 mm (4 in.). Determine the required wall thickness.
b. The mast is guyed at midheight and at the top in such a manner that the guy wires place a negligible axial load on the mast and prevent all lateral motion. The bottom end is pin-connected. What is the required wall thickness?

4-30 Select a low-strength S-section for the jib crane shown in the illustration. The crane must be able to lift a mass of 1000 kg (2200 lb) safely. Assume pinned ends at A and B.

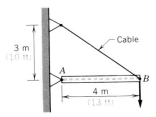

PROBLEM 4-30

4-31 Calculate the safe load when

    a. The column is as shown.

    b. The base of the column is built-in.

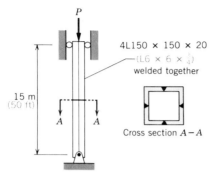

PROBLEM 4-31

4-32 a. Calculate the safe load.

    b. If the load is 1500 kN (340 kip), find the lightest S-shape that will safely carry the load.

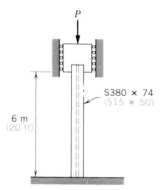

PROBLEM 4-32

4-33 Determine the safe load, P:

    a. When the structure is as shown.

    b. When the lower ends of the columns are built-in.

    c. When the lower ends of the columns are built-in and the sway-bracing has been removed.

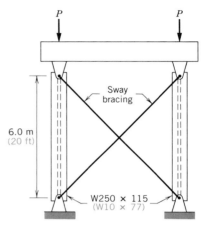

PROBLEM 4-33

Advanced Problems

4-A A 20-mm (0.8-in.) steel rod is 0.75 m (30 in.) long. It is mounted in a test machine similar to that shown in Fig. 4-2*b*.

    a. How far can the load screw be advanced before buckling begins?

    b. If the screw is advanced a further 0.6 mm (0.024 in.) determine the lateral deflection at midlength.

4-B For the column in part (b) of Prob. 4-A, draw a diagram showing the stress profile for points on a diameter of the rod at midlength.

4-C Discuss the validity of the solutions to Probs. 4-A and 4-B when the rod material is

    a. AISI 1020.

    b. AISI 4063.

4-D Curve *B* in Fig. 4-2*c* shows that as displacement increases, the force remains constant at the level of the critical load up to a point and then begins to decrease. Determine the displacement at which the drop-off occurs for an 88-mm ($3\frac{1}{2}$-in.)-pipe column 5 m (16 ft) long. The pipe material is AISI 4340 and the ends are pivoted.

4-E Working from first principles, determine the buckling load and hence confirm the effective length factor for the column shown in

a. Fig. 4-9*d*.
b. Fig. 4-9*b*.

ANALYSIS OF
STRESS

5-1 INTRODUCTION

The design of truss members is based on the stress on a plane normal to the axis of the member, the allowable stress having been determined from the stress on a plane normal to the axis of a test specimen. There seems to be no motivation for analyzing the stresses on planes inclined to the axis in either the truss member or the test specimen; however, such an analysis will help us account for the shape of the fracture surface of the test specimen shown in Fig. 1-25c and for other peculiar phenomena.

When considering stress in the web of a rolled section or a plate girder, we calculated the shearing stress on the vertical plane to make certain that it did not exceed an allowable value. However, when failure took place because of excessive shear, it did not occur by sliding along a vertical plane but rather by an unexpected action taking place on inclined planes, as seen in Fig. 2-15c and d. We will be able to account for these failure modes when we are able to analyze stresses on inclined planes.

In pressure vessel design we were able to design a welded vessel by considering only circumferential and longitudinal stresses; these are sufficient for vessels and pipes that have only longitudinal and circumferential welds. However, large-diameter pipe is frequently fabricated by forming a steel strip in such a way that a spiral weld is required as shown in Fig. 5-1. In this case the direction of the stress that tends to separate the two welded edges is neither circumferential nor longitudinal. A knowledge of stresses on inclined planes is necessary if the safe pressure in such a pipe is to be determined.

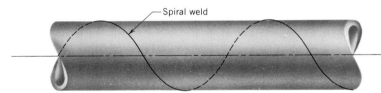

FIG. 5-1 Spiral-wound pipe.

5-2 STRESS ON AN INCLINED PLANE IN A UNIAXIAL STRESS FIELD

Before developing the equations that will give stress on an inclined plane, we will work a numerical example in order to gain some insight into the behavior of stressed materials. The simple member in tension, that is, one in a state of uniaxial stress, will be considered in the following example.

Example 5-1

For the tensile member given in Fig. 5-2a:

  a. Determine the stresses on the plane AB for $\theta = 30°$.

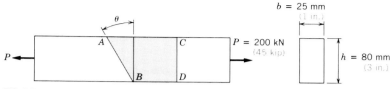

FIG. 5-2a

b. Repeat for various θ between $-90°$ and $90°$ and plot the results on a graph with normal stress as the abscissa (*x*-axis) and shear stress as the ordinate (*y*-axis).

Solution

a. For the case of $\theta = 30°$, the stresses and forces on a free body *ABCD* are shown in Fig. 5-2*b* and *c*. For equilibrium

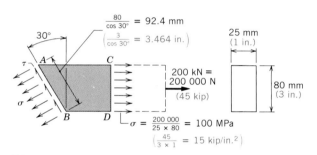

FIG. 5-2b Stresses on a free body.

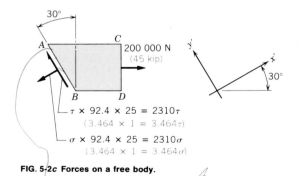

FIG. 5-2c Forces on a free body.

$$\sum F_{x'} = 0$$

$$-2310\sigma + 200\,000 \cos 30° = 0$$

$$\sigma = \frac{200\,000 \cos 30°}{2310} = \textbf{75.0 MPa}$$

Hand-written notes (right margin):

$$\sum F = 0 \rightarrow$$

$$-\left[\sigma\left(\frac{A}{\cos\theta}\right)\right]\cos\theta - \left[\tau\left(\frac{A}{\cos\theta}\right)\right]\sin\theta + \sigma_0 A = 0$$

$$(I) \quad -\sigma - \tau\tan\theta + \sigma_0 = 0$$

$$\sum F = 0 \uparrow$$

$$-\left[\sigma\left(\frac{A}{\cos\theta}\right)\right]\sin\theta + \left[\tau\left(\frac{A}{\cos\theta}\right)\right]\cos\theta = 0$$

$$(II) \quad -\sigma\tan\theta + \tau = 0$$

$$\sigma = \frac{\sigma_0}{1 + \tan^2\theta} = \frac{\sigma_0}{\sec^2\theta} = \sigma_0 \cos^2\theta$$

$$\tau = \sigma_0 \tan\theta \cos^2\theta$$

and

$$\sum F_{y'} = 0$$

$$2310\tau - 200\,000 \sin 30° = 0$$

$$\tau = \frac{200\,000 \sin 30°}{2310} = \textbf{43.3 MPa}$$

b. Repeating the method used in (a) for various θ's will give the following tabulated values:

Plane no.	θ (°)	σ (MPa)	τ (MPa)
1	−90	0.0	0.0
2	−75	6.7	−25.0
3	−60	25.0	−43.3
4	−45	50.0	−50.0
5	−30	75.0	−43.3
6	−15	93.3	−25.0
7	0	100.0	0.0
8	15	93.3	25.0
9	30	75.0	43.3
10	45	50.0	50.0
11	60	25.0	43.3
12	75	6.7	25.0
13	90	0.0	0.0

Based on the tabulated values, the graph of Fig. 5-2*d* is obtained.

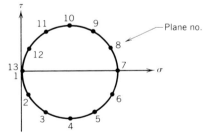

FIG. 5-2d

Example 5-1

For the tensile member given in Fig. 5-2*a*:
  a. Determine the stresses on the plane *AB* for $\theta = 30°$.
  b. Repeat for various θ between $-90°$ and $90°$ and plot the results on a graph with normal stress as the abscissa (*x*-axis) and shear stress as the ordinate (*y*-axis).

Solution

a. For the case of $\theta = 30°$, the stresses and forces on a free body are shown in Fig. 5-2*b* and *c*. For equilibrium

$$\sum F_{x'} = 0$$

$$-3.464\sigma + 45 \cos 30° = 0$$

$$\sigma = \frac{45 \cos 30°}{3.464} = \textbf{11.25 kip/in.}^2$$

and

$$\sum F_{y'} = 0$$

$$3.464\tau - 45 \sin 30° = 0$$

$$\tau = \frac{45 \sin 30°}{3.464} = \textbf{6.50 kip/in.}^2$$

b. Repeating the method used in (a) for various θ's will give the following tabulated values:

Plane no.	θ (°)	σ (kip/in.^{2})	τ (kip/in.^{2})
1	−90	0.0	0.0
2	−75	1.00	−3.75
3	−60	3.75	−6.50
4	−45	7.50	−7.50
5	−30	11.25	−6.50
6	−15	14.00	−3.75
7	0	15.00	0.0
8	15	14.00	3.75
9	30	11.25	6.50
10	45	7.50	7.50
11	60	3.75	6.50
12	75	1.00	3.75
13	90	0.0	0.0

Based on the tabulated values, the graph of Fig. 5-2d is obtained.

Several important observations can be made by an inspection of the table and the graph.

1. The normal stress on the various planes ranges from zero to 100.0 MPa (15 ksi) with the maximum occurring on the plane at right angles to the axis of the member. If a material fails because the normal stress is excessive, the failure, described as *cleavage*, will occur by the material parting along a plane normal to the axis. Cast iron and other brittle materials fail in this manner.

2. The shearing stress varies from zero to a maximum of 50.0 MPa (7.5 ksi). The maximum shear stress is just half the maximum normal stress and acts on a plane at 45° to the axis of the member. If a material happened to have a low shear strength, relative to its normal strength, it might fail in shear. The shear failure would be expected to occur on a plane at 45° to the axis. Low-strength steel and many other materials are weak in shear and consequently fail by shearing along planes at 45° to the axis of the load. These planes of shear failure can be seen near the surface of the specimen shown in Fig. 1-26c. In the interior a different mechanism of failure begins to act after an appreciable amount of necking, and the specimen ruptures by cleavage in that region. The combined modes of failure produce a cup-and-cone separation, which is characteristic of steel and other ductile materials.

Because shear failure causes particles to slide over one another, there is an opportunity for the atoms to again form a bond across the slip plane after some shear deformation has taken place. This rebonding of atoms actually happens and accounts for the deformation that takes place in ductile materials before the material breaks. In cleavage failure the atoms separate when a bond is broken and there is no opportunity to rebond; hence materials that fail by cleavage do not flow before failure and give no warning of impending failure. This mechanism accounts for the brittle nature of the failure.

3. The graph of stresses shows an orderly arrangement of points, which seem to fall on a circle. If the equations for σ and τ had been worked out in terms of θ, it would have been possible to prove that the points actually do lie on a circle. Later this will be explored further for a more general state of stress. It is interesting to observe that if a plane is thought of as rotating from the position of plane 1 to that of plane 13, that is, rotating through 180°, the points on the stress graph make a complete turn around the circle, that is, through 360°. This apparent two-to-one relationship in angle will be verified and found most useful in later analysis.

Problems 5-1 to 5-2

5-3 STRESSES ON AN INCLINED PLANE IN A GENERAL PLANE STRESS FIELD

When a plate is subjected to in-plane loads, for example, the web of an S-shape or a small piece of a boiler shell, the stresses on two planes at right angles are usually calculated first. These planes are frequently normal to the x- and y-axis with the normal stresses designated as σ_x and σ_y and shear stress as τ_{xy}. The stresses have had sign conventions such that positive stresses act in the directions indicated in Fig. 5-3a.

To define positive shearing stress on a vertical plane as being in the direction that results from a positive shearing load (Section 1-5) is adequate for stress on a vertical plane. However, this convention is not always adequate and we will now adopt a sign convention that is compatible with the above convention but is not so restricted. Henceforth, we will say that *shear stress is positive if that stress, when acting on an element, tends to rotate the element in a clockwise direction.*

This sign convention makes it possible to restate (2-21) in broader terms as: *Whenever there is shearing stress on a plane, there is also shearing stress of equal magnitude, but opposite sign, on a second plane, which is at right angles to the first plane.* For the state of stress indicated in Fig. 5-3a, equilibrium is not satisfied and stresses are required on the horizontal surfaces as shown in Fig. 5-3b.

The direction of the horizontal shear is as shown and its magnitude is 40 MPa (5.8 ksi). If we were stating this without the use of arrows to show the sense, we would say that the horizontal shear stress is -40 MPa (-5.8 ksi).

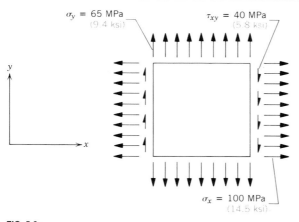

FIG. 5-3a

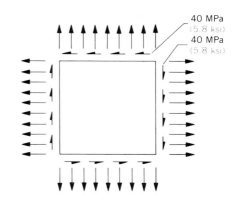

FIG. 5-3b

Consider a plate subjected to in-plane loads that induce stresses in the x- and y-directions and shear stresses on planes normal to the x- and y-axes. This arrangement of stress is known as *plane stress* or *biaxial stress*. The state of stress will be completely defined if values of σ_x, σ_y and τ_{xy} are known; our present objective is to be able to determine, from these known values, the normal and shear stress on any plane that is inclined to the coordinate axis. Using the known σ_x, σ_y, and τ_{xy} in Fig. 5-4a, we want to calculate σ_θ and τ_θ, the normal and shear stress on a plane that is normal to a line rotated through a counterclockwise angle, θ, from the x-axis. This can be accomplished by selecting a wedge-shaped

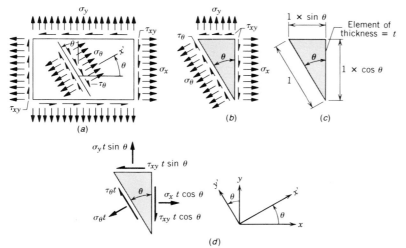

FIG. 5-4 Elements for the derivation of stress formulas on an inclined plane. (*a*) Stressed plate. (*b*) Stresses on wedge element. (*c*) Dimensions of wedge element. (*d*) Forces on wedge element.

element as shown in Fig. 5-4b. If the stresses varied throughout the plate, it would be necessary to use an infinitesimal element and we would then be dealing with stress at a point. Using constant stress enables us to arbitrarily choose an element of any convenient size, in this case the length of the inclined face is taken as 1 length unit. The final results for a constant and for a variable stress field are identical but by dealing with a constant stress field the manipulations have been made easier to understand.

The stresses in Fig. 5-4b are combined with the dimensions shown in Fig. 5-4c to give the forces in Fig. 5-4d. From the equilibrium equation

$$\sum F_{x'} = 0$$

$$-\sigma_\theta t + \sigma_x t \cos\theta \cos\theta + \sigma_y t \sin\theta \sin\theta - \tau_{xy} t \cos\theta \sin\theta$$

$$-\tau_{xy} t \sin\theta \cos\theta = 0$$

$$\boxed{\sigma_\theta = \cos^2\theta\,\sigma_x + \sin^2\theta\,\sigma_y - 2\sin\theta\cos\theta\,\tau_{xy}} \qquad \textbf{(5-1)}$$

and from

$$\sum F_{y'} = 0$$

$$\tau_\theta t - \sigma_x t \cos\theta \sin\theta + \sigma_y t \sin\theta \cos\theta - \tau_{xy} t \cos\theta \cos\theta$$

$$+\tau_{xy} t \sin\theta \sin\theta = 0$$

$$\tau_\theta = \sin\theta\cos\theta\,\sigma_x - \sin\theta\cos\theta\,\sigma_y - (\sin^2\theta - \cos^2\theta)\tau_{xy}$$

$$\boxed{\tau_\theta = \sin\theta\cos\theta(\sigma_x - \sigma_y) + (\cos^2\theta - \sin^2\theta)\tau_{xy}} \qquad \textbf{(5-2)}$$

Equations (5-1) and (5-2) enable us to calculate normal and shear stresses on any inclined plane without repeating the process used in their derivation. In these equations counterclockwise rotation is indicated by a positive θ and clockwise rotation by a negative θ. The following example illustrates their application.

Example 5-2

For the plane stress given in Fig. 5-5a:
    a. Calculate the stresses on plane A-B.
    b. Repeat part (a) for $\theta = 0$, 15, 30, 45, ..., 180° and plot points on a σ-τ diagram.

    Solution.   From Fig. 5-5a, the stress values are

$$\sigma_x = 80 \text{ MPa}$$

$$\sigma_y = 30 \text{ MPa}$$

$$\tau_{xy} = 40 \text{ MPa}$$

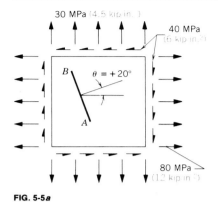

FIG. 5-5a

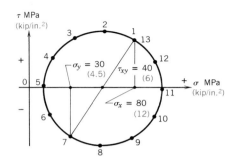

FIG. 5-5b

a. Substituting these stresses and $\theta = 20°$ into (5-1) and (5-2), we obtain

$$\sigma_\theta = \textbf{48.44 MPa}$$

$$\tau_\theta = \textbf{46.71 MPa}$$

b. Repeating the substitution process for various θ's, the tabulated results and corresponding graphical plot of Fig. 5-5b are obtained.

Plane Number	θ (°)	σ_θ (MPa)	τ_θ (MPa)
1	0	80.0	40.0
2	15	56.7	47.1
3	30	32.9	41.7
4	45	15.0	25.0
5	60	7.9	1.7
6	75	13.4	−22.1
7	90	30.0	−40.0
8	105	53.4	−47.1
9	120	77.1	−41.7
10	135	95.0	−25.0
11	150	102.1	−1.7
12	165	96.7	+22.1
13	180	80.0	+40.0

Example 5-2

For the plane stress given in Fig. 5-5a:
    a. Calculate the stresses on plane A-B.
    b. Repeat (a) for $\theta = 0, 15, 30, 45, \ldots, 180°$ and plot points on a σ-τ diagram.

Solution. From Fig. 5-5a, the stress values are

$$\sigma_x = 12 \text{ kip/in.}^2$$

$$\sigma_y = 4.5 \text{ kip/in.}^2$$

$$\tau_{xy} = 6 \text{ kip/in.}^2$$

a. Substituting these stresses and $\theta = 20°$ into (5-1) and (5-2), we obtain

$$\sigma_\theta = \textbf{7.27 kip/in.}^2$$

$$\tau_\theta = \textbf{7.00 kip/in.}^2$$

b. Repeating the substitution process for various θ's, the tabulated results and corresponding graphical plot of Fig. 5-5b are obtained.

Plane Number	θ (°)	σ_θ (kip/in.^{2})	τ_θ (kip/in.^{2})
1	0	12.00	6.00
2	15	8.50	7.07
3	30	4.93	6.25
4	45	2.25	3.75
5	60	1.18	0.25
6	75	2.00	−3.32
7	90	4.50	−6.00
8	105	8.00	−7.07
9	120	11.57	−6.25
10	135	14.25	−3.75
11	150	15.32	−0.25
12	165	14.50	3.32
13	180	12.00	7.07

An examination of the plotted points shows that certain patterns that were observed in Example 5-1 are repeated. These are (1) that the points fall on a circle, and (2) that there is a two-to-one relationship between the angular position of points on the circle and the angle of rotation of the plane on which the stresses act. Equations (5-1) and (5-2) can be manipulated to prove that these observations are correct. Otto Mohr (1835–1918) developed this pictorial representation of the equations; consequently, the circle relating σ_θ and τ_θ is referred to as *Mohr's circle for stress.* Through Mohr's circle we can achieve a much better understanding of stress than we can through a study of the equations. The circle could also be used to obtain graphical solutions but this is seldom done. In actual practice Mohr's circle is most useful in providing a visual guide to the combination of numerical values that is required to solve this type of problem. Mohr's circle provides an alternative to blind substitutions into (5-1) and (5-2). The following example illustrates how a Mohr's circle is constructed and points out some useful information that can be taken from the circle.

Example 5-3

The following stresses exist in a plate

$$\sigma_x = 100 \text{ MPa} \qquad \sigma_y = 20 \text{ MPa} \qquad \tau_{xy} = 30 \text{ MPa}$$

a. Draw Mohr's circle for stress.
b. Determine the maximum value of σ and the plane on which it acts.
c. Determine the maximum value of τ and the plane on which it acts.

Solution.   From the sign convention discussed in Section 5-3, the stresses on an element of the plate have the directions indicated in Fig. 5-6a. On a vertical plane, the given stresses of $\sigma = +100$ and $\tau = +30$ give the point labeled V on the σ-τ diagram of Fig. 5-6b.

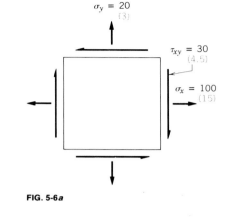

FIG. 5-6a

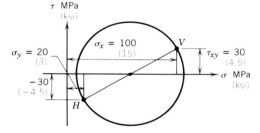

FIG. 5-6b

On a horizontal plane, $\sigma = +20$ and $\tau = -30$, give point H on the graph. Since the planes on which the stresses act are 90° apart in the plate, the points showing the state of stress on these planes will be 180° apart on the σ-τ diagram. Consequently, points V and H are on opposite ends of a diameter of the Mohr's circle. We can then draw the diameter and find the center of the circle, which is at the midpoint of the diameter. Note that because the shearing stresses on planes at right angles are always equal in magnitude but opposite in sign, the midpoint of the diameter will always fall on the σ-axis and *the center of Mohr's circle will always be on the σ-axis*. Also, the center will always be at

$$\sigma = \frac{\sigma_x + \sigma_y}{2}$$

The circle can then be drawn by finding its center and radius as shown in Fig. 5-6c.
From the circle we can see that no value of σ can exceed $(60 + 50 =)$ 110 MPa, which occurs at point P on the Mohr's circle. Such a stress is referred to as the *maximum principal stress* and is represented by σ_1. Note that from the nature of the circle the maximum principal stress must act on a plane that has zero shear stress. The minimum value of σ at Q is referred to as the *minimum principal stress*, σ_2, and again acts on a plane of zero shear.
To find the direction of the plane on which σ_1 acts, we need to note that P is 36.8° clockwise from V on the Mohr's circle. This means that in the plate the plane on which σ_1 acts is $(36.8/2 =)$ 18.4° clockwise from V as shown in Fig. 5-6d.

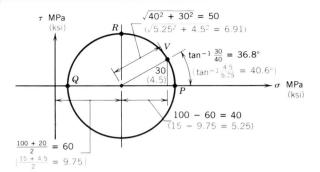

FIG. 5-6c

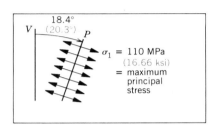

FIG. 5-6d

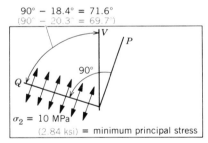

FIG. 5-6e

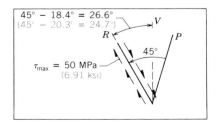

FIG. 5-6f

The minimum principal stress, σ_2, is $(60 - 50 =)$ 10 MPa. Points Q and P are at opposite ends of a diameter and consequently are 180° apart on the circle. Hence σ_2 acts on a plane at 90° to the plane of σ_1 as indicated in Fig. 5-6e.

The maximum value the shear stress can have is given by τ at the top of Mohr's circle, point R; then τ_{max} is 50 MPa. Point R is 90° on the circle from point P, therefore, the plane of maximum shear stress is at $(90/2 =)$ 45° to the plane of the maximum principal stress as shown in Fig. 5-6f.

On the negative side of the circle the largest negative shearing stress will be equal in magnitude to the maximum shearing stress but opposite in sign and, in the plate, 90° away. This is consistent with the relationships between shearing stresses that were established earlier.

Example 5-3

The following stresses exist in a plate

$$\sigma_x = 15 \text{ kip/in.}^2 \qquad \sigma_y = 3 \text{ kip/in.}^2 \qquad \tau_{xy} = 4.5 \text{ kip/in.}^2$$

a. Draw Mohr's circle for stress.
b. Determine the maximum value of σ and the plane on which it acts.
c. Determine the maximum value of τ and the plane on which it acts.

Solution. From the sign convention discussed in Section 5-3, the stresses on an element of the plate have the directions indicated in Fig. 5-6a. On a vertical plane, the given stresses of $\sigma = +15$ and $\tau = +4.5$ give the point labeled V on the σ-τ diagram of Fig. 5-6b.

On a horizontal plane, $\sigma = +3$ and $\tau = -4.5$, give point H on the graph. Since the planes on which the stresses act are 90° apart in the plate, the points showing the state of stress on these planes will be 180° apart on the σ-τ diagram. Consequently, points V and H are on opposite ends of a diameter of the Mohr's circle. We can then draw the diameter and find the center of the circle, which is at the midpoint of the diameter. Note that because the shearing stresses on planes at right angles are always equal in magnitude but opposite in sign, the midpoint of the diameter will always fall on the σ-axis and *the center of Mohr's circle will always be on the σ-axis*. Also, the center will always be at

$$\sigma = \frac{\sigma_x + \sigma_y}{2}$$

The circle can then be drawn by finding its center and radius as shown in Fig. 5-6c.

From the circle we can see that no value of σ can exceed $(9.75 + 6.91 =)$ 16.66 ksi, which occurs at point P on the Mohr's circle. Such a stress is referred to as the *maximum principal stress* and is represented by σ_1. Note that from the nature of the circle the maximum principal stress must act on a plane that has zero shear stress. The minimum value of σ at Q is referred to as the *minimum principal stress*, σ_2, and again acts on a plane of zero shear.

To find the direction of the plane on which σ_1 acts, we need to note that P is 40.6° clockwise from V on the Mohr's circle. This means that in the plate the plane on which σ_1 acts is $(40.6/2 =)$ 20.3° clockwise from V as shown in Fig. 5-6d.

The minimum principal stress, σ_2, is $(9.75 - 6.91 =)$ 2.84 ksi. Points Q and P are at opposite ends of a diameter and consequently are 180° apart on the circle. Hence, σ_2 acts on a plane at 90° to the plane of σ_1 as indicated in Fig. 5-6e.

The maximum value the shear stress can have is given by τ at the top of Mohr's circle, point R; then τ_{max} is 6.91 ksi. Point R is 90° on the circle from point P, therefore, the plane of maximum shear stress is at $(90/2 =)$ 45° to the plane of the maximum principal stress as shown in Fig. 5-6f.

On the negative side of the circle the largest negative shearing stress will be equal in magnitude to the maximum shearing stress but opposite in sign and, in the plate, 90° away. This is consistent with the relationships between shearing stresses that were established earlier.

Problems 5-3 to 5-10

With our understanding of Mohr's circle it is now possible to account for the peculiar nature of the web failures shown in Fig. 2-15c and d. In both cases the web can be taken as carrying a vertical shearing load and hence to be subjected to shearing stress on a vertical plane. Let us consider the left end of the members only and say that the vertical shear stress is S. There will also be normal stress on the vertical plane due to bending, but if we take a point on the neutral axis, the bending stress will be zero and the analysis will be much simplified while the conclusions will still be valid. As we have observed before, and as shown in Fig. 5-7a, there is a horizontal shear stress as well as the vertical shear stress. Fig. 5-7b shows Mohr's circle for this state of stress. We see that the

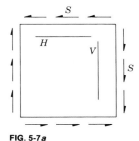

FIG. 5-7a

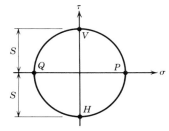

FIG. 5-7b

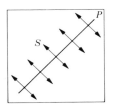

FIG. 5-7c Diagonal tension.

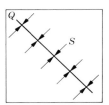

FIG. 5-7d Diagonal compression.

principal stresses are S and $-S$ at points P and Q and these act on the planes indicated in Fig. 5-7c and d.

Considering the concrete member of Fig. 2-15d, we see that although the load is a vertical shear load, tension on an inclined plane as shown in Fig. 5-7c is present. This stress is usually referred to as *diagonal tension*. Since concrete is weak in tension, it is in danger of failing by cracking on diagonal planes and will fail in this manner at quite low stress values. The concrete can be strengthened by placing diagonal bars or "shear reinforcement" in the direction of the tensile stress. The concrete is also subjected to *diagonal compression*, as indicated in Fig. 5-7d, but since it is a material that is strong in compression, this stress rarely leads to failure and hence does not usually govern the design.

The web of the plate girder of Fig. 2-15c is also subjected to the same stress patterns of Fig. 5-7c and d. In this case the material is strong in tension, and failure due to diagonal tension is unlikely. However, the web plate is usually thin and the diagonal compression along line M-L in Fig. 2-15c may cause the plate to fail by buckling. Webs that do fail by excessive shear load display the ripple pattern of Fig. 2-15c, which indicates a buckling failure resulting from excessive compression on the diagonal.

Problems 5-11 to 5-14

As well as being useful for determining principal stresses and maximum shear stress, Mohr's circle enables us to determine the stress on any given inclined plane. The stress on such a plane may be of interest when there is an inclined plane of weakness resulting from a weld, a flaw, wood grain, or another source of directional weakness. The relevant stresses are determined by the method illustrated in the following example.

Example 5-4

At a point in a plate the stresses are

$$\sigma_x = 100 \text{ MPa}$$

$$\sigma_y = 0$$

$$\tau_{xy} = -60 \text{ MPa}$$

A weld runs across the plate at $\theta = 45°$ (see Fig. 5-2a). The allowable tensile stress in the plate is 130 MPa and the weld has an efficiency of 75% with respect to tensile stress.

    a. Disregarding the weld, is the plate safe?

    b. Is the weld safe?

    c. If not, what weld efficiency would be required to make it safe?

Solution

a. Mohr's circle for the given state of stress is shown in Fig. 5-8*a*. Since the maximum principal stress, σ_1, does not exceed the allowable tensile stress of 130 MPa, **the plate is safe** at points away from the weld.

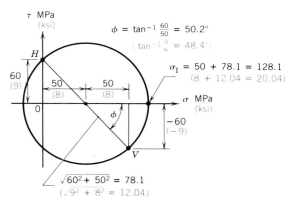

$$\phi = \tan^{-1}\frac{60}{50} = 50.2°$$

$$(\tan^{-1}\frac{9}{8} = 48.4°)$$

$$\sigma_1 = 50 + 78.1 = 128.1$$

$$(8 + 12.04 = 20.04)$$

$$\sqrt{60^2 + 50^2} = 78.1$$

$$(\sqrt{9^2 + 8^2} = 12.04)$$

FIG. 5-8*a*

b. The state of stress on the plane of the weld is given by the coordinates of *W* in Fig. 5-8*b*.

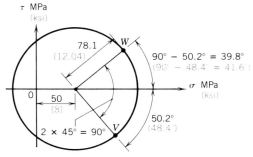

$$90° - 50.2° = 39.8°$$

$$(90° - 48.4° = 41.6°)$$

$$2 \times 45° = 90°$$

$$50.2°$$

$$(48.4°)$$

FIG. 5-8*b*

normal stress on weld, $\sigma_w = 50 + 78.1 \cos 39.8°$

$$= 50 + 60.0 = 110 \text{ MPa}$$

allowable normal stress on weld $= \dfrac{75}{100} \times 130 = 97.5 \text{ MPa}$

This means that **the weld is not safe.**

c. Required weld efficiency $= \frac{110}{130} \times 100 = $ **84.6%.**

Example 5-4

At a point in a plate the stresses are

$$\sigma_x = 16 \text{ kip/in.}^2$$

$$\sigma_y = 0$$

$$\tau_{xy} = -9 \text{ kip/in.}^2$$

A weld runs across the plate at $\theta = 45°$ (see Fig. 5-2a). The allowable tensile stress in the plate is 22 ksi and the weld has an efficiency of 75% with respect to tensile stress.

    a. Disregarding the weld, is the plate safe?
    b. Is the weld safe?
    c. If not, what weld efficiency would be required to make it safe?

Solution

    a. Mohr's circle for the given state of stress is shown in Fig. 5-8a. Since the maximum principal stress, σ_1, does not exceed the allowable tensile stress of 22 ksi, **the plate is safe** at points away from the weld.
    b. The state of stress on the plane of the weld is given by the coordinates of W in Fig. 5-8b.

$$\text{normal stress on weld, } \sigma_w = 8 + 12.04 \cos 41.6°$$

$$= 8 + 9 = 17 \text{ kip/in.}^2$$

$$\text{allowable normal stress on weld} = \frac{75}{100} \times 22 = 16.5 \text{ kip/in.}^2$$

This means that **the weld is not safe.**
    c. Required weld efficiency $= \frac{17}{22} \times 100 = $ **77.3%.**

Problems 5-15 to 5-18

5-4 MOHR'S CIRCLE AND PRINCIPAL STRESS EQUATIONS

The stresses on an inclined plane can be found from (5-1) and (5-2), which are based on a rigorous analysis, or we can use the more convenient Mohr's circle. Mohr's circle was drawn on the basis of observations made in Example 5-2 and must be proved before it can be used with confidence. We will provide a proof by deriving equations from the circle that can be turned into (5-1) and (5-2). In the process we will also derive convenient equations for the principal stresses. Referring to Fig. 5-9, if the point V on Mohr's circle represents the state of stress on the vertical plane, the stress on any arbitrary plane making an angle θ with the vertical plane is given by the coordinates of point P. From Mohr's circle:

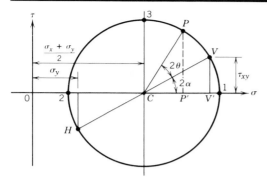

FIG. 5-9 Mohr's circle of stress.

$$\sigma_\theta = OC + CP'$$

$$= OC + CP \cos(2\alpha + 2\theta)$$

$$= OC + CV \cos(2\alpha + 2\theta)$$

$$= OC + CV \cos 2\alpha \cos 2\theta - CV \sin 2\alpha \sin 2\theta$$

But

$$CV' = CV \cos 2\alpha \qquad \text{and} \qquad VV' = CV \sin 2\alpha$$

Hence,

$$\sigma_\theta = OC + CV' \cos 2\theta - VV' \sin 2\theta$$

Substituting

$$OC = \frac{\sigma_x + \sigma_y}{2}$$

$$CV' = \frac{\sigma_x - \sigma_y}{2}$$

and

$$VV' = \tau_{xy}$$

gives

$$\sigma_\theta = \frac{\sigma_x + \sigma_y}{2} + \frac{\sigma_x - \sigma_y}{2} \cos 2\theta - \tau_{xy} \sin 2\theta$$

$$= \frac{\sigma_x + \sigma_y}{2} + \frac{\sigma_x - \sigma_y}{2}(2 \cos^2 \theta - 1) - 2 \sin \theta \cos \theta \, \tau_{xy}$$

$$= \cos^2 \theta \, \sigma_x + \sin^2 \theta \, \sigma_y - 2 \sin \theta \cos \theta \, \tau_{xy}$$

Note that this equation and (5-1) are identical. From the circle we see that:

$$\tau_\theta = PP_1 = CP \sin (2\alpha + 2\theta)$$

$$= CV \sin (2\alpha + 2\theta)$$

$$= CV \cos 2\alpha \sin 2\theta + CV \sin 2\alpha \cos 2\theta$$

$$= CV' \sin 2\theta + VV' \cos 2\theta$$

$$= \frac{\sigma_x - \sigma_y}{2} 2 \sin \theta \cos \theta + \tau_{xy}(\cos^2 \theta - \sin^2 \theta)$$

$$= \sin \theta \cos \theta (\sigma_x - \sigma_y) + (\cos^2 \theta - \sin^2 \theta) \tau_{xy}$$

We see that this equation and (5-2) are identical. We may therefore conclude that *Mohr's circle is a valid representation of stress on all planes through a stressed point*.

In Fig. 5-9, points 1 and 2 represent the principal stresses which may be determined from (5-1) provided the angle α is known. A more direct approach is found by considering Mohr's circle as follows:

$$\sigma_1 = OC + C1 = OC + CV$$

$$= OC + \sqrt{(CV')^2 + (VV')^2}$$

$$= \frac{\sigma_x + \sigma_y}{2} + \sqrt{\left(\frac{\sigma_x - \sigma_y}{2}\right)^2 + (\tau_{xy})^2}$$

$$\sigma_2 = OC - C2 = OC - CV$$

$$= \frac{\sigma_x + \sigma_y}{2} - \sqrt{\left(\frac{\sigma_x - \sigma_y}{2}\right)^2 + \tau_{xy}^2}$$

These two equations for the principal stresses can be combined to obtain

$$\boxed{\sigma_1, \sigma_2 = \frac{\sigma_x + \sigma_y}{2} \pm \sqrt{\left(\frac{\sigma_x - \sigma_y}{2}\right)^2 + \tau_{xy}^2}} \tag{5-3}$$

The maximum shear stress is found from Mohr's circle as

$$\tau_{max} = C3 = CV$$

$$\boxed{\tau_{max} = \sqrt{\left(\frac{\sigma_x - \sigma_y}{2}\right)^2 + \tau_{xy}^2}} \tag{5-4}$$

The direction of the maximum principal stress measured counterclockwise from the x-axis is given by

$$\boxed{\alpha = -\frac{1}{2} \tan^{-1}\left(\frac{\tau_{xy}}{\dfrac{\sigma_x - \sigma_y}{2}}\right)} \tag{5-5}$$

Substitution into (5-3) and (5-4) is always straightforward. However, care must be taken in using (5-5) to avoid erroneous results. The difficulty arises from the fact that for angles in the range 0 to 360°, there are two values of the antitangent corresponding to any given argument. The values differ by 180°, and hence there are always two angles given by (5-5), which differ by 90°. The ambiguity can be resolved by following certain rules, but the safest procedure is to sketch Mohr's circle and choose the angle that is given by the equation and is consistent with the sketch. The use of the equations is demonstrated in the following example.

Example 5-5

For the given stresses, determine the principal stresses, the maximum shear stress, and the direction of the maximum principal stress.

$$\sigma_x = 200 \text{ MPa}$$

$$\sigma_y = 50 \text{ MPa}$$

$$\tau_{xy} = 40 \text{ MPa}$$

Solution.   Substituting into (5-3) gives

$$\sigma_1, \sigma_2 = \frac{200 + 50}{2} \pm \sqrt{\left(\frac{200 - 50}{2}\right)^2 + 40^2}$$

$$= 125 \pm 85$$

$$\sigma_1 = \textbf{210 MPa}$$

$$\sigma_2 = \textbf{40 MPa}$$

into (5-4),

$$\tau_{max} = \sqrt{\left(\frac{200 - 50}{2}\right)^2 + 40^2} = \textbf{85 MPa}$$

and into (5-5),

$$\alpha = -\frac{1}{2} \tan^{-1}\left(\frac{40}{\dfrac{200 - 50}{2}}\right)$$

$$= -\frac{1}{2} \tan^{-1} 0.533 = -\frac{1}{2}(28.1°) \quad \text{or} \quad -\frac{1}{2}(208.1°)$$

$$= -14° \quad \text{or} \quad -104°$$

Upon sketching Mohr's circle as shown in Fig. 5-10a, it is evident that **−14°** is the correct answer.

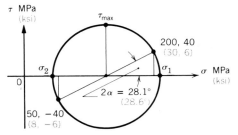

FIG. 5-10a

Example 5-5

For the given stresses, determine the principal stresses, the maximum shear stress, and the direction of the maximum principal stress.

$$\sigma_x = 30 \text{ kip/in.}^2$$

$$\sigma_y = 8 \text{ kip/in.}^2$$

$$\tau_{xy} = 6 \text{ kip/in.}^2$$

Solution.   Substituting into (5-3) gives

$$\sigma_1, \sigma_2 = \frac{30 + 8}{2} \pm \sqrt{\left(\frac{30 - 8}{2}\right)^2 + 6^2}$$

$$= 19 \pm 12.5$$

$$\sigma_1 = \textbf{31.5 kip/in.}^2$$

$$\sigma_2 = \textbf{6.5 kip/in.}^2$$

into (5-4),

$$\tau_{max} = \sqrt{\left(\frac{30 - 8}{2}\right)^2 + 6^2} = \textbf{12.5 kip/in.}^2$$

and into (5-5),

$$\alpha = -\frac{1}{2} \tan^{-1}\left(\frac{6}{\dfrac{30 - 8}{2}}\right)$$

$$= -\frac{1}{2} \tan^{-1} 0.545 = -\frac{1}{2}(28.6°) \quad \text{or} \quad -\frac{1}{2}(208.6°)$$

$$= -14° \quad \text{or} \quad -104°$$

Upon sketching Mohr's circle as shown in Fig. 5-10a, it is evident that **−14°** is the correct answer.

Note that the other angle applies to the minimum principal stress. This is always the case, so the purpose of the sketch is to sort out which stress is associated with each angle. Be careful not to draw some conclusions from this example that may lead you to believe that the sketch is unnecessary. There is a good chance that your conclusions will be oversimplified and will lead to errors under certain conditions.

Figures 5-10b and c clearly show the planes on which the principal stresses and the maximum shear stress act. The two sets of sketches show the same information to acquaint you with two methods of presentation. Figure 5-10b follows the presentation of Example 5-3 whereas Fig. 5-10c can be thought of as an element in a biaxially stressed plate that is rotated through various angles to indicate the critical planes on which the principal stresses and maximum shear stress act.

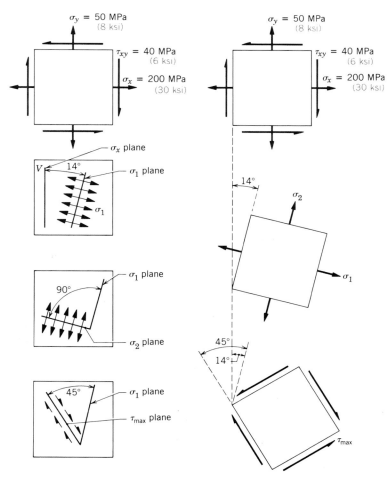

FIG. 5-10b   FIG. 5-10c

Problem 5-19

5-5 STRESSES IN THREE DIMENSIONS

Up to this point we have considered stresses in the *x-y* plane only and have ignored the fact that all engineering bodies have, in fact, a third dimension. A two-dimensional approach is adequate for the greater part of all engineering design but may lead to erroneous results in some cases. Even in the case of a simple plate, which seems to be an obvious two-dimensional case, it is sometimes essential that the third dimension be taken into account.

Let us start our look at three-dimensional or *triaxial stresses* by considering a solid, as in Fig. 5-11, that is loaded in such a way that there are stresses in the *x-y* plane only.

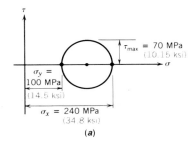

(a)

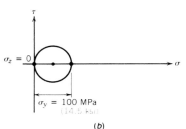

(b)

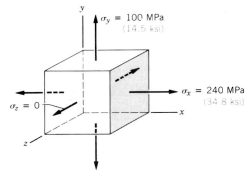

FIG. 5-11

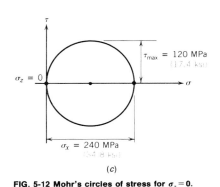

(c)

FIG. 5-12 Mohr's circles of stress for $\sigma_z = 0$.
(a) Mohr's circle of stress in the *x-y* plane.
(b) Mohr's circle of stress in the *y-z* plane.
(c) Mohr's circle of stress in the *z-x* plane.

We may feel that the whole stress picture has been presented when we draw Mohr's circle for stresses in the *x-y* plane, as in Fig. 5-12*a*. To see if anything significant has been omitted, let us consider stresses in the other coordinate planes, *y-z* and *z-x*, as shown in Fig. 5-12*b* and *c*.

Suppose that maximum shear stress is of particular interest. If we limited our attention to the *x-y* plane, we would conclude that the shear stress does not exceed 70 MPa. However, an examination of the *z-x* plane reveals a shear stress of 120 MPa. If the material is such that it can be stressed safely to 100 MPa in shear, then investigating the stresses in the *x-y* plane only would lead us to conclude that the system is safe although, in fact, the material is overloaded and would be in danger of failing by shear on the diagonal plane shown in Fig. 5-13. If, as a result of all the loads being in the *x-y* plane, we had focused our attention on that plane only, we would have made a serious error. This case points out the need to keep all three dimensions in mind even when all the action seems to be two-dimensional.

The circles in Fig. 5-12 were drawn in three separate diagrams to facilitate understanding. The usual practice is to draw all circles in a

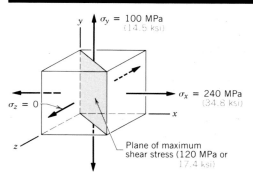

FIG. 5-13

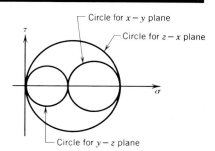

FIG. 5-14

single diagram such as Fig. 5-14. The stresses depicted in Fig. 5-14 are special in that one stress, σ_z, is zero. In general, Mohr's circles are more likely to be as shown in Fig. 5-15. However, these circles give only a limited view of stress in that they show stresses on various planes, but the planes are restricted to those that have their normal in the x-y plane, the y-z plane, or the z-x plane.

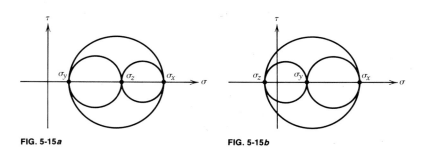

FIG. 5-15a

FIG. 5-15b

Furthermore, the given stresses were far from general because there were no shear stresses. If there had been shear stresses, the circles would not have been nested in the way that we have seen in the figures and they would have been impossible to draw. In cases where shears actually exist on the coordinate planes, the three principal stresses and their directions are determined by methods that are beyond the scope of this book. The principle directions are at right angles to one another, and these directions are used as the directions of a new set of axes. The principal axes are numbered 1, 2, and 3 so that the 1-axis is parallel to the largest principal stress, σ_1, and the 3-axis is parallel to the smallest principal stress, σ_3. *As in the two-dimensional analysis, there is no shear on the planes on which the three principal stresses act.* Consequently, a meaningful set of Mohr's circles for the planes containing the principal axes can be drawn and will be of the form of Fig. 5-16a. A view of the body stressed along the three principal axes is shown in Fig. 5-16b. Fig. 5-16a shows one negative principal stress, but it is possible to have any number of the principal stresses negative or zero depending on the nature of the loading.

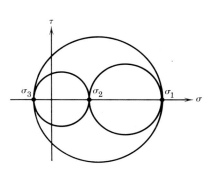

FIG. 5-16a

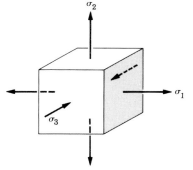

FIG. 5-16b

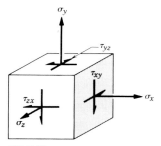

FIG. 5-17a

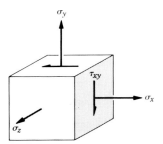

FIG. 5-17b

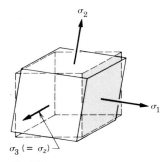

FIG. 5-17c

It is recognized that a general system of three-dimensional stresses, as shown in Fig. 5-17a, is difficult to analyze and is usually handled by a specialist. However, problems do arise in engineering design that are three-dimensional, and many are sufficiently straightforward so that the nonspecialist can deal with them. Without additional training we cannot treat cases that are more complex than that shown in Fig. 5-17b. In that case we change the stresses in the x-y plane into principal stresses, as shown in Fig. 5-17c. Part of the art of engineering is recognizing problems that are beyond your training and hence require the attention of a specialist.

Problems 5-20 to 5-24

5-6 FAILURE THEORIES

In the design of a member subjected to a uniaxial load, the stress was compared with the stress to cause failure in test specimens that had also been subjected to uniaxial load. This is the simplest of all design problems; the method is quite adequate, since the nature of the loads and the stresses in the test and in the part being designed are identical. However, we soon encounter cases where the member being designed is not so simple and the stresses are not uniaxial; consider, for example, the stresses in the web of a beam or in a pressure vessel. In these cases we know that the stress is two-dimensional or biaxial and it may, in other cases, be three-dimensional, or triaxial. For a structure having biaxial or triaxial stresses, how should we check the safety of the design? The most obvious way would be to conduct tests in which specimens are stressed to failure in the same multiaxial manner as in the structure; the allowable multiaxial stress can then be determined by the application of an adequate safety factor. However, this would require a group of tests for every new set of multiaxial stresses that occurred in design. Such tests are difficult to perform, and the cost of performing them in the required numbers would be prohibitive. Consequently, we need a theory by which the results of the standard uniaxial test can be used to predict the failure of a part made of the same material when the stresses are multiaxial. In other words, we need a *failure theory*.

To illustrate the need for a failure theory, let us consider a cylindrical pressure vessel. To avoid unnecessary complications, we will consider that all welds are 100% efficient and that the walls are thin. Under internal pressure the main stresses are circumferential and longitudinal, and it was implied in the problems in Chapter 1 that only the circumferential stress, because it is larger than the longitudinal stress, needs to be considered in judging the adequacy of the design. In this approach we tacitly assumed that the maximum stress could be treated as a uniaxial stress and that it alone determined the safety of the design. The longitudinal stress was not considered although it may, without our knowledge, have had an influence on strength. It happens that our approach in this case is acceptable, but, in a biaxial state of stress, the second stress is not always inconsequential and an understanding of failure theory is necessary in order to avoid making some serious errors.

Unfortunately, as we will discover, no single theory will be found to apply in all cases; for example, theories that are satisfactory for *ductile* materials are not acceptable for *brittle* materials. We will also find that one of the best theories is too complex for everyday use and that most designers prefer a simpler theory that introduces a small but safeside error.

In developing the various failure theories, we cannot avoid three-dimensional effects, but we will treat only those cases in which one of the stresses is zero, thus avoiding complications that would tend to obscure the important part of the theories. This is not a serious limitation, since in engineering practice most problems are reduced to the biaxial stress state for design. When shear stresses occur along with normal stresses, the principal stresses are determined. Thus, for practical purposes, we need to consider failure in a material subjected to two nonzero normal stresses while the third normal stress is zero. For ease in designating those principal stresses we will use numerical subscripts; σ_1 and σ_2 being the nonzero stresses and σ_3 being zero. *In this section the numbers do not indicate the relative sizes as they did in the previous section.*

We cannot discuss failure theory until we have defined failure. We might take the obvious definition that a material has failed when it has broken into two or more parts. However, it has already been pointed out that in most applications a member would be unserviceable due to excessive distortion long before it actually ruptured. Consequently, we will relate failure to yielding and consider that a material has failed when it will no longer return to its original shape upon release of the loads. In a simple tensile test we would then say that a *ductile* material has failed when the material begins to yield. Then for uniaxial stress, failure occurs when the stress reaches the yield stress, σ_y, in either tension or compression.

Brittle materials fail by a different mechanism and will be discussed after the theories for ductile materials have been presented.

Note: We have used the symbol σ_y to represent the yield stress and also the normal stress in the *y*-direction. Generally, the symbol does not appear in the same problem representing both quantities; however, failure theory is exceptional in this respect. The adoption of a new symbol was considered but rejected because the present usage is that found in engineering practice. Consequently, caution must be exercised to avoid using the wrong value when σ_y has two meanings within a single problem.

Maximum Principal Stress Theory

This theory predicts that failure will occur when the maximum principal stress in either tension or compression reaches the level of the maximum principal tensile stress in the test specimen at failure. The test specimen being subjected to a *uniaxial* load has a maximum principal tensile stress at failure that is simply σ_y. Expressed mathematically, failure will occur when either

$$\sigma_1 = \pm \sigma_y \qquad \text{or} \qquad \sigma_2 = \pm \sigma_y$$

These failure equations can be rearranged as

$$|\sigma_1| = \sigma_y \qquad |\sigma_2| = \sigma_y$$

or as

$$\boxed{\left|\frac{\sigma_1}{\sigma_y}\right| = 1 \qquad\qquad \left|\frac{\sigma_2}{\sigma_y}\right| = 1} \qquad\qquad \textbf{(5-6a)}$$

For the design to be *safe* both

$$\left|\frac{\sigma_1}{\sigma_y}\right| < 1 \qquad \text{and} \qquad \left|\frac{\sigma_2}{\sigma_y}\right| < 1 \qquad\qquad \textbf{(5-6b)}$$

must be satisfied. The conditions for safe design are shown graphically in Fig. 5-18.

The four lines show the stresses given by (5-6a), while the region inside the square represents stresses that satisfy the inequalities (5-6b) and hence represents safe stresses. We will refer to this region as the *safe region* and the boundary around the area as the *theoretical envelope*. Values of σ_1 and σ_2 that give points inside the square are safe to varying degrees: The further a point is from the envelope, the larger is the safety factor. Values of σ_1 and σ_2 that give points outside the square are, according to the maximum principal stress theory, unacceptable because the material has failed.

In our studies of failure theories, we will not be concerned with the safety factor but merely with the envelope that forms the boundary between the safe region and the failed region. For each theory we will compare the theoretical envelope with the results of biaxial stress tests. The comparison will be made by plotting σ_1/σ_y and σ_2/σ_y at failure for a series of imaginary biaxial tests. Figure 5-19 shows the theoretical envelope and experimental points that should, according to maximum principal stress theory, have fallen on the envelope.

In any experimental work we expect some scatter in the results due to variation in the material from one specimen to the next and we are accustomed to drawing averaging lines through groups of scattered points. Such an average line for the test results in Fig. 5-19 would be roughly elliptical in shape and does not agree well with the theoretical envelope. For example, stresses corresponding to point A would be unsafe according to the maximum principal stress theory but are shown to be safe by the experimental results. The theory is in error in the first and third quadrants, but here it errs on the safe side by indicating that certain combinations of stress will cause failure when according to the tests they are safe.

At a point such as B the stresses are within the theoretical envelope and therefore are declared safe by the theory although the experiments reveal that they would be unsafe. It is evident that the theory errs on the

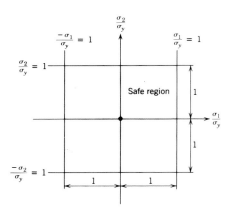

FIG. 5-18 Theoretical failure envelope and safe region for maximum principal stress theory.

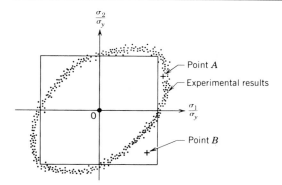

FIG. 5-19 Comparison of experimental results with maximum principal stress theory.

unsafe side in rather large areas in both the second and fourth quadrants; it therefore is unacceptable. Consequently, the maximum principal stress theory does not closely represent the failure behavior of ductile materials and is therefore seldom applied to this class of materials.

Example 5-6

A machine part, made of AISI 1020, is subjected to a load that gives principal stresses; $\sigma_3 = 0$ and $\sigma_2 = 210$ MPa. Determine the third principal stress, σ_1, that will cause failure according to the maximum principal stress theory.

Solution. From Table A, for AISI 1020,

$$\sigma_y = 250 \text{ MPa}$$

Substituting into the first of equation (5-6a), we obtain

$$\left| \frac{\sigma_1}{250} \right| = 1$$

$$|\sigma_1| = 250$$

At failure

$$\sigma_1 = \textbf{250 MPa tension}$$

or

$$\sigma_1 = \textbf{250 MPa compression}$$

Example 5-6

A machine part, made of AISI 1020, is subjected to a load that gives principal stresses; $\sigma_3 = 0$ and $\sigma_2 = 30$ kip/in.^{2}. Determine the third principal stress, σ_1, that will cause failure according to the maximum principal stress theory.

Solution. From Table A, for AISI 1020,

$$\sigma_y = 36 \text{ ksi}$$

Substituting into the first of equation (5-6a), we obtain

$$\left|\frac{\sigma_1}{36}\right| = 1$$

$$|\sigma_1| = 36$$

At failure

$$\sigma_1 = \textbf{36 kip/in.}^2 \textbf{ tension}$$

or

$$\sigma_1 = \textbf{36 kip/in.}^2 \textbf{ compression}$$

Problems 5-25 to 5-26

Maximum Principal Strain Theory

Since the failure theory based on maximum principal stress gives an envelope that is of the same general shape as the experimental envelope, it might seem reasonable to expect the closely related, maximum principal strain to form the basis of an acceptable theory.

To develop the equation for a maximum principal strain theory, we need to relate strain to stress under stress conditions that have not yet been treated. Up to this point we know that stress and strain are related by

$$\varepsilon = \frac{\sigma}{E} \qquad\qquad \textbf{(1-12b)}$$

But this applies only in a *uniaxial* stress system and cannot be used in the present case where there are two nonzero stresses, σ_1 and σ_2. We will see in Chapter 6 that (6-8) gives the strain, in the direction of principal stress, σ_1, as

$$\varepsilon_1 = \frac{1}{E}(\sigma_1 - v\sigma_2)$$

Note that stress and strain are related, as expected, through E. However, another constant, v, appears in the equation. It is also an elastic constant, Poisson's ratio, and will be discussed at length in Chapter 6.

In the test specimen with uniaxial stress, (1-12b) applies and at failure

$$\varepsilon_y = \frac{1}{E}\sigma_y$$

The maximum principal strain theory postulates that failure occurs when ε_1 in the *biaxial* stress system reaches the level of ε_y in the test specimen, that is, when

$$\frac{1}{E}(\sigma_1 - v\sigma_2) = \frac{1}{E}\sigma_y \qquad \text{or} \qquad \sigma_1 - v\sigma_2 = \sigma_y$$

If the same failure criterion applies in both tension and compression, failure occurs when

$$\sigma_1 - v\sigma_2 = \pm\sigma_y$$

Failure due to the limiting strain in the direction of σ_2 occurs when

$$\sigma_2 - v\sigma_1 = \pm\sigma_y$$

These equations in dimensionless form become

$$\boxed{\frac{\sigma_1}{\sigma_y} - v\frac{\sigma_2}{\sigma_y} = \pm 1} \qquad \boxed{\frac{\sigma_2}{\sigma_y} - v\frac{\sigma_1}{\sigma_y} = \pm 1} \qquad \textbf{(5-7a)}$$

For a design to be *safe*, the stresses must be such that the left side of both equations lies in the range -1 to $+1$; hence both

$$\left|\frac{\sigma_1}{\sigma_y} - v\frac{\sigma_2}{\sigma_y}\right| < 1 \qquad \text{and} \qquad \left|\frac{\sigma_2}{\sigma_y} - v\frac{\sigma_1}{\sigma_y}\right| < 1 \qquad \textbf{(5-7b)}$$

must be satisfied.

The theoretical envelope for the maximum principal strain theory is plotted in Fig. 5-20 for a Poisson's ratio of 0.3. It is evident that the general shape of the theoretical envelope matches more nearly that of the test results, and the regions in the second and fourth quadrants, where the theory errs on the unsafe side, are somewhat diminished but not entirely eliminated. However, new unsafe regions appear in the first and the third quadrants where previous theory erred on the safe side.

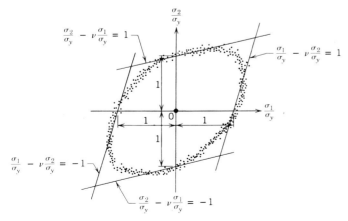

FIG. 5-20 Theoretical failure envelope for maximum principal strain theory as compared to experimental evidence.

It is evident that the maximum principal strain theory has no more merit than the maximum principal stress theory and is more complicated; consequently, it is rarely used in practice.

Example 5-7

Determine the stress to cause failure in Example 5-6 according to the maximum principal strain theory.

Solution

$$\frac{\sigma_2}{\sigma_y} = \frac{210}{250} = 0.840$$

$$v = 0.3 \qquad \text{(Table A)}$$

From Fig. 5-20 it is evident that failure is indicated by a point on the line

$$\frac{\sigma_1}{\sigma_y} - v\frac{\sigma_2}{\sigma_y} = 1$$

and another point on the line

$$\frac{\sigma_2}{\sigma_y} - v\frac{\sigma_1}{\sigma_y} = 1$$

Substituting into these equations gives

$$\frac{\sigma_1}{\sigma_y} = 1 + v\frac{\sigma_2}{\sigma_y} = 1 + 0.3 \times 0.84 = 1.252$$

$$\sigma_1 = 1.252 \times \sigma_y = 1.252 \times 250 = \textbf{313 MPa tension}$$

and

$$\frac{\sigma_1}{\sigma_y} = \frac{1}{v}\left(-1 + \frac{\sigma_2}{\sigma_y}\right) = \frac{1}{0.3}(-1 + 0.84) = -0.533$$

$$\sigma_1 = -0.533 \times \sigma_y = -0.533 \times 250 = -133 \text{ MPa} = \textbf{133 MPa compression}$$

Example 5-7

Determine the stress to cause failure in Example 5-6 according to the maximum principal strain theory.

Solution

$$\frac{\sigma_2}{\sigma_y} = \frac{30}{36} = 0.83$$

$$v = 0.3 \qquad \text{(Table A)}$$

From Fig. 5-20 it is evident that failure is indicated by a point on the line

$$\frac{\sigma_1}{\sigma_y} - v\frac{\sigma_2}{\sigma_y} = 1$$

and another point on the line

$$\frac{\sigma_2}{\sigma_y} - v\frac{\sigma_1}{\sigma_y} = 1$$

Substituting into these equations gives

$$\frac{\sigma_1}{\sigma_y} = 1 + v\frac{\sigma_2}{\sigma_y} = 1 + 0.3 \times 0.83 = 1.249$$

$$\sigma_1 = 1.249 \times \sigma_y = 1.249 \times 36 = \textbf{45 kip/in.}^2 \textbf{ tension}$$

and

$$\frac{\sigma_1}{\sigma_y} = \frac{1}{v}\left(-1 + \frac{\sigma_2}{\sigma_y}\right) = \frac{1}{0.3}(-1 + 0.83) = -0.567$$
$$\sigma_1 = -0.567 \times \sigma_y = -0.567 \times 36 = -20 \text{ kip/in.}^2 = \textbf{20 kip/in.}^2 \textbf{ compression}$$

Problems 5-27 to 5-28

Maximum Shear Stress Theory

We have already observed that *uniaxial* test specimens of ductile material fail in shear on planes at 45° to the axis of the applied stress; this means that they fail in shear at a stress that is half the yield stress:

$$\tau_y = \tfrac{1}{2}\sigma_y$$

This can be seen in Fig. 5-21, which shows the Mohr's circle for a uniaxial test specimen stressed to yield.

In a general triaxial stress system there are six components of stress: three normal and three shear stresses. Such a system can, by methods that are beyond our range of interest, always be reduced to three principal stresses σ_1, σ_2, and σ_3. The directions of these stresses are at right angles to one another in space just as the two principal stresses are at right angles in a plane.

In two dimensions the maximum shear stress was found to be one-half the difference between the principal stresses. However, in three dimensions the maximum is not always determined by the same combination of principal stresses. The combination for maximum shear–stress depends on the relative sizes of σ_1, σ_2, and σ_3 and may lead to any one of the six formulas shown in Fig. 5-22.

According to the maximum shear stress theory, failure occurs when the maximum shear stress in the triaxial stress system reaches the level of the maximum shear stress in the test specimen at failure. Failure occurs when τ_{max} as given in Fig. 5-22 is equal to τ_y in Fig. 5-21, that is, when

$$\tau_{max} = \frac{\sigma_y}{2} \tag{5-8}$$

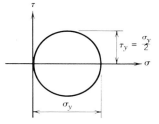

FIG. 5-21

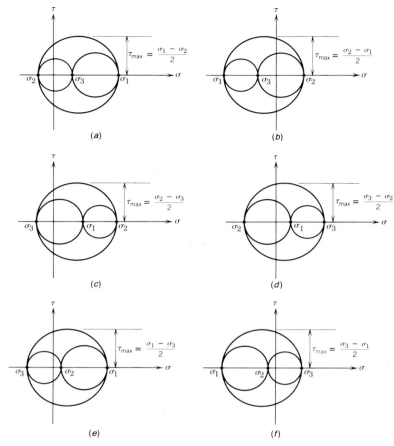

FIG. 5-22 Determination of τ_{max} by means of Mohr's stress circles for the case of three principal stresses.

For the relative sizes in Fig. 5-22a, failure occurs when

$$\frac{\sigma_1 - \sigma_2}{2} = \frac{\sigma_y}{2}$$

which gives

$$\sigma_1 - \sigma_2 = \sigma_y \qquad \textbf{(5-9a)}$$

Treating the maximum shear stress formulas from the other diagrams in Fig. 5-22 in a similar manner gives

$$\sigma_2 - \sigma_1 = \sigma_y \qquad \textbf{(5-9b)}$$

$$\sigma_2 - \sigma_3 = \sigma_y \qquad \textbf{(5-9c)}$$

$$\sigma_3 - \sigma_2 = \sigma_y \qquad \textbf{(5-9d)}$$

$$\sigma_1 - \sigma_3 = \sigma_y \qquad \text{(5-9e)}$$

$$\sigma_3 - \sigma_1 = \sigma_y \qquad \text{(5-9f)}$$

It is interesting to note that in the form given in (5-9) there is nothing to indicate that shear plays a part in the failure criteria. In fact, (5-9) could be stated as predicting that failure will occur whenever the maximum principal stress *difference* is equal to the tensile yield stress under uniaxial loading. Consequently, this theory is sometimes referred to as the maximum stress difference theory.

We will deal with the plane stress only. That is the case where $\sigma_3 = 0$. Substituting this value into (5-9) and rearranging, we get

$$\frac{\sigma_1}{\sigma_y} - \frac{\sigma_2}{\sigma_y} = 1 \qquad \text{(5-10a)}$$

$$\frac{\sigma_1}{\sigma_y} - \frac{\sigma_2}{\sigma_y} = -1 \qquad \text{(5-10b)}$$

$$\frac{\sigma_2}{\sigma_y} = 1 \qquad \text{(5-10c)}$$

$$\frac{\sigma_2}{\sigma_y} = -1 \qquad \text{(5-10d)}$$

$$\frac{\sigma_1}{\sigma_y} = 1 \qquad \text{(5-10e)}$$

$$\frac{\sigma_1}{\sigma_y} = -1 \qquad \text{(5-10f)}$$

These equations when plotted give six lines that form the theoretical envelope around the safe region as shown in Fig. 5-23.

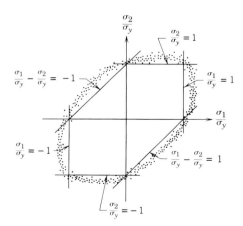

FIG. 5-23 Comparison of experimental results with theoretical failure envelope for maximum shear stress theory.

To be in the *safe* region the stresses must satisfy all of the following inequalities

$$\left| \frac{\sigma_1}{\sigma_y} - \frac{\sigma_2}{\sigma_y} \right| < 1 \qquad \text{(5-11a)}$$

$$\left| \frac{\sigma_2}{\sigma_y} \right| < 1 \qquad \text{(5-11b)}$$

$$\left| \frac{\sigma_1}{\sigma_y} \right| < 1 \qquad \text{(5-11c)}$$

It is evident from a comparison of the theoretical envelope with the test results that there is good agreement between theory and experiment. An average curve drawn through the experimental points would not encroach on the region that is safe according to theory. The theory then errs on the safe side but not to such an extent that it is too conservative. Because of its simplicity and its small safe-side error, the maximum shear stress theory is the most commonly used failure theory for ductile materials.

Example 5-8

Determine the stress to cause failure in Example 5-6 according to the maximum shear stress theory.

Solution

$$\frac{\sigma_2}{\sigma_y} = \frac{210}{250} = 0.84$$

From Fig. 5-23 it is evident that failure is indicated by a point on line

$$\frac{\sigma_1}{\sigma_y} = 1 \qquad \text{(1)}$$

and another point on the line

$$\frac{\sigma_1}{\sigma_y} - \frac{\sigma_2}{\sigma_y} = -1 \qquad \text{(2)}$$

From the first equation

$$\sigma_1 = \sigma_y = \textbf{250 MPa tension}$$

Substituting known values into the second equation gives

$$\frac{\sigma_1}{\sigma_y} = -1 + \frac{\sigma_2}{\sigma_y} = -1 + 0.84 = -0.16$$

$$\sigma_1 = -0.16\sigma_y = -0.16 \times 250 = -40 \text{ MPa}$$

$$= \textbf{40 MPa compression}$$

Example 5-8

Determine the stress to cause failure in Example 5-6 according to the maximum shear stress theory.

Solution

$$\frac{\sigma_2}{\sigma_y} = \frac{30}{36} = 0.83$$

From Fig. 5-23 it is evident that failure is indicated by a point on line

$$\frac{\sigma_1}{\sigma_y} = 1 \qquad\qquad (1)$$

and another point on the line

$$\frac{\sigma_1}{\sigma_y} - \frac{\sigma_2}{\sigma_y} = -1 \qquad\qquad (2)$$

From the first equation

$$\sigma_1 = \sigma_y = \textbf{36 kip/in.}^2 \textbf{ tension}$$

Substituting known values into the second equation gives

$$\frac{\sigma_1}{\sigma_y} = -1 + \frac{\sigma_2}{\sigma_y} = -1 + 0.83 = -0.17$$

$$\sigma_1 = -0.17\sigma_y = -0.17 \times 36 = -6 \text{ kip/in.}^2$$

$$= \textbf{6 kip/in.}^2 \textbf{ compression}$$

Problems 5-29 to 5-30

Total Strain Energy Theory

In a later chapter we will be developing energy methods for dealing with problems that would otherwise be quite difficult to solve. Until then it will not be possible for you to derive the formulas that we will encounter in the following energy theories. These theories are based on the energy stored in material that is stressed and, of course, strained. If you hold the ends of a piece of rubber tubing, you can stretch the tube and keep it in the stretched position. In this state the tube contains a certain amount of strain energy, which is equal to the work that you did in stretching the tube. When you release one end of the tube, the strain energy is recovered, sometimes with painful results. Strain energy could also be stored in the tube by twisting one end and distorting the material in the tube. For the present it will be sufficient if you will accept the idea that energy is stored in an elastic material when it is acted upon by forces that (1) change the volume of the material, (2) distort the material, or (3) simultaneously change the volume and distort the material. Condition (3) is the most common state of loading and will be dealt with first.

The total strain energy theory predicts that a material will fail when the stored strain energy per unit of volume, due to both change in volume and distortion, is equal to the total strain energy per unit volume of a *uniaxial* test specimen at failure. In terms of principal stresses the theory predicts that failure will occur when

$$\left(\frac{\sigma_1}{\sigma_y}\right)^2 - 2v\left(\frac{\sigma_1}{\sigma_y}\right)\left(\frac{\sigma_2}{\sigma_y}\right) + \left(\frac{\sigma_2}{\sigma_y}\right)^2 = 1 \qquad \textbf{(5-12)}$$

This envelope is shown in Fig. 5-24 for $v = 0.3$, where it is seen as fitting the experimental points quite well but is still somewhat short of perfection.

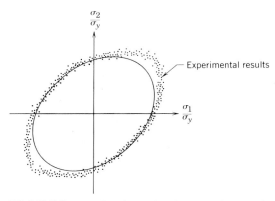

FIG. 5-24 Failure envelope for total strain energy theory evaluated at $v = 0.3$.

Example 5-9

Determine the stress to cause failure in Example 5-6 according to the total strain energy theory.

Solution

$$\frac{\sigma_2}{\sigma_y} = \frac{210}{250} = 0.84$$

Substituting into the failure equation (5-12) gives

$$\left(\frac{\sigma_1}{\sigma_y}\right)^2 - 2 \times 0.3\left(\frac{\sigma_1}{\sigma_y}\right)0.84 + 0.84^2 = 1$$

$$\left(\frac{\sigma_1}{\sigma_y}\right)^2 - 0.504\left(\frac{\sigma_1}{\sigma_y}\right) - 0.294 = 0$$

By quadratic formula

$$\frac{\sigma_1}{\sigma_y} = \frac{+0.504 \pm \sqrt{(-0.504)^2 - 4 \times (-0.294)}}{2}$$

$$= \frac{0.504 \pm 1.196}{2} = 0.850 \quad \text{or} \quad -0.346$$

$$\sigma_1 = 0.850 \times 250 \text{ MPa} = \textbf{213 MPa tension}$$

or

$$\sigma_1 = -0.346 \times 250 \text{ MPa} = -87 \text{ MPa} = \textbf{87 MPa compression}$$

Example 5-9

Determine the stress to cause failure in Example 5-6 according to the total strain energy theory.

Solution

$$\frac{\sigma_2}{\sigma_y} = \frac{30}{36} = 0.83$$

Substituting into the failure equation (5-12) gives

$$\left(\frac{\sigma_1}{\sigma_y}\right)^2 - 2 \times 0.3\left(\frac{\sigma_1}{\sigma_y}\right)0.83 + 0.83^2 = 1$$

$$\left(\frac{\sigma_1}{\sigma_y}\right)^2 - 0.498\left(\frac{\sigma_1}{\sigma_y}\right) - 0.311 = 0$$

By quadratic formula

$$\frac{\sigma_1}{\sigma_y} = \frac{+0.498 \pm \sqrt{(-0.498)^2 - 4 \times (-0.311)}}{2}$$

$$= \frac{0.498 \pm 1.221}{2} = 0.860 \quad \text{or} \quad -0.362$$

$$\sigma_1 = 0.860 \times 36 \text{ kip/in.}^2 = \textbf{31 kip/in.}^2 \textbf{ tension}$$

or

$$\sigma_1 = -0.362 \times 36 \text{ kip/in.}^2 = -13 \text{ kip/in.}^2 = \textbf{13 kip/in.}^2 \textbf{ compression}$$

Problems 5-31 to 5-32

Distortion Strain Energy Theory

In this theory only the energy required to distort the material is considered. The distortion energy in the loaded material is equated to the distortion energy in the test specimen at failure. This gives the equation of the failure envelope as

$$\left(\frac{\sigma_1}{\sigma_y}\right)^2 - \left(\frac{\sigma_1}{\sigma_y}\right)\left(\frac{\sigma_2}{\sigma_y}\right) + \left(\frac{\sigma_2}{\sigma_y}\right)^2 = 1$$ (5-13)

When this equation is plotted, as in Fig. 5-25, it is evident that, of all the theories we have investigated for ductile materials, the distortion strain energy theory gives the best agreement with the test results. You might then expect that this theory would always be used in the design of members made of ductile materials. However, it is an inconvenient formula to solve and most practicing engineers prefer to use the maximum shear stress theory for design and to accept the error thus introduced because the error is small and is on the side of safety.

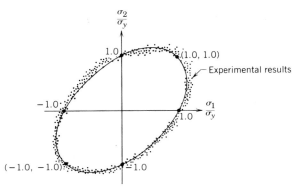

FIG. 5-25 Failure envelope for distortion strain energy theory.

Example 5-10

Determine the stress to cause failure in Example 5-6 according to the distortion strain energy theory.

Solution

$$\frac{\sigma_2}{\sigma_y} = \frac{210}{250} = 0.84$$

Substituting into the failure equation (5-13) gives

$$\left(\frac{\sigma_1}{\sigma_y}\right)^2 - \left(\frac{\sigma_1}{\sigma_y}\right)0.84 + 0.84^2 = 1$$

$$\left(\frac{\sigma_1}{\sigma_y}\right)^2 - 0.84\left(\frac{\sigma_1}{\sigma_y}\right) - 0.294 = 0$$

By the quadratic formula

$$\frac{\sigma_1}{\sigma_y} = \frac{0.84 \pm \sqrt{-0.84^2 - 4 \times (-0.294)}}{2}$$

$$= \frac{0.84 \pm 1.372}{2} = 1.106 \quad \text{or} \quad -0.266$$

$$\sigma_1 = 1.106 \times 250 \text{ MPa} = \textbf{276 MPa tension}$$

or

$$\sigma_1 = -0.266 \times 250 \text{ MPa} = -66 \text{ MPa} = \textbf{66 MPa compression}$$

Example 5-10

Determine the stress to cause failure in Example 5-6 according to the distortion strain energy theory.

Solution

$$\frac{\sigma_2}{\sigma_y} = \frac{30}{36} = 0.83$$

Substituting into the failure equation (5-13) gives

$$\left(\frac{\sigma_1}{\sigma_y}\right)^2 - \left(\frac{\sigma_1}{\sigma_y}\right)0.83 + 0.83^2 = 1$$

$$\left(\frac{\sigma_1}{\sigma_y}\right)^2 - 0.83\left(\frac{\sigma_1}{\sigma_y}\right) - 0.311 = 0$$

By the quadratic formula

$$\frac{\sigma_1}{\sigma_y} = \frac{0.83 \pm \sqrt{-0.83^2 - 4 \times (-0.311)}}{2}$$

$$= \frac{0.83 \pm 1.390}{2} = 1.110 \quad \text{or} \quad -0.280$$

$$\sigma_1 = 1.110 \times 36 \text{ kip/in.}^2 = \textbf{40 kip/in.}^2 \textbf{ tension}$$

or

$$\sigma_1 = -0.280 \times 36 \text{ kip/in.}^2 = -10 \text{ kip/in.}^2 = \textbf{10 kip/in.}^2 \textbf{ compression}$$

It is often instructive to compare the results obtained by applying the various theories. In Examples 5-6 to 5-10, for a material having a yield stress of 250 MPa (36 ksi), we found the value of the third principal stress to cause failure when the other principal stresses were zero and 210 MPa (30 ksi). These were

Maximum principal stress theory:	250 or	−250 MPa (36 or −36 ksi)
Maximum principal strain theory:	313 or	−133 MPa (45 or −20 ksi)
Maximum shear stress theory:	250 or	−40 MPa (36 or −6 ksi)
Total strain energy theory:	213 or	−87 MPa (31 or −13 ksi)
Distortion strain energy theory:	276 or	−66 MPa (40 or −10 ksi)

These figures show that products supposedly having the same safety factor may, in fact, have vastly different margins of safety depending on the theory to which the designers happen to subscribe.

Problems 5-33 to 5-35

Failure Theory for Brittle Materials

Up to this point all discussion of failure theories has been limited to ductile materials. Although ductile materials are preferred for most engineering applications, there are many brittle materials that are used because they possess other desired properties. Concrete, gray cast iron, bricks, and other ceramics are among the commonly used brittle materials.

Because ductile materials have approximately the same yield strength in compression and in tension, the test points in the diagrams that we have seen have a certain symmetry which aids in fitting theoretical curves to the test points. This symmetry is lacking in brittle materials and is evident in the well-known difference between tensile and compressive strengths of gray cast iron (see Table A) and other brittle materials. This difference is obvious in the set of typical experimental points shown in Fig. 5-26.

In the first and third quadrants the position of the experimental points agrees well with maximum principal stress theory provided that we use a failure stress in compression, σ_{fc}, that differs from the failure stress in tension, σ_{ft}. The theoretical envelope would then be as shown in Fig. 5-27.

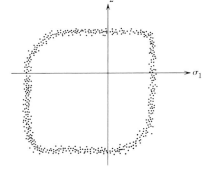

FIG. 5-26 Typical experimental points for a brittle material.

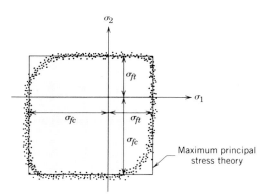

FIG. 5-27

Although the theory agrees well with the experimental points in the first and third quadrants, it errs considerably, and on the unsafe side, in the second and fourth quadrants. This is not an acceptable error, but unfortunately no single theory can be found that is satisfactory in all four quadrants. Consequently, it is common practice to follow the suggestion that was made in 1885 by C. Duguet and modify the envelope by arbitrary straight lines as shown in Fig. 5-28.

Failure theories have been considered for the special case where one of the principal stresses is zero. That is, we have dealt with failure in the plane stress or biaxial stress case. Having seen the complexity of the theories for plane stress, we can easily imagine the difficulties involved in extending them to the general triaxial case. Triaxial failure theory is beyond the scope of this book and, fortunately, is very rarely required in design practice. In fact, even biaxial theory is not called upon as frequently as you might expect, and many designers manage to succeed with only the simplistic view of failure that we had before we encountered the failure theories in this chapter. Such a limited view causes no serious error as long as the principal stresses are both of the same sign, in which case the numerically larger principal stress alone is a very good indicator of the adequacy of the design. However, when there is a difference in sign, an unsafe error can be made by focusing attention on one stress only. Since the consequences of such an error can be disastrous, we must watch for situations that require the applications of failure theories.

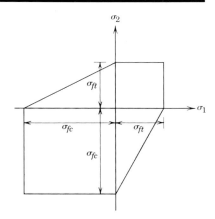

FIG. 5-28 Modified maximum principal stress theory for brittle materials.

PROBLEMS

5-1 A rectangular bar, 16×12 mm (0.6×0.5 in.), is subjected to an axial load of 20.0 kN (4.5 kip). Draw a free-body diagram and use equilibrium to determine the normal and shear stresses on a plane having θ (see Example 5-1) equal to

   a. $0°$.
   b. $30°$.
   c. $45°$.
   d. $90°$.

5-2 A rectangular bar, 20×15 mm (0.8×0.6 in.), is subjected to an axial load of 45 kN (10 kip).

   a. Draw a free-body diagram and use equilibrium to determine the normal and shear stresses on a plane having $\theta = 0$, 45, and $90°$ (see Example 5-1).
   b. Plot the answers to part (a) on σ-τ axes.
   c. Complete the σ-τ diagram and from it determine the normal and shear stresses on a plane having θ equal to $30°$.

5-3 Use equations (5-1) and (5-2) to solve Prob. 5-1.

5-4 For the plane stresses given in Fig. 5-3a, use equations (5-1) and (5-2) to determine the normal and shear stresses acting on a plane having

   a. $\theta = 20°$.
   b. $\theta = 110°$.
   c. $\theta = 200°$.
   d. $\theta = -20°$.

5-5 Given the plane stresses

$$\sigma_x = 150 \text{ MPa (30 ksi)}$$

$$\sigma_y = 30 \text{ MPa (6 ksi)}$$

$$\tau_{xy} = 60 \text{ MPa (12 ksi)}$$

   a. Determine the principal stresses.
   b. Show, by a sketch, the planes on which the principal stresses act.
   c. Determine the maximum shear stress.
   d. Show, by a sketch, the plane on which the maximum shear stress acts.

5-6 Given the plane stress conditions

$$\sigma_x = 40 \text{ MPa (8 ksi)}$$

$$\sigma_y = -100 \text{ MPa } (-20 \text{ ksi})$$

$$\tau_{xy} = -50 \text{ MPa } (-10 \text{ ksi})$$

a. Determine the principal stresses.
b. Show, by a sketch, the planes on which the principal stresses act.
c. Determine the maximum shear stress.
d. Show, by a sketch, the plane on which the maximum shear stress acts.

5-7 Given the plane stress conditions

$$\sigma_x = -150 \text{ MPa } (-30 \text{ ksi})$$

$$\sigma_y = -20 \text{ MPa } (-4 \text{ ksi})$$

$$\tau_{xy} = 90 \text{ MPa (18 ksi)}$$

a. Determine the principal stresses.
b. Show, by a sketch, the planes on which the principal stresses act.
c. Determine the maximum shear stress.
d. Show, by a sketch, the plane on which the maximum shear stress acts.

5-8 A pipeline contains steam at a pressure of 5000 kPa (725 psi). The pipe has an outside diameter of 203 mm (8 in.) and a wall thickness of 4 mm (0.157 in.). In addition to the circumferential and longitudinal stresses due to pressure, there is a stress caused by a torque load of 20 kN·m (14.7 kip·ft) applied to the pipe. (a) Determine the maximum tensile stress in the pipe wall. (b) If the allowable normal stress is 140 MPa (20.3 ksi), is the pipe safe?

5-9 A steel pipeline of 350 mm (13.8 in.) outside diameter and 3.5 mm (0.138 in.) thickness is subjected to an internal pressure of 4000 kPa (580 psi). An axial tensile force of 80 kN (18 kip) is also applied in addition to the axial pressure load.

a. Determine the principal stresses.
b. Determine the maximum shear stress.
c. Determine the normal and shear stresses acting on a plane at 20° to the longitudinal axis of the pipeline.
d. If the allowable normal stress is 220 MPa (31.9 ksi), determine the maximum torque load that may act on the pipeline in addition to the given loads.

5-10 A piece of cast iron is loaded in such a way that it is subjected to stresses of

$$\sigma_x = 60 \text{ MPa (12 ksi)}$$

$$\sigma_y = -10 \text{ MPa } (-2 \text{ ksi})$$

$$\tau_{xy} = 40 \text{ MPa (8 ksi)}$$

a. If the allowable stress in tension is 70 MPa (14 ksi), is the material overloaded?
b. If the material fails because of overload, show the direction of the cracks that you would expect to appear.

5-11 a. Using the allowable stresses given in Table A for high-strength concrete, determine the safe load, W, considering shear only.
b. Determine the maximum diagonal tension in the concrete when the load in part (a) is applied.

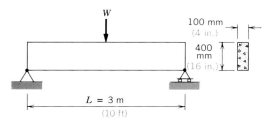

PROBLEM 5-11

5-12 A Western hemlock beam is 100 mm (4 in.) wide, 400 mm (16 in.) deep, and 3 m (10 ft) long. It is simply supported at the ends and carries a uniformly distributed load. Using the allowable stresses given in Table A, determine the safe load considering

a. Bending stress only.
b. Shear stress only.
c. Both bending and shear stress.

5-13 The given beam must carry a total load of 200 kN (50 kip). Using the allowable shear stress given in Table A for high-strength concrete, and ignoring bending stresses, determine the regions where the allowable stress is exceeded. Show the overstressed region in a sketch, giving the dimensions of the region, and indicate the direction of reinforcing bars that would prevent shear failure. Consider the load to be:

a. Applied half on point *A* and half on *B*.
b. Uniformly distributed over the length of beam.

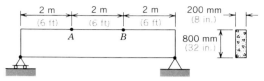

PROBLEM 5-13

5-14 For the given beam, determine the maximum principal stress at *A*. Show the direction of the stress on a sketch.

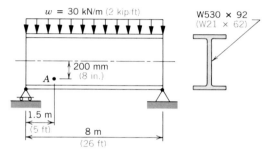

PROBLEM 5-14

5-15 In Example 5-4, by how much would the angle θ have to be increased to permit the use of the 75% efficient weld.

5-16 A pipe having an outside diameter of 305 mm (12 in.) and a wall thickness of 4.75 mm (0.187 in.) is subjected to a bending moment of 40 kN·m (29.5 kip·ft) and a torque of 35 kN·m (25.8 kip·ft). The material has the same properties as low-strength structural steel.

a. Determine the maximum principal stress.
b. Is the pipe safe?

5-17 A pipeline contains gas at a pressure of 7000 kPa (1015 psi). The pipe has an inside diameter of 1200 mm (48 in.) and a wall thickness of 30 mm (1.2 in.). The line is made of spiral-wound pipe (Fig. 5-1) with a pitch angle of 20° ($\theta = 20°$ in Fig. 5-4). Determine the normal stress on the spiral weld.

5-18 Pipe having an outside diameter of 900 mm (35.4 in.) and a wall thickness of 16 mm (0.63 in.) is used for a pipeline to transmit gas at a pressure of 7500 kPa (1090 psi). In addition to the gas pressure the pipe is subjected to a torque load of 750 kN·m (550 kip·ft).

The allowable stress in the pipe is 220 MPa (31.9 ksi) and the weld is 70% efficient. Calculate the normal stress on the weld and determine if the pipe is safe when the torque load is:

a. As shown.
b. In the opposite direction.

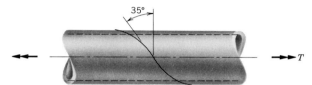

PROBLEM 5-18

5-19 Use equations (5-3), (5-4), and (5-5) to solve:

a. Prob. 5-5.
b. Prob. 5-6.
c. Prob. 5-7.
d. Prob. 5-10.

5-20 Draw Mohr's circles of stress and determine the maximum shear stress for each set of principal stresses:

	σ_1	σ_2	σ_3
a.	200 MPa	150 MPa	50 MPa
	(40 ksi)	(30 ksi)	(10 ksi)
b.	200 MPa	150 MPa	0 MPa
	(40 ksi)	(30 ksi)	(0 ksi)
c.	200 MPa	20 MPa	0 MPa
	(40 ksi)	(4 ksi)	(0 ksi)
d.	200 MPa	0 MPa	0 MPa
	(40 ksi)	(0 ksi)	(0 ksi)
e.	200 MPa	0 MPa	−50
	(40 ksi)	(0 ksi)	(−10 ksi)
f.	200 MPa	−50 MPa	−50 MPa
	(40 ksi)	(−10 ksi)	(−10 ksi)
g.	200 MPa	−50 MPa	−100 MPa
	(40 ksi)	(−10 ksi)	(−20 ksi)
h.	100 MPa	−50 MPa	−100 MPa
	(20 ksi)	(−10 ksi)	(−20 ksi)
i.	0 MPa	−50 MPa	−100 MPa
	(0 ksi)	(−10 ksi)	(−20 ksi)
j.	−20 MPa	−50 MPa	−100 MPa
	(−4 ksi)	(−10 ksi)	(−20 ksi)
k.	−50 MPa	−50 MPa	−100 MPa
	(−10 ksi)	(−10 ksi)	(−20 ksi)

5-21 A body of the dimensions given in the illustration is acted upon by forces in three directions.

    a. Draw Mohr's circles of stress and determine the maximum shear stress.

    b. The maximum allowable shear stress is 120 MPa (24 ksi). If P_2 and P_3 have the given values, determine the limiting values of P_1. (Note that P_1 may change its direction.)

    c. The maximum allowable shear stress is 120 MPa (24 ksi). If P_1 and P_3 have the given values, determine the limiting values of P_2.

    d. The maximum allowable shear stress is 120 MPa (24 ksi). If P_1 and P_2 have the given values, determine the limiting values of P_3.

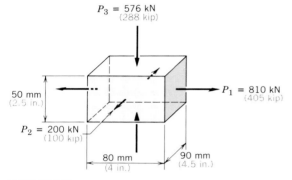

PROBLEM 5-21

5-22 For an allowable stress in shear of 100 MPa (20 ksi) and $\sigma_z = 0$, fill in the limiting values of σ_y in the table.

	σ_x		Limits in σ_y	
Case	(MPa)	(ksi)	Upper	Lower
a	200	(40)		
b	150	(30)		
c	100	(20)		
d	50	(10)		
e	0	(0)		
f	−50	(−10)		
g	−100	(−20)		
h	−150	(−30)		
i	−200	(−40)		

5-23 For each case in Prob. 5-22 plot two points on a diagram having σ_x as the abscissa and σ_y as the ordinate. Join the points to form a closed curve and show the regions corresponding to a safe state of stress and the unsafe state.

5-24 Draw Mohr's circles and determine the maximum shear stress for each case given below.

	σ_x		σ_y		σ_z		τ_{xy}	
Case	(MPa)	(ksi)	(MPa)	(ksi)	(MPa)	(ksi)	(MPa)	(ksi)
a	−40	(−8)	0	(0)	180	(36)	0	(0)
b	40	(8)	0	(0)	180	(36)	0	(0)
c	80	(16)	180	(36)	180	(36)	60	(12)
d	80	(16)	180	(36)	−60	(−12)	60	(12)
e	0	(0)	120	(24)	100	(20)	−80	(−16)
f	0	(0)	120	(24)	180	(36)	−80	(−16)

5-25 A machine part, made of AISI 1090, is subjected to the stresses given below. Use Fig. 5-19 to determine in each case if the part is safe according to

    (i) Maximum principal stress theory.

    (ii) Experimental evidence.

    a. $\sigma_x = 300$ MPa (43.5 ksi)
        $\sigma_y = 400$ MPa (58 ksi)
        $\tau_{xy} = 0$ MPa (0 ksi)

    b. $\sigma_x = 300$ MPa (43.5 ksi)
        $\sigma_y = 500$ MPa (72.5 ksi)
        $\tau_{xy} = 0$ MPa (0 ksi)

c. $\sigma_x = -300$ MPa (-43.5 ksi)
$\sigma_y = 200$ MPa (29.0 ksi)
$\tau_{xy} = 0$ MPa (0 ksi)
d. $\sigma_x = -300$ MPa (-43.5 ksi)
$\sigma_y = 400$ MPa (58.0 ksi)
$\tau_{xy} = 0$ MPa (0 ksi)
e. $\sigma_x = -300$ MPa (-43.5 ksi)
$\sigma_y = 500$ MPa (72.5 ksi)
$\tau_{xy} = 0$ MPa (0 ksi)
f. $\sigma_x = 500$ MPa (72.5 ksi)
$\sigma_y = 500$ MPa (72.5 ksi)
$\tau_{xy} = 0$ MPa (0 ksi)
g. $\sigma_x = 0$ MPa (0 ksi)
$\sigma_y = 0$ MPa (0 ksi)
$\tau_{xy} = 400$ MPa (58.0 ksi)
h. $\sigma_x = -90$ MPa (-13.1 ksi)
$\sigma_y = 270$ MPa (39.1 ksi)
$\tau_{xy} = 240$ MPa (34.8 ksi)

5-26 A straight piece of pipe is made of AISI 4340. It has an internal diameter of 760 mm (30 in.) and a wall thickness of 20 mm (0.8 in.). There are axial loads of such a nature that the longitudinal stress is always 324 MPa (67 ksi) but may be either tension or compression. Determine the internal pressure that will cause failure according to the maximum principal stress theory when the longitudinal stress is

a. Tensile.
b. Compressive.

5-27 Solve Prob. 5-25 using the maximum principal strain theory.

5-28 Solve Prob. 5-26 using the maximum principal strain theory.

5-29 Solve Prob. 5-25 using the maximum shear stress theory.

5-30 Solve Prob. 5-26 using the maximum shear stress theory.

5-31 Solve Prob. 5-25 using the total strain energy theory.

5-32 Solve Prob. 5-26 using the total strain energy theory.

5-33 Solve Prob. 5-25 using the distortion strain energy theory.

5-34 Solve Prob. 5-26 using the distortion strain energy theory.

5-35 A cylindrical pressure vessel has an inside diameter of 2200 mm (87 in.) and a wall thickness of 30 mm (1.2 in.). The material is AISI 4340. Determine the pressure at which failure would occur according to

a. Maximum principal stress theory.
b. Maximum principal strain theory.
c. Maximum shear stress theory
d. Total strain energy theory.
e. Distortion strain energy theory.

Advanced Problems

5-A A pipeline is being designed to carry gas at a pressure of 6500 kPa (960 psi). The pipe is to have an inside diameter of 1500 mm (59.0 in.) and is to be formed by spiral welding plate that is 2000 mm (78.7 in.) wide and 25 mm (1 in.) thick.

a. Determine the normal stress on the weld.

For an allowable stress of 200 MPa (29 ksi), determine the required efficiency

b. In the spiral weld.
c. In the circumferential welds that are used in the field to join lengths of pipe.

If the pipe had been made of 25-mm (1 in.) plate with a single longitudinal weld,

d. How wide would the plate have to be?
e. What weld efficiency would be required?

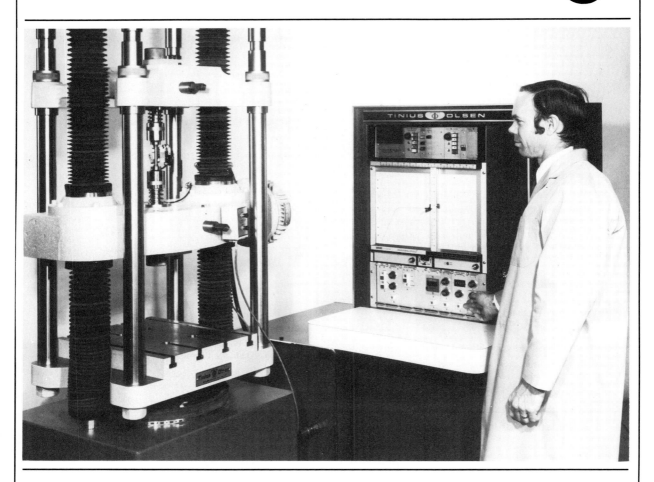

6

STRESS–STRAIN RELATIONSHIPS

6-1 INTRODUCTION

The relationship between stress and strain within the proportional limit given in (1-12b)

$$\varepsilon = \frac{\sigma}{E}$$

has been found to be useful in determining:

1. The change in length of members under axial load.
2. The solution to certain statically indeterminate problems.
3. The stress distribution on a cross section subjected to a bending moment.

The relationship was derived from observations made on test specimens that were loaded with an axial force only and, consequently, is applicable only to states of *uniaxial* stress. If it is applied to cases in which there is stress in more than one direction, that is, biaxial or triaxial states of stress, the results will be in error. For example, it cannot be used to calculate the change in dimensions of a plate such as that of Fig. 6-1, which is in a state of *biaxial stress* or *plane stress*. In this case biaxial stress–strain relationships must be established before dimensional changes can be calculated. The relationships are also needed so that biaxial stresses can be determined from test measurements. Unfortunately, there is no practical method for directly measuring stress; however, strains can be measured and the corresponding stresses calculated by means of the stress–strain relationships.

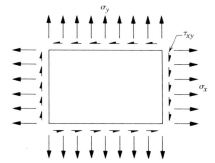

FIG. 6-1 Biaxial or plane stress.

6-2 STRAINS FROM KNOWN STRESSES

When strength tests were described in Chapter 1, we were interested in the axial stress and, consequently, focused our attention on the axial strain. The change in diameter that can be seen in Fig. 1-26b, which takes place during yielding, was rather obvious. If careful measurements of diameter had been taken while the material was still in the elastic range, a small change in diameter would have been found. This decrease in lateral dimension was predicted on the basis of molecular theory by S. D. Poisson in 1828 and hence is referred to as the *Poisson effect*. If the diametral changes were measured and the diametral strain calculated at various loads and compared with the axial strain at the corresponding load, it would be found that the two strains were always in a fixed ratio. This ratio is designated as *Poisson's ratio* and is approximately 0.3 for all types of steel. In a plate the strains could be seen by the change in length of lines scribed on the surface such as *A-B* and *A-C* in Fig. 6-2a. If such a plate were loaded *uniaxially in the x-direction* as shown in Fig. 6-2b and measurements were made, the lengths of the lines would be found to change, but the lines would remain at right angles. It would be found that the stress, σ_x, caused *A-B* to increase in length and, if the change in

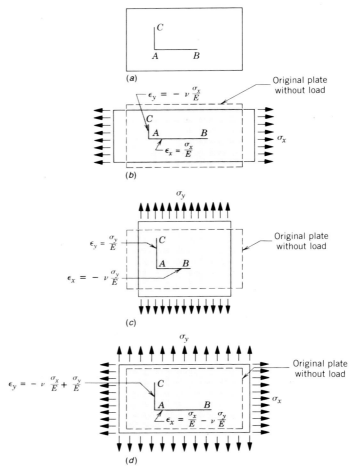

FIG. 6-2 Poisson effect. (*a*) Plate without load. (*b*) Plate stressed in *x*-direction only. (*c*) Plate stressed in *y*-direction only. (*d*) Plate stressed in *x*- and *y*-direction.

length were used to calculate strain, it would agree with (1-12b). To indicate the direction of the strain and stress, we rewrite (1-12b) as

$$\varepsilon_x = \frac{\sigma_x}{E} \qquad \textbf{(6-1)}$$

Observations on the line *A-C* would show that its length decreases and that the value of the strain, which is compressive, is related to the *x*-strain by a constant denoted by *v*; thus

$$v = -\frac{\text{lateral strain}}{\text{axial strain}} = -\frac{\varepsilon_y}{\varepsilon_x} \qquad \textbf{(6-2)}$$

For commonly used engineering materials, the constant v, *Poisson's ratio*, ranges from about 0.15 for concrete to slightly less than 0.5 for rubber.

Substituting (6-1) into (6-2) and solving for ε_y gives

$$\varepsilon_y = -v\frac{\sigma_x}{E} \qquad \text{(6-3)}$$

If measurements had been made on the thickness of the plate during loading, that is, in the *z*-direction, it would have been observed that

$$\varepsilon_z = -v\frac{\sigma_x}{E} \qquad \text{(6-4)}$$

In a similar manner, the plate of Fig. 6-2c, loaded *uniaxially in the y-direction* only would be observed to experience strains given by

$$\varepsilon_y = \frac{\sigma_y}{E} \qquad \text{(6-5)}$$

$$\varepsilon_x = -\frac{v\sigma_y}{E} \qquad \text{(6-6)}$$

$$\varepsilon_z = -\frac{v\sigma_y}{E} \qquad \text{(6-7)}$$

When stresses in the *x*-direction and the *y*-direction are applied simultaneously, that is, when we are dealing with a *biaxial* stress condition, as shown in Fig. 6-2d, the resultant strains can be found by superimposing the strains from each separate stress, thus:

$$\varepsilon_x = \frac{\sigma_x}{E} - v\frac{\sigma_y}{E} = \frac{1}{E}(\sigma_x - v\sigma_y) \qquad \text{(6-8)}$$

$$\varepsilon_y = -v\frac{\sigma_x}{E} + \frac{\sigma_y}{E} = \frac{1}{E}(-v\sigma_x + \sigma_y) \qquad \text{(6-9)}$$

$$\varepsilon_z = -v\frac{\sigma_x}{E} - v\frac{\sigma_y}{E} = \frac{1}{E}(-v\sigma_x - v\sigma_y) \qquad \text{(6-10)}$$

Since we are usually interested in strains in the plane of the stress only, (6-10) is seldom required.

Problems 6-1 to 6-8

Consider next the rectangular plate of Fig. 6-3 that, as before, has lines *A-B* and *A-C* scribed on its surface. The lines are of a certain length and are at right angles to one another. If a *shear stress* is applied to the plate, we would find that the plate is no longer rectangular, the angles changing slightly from right angles while the sides remain unchanged in

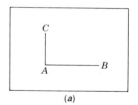

(a)

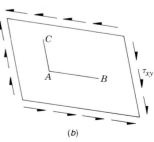

(b)

FIG. 6-3 Shear strain. (*a*) Plate without load. (*b*) Plate with shear load only.

length. The angle CAB will become slightly larger than $90°$ but the lengths of the lines A-B and A-C do not change. The shear stress then makes no contribution to ε_x or to ε_y but changes the shape of the plate by giving it a *shear strain*. The shear strain is defined as the change in the angle CAB in radians and is represented by the symbol γ_{xy}. A series of tests that keep all stresses within the proportional limit would show a linear relationship between the shear stress and the shear strain; thus

$$\boxed{\tau_{xy} = G\gamma_{xy}} \tag{6-11a}$$

or

$$\boxed{\gamma_{xy} = \frac{1}{G}\tau_{xy}} \tag{6-11b}$$

The experimental results would show that G is a very large number and the similarity of (6-11a) to (1-12c) might suggest that G is Young's modulus. However, it would be found to have a different value, although of the same order of magnitude. Because of its relationship with shear, G is called the *shear modulus*.

The observation that the shear stress contributes nothing to the axial strains and that the axial stress does not contribute to the shear strain can be expressed in the equation by writing (6-8), (6-9), and (6-11b) as:

$$\varepsilon_x = \frac{1}{E}(\sigma_x - v\sigma_y + 0\tau_{xy})$$

$$\varepsilon_y = \frac{1}{E}(-v\sigma_x + \sigma_y + 0\tau_{xy}) \tag{6-12a}$$

$$\gamma_{xy} = \left(0\sigma_x + 0\sigma_y + \frac{1}{G}\tau_{xy}\right)$$

These can be written in matrix format as

$$
\begin{Bmatrix} \varepsilon_x \\ \varepsilon_y \\ \gamma_{xy} \end{Bmatrix} = \frac{1}{E}
\begin{bmatrix} 1 & -v & 0 \\ -v & 1 & 0 \\ 0 & 0 & \dfrac{E}{G} \end{bmatrix}
\begin{Bmatrix} \sigma_x \\ \sigma_y \\ \tau_{xy} \end{Bmatrix}
\tag{6-12b}
$$

It appears from (6-12) that we have three *elastic constants* relating strain to stress: E, Young's modulus; v, Poisson's ratio; and G, the shear modulus. We will now check to see if the constants are independent of each other by attempting to solve a shear strain problem without the use of the shear modulus.

Consider a square plate element as shown in Fig. 6-4a with all sides of unit length. Before loading, the angle at the lower left corner is $90°$, after loading it is greater by a small angle that is defined as the shear

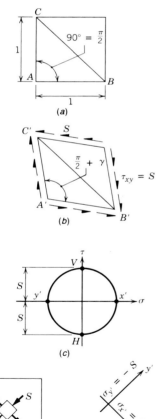

FIG. 6-4 (*a*) Unloaded element. (*b*) Stressed element. (*c*) Mohr's circle of stress. (*d*) Stresses in direction of diagonals.

strain. The state of stress on any plane can be found from Mohr's circle in Fig. 6-4c; consequently, the stresses in the direction of the diagonals are as shown in Fig. 6-4d. Rotated axes are used to simplify the expression of this state of stress. For the rotated axes, (6-8) can be written as

$$\varepsilon'_x = \frac{1}{E}(\sigma'_x - v\sigma'_y)$$

When the known stress values are substituted

$$\varepsilon'_x = \frac{1}{E}[s - v(-s)] = \frac{1}{E}(1 + v)\,s$$

The unstressed length of the diagonal B-C was $1 \times \sqrt{2}$, and consequently, its increase in length will be, from (1-11c)

$$\Delta L = L \times \varepsilon$$

$$= 1 \times \sqrt{2} \times \varepsilon'_x$$

$$= \sqrt{2}\,\frac{1}{E}(1 + v)\,s$$

If the triangle $A'B'C'$ from Fig. 6-4b is placed on top of ABC from Fig. 6-4a so that $A'C'$ coincides with A-C, we will have Fig. 6-5a. The lines B-B'' and B-B' are circular arcs but, since the displacements and angles in most engineering materials are very small, we can replace the arcs by straight lines and get the triangle in Fig. 6-5b.

The shear strain is then given by

$$\gamma = \frac{\sqrt{2}\,\Delta L}{1}$$

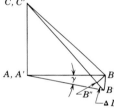

FIG. 6-5a

But ΔL is known so that

$$\gamma = \sqrt{2}\sqrt{2}\,\frac{1}{E}(1 + v)\,s$$

$$= \frac{2(1 + v)}{E}\,s$$

$$\gamma_{xy} = \frac{2(1 + v)}{E}\,\tau_{xy}$$

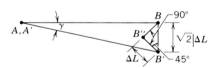

FIG. 6-5b Enlarged detail from Fig. 6-5a.

Comparing this with (6-11b) shows that

$$\frac{1}{G} = \frac{2(1 + v)}{E}$$

or

$$\boxed{G = \frac{E}{2(1 + v)}}$$

(6-13)

Consequently, the shear modulus is not a new and independent constant but merely a combination of Young's modulus and Poisson's ratio which appear to be the basic elastic constants. Another approach to stress–strain relations makes G and the *bulk modulus*,

$$K = \frac{E}{1 - 2v}$$

appear to be the fundamental constants. Yet another point of view shows G and *Lamé's constant*,

$$\lambda = \frac{vE}{(1 + v)(1 - 2v)}$$

as the elastic constants. Since E and v are most readily related to the physical changes in a stressed material, designers usually treat them as the *fundamental elastic constants*.

Now that G is known in terms of E and v, the relationships in (6-12) can be written as:

$$\varepsilon_x = \frac{1}{E}(\sigma_x - v\sigma_y)$$

$$\varepsilon_y = \frac{1}{E}(- v\sigma_x + \sigma_y) \tag{6-14a}$$

$$\gamma_{xy} = \frac{2(1 + v)}{E}\tau_{xy}$$

$$\begin{Bmatrix} \varepsilon_x \\ \varepsilon_y \\ \gamma_{xy} \end{Bmatrix} = \frac{1}{E} \begin{bmatrix} 1 & -v & 0 \\ -v & 1 & 0 \\ 0 & 0 & 2(1+v) \end{bmatrix} \begin{Bmatrix} \sigma_x \\ \sigma_y \\ \tau_{xy} \end{Bmatrix} \tag{6-14b}$$

When a three-dimensional solid is considered, there will be an additional normal stress and two more shear stresses making a total of six components of stress. There will also be six components of strain. These can be related by extending (6-14) to give

$$\begin{Bmatrix} \varepsilon_x \\ \varepsilon_y \\ \varepsilon_z \\ \gamma_{xy} \\ \gamma_{yz} \\ \gamma_{zx} \end{Bmatrix} = \frac{1}{E} \begin{bmatrix} 1 & -v & -v & 0 & 0 & 0 \\ -v & 1 & -v & 0 & 0 & 0 \\ -v & -v & 1 & 0 & 0 & 0 \\ 0 & 0 & 0 & 2(1+v) & 0 & 0 \\ 0 & 0 & 0 & 0 & 2(1+v) & 0 \\ 0 & 0 & 0 & 0 & 0 & 2(1+v) \end{bmatrix} \begin{Bmatrix} \sigma_x \\ \sigma_y \\ \sigma_z \\ \tau_{xy} \\ \tau_{yz} \\ \tau_{zx} \end{Bmatrix} \tag{6-15}$$

In engineering practice three-dimensional stress problems are rarely solved; hence (6-15) is seldom used.

Problems 6-9 to 6-11

6-3 STRESSES FROM KNOWN STRAINS

The relationships expressed in (6-14) are useful in determining the strains when stresses are known. In many practical problems the strains are known, having been determined from strain gage measurements, and the stresses are required. Under these conditions the equations of (6-14) constitute three simultaneous linear equations in three unknowns. It would be wasteful of effort to solve these equations each time stresses are required from known strains. That is not necessary, since it is possible to write a general solution by inverting the matrix in (6-14) to give:

$$\begin{Bmatrix} \sigma_x \\ \sigma_y \\ \tau_{xy} \end{Bmatrix} = \frac{E}{1-v^2} \begin{bmatrix} 1 & v & 0 \\ v & 1 & 0 \\ 0 & 0 & \dfrac{1-v}{2} \end{bmatrix} \begin{Bmatrix} \varepsilon_x \\ \varepsilon_y \\ \gamma_{xy} \end{Bmatrix} \qquad \textbf{(6-16a)}$$

It is often more convenient to take advantage of the fact that the zeros in the matrix permit the equations to be separated into:

$$\begin{Bmatrix} \sigma_x \\ \sigma_y \end{Bmatrix} = \frac{E}{1-v^2} \begin{bmatrix} 1 & v \\ v & 1 \end{bmatrix} \begin{Bmatrix} \varepsilon_x \\ \varepsilon_y \end{Bmatrix} \qquad \textbf{(6-16b)}$$

or

$$\sigma_x = \frac{E}{1-v^2}(\varepsilon_x + v\varepsilon_y)$$

$$\sigma_y = \frac{E}{1-v^2}(v\varepsilon_x + \varepsilon_y) \qquad \textbf{(6-16c)}$$

and

$$\tau_{xy} = \frac{E}{1-v^2}\frac{1-v}{2}\gamma_{xy} = \frac{E}{2(1+v)}\gamma_{xy} \qquad \textbf{(6-16d)}$$

It is obvious that three strains must be known to calculate the three stresses that exist in the general plane stress state. It might seem that in experimental work we would measure the three strains and put them into (6-16) along with the elastic constants to give the required stresses. The axial strains can be measured by various types of *strain gages*, the most convenient being *electric resistance strain gages* such as the one shown in Fig. 6-6a. This gage consists of a fine wire attached to a backing strip. When the backing strip is bonded to a part that is then subjected to a load, the wires are strained to the same extent as the test part. By measuring the change in resistance of the gage, the strain in the direction of the axis

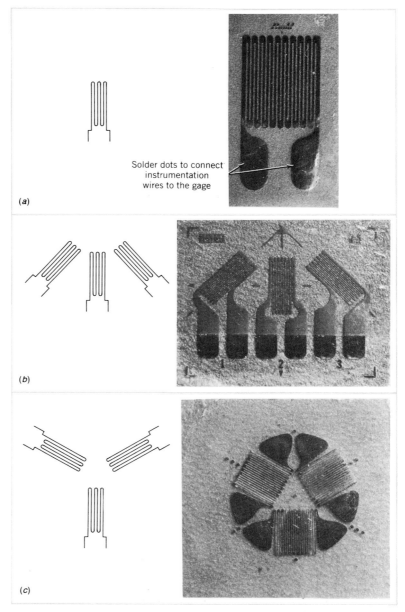

Solder dots to connect instrumentation wires to the gage

(a)

(b)

(c)

FIG. 6-6 Strain gages showing the principle of each gage on the left and a photograph of a typical gage on the right. (a) Electric resistance strain gage. (b) Rectangular strain rosette. (c) Equiangular or delta rosette.

of the wires can be calculated. Thus axial strains can be determined accurately. However, the same gage cannot give shear strain directly and an indirect method must be found to determine shear stresses. This can be done by using a *strain rosette* such as that of Fig. 6-6*b* or *c*. These gages give three pieces of strain data from which we are able to calculate three stress components. The calculation of the stresses is not a simple process; we need another strain equation before we are ready to undertake the processing of strain gage data. This equation is one that enables us to calculate the strain along an axis in the θ-direction, in Fig. 6-7*a*, in terms of ε_x, ε_y, and γ_{xy}.

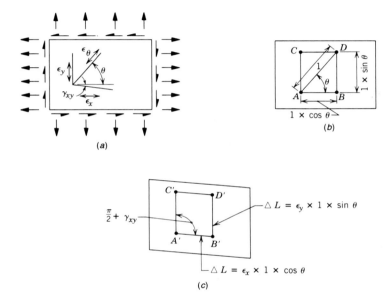

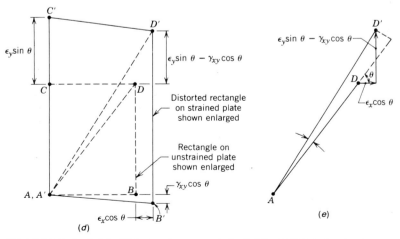

FIG. 6-7 (*a*) Strained plate. (*b*) Rectangle on an unstrained plate. (*c*) Distorted rectangle on a strained plate. (*d*) Original and distorted rectangles. (*e*) Changes in line *AD*.

Consider a rectangle *ABCD* drawn on the plate, as in Fig. 6-7*b*, before the loads are applied. The diagonal *A-D*, which is in the direction in which we wish to calculate the strain, ε_θ, is arbitrarily made one unit in length. Then the sides have the dimensions shown in Fig. 6-7*b*.

After the loads are applied, the rectangle distorts as shown in Fig. 6-7*c*. The distorted rectangle is compared with the original rectangle in Fig. 6-7*d* where *A'B'D'C'* is superimposed on *ABDC* with *A'* coinciding with *A* and the side *A'C'* falling along *A-C*. The point *D* is seen to move upward by

$$\varepsilon_y \sin \theta - \gamma_{xy} \cos \theta$$

and to the right by

$$\varepsilon_x \cos \theta$$

These movements, shown in Fig. 6-7*e*, may be broken into components along *A-D* and normal to *A-D*. The motion of *D* in the direction of *A-D*, which is the θ-direction, is

$$(\varepsilon_x \cos \theta) \cos \theta + (\varepsilon_y \sin \theta - \gamma_{xy} \cos \theta) \sin \theta$$

This is the change in length of *A-D*:

$$\Delta L = \cos^2 \theta \, \varepsilon_x + \sin^2 \theta \, \varepsilon_y - \sin \theta \cos \theta \, \gamma_{xy}$$

The strain along *A-D* is the strain ε_θ and is given by

$$\varepsilon_\theta = \frac{\Delta L}{L} = \frac{\Delta L}{1}$$

or

$$\boxed{\varepsilon_\theta = \cos^2 \theta \, \varepsilon_x + \sin^2 \theta \, \varepsilon_y - \sin \theta \cos \theta \, \gamma_{xy}} \qquad \textbf{(6-17)}$$

The motion of *D* normal to the direction of *A-D* is

$$-\varepsilon_x \cos \theta \sin \theta + (\varepsilon_y \sin \theta - \gamma_{xy} \cos \theta) \cos \theta$$

and, hence, the line *A-D* rotates through an angle given by

$$\phi = \cos \theta \sin \theta (-\varepsilon_x + \varepsilon_y) - \cos^2 \theta \, \gamma_{xy} \qquad \textbf{(6-18)}$$

If, in the original diagram, we had included a line *A-E* normal to the diagonal as in Fig. 6-8*a*, we would have had lines in a position to indicate the shear strain associated with the θ-direction. The increase in the angle *EAD* is a measure of this shear strain, γ_θ. From Fig. 6-8*b*

$$\gamma_\theta = \psi - \phi \qquad \textbf{(6-19)}$$

We already have an equation for ϕ in (6-18). Since this is a general equation derived without restriction on the angle, it can be used to evaluate ψ by substituting $\theta + 90°$ in place of θ. Then

$$\psi = \cos(\theta + 90) \sin(\theta + 90)(-\varepsilon_x + \varepsilon_y) - \cos^2(\theta + 90)\gamma_{xy}$$

$$= -\sin \theta \cos \theta (-\varepsilon_x + \varepsilon_y) - \sin^2 \theta \, \gamma_{xy} \qquad \textbf{(6-20)}$$

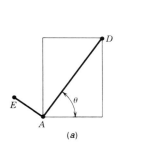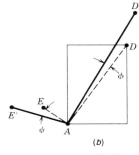

FIG. 6-8 (a) Unstrained plate with shear measuring lines. (b) Strained plate showing shear strain.

Substituting (6-20) and (6-18) into (6-19) gives

$$\gamma_\theta = -\sin\theta\cos\theta\,(-\varepsilon_x + \varepsilon_y) - \sin^2\theta\,\gamma_{xy}$$
$$-[\cos\theta\sin\theta\,(-\varepsilon_x + \varepsilon_y) - \cos^2\theta\,\gamma_{xy}]$$
$$= 2\sin\theta\cos\theta\,(\varepsilon_x - \varepsilon_y) + (\cos^2\theta - \sin^2\theta)\,\gamma_{xy}$$

or

$$\frac{\gamma_\theta}{2} = \sin\theta\cos\theta\,(\varepsilon_x - \varepsilon_y) + (\cos^2\theta - \sin^2\theta)\frac{\gamma_{xy}}{2} \qquad \textbf{(6-21)}$$

If equation (6-17) is compared with (5-1), and (6-21) with (5-2), a similarity is obvious: wherever σ_x occurred in (5-1) and (5-2), ε_x occurs in (6-17) and (6-21). Similarly, σ_y is replaced by ε_y. A change in the replacement pattern for shear is noted; in place of τ_{xy} and τ_θ in (5-1) and (5-2) we find $\gamma_{xy}/2$ and $\gamma_\theta/2$ in (6-17) and (6-21). Consequently, all that was observed with respect to the stress relationships can be applied to the strain relationships except that where in stress we had τ, in strain we have $\gamma/2$. We know that the stress equations can be represented graphically as a Mohr's circle of stress, and consequently, the strain equations can be represented graphically as a *Mohr's circle of strain* provided that the ordinates are taken as one-half the shear strain. Thus Mohr's circle, which has the advantages of giving better insight into variations with a change in angle, is available for studies in strain as well as stress.

Example 6-1

Determine the strain along a line at 12° ccw to the x-axis, and the shear strain associated with the line, when

$$\varepsilon_x = 400 \times 10^{-6} \qquad \varepsilon_y = 100 \times 10^{-6} \qquad \gamma_{xy} = 250 \times 10^{-6}$$

Also determine the maximum strain and show its direction on a sketch.

Solution. Using the strains ε_x and $\gamma_{xy}/2$ we plot point x shown in Fig. 6-9a. Similarly, using ε_y and $-\gamma_{xy}/2$, point y is obtained. Joining the two points by a straight line locates the center of Mohr's circle of strain on the ε-axis.

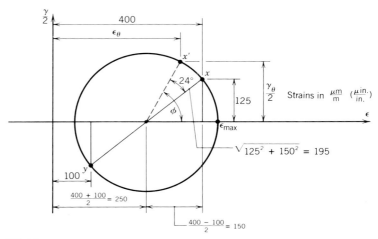

FIG. 6-9a

$$\phi = 24 + \tan^{-1}\tfrac{125}{150}$$

$$= 24 + 39.8 = 63.8°$$

We will refer to the given line as the x'-axis. Since this axis is 12° ccw from the x-axis, the state of strain is given by a point on the circle that is 24° ccw from x. The point is shown as x'. The required strains can now be calculated from the dimensions given on the circle.

$$\varepsilon_\theta = 250 + 195 \cos 63.8$$

$$= 250 + 86 = 336 \frac{\mu m}{m}$$

The required strain along the given line is **336×10^{-6}**.

$$\frac{\gamma_\theta}{2} = 195 \sin 63.8 = 175 \frac{\mu m}{m}$$

$$\gamma_\theta = 2 \times 175 = 350 \frac{\mu m}{m}$$

The shear strain associated with the given line is **350×10^{-6}**. The maximum strain is

$$\varepsilon_{max} = 250 + 195 = 445 \frac{\mu m}{m} = \mathbf{445 \times 10^{-6}}$$

and its direction is as indicated on Fig. 6-9b. Note that the clockwise rotation from the x-axis in Fig. 6-9b is one-half the clockwise rotation from x to ε_{max} on Mohr's circle of strain.

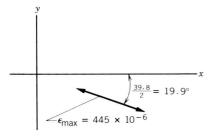

FIG. 6-9b

Example 6-1

Determine the strain along a line at 12° ccw to the x-axis, and the shear strain associated with the line, when

$$\varepsilon_x = 400 \times 10^{-6} \qquad \varepsilon_y = 100 \times 10^{-6} \qquad \gamma_{xy} = 250 \times 10^{-6}$$

Also determine the maximum strain and show its direction on a sketch.

Solution.   Using the strains ε_x and $\gamma_{xy}/2$ we plot point x shown in Fig. 6-9a. Similarly, using ε_y and $-\gamma_{xy}/2$, point y is obtained. Joining the two points by a straight line locates the center of Mohr's circle of strain on the ε-axis.

$$\phi = 24 + \tan^{-1} \tfrac{125}{150}$$

$$= 24 + 39.8 = 63.8°$$

We will refer to the given line as the x'-axis. Since this axis is 12° ccw from the x-axis, the state of strain is given by a point on the circle that is 24° ccw from x. The point is shown as x'. The required strains can now be calculated from the dimensions given on the circle.

$$\varepsilon_\theta = 250 + 195 \cos 63.8$$

$$= 250 + 86 = 336 \, \frac{\mu\text{in.}}{\text{in.}}$$

The required strain along the given line is **336 × 10⁻⁶**.

$$\frac{\gamma_\theta}{2} = 195 \sin 63.8 = 175 \, \frac{\mu\text{in.}}{\text{in.}}$$

$$\gamma_\theta = 2 \times 175 = 350 \, \frac{\mu\text{in.}}{\text{in.}}$$

The shear strain associated with the given line is **350 × 10⁻⁶**. The maximum strain is

$$\varepsilon_{\max} = 250 + 195 = 445 \, \frac{\mu\text{in.}}{\text{in.}} = \textbf{445} \times \textbf{10}^{-6}$$

and its direction is as indicated on Fig. 6-9b. Note that the clockwise rotation from the x-axis in Fig. 6-9b is one-half the clockwise rotation from x to $\varepsilon_{\max}$ on Mohr's circle of strain.

Problems 6-12 to 6-14

6-4   STRESSES FROM STRAIN GAGE DATA

When stresses must be determined experimentally, strains are measured and used to calculate stresses. Unfortunately, it is impractical to measure shear strain directly. Consequently, in the general case, the three strains ε_x, ε_y, and γ_{xy} must be deduced from three values of axial strain measured, usually, by strain rosettes such as those of Fig. 6-6b and c. The strain readings enable us to write (6-17) three times as three simultaneous equations in the three unknowns, ε_x, ε_y, and γ_{xy}. Solving three simultaneous equations may be a rather laborious task, however, if the gages are oriented relative to the coordinate axes in the most advantageous manner, the equations can be simplified. The best orientation will have one gage or, if possible, two gages, parallel to the x-axis or the y-axis.

Example 6-2

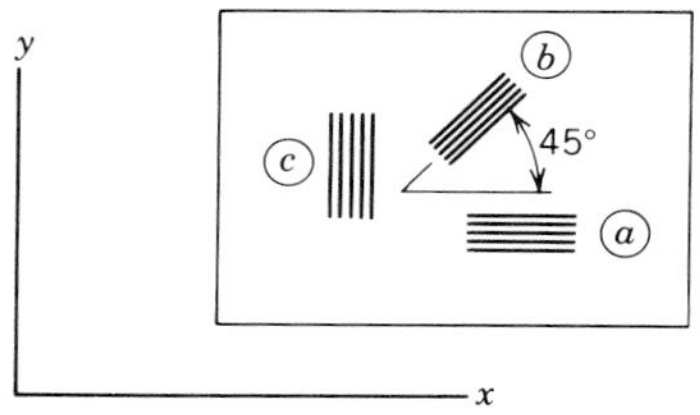

FIG. 6-10

A rectangular strain rosette mounted on a steel plate as shown in Fig. 6-10 provided the following strain gage data:

$$\varepsilon_a = 450 \times 10^{-6}$$

$$\varepsilon_b = -300 \times 10^{-6}$$

$$\varepsilon_c = 200 \times 10^{-6}$$

Determine the strains ε_x, ε_y, γ_{xy} and the stresses σ_x, σ_y, τ_{xy}.

Solution. Elastic constants for steel from Table A are

$$E = 200\ 000 \text{ MPa}$$

$$v = 0.3$$

$$G = 77\ 000 \text{ MPa}$$

$$\frac{E}{1 - v^2} = \frac{0.2 \times 10^6}{1 - 0.3^2} = 0.220 \times 10^6 \text{ MPa}$$

Converting strains to units of μm/m gives

$$\varepsilon_a = 450 \text{ at } \theta = 0°$$

$$\varepsilon_b = -300 \text{ at } \theta = 45°$$

$$\varepsilon_c = 200 \text{ at } \theta = 90°$$

Substituting into (6-17), data from each gage in turn:

Gage a:

$$\cos^2 0\ \varepsilon_x + \sin^2 0\ \varepsilon_y - \sin 0 \cos 0\ \gamma_{xy} = 450$$

$$1\ \varepsilon_x + 0\ \varepsilon_y - 0\ \gamma_{xy} = 450$$

$$\varepsilon_x = 450\ \frac{\mu m}{m} = \mathbf{450 \times 10^{-6}}$$

Gage c:

$$\cos^2 90\ \varepsilon_x + \sin^2 90\ \varepsilon_y - \sin 90 \cos 90\ \gamma_{xy} = 200$$

$$0 \; \varepsilon_x + 1 \; \varepsilon_y - 1 \times 0 \; \gamma_{xy} = 200$$

$$\varepsilon_y = 200 \, \frac{\mu m}{m} = \textbf{200} \times \textbf{10}^{-6}$$

Gage b:

$$\cos^2 45 \; \varepsilon_x + \sin^2 45 \; \varepsilon_y - \sin 45 \cos 45 \; \gamma_{xy} = -300$$

$$0.5 \; \varepsilon_x + 0.5 \; \varepsilon_y - 0.5 \; \gamma_{xy} = -300$$

$$0.5 \times 450 + 0.5 \times 200 - 0.5 \times \gamma_{xy} = -300$$

$$\gamma_{xy} = 600 + 450 + 200$$

$$= 1250 \, \frac{\mu m}{m} = \textbf{1250} \times \textbf{10}^{-6}$$

From these strains and the known elastic constants, the stresses can be determined.

$$\sigma_x = \frac{E}{1 - v^2} (\varepsilon_x + v\varepsilon_y)$$

$$= 0.220 \times 10^6 (450 + 0.30 \times 200) \, 10^{-6} = \textbf{112 MPa}$$

$$\sigma_y = \frac{E}{1 - v^2} (v\varepsilon_x + \varepsilon_y)$$

$$= 0.220 \times 10^6 (0.30 \times 450 + 200) \times 10^{-6} = \textbf{74 MPa}$$

$$\tau_{xy} = G \, \gamma_{xy} = 0.077 \times 10^6 \times 1250 \times 10^{-6} = \textbf{96 MPa}$$

Example 6-2

A rectangular strain rosette mounted on a steel plate as shown in Fig. 6-10 provided the following strain gage data:

$$\varepsilon_a = 450 \times 10^{-6}$$

$$\varepsilon_b = -300 \times 10^{-6}$$

$$\varepsilon_c = 200 \times 10^{-6}$$

Determine the strains ε_x, ε_y, γ_{xy} and the stresses σ_x, σ_y, τ_{xy}.

Solution. Elastic constants for steel from Table A are

$$E = 29 \times 10^3 \text{ ksi}$$

$$v = 0.3$$

$$G = 11 \times 10^3 \text{ ksi}$$

$$\frac{E}{1 - v^2} = \frac{29 \times 10^3}{1 - 0.3^2} = 31.87 \times 10^3 \text{ kip/in.}^2$$

Converting strains to units of μin./in. gives

$$\varepsilon_a = 450 \text{ at } \theta = 0°$$

$$\varepsilon_b = -300 \text{ at } \theta = 45°$$

$$\varepsilon_c = 200 \text{ at } \theta = 90°$$

Substituting into (6-17), data from each gage in turn:

Gage a:

$$\cos^2 0 \, \varepsilon_x + \sin^2 0 \, \varepsilon_y - \sin 0 \cos 0 \, \gamma_{xy} = 450$$

$$1 \, \varepsilon_x + 0 \, \varepsilon_y - 0 \, \gamma_{xy} = 450$$

$$\varepsilon_x = 450 \, \frac{\mu\text{in.}}{\text{in.}} = \mathbf{450 \times 10^{-6}}$$

Gage c:

$$\cos^2 90 \, \varepsilon_x + \sin^2 90 \, \varepsilon_y - \sin 90 \cos 90 \, \gamma_{xy} = 200$$

$$0 \, \varepsilon_x + 1 \, \varepsilon_y - 1 \times 0 \, \gamma_{xy} = 200$$

$$\varepsilon_y = 200 \, \frac{\mu\text{in.}}{\text{in.}} = \mathbf{200 \times 10^{-6}}$$

Gage b:

$$\cos^2 45 \, \varepsilon_x + \sin^2 45 \, \varepsilon_y - \sin 45 \cos 45 \, \gamma_{xy} = -300$$

$$0.5 \, \varepsilon_x + 0.5 \, \varepsilon_y - 0.5 \, \gamma_{xy} = -300$$

$$0.5 \times 450 + 0.5 \times 200 - 0.5 \times \gamma_{xy} = -300$$

$$\gamma_{xy} = 600 + 450 + 200$$

$$= 1250 \, \frac{\mu\text{in.}}{\text{in.}} = \mathbf{1250 \times 10^{-6}}$$

From these strains and the known elastic constants, the stresses can be determined.

$$\sigma_x = \frac{E}{1 - v^2} (\varepsilon_x + v\varepsilon_y)$$

$$= 31.87 \times 10^3 (450 + 0.30 \times 200) \, 10^{-6} = \mathbf{16.25 \ kip/in.^2}$$

$$\sigma_y = \frac{E}{1 - v^2} (v\varepsilon_x + \varepsilon_y)$$

$$= 31.87 \times 10^3 (0.30 \times 450 + 200) \times 10^{-6} = \mathbf{10.68 \ kip/in.^2}$$

$$\tau_{xy} = G \, \gamma_{xy} = 11 \times 10^3 \times 1250 \times 10^{-6} = \mathbf{13.75 \ kip/in.^2}$$

Problems 6-15 to 6-17

6-5 PRINCIPAL STRAINS USING STRAIN GAGE ROSETTES

The equations giving the *principal strains* ε_1 and ε_2 may be obtained by using the similarity between stress and strain as noted in Section 6-3. That is, by replacing σ_x with ε_x, σ_y with ε_y, τ_{xy} with $\gamma_{xy}/2$, and τ_θ with $\gamma_\theta/2$ in (5-3), (5-4), and (5-5), we obtain

$$\varepsilon_1, \varepsilon_2 = \frac{\varepsilon_x + \varepsilon_y}{2} \pm \sqrt{\left(\frac{\varepsilon_x - \varepsilon_y}{2}\right)^2 + \left(\frac{\gamma_{xy}}{2}\right)^2} \qquad \textbf{(6-22)}$$

$$\frac{\gamma_{max}}{2} = \sqrt{\left(\frac{\varepsilon_x - \varepsilon_y}{2}\right)^2 + \left(\frac{\gamma_{xy}}{2}\right)^2} \qquad \textbf{(6-23)}$$

$$\alpha = -\frac{1}{2}\tan^{-1}\left(\frac{\gamma_{xy}}{\varepsilon_x - \varepsilon_y}\right) \qquad \textbf{(6-24)}$$

These equations require the determination of ε_x, ε_y, and γ_{xy} as a preliminary step. A direct determination of ε_1 and ε_2 from the three measured ε_θ values would make this step unnecessary; the equations for such a direct determination will now be developed.

Equation (6-17) gives the strain, ε_θ, in terms of the normal and shear strains in the x and y coordinate directions. The angular relationships between axes, principal strain directions, and the three strains ε_a, ε_b, and ε_c, are shown in Fig. 6-11. Suppose now that the x-direction, being arbitrary, is made to coincide with the direction of the principal stress, 1, and that strain has been measured in the three directions a, b, and c. Equation (6-17) then gives

$$\varepsilon_a = \varepsilon_1 \cos^2\alpha_a + \varepsilon_2 \sin^2\alpha_a - 0 \sin\alpha_a \cos\alpha_a$$

$$\varepsilon_b = \varepsilon_1 \cos^2(\alpha_a + \theta_b) + \varepsilon_2 \sin^2(\alpha_a + \theta_b) - 0 \sin(\alpha_a + \theta_b) \cos(\alpha_a + \theta_b)$$

$$\varepsilon_c = \varepsilon_1 \cos^2(\alpha_a + \theta_c) + \varepsilon_2 \sin^2(\alpha_a + \theta_c) - 0 \sin(\alpha_a + \theta_c) \cos(\alpha_a + \theta_c)$$

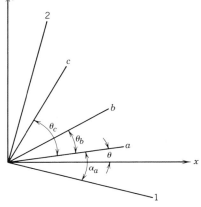

FIG. 6-11

or

$$\varepsilon_a = \varepsilon_1 \cos^2\alpha_a + \varepsilon_2 \sin^2\alpha_a \qquad \textbf{(6-25a)}$$

$$\varepsilon_b = \varepsilon_1 \cos^2(\alpha_a + \theta_b) + \varepsilon_2 \sin^2(\alpha_a + \theta_b) \qquad \textbf{(6-25b)}$$

$$\varepsilon_c = \varepsilon_1 \cos^2(\alpha_a + \theta_c) + \varepsilon_2 \sin^2(\alpha_a + \theta_c) \qquad \textbf{(6-25c)}$$

Solving these equations for ε_1, ε_2, and α_a would give the principal strains and their direction directly in terms of the measured strains and the known angles θ_b and θ_c. However, it is not practical to solve the general equations, but they can be solved for the particular angles used in rosettes.

Two commonly used rosettes are shown in Figs. 6-6b and 6-6c, the rectangular rosette and the delta rosette. The angles as defined by Fig. 6-12 give:

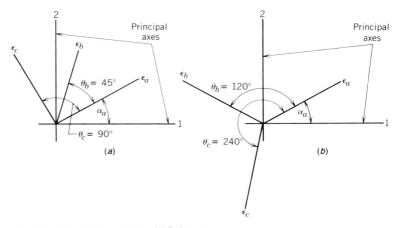

FIG. 6-12 (a) Rectangular rosette. (b) Delta rosette.

for the rectangular rosette: $\theta_b = 45°$, $\theta_c = 90°$

for the delta rosette: $\theta_b = 120°$, $\theta_c = 240°$

Solving (6-25) for ε_1, ε_2, and α_a for the rectangular rosette gives:

$$\varepsilon_1, \varepsilon_2 = \frac{\varepsilon_a + \varepsilon_c}{2} \pm \sqrt{\frac{(\varepsilon_a - \varepsilon_b)^2 + (\varepsilon_c - \varepsilon_b)^2}{2}} \tag{6-26}$$

$$2\alpha_a = \tan^{-1}\left(\frac{\varepsilon_a - 2\varepsilon_b + \varepsilon_c}{\varepsilon_a - \varepsilon_c}\right) \tag{6-27}$$

For the *delta* rosette solving for ε_1, ε_2, and α_a gives:

$$\varepsilon_1, \varepsilon_2 = \frac{\varepsilon_a + \varepsilon_b + \varepsilon_c}{3} \pm \sqrt{\frac{(2\varepsilon_a - \varepsilon_b - \varepsilon_c)^2}{9} + \frac{(\varepsilon_b - \varepsilon_c)^2}{3}} \tag{6-28}$$

$$2\alpha_a = \tan^{-1}\frac{\sqrt{3}(\varepsilon_b - \varepsilon_c)}{2\varepsilon_a - \varepsilon_b - \varepsilon_c} \tag{6-29}$$

Since the 1-axis and the 2-axis are at right angles, (6-16b) can be used to relate principal stress to principal strain giving

$$\sigma_1 = \frac{E}{1 - v^2}(\varepsilon_1 + v\varepsilon_2) \tag{6-30a}$$

$$\boxed{\sigma_2 = \frac{E}{1-v^2}(v\varepsilon_1 + \varepsilon_2)} \qquad \text{(6-30b)}$$

Thus the principal stresses are determined from the three measured values of strain by substitution into (6-26) or (6-28) and (6-30).

Using (6-27) or (6-29) to determine the direction of the principal strains can often give results that are in error by 90°. This also applies to the determination of the direction of the principal stress by (5-5). Equations can be devised to avoid this error, but it is recommended that when directions are required, Mohr's circle be used to obtain the correct direction.

Example 6-3

A rectangular strain gage rosette is mounted in an arbitrary direction on a steel plate. The strains are measured as

$$\varepsilon_a = 300 \times 10^{-6} \qquad \varepsilon_b = 400 \times 10^{-6} \qquad \varepsilon_c = -350 \times 10^{-6}$$

Determine the principal strains and their direction as well as the principal stresses.

Solution

$$\varepsilon_1, \varepsilon_2 = \frac{\varepsilon_a + \varepsilon_c}{2} \pm \sqrt{\frac{(\varepsilon_a - \varepsilon_b)^2 + (\varepsilon_c - \varepsilon_b)^2}{2}}$$

$$= \frac{(300 - 350)}{2}10^{-6} \pm \sqrt{\frac{(300 - 400)^2 + (-350 - 400)^2}{2}}10^{-6}$$

$$= -25 \times 10^{-6} \pm 535 \times 10^{-6}$$

$$\varepsilon_1 = \mathbf{510 \times 10^{-6}}$$

$$\varepsilon_2 = -\mathbf{560 \times 10^{-6}}$$

$$2\alpha_a = \tan^{-1}\left(\frac{\varepsilon_a - 2\varepsilon_b + \varepsilon_c}{\varepsilon_a - \varepsilon_c}\right)$$

$$= \tan^{-1}\left(\frac{300 - 2 \times 400 - 350}{300 - (-350)}\right)\frac{10^{-6}}{10^{-6}}$$

$$= \tan^{-1}\left(\frac{-850}{650}\right) = \tan^{-1} - 1.308$$

$$= 52.6°$$

$$\alpha_a = 26.3°$$

The maximum principal strain occurs along an axis **26.3°** **counterclockwise** from the *a*-axis. Next, the principal stresses are obtained by substitution into (6-16c).

$$\sigma_1 = \frac{E}{1 - v^2}(\varepsilon_1 + v\varepsilon_2)$$

$$= \frac{200 \times 10^3}{1 - 0.3^2}[510 + 0.3 \times (-560)] \times 10^{-6} = \textbf{75.2 MPa}$$

$$\sigma_2 = \frac{E}{1 - v^2}(v\varepsilon_1 + \varepsilon_2) = \frac{200 \times 10^3}{1 - 0.3^2}(0.3 \times 510 - 560) \times 10^{-6}$$

$$= \textbf{-89.5 MPa}$$

Example 6-3

A rectangular strain gage rosette is mounted in an arbitrary direction on a steel plate. The strains are measured as

$$\varepsilon_a = 300 \times 10^{-6} \qquad \varepsilon_b = 400 \times 10^{-6} \qquad \varepsilon_c = -350 \times 10^{-6}$$

Determine the principal strains and their direction as well as the principal stresses.

Solution

$$\varepsilon_1, \varepsilon_2 = \frac{\varepsilon_a + \varepsilon_c}{2} \pm \sqrt{\frac{(\varepsilon_a - \varepsilon_b)^2 + (\varepsilon_c - \varepsilon_b)^2}{2}}$$

$$= \frac{(300 - 350)}{2} 10^{-6} \pm \sqrt{\frac{(300 - 400)^2 + (-350 - 400)^2}{2}} 10^{-6}$$

$$= -25 \times 10^{-6} \pm 535 \times 10^{-6}$$

$$\varepsilon_1 = \textbf{510} \times \textbf{10}^{-6}$$

$$\varepsilon_2 = -\textbf{560} \times \textbf{10}^{-6}$$

$$2\alpha_a = \tan^{-1}\left(\frac{\varepsilon_a - 2\varepsilon_b + \varepsilon_c}{\varepsilon_a - \varepsilon_c}\right)$$

$$= \tan^{-1}\left(\frac{300 - 2 \times 400 - 350}{300 - (-350)}\right)\frac{10^{-6}}{10^{-6}}$$

$$= \tan^{-1}\left(\frac{-850}{650}\right) = \tan^{-1} - 1.308$$

$$= 52.6°$$

$$\alpha_a = 26.3°$$

The maximum principal strain occurs along an axis **26.3°** **counterclockwise** from the *a*-axis. Next, the principal stresses are obtained by substitution into (6-16c).

$$\sigma_1 = \frac{E}{1 - v^2} (\varepsilon_1 + v\varepsilon_2)$$

$$= \frac{29 \times 10^3}{1 - 0.3^2} [510 + 0.3 \times (-560)] \times 10^{-6} = \textbf{10.90 kip/in.}^2$$

$$\sigma_2 = \frac{E}{1 - v^2} (v\varepsilon_1 + \varepsilon_2) = \frac{29 \times 10^3}{1 - 0.3^2} (0.3 \times 510 - 560) \times 10^{-6}$$

$$= \textbf{-12.97 kip/in.}^2$$

Problems 6-18 to 6-20

6-6 SUMMARY

The material presented in this chapter shows how stress and strain are connected and permits the calculation of the distortion of a body under a given state of stress. It has been shown that strain exists in a direction perpendicular to the direction of an applied load and that this strain is equal to Poisson's ratio times the strain in the direction of the applied load but of opposite sign. A complete picture of the strain field can be obtained graphically by means of Mohr's circle of strain. An alternative solution, involving any problem where strains are measured using rectangular or delta rosettes and where the principal strains are of prime interest, can be obtained using the equations given in Section 6-5. Using these equations, the principal strains and principal stresses may be obtained directly from the measured strain data. However, some ambiguity exists in the solutions thus obtained. This arises because there are two angles, within the range 0 to 360°, having any given tangent. These angles differ by 180° and hence the half-angles differ by 90°. If we are looking for the direction of the maximum principal strain, equation (6-24) gives two angles. A sketch of Mohr's circle will reveal which angle is correct.

PROBLEMS

6-1 An L200 × 150 × 25 (L8 × 6 × 1) is 5 m (16 ft) long and carries an axial tensile load of 1600 kN (360 kip). Determine the change in each of the four given dimensions due to the load.

6-2 A W690 × 140 (W27 × 94) beam, made of high-strength structural steel, is subjected to the maximum allowable bending stress. Determine the changes in the width and thickness of the compression flange.

6-3 Calculate the changes in the dimensions of the given steel plate when a tensile load of 750 kN (175 kip) is applied in the x-direction and, simultaneously, a tensile load of 1440 kN (335 kip) in the y-direction. Assume that the loads give uniform stresses throughout the plate.

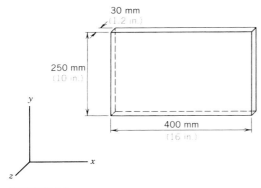

PROBLEM 6-3

6-4 Calculate the changes in dimensions required in Prob. 6-3 when the load in the y-direction is compressive.

6-5 A 300-mm (12-in.) steel pipe is subjected to an axial tensile load of 2000 kN (450 kip). Determine the change in (a) the circumference, (b) the diameter, and (c) the wall thickness due to the load.

6-6 A 300-mm (12-in.) steel pipe has capped ends and contains fluid under a pressure of 12 500 kPa (1800 lb/in.²). Determine the change in diameter due to the fluid pressure.

6-7 A steel gas line has an inside diameter of 1.20 m (4 ft) and an outside diameter of 1.26 m (4.2 ft). When gas under a pressure of 12 500 kPa (1800 psi) is admitted to the pipe, determine the change in the following:

a. The circumference.
b. The diameter.
c. The distance between two points that were initially 100 m (325 ft) apart measured along the length of the pipe.

6-8 A steel pressure vessel has a cylindrical central portion that is 12 m (40 ft) long, has an outside diameter of 1.3 m (4.25 ft), and a wall thickness of 45 mm (1.77 in.). The ends are capped. By how much will the diameter and the length of the cylindrical portion of the vessel change when it contains fluid at 10 000 kPa (1450 psi)?

6-9 Determine the changes in the given dimensions (including the angle) when loads are applied that give uniform stresses throughout the plate of

$$\sigma_x = 250 \text{ MPa (36 ksi)}$$

$$\sigma_y = -150 \text{ MPa } (-22 \text{ ksi})$$

$$\tau_{xy} = 200 \text{ MPa (29 ksi)}$$

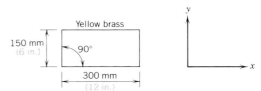

PROBLEM 6-9

6-10 Determine the strains at point A in the web of the given beam.

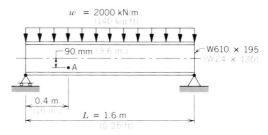

PROBLEM 6-10

6-11 The given aluminum block is subjected to tensile loads of:

90 kN (21 kip) in the *x*-direction

60 kN (14 kip) in the *y*-direction

180 kN (42 kip) in the *z*-direction

Assuming that the stresses are uniformly distributed, calculate the change in

a. The dimensions.
b. The volume.

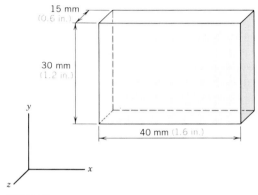

PROBLEM 6-11

6-12 Two lines, drawn on an unstressed aluminum alloy plate, have the given dimensions. After loading, the horizontal line increases in length by 0.50 mm (0.02 in.) while the vertical line decreases by 0.10 mm (0.004 in.). The angle increases to 90.10 degrees. Determine the magnitude and direction of the maximum principal strain and the maximum principal stress.

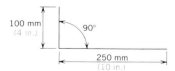

PROBLEM 6-12

6-13 A block of magnesium alloy is subjected to the given loads, which are applied in such a manner that the stresses are uniform throughout the block. By means of a Mohr's circle of strain, determine the strain in the elements of

a. The delta rosette (see Fig. 6-6).
b. The rectangular rosette.

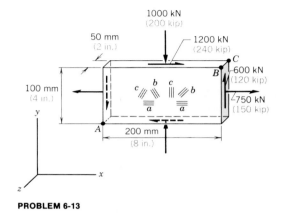

PROBLEM 6-13

6-14 For the material and loads given in Prob. 6-13, determine the change in length of

a. The diagonal *A-B*.
b. The edge *B-C*.

6-15 An equilateral triangular strain rosette attached to a steel plate gives strain values of:

$$\varepsilon_a = 50 \times 10^{-6}$$

$$\varepsilon_b = 100 \times 10^{-6}$$

$$\varepsilon_c = 40 \times 10^{-6}$$

Determine the maximum principal strain and the maximum principal stress. Draw a sketch showing the direction of the maximum principal strain.

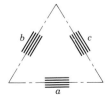

PROBLEM 6-15

6-16 Strain gages are placed as shown on a steel plate and the given strains are measured.

a. Determine the maximum principal stress and show its direction on a sketch.

b. If the material is medium-strength structural steel, is it overloaded?

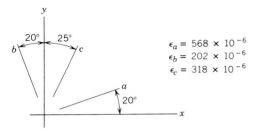

$$\epsilon_a = 568 \times 10^{-6}$$
$$\epsilon_b = 202 \times 10^{-6}$$
$$\epsilon_c = 318 \times 10^{-6}$$

PROBLEM 6-16

6-17 Strain gages are placed as shown on a steel plate and the given strains are measured.

a. Determine the maximum principal stress and show its direction on a sketch.

b. If the material is low-strength structural steel, is it overloaded?

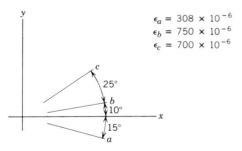

$$\epsilon_a = 308 \times 10^{-6}$$
$$\epsilon_b = 750 \times 10^{-6}$$
$$\epsilon_c = 700 \times 10^{-6}$$

PROBLEM 6-17

6-18 For each of the following rectangular rosette problems, determine the magnitude of the principal strains and their directions relative to the *a*-axis. All strains are in μm/m (μin./in.).

(a)	$\varepsilon_a = 600$	$\varepsilon_b = -300$	$\varepsilon_c = -100$
(b)	$\varepsilon_a = -550$	$\varepsilon_b = 1240$	$\varepsilon_c = 800$
(c)	$\varepsilon_a = 720$	$\varepsilon_b = 900$	$\varepsilon_c = -460$

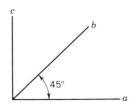

PROBLEM 6-18

6-19 A rectangular rosette, attached to a steel plate, furnished the following strain data in the three directions indicated in the illustration:

$$\varepsilon_a = 425 \times 10^{-6}$$

$$\varepsilon_b = 550 \times 10^{-6}$$

$$\varepsilon_c = 75 \times 10^{-6}$$

Determine the principal strains, principal stresses, and the directions in which they act. On a sketch show the direction in which the maximum principal stress acts relative to the *a*-axis.

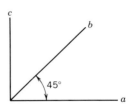

PROBLEM 6-19

6-20 The ends of the given pipe are capped and fluid inside the pipe is under pressure, *p*. In addition, an axial force, *P*, and a torque, *T*, are applied. There is no bending moment in the pipe. Calculate *p*, *P*, and *T* if the following strains are measured on the surface:

$$\varepsilon_a = 40 \times 10^{-6}$$

$$\varepsilon_b = 140 \times 10^{-6}$$

$$\varepsilon_c = 100 \times 10^{-6}$$

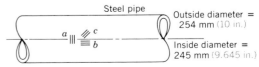

PROBLEM 6-20

Advanced Problems

6-A A block of elastic material, measuring $A \times B \times C$, is submerged in fluid and subjected to pressure p.

   a. Determine the change in volume of the block in terms of the pressure, dimensions, and elastic constants.

   b. Determine the change in volume if the material is rubber with a Poisson's ratio of 0.5.

   c. Determine the change in volume if the material has a Poisson's ratio of 0.6.

6-B The pipe described in Prob. 6-20 is loaded by internal pressure only. The pressure is not known but can be deduced from the strain measured by a single gage placed on the surface and oriented in any known direction. Determine the pressure if the gage is in the direction of:

   a. Gage a and indicates a strain of 500×10^{-6}.

   b. Gage b and indicates a strain of 150×10^{-6}.

   c. Gage c and indicates a strain of 200×10^{-6}.

6-C From the experimental data, determine (a) the load P and (b) the distance X.

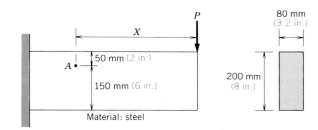

Strain Measurements

$\epsilon_a = 249 \times 10^{-6}$

$\epsilon_b = -228 \times 10^{-6}$

$\epsilon_c = 422 \times 10^{-6}$

Strain Rosette at A

PROBLEM 6-C

TORSION

7-1 INTRODUCTION

In many machines power is transmitted from one part to another by a rotating shaft subjected to torque. When the shaft carries only a torque load, the cross section can be shaped to optimize its ability to carry that load. The least amount of material will be required if the shaft has a circular cross section. We will see later that a hollow, circular cross section is superior to a solid section; however, the added cost of forming the hollow section may more than offset the saving in material.

The rolled-steel sections, which are commonly used for beams, are proportioned to give good performance in bending but, as we will see in this chapter, are not well shaped for carrying torsional loads. In the beams that have been considered up to this point, the loads were always in the plane of symmetry of the member and the familiar bending-stress and shear-stress formulas were applicable. If the loads on an S-shape do not lie in the plane of symmetry of the section, the member will be subjected to torsion as well as bending and both must be taken into account in the design. This load arrangement is avoided where possible and does not occur frequently. However, practical considerations in the design of tall buildings often lead to a support system for the brick veneer that subjects a member to both types of load. The brick work is usually supported on angle brackets at each floor level. This imposes a load on the angle that results in torque combined with bending and complicates the design of the brackets.

7-2 TORSIONAL STRESS IN A CYLINDRICAL ROD

Imagine that we have a cylindrical rod made of styrofoam or other very flexible material and that we make some observations on its behavior when subjected to torque. To make its behavior more readily apparent, suppose that we draw lines, such as those shown in Fig. 7-1a, on the surface. The markings consist of a set of longitudinal lines parallel to the axis and a set of circumferential lines. After a torsional or twisting load is applied, the cylinder would take on the appearance of Fig. 7-1b. If we were able to make the necessary measurements, we would find that all circular segments remained unchanged in length and merely moved a small distance in the circumferential direction. The amount of motion would be found to vary linearly from zero at the fixed end to a maximum at the loaded end.

The segments of the longitudinal lines would also be found to be unchanged in length but to be no longer parallel to the centerline of the cylinder. From these observations it would seem reasonable to assume, as we did with beams in bending, that initially plane sections remain plane after deformation. In this case, however, *planes that are initially parallel remain parallel but are rotated, relative to one another, about the cylinder centerline, by a small angle.*

A small rectangle *ABDC*, see Fig. 7-1, on the surface will change to a parallelogram *A'B'D'C'* without any alteration in the lengths of the sides. The angle *A'B'D'* will be larger than a right angle by a small amount, γ.

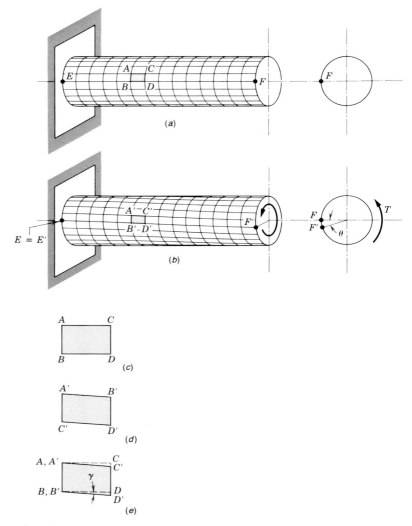

FIG. 7-1 Cylindrical rod subjected to torsion. (*a*) Surface markings on unloaded cylinder. (*b*) Surface markings on loaded cylinder. (*c*) Surface rectangle before loading. (*d*) Surface rectangle after loading. (*e*) Superimposed rectangles.

From these observations it is evident that there exists no axial or circumferential strain but only shear strain. The rod has complete polar symmetry, and the same observations would be made regardless of the surface position from which the rectangle was taken.

Examining the action of the line *EF* under load, we see in Fig. 7-2 that for small γ (in radians), the displacement of the point *F* is given by

$$FF' = L\gamma \tag{7-1}$$

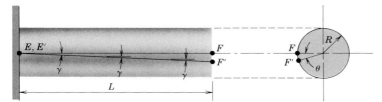

FIG. 7-2

and the rotation of the end (in radians) is

$$\theta = \frac{F'F}{R}$$

$$= \frac{L\gamma}{R} \qquad (7\text{-}2)$$

Now consider the internal lines shown in Fig. 7-3a. The line *GJH*, which in the unloaded state is parallel to the axis and at a distance *r* from it, moves to *GJ'H'* with the distance *HH'* being given by

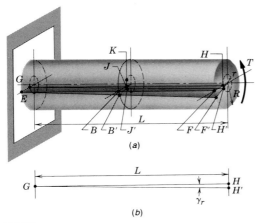

FIG. 7-3

$$HH' = FF'\frac{r}{R}$$

The inclination of the line *GJ'H'* will then be

$$\frac{HH'}{L} = \frac{FF'r}{LR} \qquad (7\text{-}3)$$

and the angle that it makes with a small circular arc such as *JK*, which was initially a right angle, will be increased by the angle of inclination given in (7-3). This increase in angle, the shearing strain at a radial distance r, can be represented by γ_r and determined from Fig. 7-3b as

$$\gamma_r = \frac{HH'}{L}$$

$$= \frac{FF'r}{LR} \tag{7-4}$$

Substituting for FF' from (7-1) gives

$$\gamma_r = \frac{L\gamma r}{LR} = \frac{\gamma r}{R} \tag{7-5}$$

We can multiply both sides of (7-5) by G to get

$$G\gamma_r = \frac{G\gamma r}{R}$$

which can be written in terms of stresses by (6-11a) as

$$\tau_r = \frac{\tau r}{R}$$

where τ_r is the shear stress at distance r from the axis and τ is the shear stress at the surface. This tells us that *shear stress in a cylindrical rod subjected to torsion varies linearly from zero at the center of the rod to a maximum at the surface.* It is interesting to note the similarity between this discovery and the conclusion reached in Chapter 2 with respect to the variation in bending stress with distance from the neutral axis. In order not to loose sight of the fact that τ is the maximum shear stress, it will be replaced by $\tau_{\max}$ and the shear stress equation will become

$$\boxed{\tau_r = \tau_{\max} \frac{r}{R}} \tag{7-6}$$

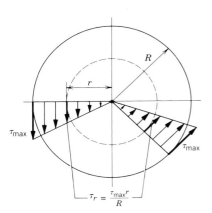

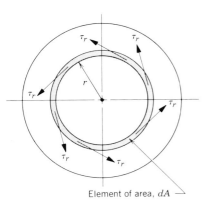

FIG. 7-4 Shear stress distribution in a circular rod.

$$\tau_r = \frac{\tau_{\max} r}{R}$$

The distribution of stress on a typical cross section is shown in Fig. 7-4. Along any radial line the stress varies linearly with r and is always normal to the radial line.

To find the relationship between torque and stress we will consider the element of area shown in Fig. 7-5. An element of this shape with small radial thickness is chosen so that all points in the element will have substantially the same stress, τ_r. The shear stress on the element gives a total force of $\tau_r\, dA$ with all elements of this force acting at normal distance r from the center of the circle. This gives a moment or torque of

$$dT = \tau_r\, dA r$$

$$= \tau_r r\, dA$$

Substituting for τ_r from (7-6) gives

$$dT = \frac{\tau_{\max}}{R} r^2\, dA$$

Element of area, dA

FIG. 7-5

The total torque on the cross section is obtained by integration as

$$T= \int_{area} \frac{\tau_{max}}{R} r^2 \, dA$$

But τ_{max} and R are constants; hence

$$T= \frac{\tau_{max}}{R} \int_{area} r^2 \, dA \qquad \textbf{(7-7)}$$

Since stress is usually the quantity to be determined, the equation is more convenient in the form

$$\tau_{max} = \frac{TR}{\int_{area} r^2 \, dA}$$

The integral is a property of the cross section and is usually evaluated separately. For this purpose it is given the name *polar moment of inertia* and the symbol *J*. Thus

$$\boxed{J= \int_{area} r^2 \, dA} \qquad \textbf{(7-8)}$$

and

$$\boxed{\tau_{max} = \frac{TR}{J}} \qquad \textbf{(7-9)}$$

This equation is known as the *torsion formula* for circular shafts. It relates the maximum shearing stress, τ_{max}, to the torque, T, and to the cross-section parameters R and J. Because the circular cross section occurs frequently in torsional problems, it is useful to derive the formula for *J* for circular shafts. To determine *J*, we select an element of area, as indicated in Fig. 7-6, such that all points are at substantially the same distance from the center of the circle. Then

$$dA = 2\pi r \, dr$$

and from (7-8)

$$J= \int_0^R r^2 \, dA$$

$$= \int_0^R r^2 2\pi r \, dr$$

$$= 2\pi \int_0^R r^3 \, dr = 2\pi \frac{r^4}{4} \Big]_0^R$$

$$\boxed{J= \frac{\pi}{2} R^4 = \frac{\pi}{32} D^4} \qquad \textbf{(7-10)}$$

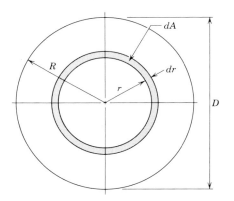

FIG. 7-6 Element for determination of the polar moment of inertia.

This formula for a *solid circular shaft* agrees with the equation listed in Table H for a circle.

Using (7-10), we can rewrite (7-9) as

$$\tau_{max} = \frac{TR}{\frac{\pi}{2}R^4}$$

$$= \frac{2}{\pi}\frac{T}{R^3}$$

$$\boxed{\tau_{max} = \frac{16}{\pi}\frac{T}{D^3}} \tag{7-9a}$$

Example 7-1

Determine the maximum shear stress in the rod of Fig. 7-7.

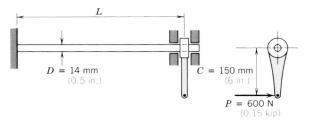

$D = 14$ mm
(0.5 in.)

$C = 150$ mm
(6 in.)

$P = 600$ N
(0.15 kip)

FIG. 7-7 Rod subjected to torsion only.

Solution. The supports at the right-hand end of the rod ensure that the rod is subjected to torque only; that is, the rod can rotate but not bend.

$$T = 600 \times 150 \text{ N} \cdot \text{mm}$$

$$= 90\,000 \text{ N} \cdot \text{mm}$$

From (7-10)

$$J = \frac{\pi}{32}D^4$$

$$= \frac{\pi}{32}14^4 \text{ mm}^4 = 3771 \text{ mm}^4$$

Substituting into (7-9) gives

$$\tau_{max} = \frac{TR}{J} = \frac{90\,000 \times 7}{3771}\frac{\text{N} \cdot \text{mm} \cdot \text{mm}}{\text{mm}^4}$$

$$= 167\frac{\text{N}}{\text{mm}^2} = \textbf{167 MPa}$$

Example 7-1

Determine the maximum shear stress in the rod of Fig. 7-7.

Solution. The supports at the right-hand end of the rod ensure that the rod is subject to torque only; that is, the rod can rotate but not bend.

$$T = 0.15 \times 6$$

$$= 0.90 \text{ kip} \cdot \text{in.}$$

From (7-10)

$$J = \frac{\pi}{2} R^4 = \frac{\pi}{2} \left(\frac{0.5}{2} \right)^4 \text{ in.}^4$$

$$= 6.14 \times 10^{-3} \text{ in.}^4$$

Substituting into (7-9)

$$\tau_{max} = \frac{TR}{J} = \frac{0.90 \times 0.25}{6.14 \times 10^{-3}} \frac{\text{kip} \cdot \text{in.} \cdot \text{in.}}{\text{in.}^4}$$

$$= \textbf{36.65 kip/in.}^2$$

Before going on to more complex torsion problems, it is useful to summarize the three key concepts used in deriving the torsion formula (7-9) for circular members:

1. Plane sections remain plane.
2. The shearing strain, γ_r, varies linearly from the centerline of the rod.
3. Shearing stress, τ_r, is related to γ_r through the shear modulus, G, and hence varies linearly from zero at the center to a maximum at the surface. This indicates that *elastic* action has been assumed.

We will later also be interested in the *inelastic* action of torsion members. All of the three key assumptions apply not only to solid circular members, but also to *hollow* circular members or *tubes*. The only fundamental difference in applying (7-9) to tubes, comes from the polar moment of inertia, J. Referring to Fig. 7-8, we see that the polar moment of inertia for a hollow circular cross section is determined by integrating between the limits R_i, inside radius, and R_o, outside radius. From (7-8)

$$J = \int_{area} r^2 \, dA$$

$$= \int_{R_i}^{R_o} r^2 2\pi r \, dr = 2\pi \int_{R_i}^{R_o} r^3 \, dr$$

$$= 2\pi \left(\frac{R_o^4}{4} - \frac{R_i^4}{4} \right)$$

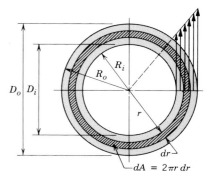

FIG. 7-8 Shear stress distribution in a hollow circular section.

$$J = \frac{\pi}{2}(R_o^4 - R_i^4) = \frac{\pi}{32}(D_o^4 - D_i^4) \tag{7-11}$$

Example 7-2

If the solid rod in Example 7-1 is replaced with a tube having an inner diameter equal to the rod's diameter of 14 mm, determine the required outer diameter if the tube is to carry the same torque without exceeding the maximum shear stress of 167 MPa.

Solution. From (7-9) and (7-11)

$$\tau_{max} = \frac{TR_o}{(\pi/2)(R_o^4 - R_i^4)}$$

and substituting for the known quantities, we obtain

$$167 = \frac{90 \times 10^3 \times R_o}{(\pi/2)(R_o^4 - 7^4)} \frac{N \cdot mm \cdot mm}{mm^4} = \frac{N}{mm^2}$$

Solving by trial and error for R_o, we determine outer radius as

$$R_o = \textbf{8.55 mm}$$

Example 7-2

If the solid rod in Example 7-1 is replaced with a tube having an inner diameter equal to the rod's diameter of 0.5 in., determine the required outer diameter if the tube is to carry the same torque without exceeding the maximum shear stress of 36.65 kip/in.^{2}.

Solution. From (7-9) and (7-11),

$$\tau_{max} = \frac{TR_o}{(\pi/2)(R_o^4 - R_i^4)}$$

and substituting for the known quantities, we obtain

$$36.65 = \frac{0.90 \times R_o}{(\pi/2)(R_o^4 - 0.25^4)} \frac{kip \cdot in. \cdot in.}{in.^4} = \frac{kip}{in.^2}$$

Solving by trial and error for R_o, we determine outer radius as

$$R_o = \textbf{0.305 in.}$$

It is of interest to compare the amounts of material used in the solid rod of Example 7-1 and the tube of Example 7-2 to transmit the same torque at the same maximum stress. From the diameters of the two members, the cross-sectional areas can be calculated. In this case the cross-sectional area of the tube is only one-half that of the rod. The fact that the tube requires less material than the rod demonstrates the inherent efficiency of hollow circular members to transmit torsional loads. The reason for this efficiency is that all material is stressed to a high level. In this case all the material in the tube is stressed to within 20% of the maximum stress. On the other hand, the material near the center of the rod has a very low stress and contributes little to the torque-carrying capacity. For tubes still thinner than the one in this example, the shear stress will become almost uniform throughout the wall thickness. It is evident that thinner tubes make more effective use of material and hence require a smaller amount of material. However, a very thin wall may fail by local buckling due to the diagonal compression in the wall. This imposes a limit on how thin the wall can be made in practice.

Problems 7-1 to 7-6

The torque, T, in the torsion formula (7-9) is the *internal torque* acting on a particular section of a shaft under investigation. In a problem we determine the internal torque caused by the applied *external torques* by equilibrium considerations using free-body diagrams. The torsional loading of the stepped shaft in Fig. 7-9a illustrates this point. A summation of the external torques T_A, T_B, T_C, and T_D according to $\sum M_x = 0$ shows that the clockwise torques are balanced by the counterclockwise torques, and hence overall equilibrium of the shaft exists. If we wanted to determine the internal torque in the shaft portion A-B, we pass a cutting plane M-M through the portion A-B and isolate the free body shown in Fig. 7-9b. From

$$\sum M_x = 0$$

external torque = internal torque

or

$$T_A = T_M = 30 \text{ N} \cdot \text{m} (22 \text{ lb} \cdot \text{ft})$$

Proceeding in a similar manner for the shaft portion B-C, the internal torque, T_N is obtained by using the free-body diagram in Fig. 7-9c. Hence

$$T_N = T_B - T_A = 40 - 30 = 10 \text{ N} \cdot \text{m} (29 - 22 = 7 \text{ lb} \cdot \text{ft})$$

By inspection, the internal torque in the shaft portion C-D is seen to be 15 N·m (11 lb·ft).

In analyzing the stepped shaft of Fig. 7-9, we now realize that the internal torque can vary from one section to another so that in working torsion problems, we may have to investigate several sections to arrive at the *critical section* that governs the design. If we are dealing with a shaft of *constant* cross-sectional area made of the *same material* throughout, the critical section will obviously be the section where the maximum torque is applied. On the other hand, for the *stepped* shaft made of the *same material* throughout, the critical section can be either

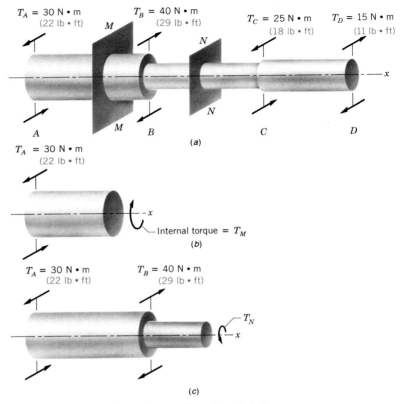

$T_A = 30 \text{ N} \cdot \text{m}$
(22 lb • ft)

$T_B = 40 \text{ N} \cdot \text{m}$
(29 lb • ft)

$T_C = 25 \text{ N} \cdot \text{m}$
(18 lb • ft)

$T_D = 15 \text{ N} \cdot \text{m}$
(11 lb • ft)

(a)

$T_A = 30 \text{ N} \cdot \text{m}$
(22 lb • ft)

Internal torque $= T_M$

(b)

$T_A = 30 \text{ N} \cdot \text{m}$
(22 lb • ft)

$T_B = 40 \text{ N} \cdot \text{m}$
(29 lb • ft)

T_N

(c)

FIG. 7-9 Determination of internal torques for a stepped shaft.

in the shaft portion *A-B, B-C* or *C-D*. Also, if the three portions of the stepped shaft are made of three different materials, any of the three portions may represent the critical section and all three would have to be investigated.

Example 7-3

The stepped shaft illustrated in Fig. 7-10 is made of two materials, steel and aluminum. If the allowable shear stresses are 170 MPa and 90 MPa for steel and aluminum, respectively, check to see if the shaft can safely carry the given loads.

Solution. The steel section of the shaft is identical to the rod of Example 7-1. Since the applied torque of 90 N · m is also identical, the maximum shear stress of 167 MPa obtained in Example 7-1 can be directly compared to the allowable stress of 170 MPa. It is seen to be safe.

From $\sum M_x = 0$, the aluminum section of the shaft is subjected to a total torque of $60 + 90 = 150 \text{ N} \cdot \text{m} = 150\,000 \text{ N} \cdot \text{mm}$. Substituting into the shear stress equation gives

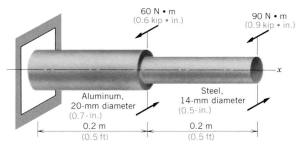

FIG. 7-10

$$\tau_{max} = \frac{16T}{\pi D^3} \quad (7\text{-}9a)$$

$$= \frac{16 \times 150\,000}{\pi(20)^3} \frac{N \cdot mm}{mm^3}$$

$$= 95.5 \frac{N}{mm^2} = 95.5 \text{ MPa}$$

The shear stress in the aluminum exceeds the allowable value of 90 MPa and the shaft is therefore **not safe** for the given loading.

Example 7-3

The stepped shaft illustrated in Fig. 7-10 is made of two materials, steel and aluminum. If the allowable shear stresses are 40 ksi and 21 ksi for steel and aluminum, respectively, check to see if the shaft can safely carry the given loads.

Solution. The steel section of the shaft is identical to the rod of Example 7-1. Since the applied torque of 0.90 kip·in. is also identical, the maximum shear stress of 36.65 kip/in.^{2} obtained in Example 7-1 can be directly compared to the allowable stress of 40 ksi. It is seen to be safe.

From $\sum M_x = 0$, the aluminum section of the shaft is subjected to a total torque of $0.60 + 0.90 = 1.50$ kip·in. Substituting into the shear stress equation gives

$$\tau_{max} = \frac{16T}{\pi D^3} \quad (7\text{-}9a)$$

$$= \frac{16 \times 1.50}{\pi(0.7)^3} \frac{kip \cdot in.}{in.^3}$$

$$= 22.3 \text{ kip/in.}^2$$

The shear stress in the aluminum exceeds the allowable value of 21 ksi and the shaft is therefore **not safe** for the given loading.

Circular torsional members are frequently used in machines as shafts to transmit power. It is useful to develop a formula that expresses the power transmitted through a shaft as a function of the applied torque.

We will denote the angular velocity of the shaft in revolutions per minute (rpm) by n. During a time interval, t(min), the torque, T(N·m), does an amount of work, U(N·m), equal to T times the angular displacement in radians or

$$U = T \times 2\pi n \times t \qquad \text{N·m} \cdot \frac{\text{rad}}{\text{rev}} \cdot \frac{\text{rev}}{\text{min}} \cdot \text{min} = \text{N·m}$$

The rate at which work is done, or the power, P, that is transmitted along the shaft is given by

$$P = \frac{U}{t} = T2\pi n \frac{\text{N·m}}{\text{min}}$$

$$= \frac{T2\pi n}{60} \frac{\text{N·m}}{\text{s}}$$

The *watt* is defined as N·m/s (see Table I) so that the power in *watts* transmitted by a rotating shaft is given by

$$P = \frac{Tn}{9.55} \qquad \textbf{(7-12a) (SI)}$$

where T is torque in newton metres (N·m) and n is the shaft speed in revolutions per minute.

Then a shaft transmitting power is subjected to a torque load given by

$$\boxed{T = \frac{9.55P}{n}} \qquad \textbf{(7-12b) (SI)}$$

Circular torsional members are frequently used in machines as shafts to transmit power. It is useful to develop a formula which expresses the power transmitted through a shaft as a function of the applied torque. We denote the angular velocity of the shaft in revolutions per minute by n. During a time interval, t(min), the torque, T(lb·ft), does an amount of work, U(lb·ft), equal to T times the angular displacement in radians or

$$U = T \times 2\pi n \times t \frac{\text{lb·ft·rad}}{\text{rev}} \cdot \frac{\text{rev}}{\text{min}} \cdot \text{min} = \text{lb·ft}$$

The rate at which work is done, or the power, P, that is transmitted along the shaft, is given by

$$P = \frac{U}{t} = T2\pi n \; \text{lb·ft/min}$$

One *horsepower* is defined as 33,000 lb·ft/min so that the power in *horsepower* transmitted by a rotating shaft is given by

$$P = \frac{T2\pi n}{33,000} \frac{\text{lb·ft/min}}{\text{lb·ft/min}} = \frac{T \times n}{5250} \qquad \textbf{(7-12a) (Imperial)}$$

where P is in horsepower, T is torque in pounds-feet, and n is the shaft speed in revolutions per minute.

Then a shaft transmitting power is subjected to a torque load given by

$$T = \frac{5250P}{n} \qquad \text{(7-12b) (Imperial)}$$

Example 7-4

For a motor operating at 5000 rpm and delivering 20 kW, design a solid steel shaft. The allowable shear stress is 130 MPa.

Solution.   Substituting into (7-12b)(SI), we find that the torque is

$$T = \frac{9.55 \times 20 \times 10^3}{5000} = 38.2 \text{ N} \cdot \text{m}$$

From (7-9a)

$$D^3 = \frac{16 \times T}{\pi \times \tau_{\text{max}}}$$

$$= \frac{16 \times 38.2 \times 10^3}{\pi \times 130} \frac{\text{N} \cdot \text{mm}}{\text{N/mm}^2} = \text{mm}^3$$

$$= 1497 \text{ mm}^3$$

$$D = \textbf{11.44 mm}$$

In practice a 12-mm shaft would be specified. It is interesting to note the effect on D if the motor were to operate at a considerably slower speed, say 500 rpm. The torque then is multiplied by a factor of 10 resulting in a shaft diameter of about 25 mm. This example demonstrates why high-speed machinery is extensively used in combination with torsional shafts.

Example 7-4

For a motor operating at 5000 rpm and delivering 25 hp, design a solid steel shaft. The allowable shear stress is 18 kip/in.^{2}.

Solution.   Substituting into (7-12b)(Imperial), we find that the torque is

$$T = \frac{5250 \times 25}{5000} = 26.25 \text{ lb} \cdot \text{ft}$$

From (7-9a)

$$D^3 = \frac{16 \times T}{\pi \times \tau_{max}}$$

$$= \frac{16 \times 26.25 \times 12}{\pi \times 18,000} \frac{\text{lb} \cdot \text{in.}}{\text{lb/in.}^2}$$

$$= 0.0891 \text{ in.}^3$$

$$D = \textbf{0.45 in.}$$

In practice a 0.5-in. shaft would be specified. It is interesting to note the effect on D if the motor were to operate at a considerably slower speed, say 500 rpm. The torque then is multiplied by a factor of 10 resulting in a shaft diameter of about 1 in. This example demonstrates why high-speed machinery is extensively used in combination with torsional shafts.

Problems 7-7 to 7-15

7-3 TORSION BARS AND HELICAL SPRINGS

The arrangement of parts in Fig. 7-7 is similar to that used in the suspension systems of some motor vehicles, where it is referred to as a *torsion-bar suspension*. In order to assess the riding characteristics of such a suspension system, the designer needs to be able to calculate the stiffness of the system; that is, it is necessary to know the relationship between the load, P, and the deflection caused by that load. To make such a calculation, and for other applications, we will need to know θ of Fig. 7-2 in terms of the load and dimensions of the system.

Recalling (7-2)

$$\theta = \frac{L\gamma}{R}$$

and substituting τ_{max}/G for γ gives

$$\theta = \frac{L}{R} \frac{\tau_{max}}{G}$$

Substituting for τ_{max} from (7-9), we obtain

$$\theta = \frac{L}{R} \frac{TR}{GJ}$$

$$\boxed{\theta = \frac{TL}{GJ}} \tag{7-13a}$$

If we substitute typical units: $N \cdot mm$ (lb $\cdot$ in.) for T, mm (in.) for L, N/mm^2 (lb/in.^{2}) for G and mm^4 (in.^{4}) for J, then θ is seen to be dimensionless or in radians.

For a circular shaft $J = (\pi/2)R^4$, this equation becomes

$$\theta = \frac{TL}{G\pi/2R^4}$$

$$= \frac{2}{\pi}\frac{TL}{GR^4}$$

$$\boxed{\theta = \frac{32}{\pi}\frac{TL}{GD^4}}$$ (7-13b)

It is interesting to note the similarity between (1-13), (2-10), and (7-13) as follows:

$$\Delta L = \frac{PL}{EA}$$ (1-13)

$$\alpha = \frac{ML}{EI}$$ (2-10)

$$\theta = \frac{TL}{GJ}$$ (7-13a)

Since ΔL, α, and θ represent the basic deformations occurring due to the associated loadings of P, M, and T (axial, bending, and torsional loadings), each of the three *deformation-load equations* can be reconstructed from the others by substituting the analogous terms. The equation for torsional deformation or angle of twist is an important and useful one in that it enables us to work out shaft problems where θ may be restricted to a certain permissible value. The product of G and J is often referred to as the *torsional stiffness* of a shaft. If, in a shaft problem, θ, T, and L, for instance, are specified, we can design the shaft for adequate stiffness.

Example 7-5

For the stepped shaft of Fig. 7-11 a maximum allowable angle of twist of 8° is specified. Check to see if the shaft satisfies this deformation criterion.

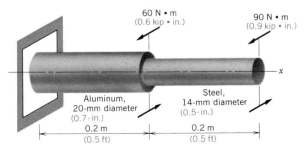

60 N • m
(0.6 kip • in.)

90 N • m
(0.9 kip • in.)

x

Aluminum,
20-mm diameter
(0.7-in.)

Steel,
14-mm diameter
(0.5-in.)

0.2 m
(0.5 ft)

0.2 m
(0.5 ft)

FIG. 7-11

Solution. Subscripts a and s refer to aluminum and steel, respectively. For the steel portion of the shaft

$$T_s = 90 \times 10^3 \text{ N} \cdot \text{mm}$$

$$L_s = 0.2 \times 10^3 \text{ mm}$$

$$G_s = 77 \times 10^3 \frac{\text{N}}{\text{mm}^2} \qquad \text{(Table A)}$$

$$J_s = \frac{\pi}{32} D^4 \qquad \text{(7-10)}$$

$$= \frac{\pi}{32} 14^4 \text{ mm}^4 = 3771 \text{ mm}^4$$

From (7-13a)

$$\theta_s = \frac{TL}{GJ} = \frac{90 \times 10^3 \times 0.2 \times 10^3}{77 \times 10^3 \times 3771} \frac{\text{N} \cdot \text{mm} \times \text{mm}}{\text{N/mm}^2 \times \text{mm}^4}$$

$$= 0.06199 \text{ radians}$$

$$= 0.06199 \times \frac{360}{2\pi} = 3.55^\circ$$

For the aluminum portion of the shaft

$$T_a = 150 \times 10^3 \text{ N} \cdot \text{mm}$$

$$L_a = 0.2 \times 10^3 \text{ mm}$$

$$G_a = 26 \times 10^3 \frac{\text{N}}{\text{mm}^2} \qquad \text{(Table A)}$$

$$J_a = \frac{\pi}{32} D^4 \qquad \text{(7-10)}$$

$$= \frac{\pi}{32} 20^4 \text{ mm}^4 = 15\,708 \text{ mm}^4$$

Hence

$$\theta_a = \frac{TL}{GJ} = \frac{150 \times 10^3 \times 0.2 \times 10^3}{26 \times 10^3 \times 15\,708} \frac{\text{N} \cdot \text{mm} \times \text{mm}}{\text{N/mm}^2 \times \text{mm}^4}$$

$$= 0.07346 \text{ radians}$$

$$= 0.07346 \times \frac{360}{2\pi} = 4.21^\circ$$

The angle of twist at the free end is

$$\theta_{\max} = \theta_s + \theta_a = \textbf{7.76}^\circ$$

This is less than the maximum permissible rotation of 8°, and the torsional stiffness of the stepped shaft therefore is satisfactory.

Example 7-5

For the stepped shaft of Fig. 7-11 a maximum allowable angle of twist of 11° is specified. Check to see if the shaft satisfies this deformation criterion.

Solution. Subscripts a and s refer to aluminum and steel, respectively. For the steel portion of the shaft

$$T_s = 0.9 \text{ kip} \cdot \text{in.}$$

$$L_s = 0.5 \times 12 = 6 \text{ in.}$$

$$G_s = 11 \times 10^3 \text{ kip/in.}^2 \qquad \text{(Table A)}$$

$$J_s = \frac{\pi}{2} R^4 \qquad \text{(7-10)}$$

$$= \frac{\pi}{2} \times 0.25^4 = 6.14 \times 10^{-3} \text{ in.}^4$$

From (7-13a)

$$\theta_s = \frac{TL}{GJ} = \frac{0.9 \times 6}{11 \times 10^3 \times 6.14 \times 10^{-3}} \frac{\text{kip} \cdot \text{in.} \times \text{in.}}{\text{kip/in.}^2 \times \text{in.}^4}$$

$$= 0.07995 \text{ radians}$$

$$= 0.07995 \times \frac{360}{2\pi} = 4.58°$$

For the aluminum portion of the shaft

$$T_a = 1.5 \text{ kip} \cdot \text{in.}$$

$$L_a = 0.5 \times 12 = 6 \text{ in.}$$

$$G_a = 4 \times 10^3 \text{ kip/in.}^2 \qquad \text{(Table A)}$$

$$J_a = \frac{\pi}{2} R^4 = \frac{\pi}{2} (0.35)^4 = 23.6 \times 10^{-3} \text{ in.}^4$$

Hence

$$\theta_a = \frac{TL}{GJ} = \frac{1.5 \times 6}{4 \times 10^3 \times 23.6 \times 10^{-3}} \frac{\text{kip} \cdot \text{in.} \times \text{in.}}{\text{kip/in.}^2 \times \text{in.}^4}$$

$$= 0.09534 \text{ radians}$$

$$= 0.09534 \times \frac{360}{2\pi} = 5.46°$$

The angle of twist at the free end is

$$\theta_{\max} = \theta_s + \theta_a = \mathbf{10.04°}$$

This is less than the maximum permissible rotation of 11°, and the torsional stiffness of the stepped shaft therefore is satisfactory.

Problems 7-16 to 7-22

The torsion bar in Fig. 7-7 was fitted with end supports arranged so that the bar was subjected to a twisting moment, a torque, only. We will now make some modifications to the system, leaving out the bearings at the right end and curving the bar. The new system is shown in Fig. 7-12. It is important that the bar be curved to a radius equal to C. That is so that the center of curvature of the bar can be made to coincide with the point, O, at which the load, P, is applied. We will call the radius of curvature $\bar{R}$. If a typical section A-B is taken and the free body, consisting of the arm and the portion of the curved rod from the arm to the cutting plane, is examined for equilibrium, it will be found that the section A-B is subjected to a torque

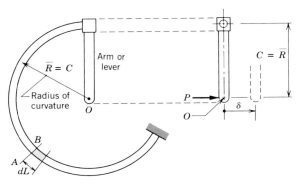

FIG. 7-12

$$T = P\bar{R}$$

and to *no bending moment*. Wherever section A-B is taken, the torque remains the same and we have in the bar substantially the same load state as for the torsion-bar arrangement. There is also a vertical shear load, equal to P, on the section, but for the proportions found in most engineering applications this can be ignored without substantial error.

A small element of the rod having length dL and an applied torque $P\bar{R}$, Fig. 7-12, will twist a small amount, which is determined by substitution into (7-13a) as

$$\theta = \frac{P\bar{R}\,dL}{GJ}$$

This will cause the whole body beyond A-B to rotate through angle θ and will cause point O at the end of the arm to move a distance

$$\Delta\delta = \theta\bar{R}$$

$$= \frac{P\bar{R}\,dL\,\bar{R}}{GJ}$$

$$= \frac{P\bar{R}^2\,dL}{GJ}$$

The element, of length dL, is typical of all regions of the bar, and its contribution to deflection is independent of its location. When all parts of the bar are taken into account, the total deflection at O will be given by

$$\delta = \frac{P\bar{R}^2 L}{GJ} \qquad \textbf{(7-14)}$$

If a larger δ is required, it can be obtained by increasing L without changing any other parameter. For the curve of Fig. 7-12, where the bar is in the plane of the figure, the bar length will reach a limit when the support conflicts with the hub of the arm. When a larger length is required the excess length can be accommodated by curving the bar out of the plane so that its centerline becomes a helix. The length can then be greatly increased and the torsion bar has evolved into the well-known *coil spring*, similar to that shown in Fig. 7-13. It is convenient in coil springs to specify the number of turns, N_c, rather than the length, L, and to give the mean diameter of the coils $\bar{D}$ rather than $\bar{R}$. These changes, and substituting for J from (7-10), alter (7-14) to the more convenient form:

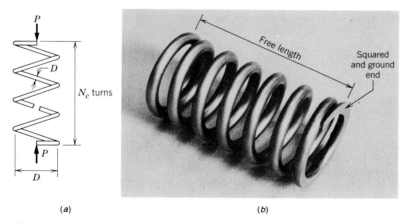

(a) (b)

FIG. 7-13 Coil spring.

$$\delta = \frac{P\left(\dfrac{\bar{D}}{2}\right)^2 N_c \pi \bar{D}}{G\dfrac{\pi}{32}D^4}$$

$$\boxed{\delta = \frac{8PN_c\bar{D}^3}{GD^4}} \qquad \textbf{(7-15)}$$

The torsional moment on the cross section of the spring is

$$T = P\frac{\bar{D}}{2}$$

Substituting into (7-9a) and converting to spring dimensions, we determine the maximum shearing stress in the rod of the coil spring as

$$\tau_{max} = \frac{16P(\bar{D}/2)}{\pi D^3}$$

$$\boxed{\tau_{max} = \frac{8P\bar{D}}{\pi D^3}} \tag{7-16}$$

Note that in (7-15) and (7-16) $\bar{D}$ refers to the mean *coil* diameter whereas D refers to the *bar* diameter.

The formula for shear stress (7-16) is based on a pure torque load and does not take into account the direct shear load, equal to P. It, therefore, is approximate. Both the stress and deflection formulas are further in error because neighboring sections, taken normal to the coil centerline, will not be parallel and consequently the stress does not vary linearly with radial distance.

These errors diminish as the ratio of $\bar{D}$ to D increases and may be neglected in many practical cases without significant error. When the extra effort is warranted, spring designers make use of correcting constants* to eliminate the above-mentioned, and other, errors.

The torsion bar of Fig. 7-7 had an arm or lever attached to the end and in the above development the arm was left in place. In coil springs such arms are never used, the end portion of the spring may be bent inward to form the equivalent of the arm. However, the most common practice is to wind the spring so that the last turn is closed. The end is then ground so that there is a plane surface normal to the helix axis, as seen in Fig. 7-13b. The load, when applied uniformly to this ground surface, is equivalent to the centrally applied load P. In calculations only the coils that are not closed, the *active turns*, are taken into consideration.

Problems 7-23 to 7-27

7-4 MEMBRANE ANALOGY

Having arrived at two useful equations, (7-9) and (7-13), for the design of shafts with circular cross sections, the next natural step would be to apply them in cases where the cross section has another shape: rectangular, for example. Although the determination of the polar moment of inertia of a rectangle poses some problems, with a little effort it can be calculated. For a given shaft the other quantities can also be determined and the stress and rotation calculated. Unfortunately, both values would be very much in error.

The error in applying (7-9) and (7-13) to any section other than circular goes back to the initial assumption that plane sections remain plane when torque is applied. The error in this assumption can be seen when a styrofoam member having surface markings as in Fig. 7-14a is

*See V. M. Faires, *Design of Machine Elements*, 4th ed., Macmillan, New York, 1965.

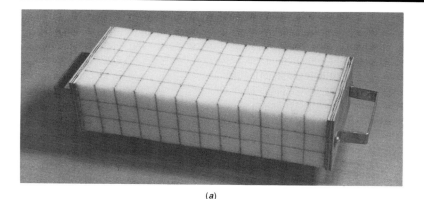

(a)

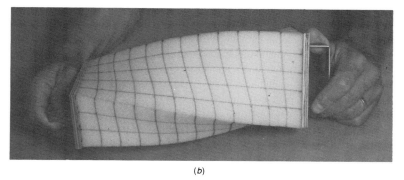

(b)

FIG. 7-14 Torsion of a rectangular section. (a) Unloaded styrofoam member. (b) Twisted styrofoam member showing warping of sections.

twisted. The lines that run circumferentially around the member do not remain straight when the torque is applied but rather distort as seen in Fig. 7-14b. Thus *plane sections do not remain plane for noncircular members* and the development of the formulas has no foundation in fact. The amount of *warping* of the plane sections forms the basis of the correct theoretical analysis that was accomplished by Saint Venant in 1855. Since this analysis requires a theoretical treatment that is well beyond the scope of this study, we will compromise by introducing an analogy which will make it possible to visualize the regions of high stress and to evaluate the stress in some important cross sections.

The analogy referred to above is the *membrane analogy**, which can be proved only after a lengthy theoretical analysis. For our purposes the following statement of the analogy will suffice. If we want the stress pattern in a shaft having any cross section, a wire frame having the shape of the periphery of the cross section is formed. A membrane is stretched with uniform tension over the frame; a soap film does this very well. A uniform pressure applied to the membrane will deflect it and then, provided the deflection is very small, the shape of the membrane has these characteristics:

*Introduced by L. Prandtl in 1903. See E. E. Sechler, *Elasticity in Engineering*, Wiley, New York, 1952.

1. The *slope* of the membrane, at any given point in a given direction, *is a measure of the shear stress* in the shaft at that point, the stress being at right angles to the slope direction.
2. The *volume* under the deflected membrane *is a measure of the torque* applied to the shaft.
3. The *pressure* required to deflect the membrane *is a measure of the twist* of the shaft.

It is obvious that the task of setting up such an experiment and taking the necessary measurements is formidable and, consequently, the analogy has not been used extensively to obtain experimental solutions. However, because the membrane is easy to visualize, the analogy can direct our attention to regions of high stress and perhaps guide us in reshaping a section so that high stresses will be reduced.

In order to become familiar with the use of the membrane analogy, let us first consider a circular section, forgetting for the moment that we know the solution to this problem. Imagine a soap film stretched across a wire loop and a small upward pressure, *p*, applied to the film. The shape of the soap bubble can be imagined to be as shown in Fig. 7-15. At any

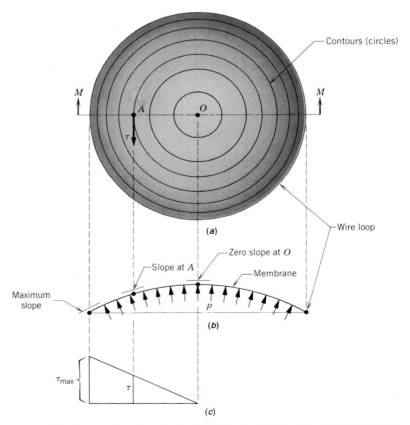

FIG. 7-15 Membrane analogy for a circular shaft in torsion. (*a*) Plan view of a circular cross section indicating contour lines. (*b*) Elevation at section *M-M*. (*c*) Shear stress distribution.

point, *A*, we can take a slope in any direction and get the stress at right angles to that direction. If the direction chosen is tangential to the contour through *A*, the slope will be zero, which tells us that the component of stress normal to the contour, and hence radial, is zero. For an extreme value of stress we should consider the maximum slope which will be in the direction normal to the contour. This slope gives the value of stress tangential to the contour, and hence in a circumferential direction. These components of stress together give a stress that is normal to the radial line *A-O*. Since *A* was an arbitrary point, we can conclude that at all points in the section the shear stress is circumferential.

The shape of the profile of the imaginary bubble makes it evident that the stress at the center is zero and that stress increases with radial distance. An analysis of the membrane shape would show that this is a linear variation, hence verifying the known solution or, more realistically, verifying the membrane analogy.

Consider now the same shaft with a longitudinal groove, such as the key-way, shown in Fig. 7-16. The purpose of a key-way is to provide a means for connecting a gear or pulley to a shaft. For most key-ways we would want the cross section to be rectangular, but let us accept for now a *semicircular* groove. At first our intuition might lead us to expect no serious disturbance in the stress pattern. But consider the membrane before reaching any conclusion. The indentation in the wire loop due to

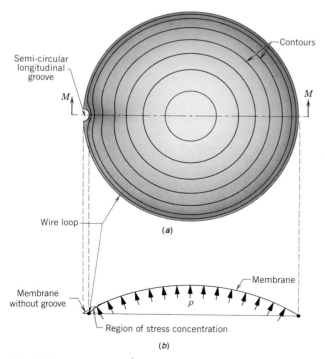

FIG. 7-16 Membrane analogy for a shaft with longitudinal groove. (*a*) Plan view. (*b*) Elevation at section *M-M* showing the steep slope at the bottom of the groove; hence high stress.

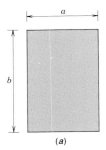

(a)

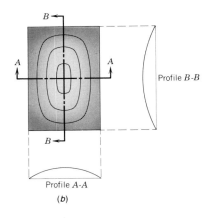

Profile B-B

Profile A-A

(b)

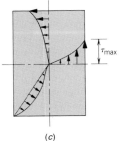

τ_{max}

(c)

FIG. 7-17 Rectangular shaft in torsion.
(a) Dimensions of cross section.
(b) Membrane analogy. (c) Shear stress distribution.

the groove will leave the shape of the film largely unchanged except in the region of the indentation. If we visualize the film in the vicinity of the groove, we will see the contours bunching up as seen in Fig. 7-16. Throughout most of the section the contours will not be substantially altered. It is easy to imagine that the steepest slope is at the bottom of the groove, and hence that the stress at the bottom of the groove is significantly greater than the maximum stress in the original shaft. In actual fact, if the semicircular groove is very small relative to the shaft, introducing the groove raises the maximum stress by a factor of two.*

If the key-way had been *rectangular*, the bunching-up of contours would have been even greater. At the sharp corners of the key-way the membrane would be vertical, which indicates an infinite stress. In practice, an infinite stress is impossible because local yielding would occur and the analysis would no longer be valid, since we tacitly assumed that all stresses would be in the elastic range. Thinking about this key-way analogy is useful because it quickly warns us that the sharp corners at the bottom of a key-way should be avoided at all cost. Instead, some fillet should be provided with as large a radius as possible, the limit being a semicircular key-way, where the *stress concentration factor*, as we have just seen, will still be two.

7-5 STRESSES IN NONCIRCULAR SECTIONS

We will now consider a rectangular shaft having the cross-sectional dimensions shown in Fig. 7-17a. By imagining a membrane analogy for this shape, we can visualize the contours and profiles of the membrane to be as shown in Fig. 7-17b. If we want actual quantities for the membrane deflection, stresses, torque, and rotation, they can be obtained by numerical methods with the aid of a computer. An inspection of the deflected membrane shows that the steepest slope is found at the ends of section A-A; hence the greatest torsional stress occurs at the midpoints of the longer sides. Note that along the shorter sides the stress also peaks at the midpoints, but this stress is not as great as that on the longer sides.

An inspection of the membrane in the region of corners would show that the slope approaches zero as the corner is approached. This means that there is zero stress in the fibers that are located the largest distance from the centroid of a rectangular cross section. The situation is vastly different from the case of a circular shaft, where we found that the extreme fibers were in fact subjected to the maximum torsional stress. It is evident from the stress variations in Fig. 7-17c that the application of the circular-shaft formulas to noncircular sections will give meaningless results.

The correct solution to the *rectangular cross section* in torsion can

---

*See F. B. Seely, and J. O. Smith, *Advanced Mechanics of Materials*, 2nd ed., Wiley, New York, 1952 for a more thorough treatment of this topic.

be found either by the membrane analogy or by using the theory of elasticity.* By either method the following formulas can be derived:

$$\tau_{max} = C_1 \frac{3T}{a^2 b}$$

(7-17)

$$\theta = C_2 \frac{3LT}{Ga^3 b}$$

(7-18)

By eliminating T, we get

$$\tau_{max} = \frac{C_1}{C_2} \frac{Ga\theta}{L} = C_3 \frac{Ga\theta}{L}$$

(7-19)

The values of C_1, C_2, and C_3 can be determined for various ratios of b to a and are given in Fig. 7-18. Note that when b/a is greater than 7, all C's are within 10% of unity. Hence, for *slender* rectangular sections having b/a greater than 7, the formulas can be reduced to

$$\tau_{max} = \frac{3T}{a^2 b}$$

(7-17a)

$$\theta = \frac{3LT}{Ga^3 b}$$

(7-18a)

$$\tau_{max} = \frac{Ga\theta}{L}$$

(7-19a)

with an error that is less than 10%.

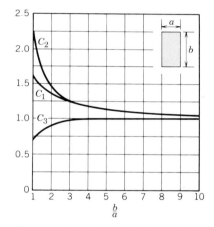

FIG. 7-18 Constants for torsional analysis of rectangular shafts.

Problems 7-28 to 7-32

Much can be learned about the usefulness of the membrane analogy by examining the solution to Prob. 7-30. Case (a), which can be solved without the help of the analogy, provides a reference for comparison between the circular section and other cross sections. Simple application of the formulas for rectangular sections gives the solutions to cases (b), (c), and (d), but the step from (d) to (e) requires some thought and imagination. To make this step, we must first recognize that the significant properties of the membranes, that is, the maximum slope and volume, are the same in both cases and that consequently the stress and

*The general solution to the problem of noncircular sections subjected to torque was developed by the French engineer B. de St. Venant in 1855.

twist are the same. After making the discovery that (7-17) and (7-18) are applicable to crescents such as case (e), it requires only a small step to see that they apply to case (f). There is no need for the analogy to solve case (g), as the circular shaft formulas apply. If the analogy is used, there is a good chance that it will be used incorrectly and that the answer will seem to be the same as in case (f). In case (g) the analogy requires that we stretch the membrane between two hoop boundaries, and we might be inclined to fix both hoops in a common plane. But, since the boundaries are not connected, there is no reason for keeping them coplanar, and if they are so constrained a large error will result. The answers to the problem show that of all the shapes considered, the hollow tube of case (g) is by far the strongest and stiffest.

The formulas for rectangular sections can be useful in certain practical problems in which the section is not a simple rectangle. Let us consider first an *angle*, having dimensions given in Fig. 7-19a, subjected to torsion. Only by extensive numerical or experimental work can the solution to this problem be obtained. However, if we straighten out the angle and square off the ends we have the familiar rectangle in torsion. It seems obvious that the profile on sections *A-A* and *B-B* will remain unaltered in the transformation. Thus formulas (7-17), (7-18), and (7-19)

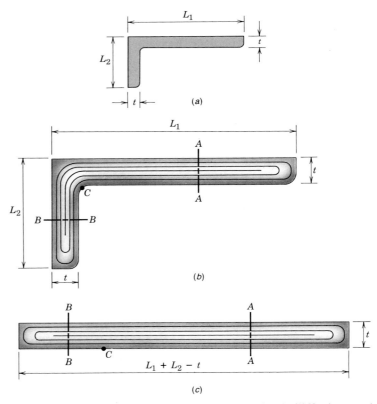

FIG. 7-19 Angle section subjected to torque. (*a*) Dimensions of angle. (*b*) Membrane contours. (*c*) Contours for simplified problem.

may be expected to apply because the volume of the bubble and its slopes seem to be the same. In fact, the volume is substantially unaltered and (7-18) will be quite accurate. Stress by (7-17) will be quite accurate at *A-A* and *B-B* but significantly in error at *C*. The reason for the error at *C* is that the tension in the membrane in the direction of the contours where they are curved tends to pull the membrane toward the center of curvature and will cause the contours to bunch together near the fillet at *C*. Thus the maximum slope of the membrane in this region is significantly greater than at the other sections and there will be a stress concentration factor at the fillet that must be applied to the stress as calculated by the formulas. The magnitude of the stress concentration factor can be diminished by increasing the radius of the fillet. Since the peak stress is highly localized, it is often not taken into consideration in design. However, as large a radius as possible is provided in the fillet in order to keep the peak stress at a reasonable level.

Problem 7-33

The torsion problem for a structural shape other than an angle is somewhat more difficult because the web and flange thicknesses are not the same. The contours for a membrane analogy solution to an *S-shape* are approximately as shown in Fig. 7-20*a*. Since the membrane in the flange spans a distance a_1, which is larger than the corresponding

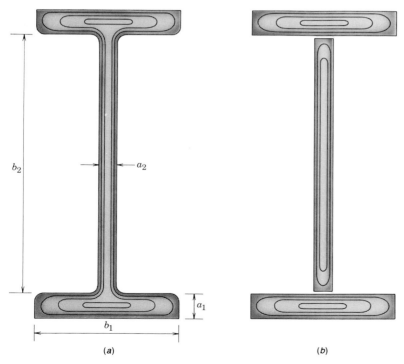

(*a*)   (*b*)

FIG. 7-20 S-shape in torsion. (*a*) Membrane contours. (*b*) Equivalent contours.

distance a_2 in the web, the bubble in the region of the flange rises to a greater height than in the region of the web. Consequently, the shear stresses at the surfaces are different, the greatest being in the flange. The contours will not be much altered if the cross section is partitioned as in Fig. 7-20b; the problem can then be solved by treating the three separate rectangular sections. This treatment is somewhat similar to the case of the straightened angle but is more difficult because of the different thicknesses in the web and flange regions. However, all regions have a common twist angle, so the best approach is to use (7-18) to find the torque on each region in terms of the twist angle, θ. Summing the torques gives the total load in terms of the twist angle, and leads to an equation for θ in terms of the load T and thence to the value of θ. Substituting the now known value of θ into (7-19) gives the maximum stress in the member. As with the angle there is a stress concentration at the junctions of the flanges with the web and a large fillet radius is provided in standard sections in order to reduce the stress concentration.

Problems 7-34 to 7-36

7-6 THIN-WALLED IRREGULAR SECTIONS

In practice members having the cross sections shown in Fig. 7-21 are often subjected to torsion. We have formulas that can deal adequately with the circular sections, but even the membrane analogy is difficult to apply in the other cases. We will start with the membrane analogy and through it meet a new concept, *shear flow*, which will enable us to get an approximate solution to shapes such as those in Fig. 7-21c and d. The approximate solution will be quite accurate, provided that the wall thickness is small relative to the other dimensions.

Let us start by considering the membrane analogy for a *solid circular shaft* in torsion. The membrane will be as in Fig. 7-22a; the volume of the bubble is known to be a measure of the torque carrying capacity of the shaft. Imagine two different regions in the shaft with an interface at radius R_i in Fig. 7-22b. The total torque on the shaft is shared between the regions in the ratio of the two volumes indicated. It is evident that the outer region is better able to carry the load, and we might conclude that little would be lost by removing the core to arrive at a hollow shaft. If we wanted to get experimental results for the *hollow shaft*, we would have to reproduce the lower portion of the bubble in some way. This can be done by having a fixed circular hoop for the outside support, as before, and a disk for the inner region with the membrane spanning the space between the hoop and the disk. The arrangement is shown in Fig. 7-23a. The pressure must be allowed to act on the disk surface, and the disk must be supported so that it is in effect weightless and constrained in a manner ensuring that its only motion is vertical translation. In this case there are no unbalanced forces tending to impart

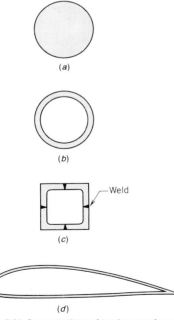

(a)

(b)

Weld

(c)

(d)

FIG. 7-21 Cross sections of torsion members.

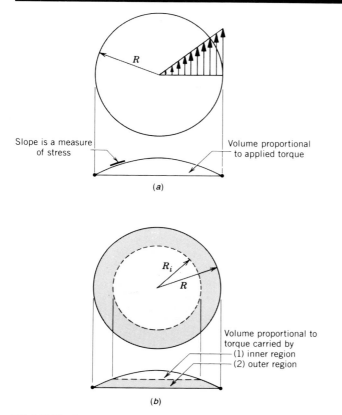

FIG. 7-22 Membrane analogy for circular shafts. (*a*) Solid shaft. (*b*) Solid shaft divided into two regions.

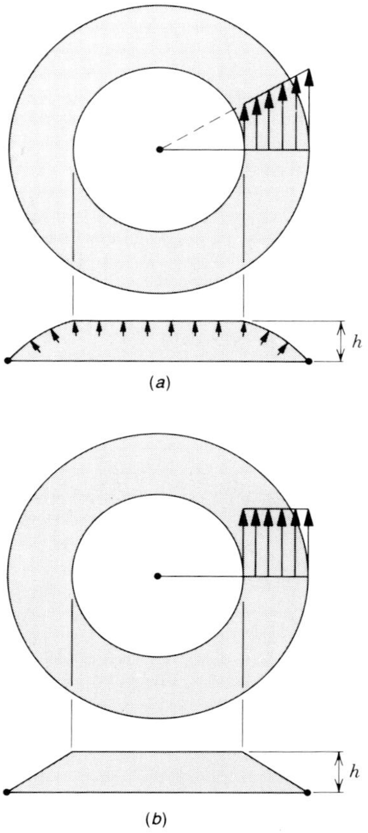

FIG. 7-23 Membrane analogy for hollow shaft.

anything but the allowed motion to the disk, and hence no constraint is required; later cases will require some constraint.

If we did not apply fluid pressure but merely raised the disk the same distance that it would move under pressure, we would have the membrane as in Fig. 7-23*b*, which indicates a constant stress on the tube wall. This is known to be incorrect, but the error diminishes with wall thickness and is insignificant in many engineering applications. Note that the wall thickness in the figures should be shown as small relative to the other dimensions but is left large so that the diagrams will be easier to read.

We will now introduce the concept of *shear flow* by again considering the approximate stress in a tube as shown in Fig. 7-24*a*. *Shear flow is defined as the force, resulting from shear stress acting on an element of cross-sectional area, per unit of length measured along the wall.*

Using the element of area shown in Fig. 7-24*a*, the force is given by

$$F = \tau t \, dL \qquad \textbf{(7-20)}$$

When divided by the length, dL, to get the force per unit of length, we have the shear flow,

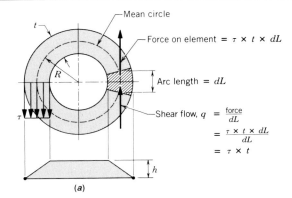

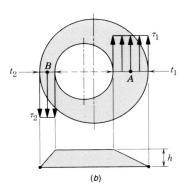

FIG. 7-24 Hollow tubes in torsion. (a) Concentric tube. (b) Eccentric tube.

$$q = \frac{F}{dL} \qquad \text{(7-21a)}$$

$$q = \tau t \qquad \text{(7-21b)}$$

In this case the shear flow is obviously the same regardless of where the area is chosen in the periphery of the tube. A more striking conclusion is reached if we vary the tube wall thickness. Let us do this by making the bore of the tube *eccentric* as in Fig. 7-24b. For the eccentric tube the membrane will be formed by moving the disk to one side. The height h must be maintained so that the volume and, hence, the torque will be the same. The membrane slopes are different at A and B; consequently, the stress is no longer uniform around the circumference of the tube. At point A

$$\tau_1 \propto \frac{h}{t_1}$$

or

$$\tau_1 = k\frac{h}{t_1}$$

At point B

$$\tau_2 = k\frac{h}{t_2}$$

At point A the *shear flow*, by (7-21b), is

$$q_1 = \tau_1 t_1$$

$$= k\frac{h}{t_1}t_1 = kh$$

At point B the *shear flow* is

$$q_2 = \tau_2 t_2$$

$$= k\frac{h}{t_2}t_2 = kh$$

This could be repeated for *any point* in the tube wall or for *any shape* of wall section, and we would always reach the conclusion that *shear flow, q, is constant around the circumference of any thin-walled member in torsion*. Note that this applies to closed sections as in Prob. 7-30(g) and not to open sections as in Prob. 7-30(f). Once we have determined the value of shear flow in a wall we can determine stress at any point, p, by dividing the shear flow by the wall thickness, t_p, at that point, that is,

$$\boxed{\tau_p = \frac{q}{t_p}} \qquad \textbf{(7-22)}$$

We could apply the idea of shear flow to the eccentric tube and get a stress formula, but let us solve the more general case of a tube with any thin-walled cross section.

Consider a member under torque load T that has a cross section as shown in Fig. 7-25. On a typical element of wall, a force F will act tangential to the wall. The moment of that force about O will contribute an increment to the torque given by

$$\Delta T = F \times a$$

Notice that the area of the shaded triangle, A_1, having base dL and altitude a is given by

$$A_1 = \frac{1}{2}a \times dL$$

or

$$a = \frac{2A_1}{dL}$$

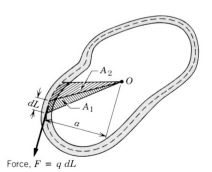

Force, $F = q\,dL$

FIG. 7-25 Thin-walled torsion member.

Rearranging (7-21a) gives

$$F = q \, dL \qquad \text{(7-21c)}$$

Substituting into the torque equation, we get

$$\Delta T = Fa$$

$$= q \, dL \frac{2A_1}{dL} = 2qA_1$$

By taking the neighboring wall region, we get another contribution to torque that has q and 2 as above, but the area term is replaced by the next triangular area, A_2. When contributions from the whole wall are summed, we have the total torque

$$T = 2q[A_1 + A_2 + \ldots + \ldots]$$

and all areas add up to the area enclosed within the dotted line. Rewriting

$$\boxed{T = 2q \ \langle \bar{A} \rangle} \qquad \text{(7-23)}$$

where $\langle \bar{A} \rangle$ is the area inside the midthickness line.

Usually, we want to calculate q, so the formula is more useful as

$$\boxed{q = \frac{T}{2 \ \langle \bar{A} \rangle}} \qquad \text{(7-24)}$$

For many practical problems relating to thin-walled torsion members the *shear flow*, which applies to the whole section, is determined by (7-24). The stress at any point of interest, most likely where the wall is thinnest, can then be determined by (7-22).

Problems 7-37 to 7-42

7-7 TORSION IN THE INELASTIC RANGE

The equations giving stress and displacement in terms of torque were developed on the assumption that the material is such that stress is proportional to strain up to the yield stress and that all particles are stressed to levels below the yield stress and hence remain elastic. In order to extend our studies into the *inelastic* range, we would like to have a formula that relates stress to strain beyond the yield stress. Curves for stress–strain in shear for various engineering materials are similar in form to those for uniaxial tension as shown in Fig. 1-30. It is evident that outside of the straight-line portion of the curves, no single formula will match the various shapes that are characteristic of different engineering materials. Furthermore, for any given material the curve is of such a nature that it is difficult to match a single mathematical curve to the

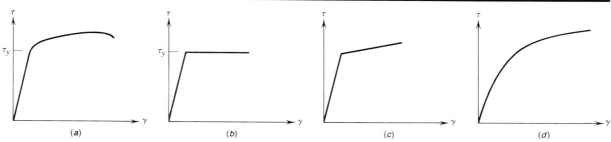

FIG. 7-26 Stress-strain relationships for shear. (a) Actual stress-strain curve. (b) Elastic—perfectly plastic. (c) Elastic—plastic, with strain hardening. (d) Empirical.

experimental curve over the full range of strain. Consequently, it is not possible to derive formulas that will give stress in terms of strain, which are applicable in the inelastic as well as the elastic range. In practice the typical curve of Fig. 7-26a is approximated by two straight lines as in Fig. 7-26b or c or a more complex empirical curve in Fig. 7-26d.

The curve in Fig. 7-26b, when compared with the other approximations, leads to the simplest calculations but gives the largest error. The curve in Fig. 7-26c represents an improvement in that it makes provision for strain hardening. The curve in Fig. 7-26d represents an empirical formula that can be made to model the actual case to any desired degree of accuracy. However, because of the nature of the experimental curve, good accuracy can only be achieved by including many terms in the empirical formula.

We will first consider a *circular* shaft, either solid or hollow, that is strained into the inelastic range. As before, we assume that plane sections remain plane during loading. This, as we have seen, leads to the conclusion that shear strain is proportional to distance from the center. However, when stress is not proportional to strain, the stress does not vary linearly with distance from the center. The variation in stress and strain is then as shown in Fig. 7-27. The shape of the portion of the stress curve in the inelastic region will depend on the inelastic stress–strain relationship.

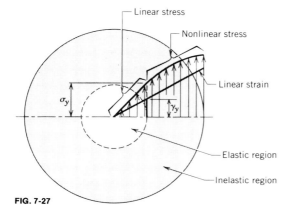

FIG. 7-27

The stress–strain approximation in Fig. 7-26b is usually selected by designers because of its simplicity and the fact that neglecting strain hardening introduces a safe-side error. The following example is typical of elastic-plastic torsion problems.

Example 7-6

A gradually increasing torque is applied to a hollow steel shaft having an outside diameter of 160 mm and an inside diameter of 60 mm. The shaft material has a shear yield stress of 200 MPa and may be taken to have a stress-strain curve similar to Fig. 7-26b. Determine the torque when the plastic region has penetrated to a depth of 20 mm.

Solution. The elastic-plastic state of stress is shown in Fig. 7-28. From similar triangles, in the elastic region

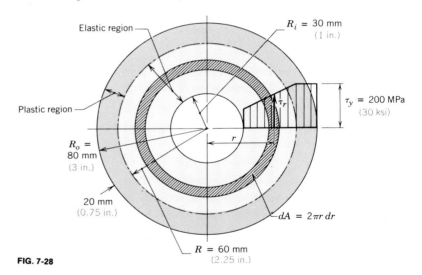

FIG. 7-28

$$\frac{\tau_r}{r} = \frac{\tau_y}{R}$$

or

$$\tau_r = \tau_y \frac{r}{R} \qquad \text{for } R_i \leqslant r \leqslant R$$

In the plastic region

$$\tau_r = \tau_y \qquad \text{for } R \leqslant r \leqslant R_o$$

The torsion on the element of area, dA, is given by

$$dT = r\tau_r \, dA$$

$$dT = r\tau_y \frac{r}{R} 2\pi r \, dr \qquad \text{for } R_i \leqslant r \leqslant R$$

and

$$dT = r\tau_y 2\pi r \, dr \qquad \text{for } R \leqslant r \leqslant R_o$$

The total torque is obtained by integrating over the whole cross section; thus

$$T = \int_{R_i}^{R} \frac{2\pi\tau_y}{R} r^3 \, dr + \int_{R}^{R_o} 2\pi\tau_y r^2 \, dr$$

$$= 2\pi\tau_y \left[\frac{1}{R} \int_{R_i}^{R} r^3 \, dr + \int_{R}^{R_o} r^2 \, dr \right]$$

$$= 2\pi\tau_y \left[\frac{1}{4R} (R^4 - R_i^4) + \frac{1}{3} (R_o^3 - R^3) \right]$$

$$= 2\pi\tau_y \left[\frac{1}{3} R_o^3 + \left(\frac{1}{4} - \frac{1}{3} \right) R^3 - \frac{1}{4} \frac{R_i^4}{R} \right]$$

$$= 2\pi\tau_y \left[\frac{1}{3} R_o^3 - \frac{1}{12} R^3 - \frac{1}{4} \frac{R_i^4}{R} \right] \tag{1}$$

For the given dimensions

$$T = 2\pi 200 \left[\frac{1}{3} 80^3 - \frac{1}{12} 60^3 - \frac{1}{4} \frac{30^4}{60} \right] \frac{N}{mm^2} \cdot mm^3$$

$$= 188 \times 10^6 \, \text{N} \cdot \text{mm}$$

$$= \mathbf{188 \, kN \cdot m}$$

Example 7-6

A gradually increasing torque is applied to a hollow steel shaft having an outside diameter of 6 in. and an inside diameter of 2 in. The shaft material has a shear yield stress of 30 ksi and may be taken to have a stress-strain curve similar to Fig. 7-26b. Determine the torque when the plastic region has penetrated to a depth of 0.75 in.

Solution. The elastic-plastic state of stress is shown in Fig. 7-28. From similar triangles, in the elastic region

$$\frac{\tau_r}{r} = \frac{\tau_y}{R}$$

or

$$\tau_r = \tau_y \frac{r}{R} \qquad \text{for } R_i \leqslant r \leqslant R$$

In the plastic region

$$\tau_r = \tau_y \qquad \text{for } R \leqslant r \leqslant R_o$$

The torsion on the element of area dA, is given by

$$dT = r\tau_r \, dA$$

$$dT = r\tau_y \frac{r}{R} 2\pi r \, dr \qquad \text{for } R_i \leqslant r \leqslant R$$

and

$$dT = r\tau_y 2\pi r \, dr \qquad \text{for } R \leqslant r \leqslant R_o$$

The total torque is obtained by integrating over the whole cross section; thus

$$T = \int_{R_i}^{R} \frac{2\pi\tau_y}{R} r^3 \, dr + \int_{R}^{R_o} 2\pi\tau_y r^2 \, dr$$

$$= 2\pi\tau_y \left[\frac{1}{R} \int_{R_i}^{R} r^3 \, dr + \int_{R}^{R_o} r^2 \, dr \right]$$

$$= 2\pi\tau_y \left[\frac{1}{4R} (R^4 - R_i^4) + \frac{1}{3}(R_o^3 - R^3) \right]$$

$$= 2\pi\tau_y \left[\frac{1}{3} R_o^3 + \left(\frac{1}{4} - \frac{1}{3} \right) R^3 - \frac{1}{4} \frac{R_i^4}{R} \right]$$

$$= 2\pi\tau_y \left[\frac{1}{3} R_o^3 - \frac{1}{12} R^3 - \frac{1}{4} \frac{R_i^4}{R} \right] \tag{1}$$

For the given dimensions

$$T = 2\pi 30 \left[\frac{1}{3} 3^3 - \frac{1}{12} 2.25^3 - \frac{1}{4} \frac{1^4}{2.25} \right] \frac{\text{kip}}{\text{in.}^2} \cdot \text{in.}^3$$

$$= 1.497 \times 10^3 \text{ kip} \cdot \text{in.}$$

$$= \mathbf{124.7 \text{ kip} \cdot \text{ft}}$$

Using the general formula (1) derived in the example, the torque required to produce varying degrees of plasticity can be readily calculated. The results of such a calculation for the shaft specified in the example are given in the following table.

Depth of Plastic Region		R		Torque	
(mm)	(in.)	(mm)	(in.)	(kN·m)	(kip·ft)
0	(0)	80	(3)	$158 = T_y$	$(104.8 = T_y)$
10	(0.4)	70	(2.6)	175	(116.8)
20	(0.8)	60	(2.2)	188	(125.7)
30	(1.2)	50	(1.8)	196	(131.6)
40	(1.6)	40	(1.2)	201	(135.8)
50	(2.0)	30	(1.0)	$203 = T_u$	$(136.2 = T_u)$

The table shows that the torque at the beginning of yielding is 158 kN·m (104.8 kip·ft) and when the section is fully plastic the torque is 203 kN·m (136.2 kip·ft). The ratio of these torques, 203:158 = 1.28 (136.2:104.8 = 1.30) in this case, is called the *shape factor for torsion* and is a useful number in design because it shows the amount by which the torque at first yielding can be increased, 28% (30%) in this case, before the shaft becomes fully plastic. In the fully plastic state, the member has developed its full capacity for carrying torque; the corresponding torque is referred to as the *ultimate torque*. Note that this load, in spite of its name, is not in any way associated with the ultimate stress.

Let us now return to noncircular cross sections and the membrane analogy. We will consider a torsion member having a *rectangular* cross section with the dimensions given in Fig. 7-29a. Corresponding to the condition of no load, a membrane stretched across a rectangular frame would have zero deflection. We will assume a stress–strain curve in shear as in Fig. 7-26b.

Imagine a small torque applied to the shaft; the corresponding membrane contours would then be as in Fig. 7-29b. The highest stresses would be at points A, but let the load be such that they are well below the yield stress. The effect of additional twist of the shaft can be simulated by raising the pressure under the membrane, which will cause the bubble to inflate more. The change in the volume under the membrane represents

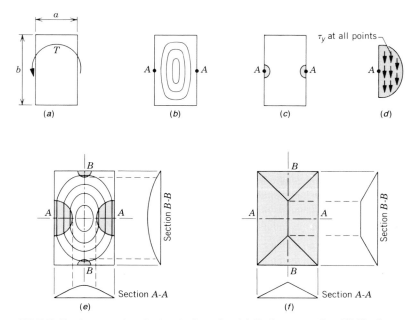

FIG. 7-29 Membrane analogy in the plastic region. (*a*) Shaft cross section. (*b*) Membrane contours for elastic stresses. (*c*) Plastic regions. (*d*) Stresses in plastic region. (*e*) Contours for partially plastic region. (*f*) Constraint roof.

the increase in the torque required to give the shaft the extra twist, and the change in the slope at *A* indicates the addition to the maximum stress. If the stress at *A* is less than the yield stress, we can be certain that all fibers are in the elastic state.

Let us now apply another increment to the twist such that the strains in the immediate area of *A* go beyond the elastic limit. We can imagine the plastic regions as shown by the shaded areas in Fig. 7-29c, where the stress will be as shown in Fig. 7-29d. The stress is constant in the region and parallel to the edge. The membrane will no longer give a correct representation of the stress in the plastic region. So let us reverse the process and establish the shape of the membrane corresponding to the yielded region. Since the stress is constant, the maximum slope will also be constant at that value which corresponds to the yield stress. The direction of the maximum slope, which we know to be at right angles to the stress, is then normal to the edge. The membrane for the plastic region is therefore a plane which slopes upward from the edge at a steepness corresponding to the yield stress. The membrane in the elastic region still has the curved shape of a bubble. With a further increase in twist, the regions at *B* will also go plastic and the contours would then be as shown in Fig. 7-29e.

It may appear to be impractical to construct such a membrane, the determination of the boundaries of the plastic regions alone appearing to be an insurmountable task. However, if we want to solve the problem experimentally, we merely erect over the membrane a set of rigid planar surfaces that will constrain the membrane wherever it tends to indicate stress above the yield. This can be done by erecting a "bungalow roof" over the membrane as in Fig. 7-29f, where the slope of the roof corresponds to the yield stress. When stresses are low, the membrane does not touch the roof and it plays no part in the stress distribution. However, when twist, or load, is such that some plastic flow would occur, the roof will constrain the membrane in areas where the stress has reached yield. The idea of a constraining roof has been developed here for a rectangular section, but it can be applied to other sections as well. For all sections the roof can be described as having the same shape as the surface of the largest sand pile that could be placed on a plate having the shape of the cross section.

There are many difficulties in the way of using the constraining roof in experimental work. However, as a concept it is invaluable in the numerical solution to elastic-plastic problems and also for determining the torque that can be carried by a shaft when it is fully plastic. By imagining the membrane pressed against all points in the roof of Fig. 7-29f, we can see that the shear stresses are as shown in Fig. 7-30. For a given set of dimensions the stress resultants can be worked out and the *ultimate torque*, T_u, for a fully plastic section calculated in terms of the yield stress. By previous methods the torque, T_y, at which yielding was just starting, can also be determined. The relationship between these torques is dependent upon the shape of the cross section, and the ratio of T_u to T_y, the *shape factor*, can be determined.

FIG. 7-30 Stress in a fully plastic rectangular section.

Example 7-7

A steel torsion member has the cross section given in Fig. 7-31a. Determine the torque, T_u, for a fully plastic cross section if the yield stress in shear is 250 MPa.

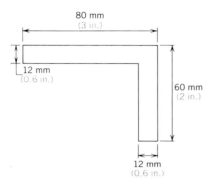

FIG. 7-31a Steel cross section subjected to torsion.

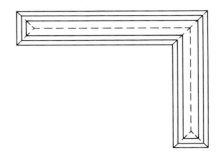

FIG. 7-31b Roof contours for angle section.

Solution. The contours of the "roof" can be sketched out as shown in Fig. 7-31b. Note that at the re-entrant angle the contours should be curved; this can be seen by visualizing the sand pile. However, curved contours would make the calculations very lengthy whereas the contours given above will provide good accuracy and err on the safe side.

Fig. 7-31c to e demonstrates the required steps to arrive at the torque calculations. The fully plastic torque is determined as:

$$T_u = 9 \times 76 + 102 \times 6 + 9 \times 56 + 72 \times 6 \text{ kN} \cdot \text{mm}$$

$$= 684 + 612 + 504 + 432 \text{ N} \cdot \text{m}$$

$$= \mathbf{2232 \ N \cdot m}$$

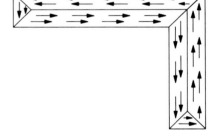

FIG. 7-31c Constant stress regions.

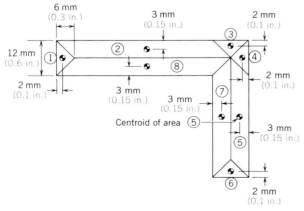

FIG. 7-31d Area subdivision.

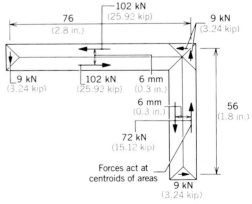

FIG. 7-31e Forces acting on areas 1 to 8.

It is interesting to note that the torque carried by the horizontal forces (612 + 504 = 1116) equals that of the vertical forces (684 + 432 = 1116). This is a principle that applies to all torsion whether elastic, plastic, or elastic-plastic and to all cross-sectional shapes.

Example 7-7

A steel torsion member has the cross section given in Fig. 7-31a. Determine the torque, T_u, for a fully plastic cross section if the yield stress in shear is 36 ksi.

Solution. The contours of the "roof" can be sketched out as shown in Fig. 7-31b. Note that at the re-entrant angle the contours should be curved; this can be seen by visualizing the sand pile. However, curved contours would make the calculations very lengthy whereas the contours given above will provide good accuracy and err on the safe side.

Fig. 7-31c to e demonstrates the required steps to arrive at the torque calculations. The fully plastic torque is determined as:

$$T_u = 3.24 \times 2.8 + 25.92 \times 0.3 + 3.24 \times 1.8 + 15.12 \times 0.3 \text{ kip} \cdot \text{in.}$$

$$= 9.072 + 7.776 + 5.832 + 4.536 \text{ kip} \cdot \text{in.}$$

$$= \textbf{27.216 kip} \cdot \textbf{in.}$$

It is interesting to note that the torque carried by the horizontal forces (7.776 + 5.832 = 13.608) equals that of the vertical forces (9.072 + 4.536 = 13.608). This is a principle that applies to all torsion whether elastic, plastic, or elastic-plastic and to all cross-sectional shapes.

Problems 7-43 to 7-47

7-8 SUMMARY

Engineers are frequently required to solve problems involving torsion, which is one of the most error-prone areas in design. For example, an engineer who does not fully appreciate the vast difference between an open and a closed section might be persuaded by the fabricator to allow the weld at A, in the member in Prob. 7-37, to be omitted. Or the designer might allow a pipe to be replaced by an angle of equal cross-sectional area and thus create an unsafe member. If torsion is all that important, you may wonder why it was not given a more rigorous treatment. The answer is that a complete treatment of the subject requires mathematics beyond the level of most undergraduate studies and the manipulations leading to rigorous solutions do little to enhance the understanding of torsion and to avoid pitfalls such as those cited above.

The treatment given here has been largely intuitive and based on the membrane analogy. However, the results, when not misapplied, are in reasonably good agreement with rigorous solutions. Furthermore, you will find that the membrane analogy leaves you with a good basic understanding of the nature of torsional stresses, their magnitudes and locations.

PROBLEMS

7-1 Calculate the maximum shear stress in the steel rod if the force P comes from supporting a mass of 50 kg (110 lb).

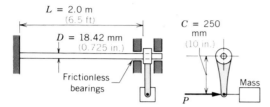

L = 2.0 m
(6.5 ft)

D = 18.42 mm
(0.725 in.)

C = 250 mm
(10 in.)

Frictionless bearings

Mass

P

PROBLEM 7-1

7-2 The solid rod in Prob. 7-1 is replaced by a pipe having an outside diameter of 25 mm (1 in.) and an inside diameter of 22 mm (0.88 in.). Compare the maximum stress and the amount of material in the two torsion members.

7-3 A 20-mm (0.8-in.) yellow brass shaft transmits a torque of 200 N·m (150 lb·ft). What strain would be recorded by a strain gage placed on the surface of the shaft at an angle of 45° to the shaft axis?

7-4 An automobile drive shaft is made by rolling and welding a 5-mm-thick (0.2-in.-thick) strip of AISI 1090 steel into a tube having an outside diameter of 75 mm (3 in.). What torque can be carried by the shaft if the longitudinal weld is 80% efficient? Use a safety factor of 1.5 based on yield and assume that the yield stress in shear is 60% of the yield stress in tension.

7-5 A 50-m-long (165-ft-long) propeller shaft for a ship has been designed with an allowable shear stress of 70 MPa (10 ksi). To safely carry the torque ioad, the solid shaft would require a diameter of 325 mm (13 in.).

It has been decided to change the design to a hollow shaft made of material having an allowable shear stress of 130 MPa (19 ksi). The outside diameter is to be retained.

a. Determine the maximum safe inside diameter of the new shaft.

b. Determine the number of tonnes (tons) of steel in each shaft.

c. If the high-strength material costs four times as much per tonne (ton), which shaft has the lower material cost?

7-6 Determine the required outside diameter of a hollow, phosphor bronze shaft, which is to carry a torsional load of 6000 N·m (4400 lb·ft) if the internal diameter of the shaft is 50 mm (2 in.). A factor of safety of 1.5 based on the yield strength in shear is required. Assume that the yield stress in shear is 57% of the yield stress in tension.

7-7 What power can a 50-mm (2-in.) brass shaft transmit when rotating at 3600 rpm if the shear stress is limited to 175 MPa (25 ksi)?

7-8 A shaft with a 200-mm (8-in.) diameter is made of AISI 1045 steel. The allowable shear stress is one-quarter of the yield stress in tension.

a. Determine the safe torque load.

b. If the shaft turns at 150 rpm, determine the power that can be safely transmitted.

7-9 What power can a 120-mm (5-in.) shaft transmit when rotating at 360 rpm if the shaft is AISI 1045 steel and a factor of safety of 1.5 based on the yield strength in shear is required? Assume that the yield stress in shear is 55% of the yield stress in tension.

7-10 A 15-mm (0.6-in.) shaft turning at 1750 rpm is used to transmit 10 kW (14 horsepower). Determine the change in the power handling capacity of the shaft caused by drilling a 5-mm (0.2-in.) hole along the shaft centerline.

7-11 A circular shaft turns at 1200 rpm and is required to transmit 7500 kW (10,000 hp). It is proposed to construct the shaft from AISI 1090 steel shafting 120 mm (5 in.) in diameter. Determine the safety factor based on yield assuming that the yield stress in shear is 60% of the yield stress in tension.

7-12 A water wheel has a rated power output of 8 MW (10,700 hp) with a 5% overload capacity. A vertical 600-mm (24-in.) shaft, turning at 60 rpm, connects the water wheel to an electric generator. Determine the shear stress in the shaft at

a. Rated power.   b. Overload.

7-13 A 75-kW (100-hp) electric motor, turning at 1750 rpm, is used to drive a hoist. The motor is connected by a shaft to a gearbox having a 30:1 speed ratio and an efficiency of 95%. A second shaft connects the gearbox to a hoisting drum. The shaft material is AISI 1045. Assuming the yield stress in shear to be 60% of the tensile yield stress, and using a safety factor of 2.25 based on yield, determine the minimum safe diameter of each shaft.

7-14 A phosphor bronze shaft is used to transmit power at constant speed through the pulleys shown. If power is input at *B* through a torque of 200 N · m (1800 lb · in.) and is extracted at *A* and *C* through torques of 75 N · m (675 lb · in.) and 125 N · m, (1125 lb · in.), respectively, determine the shear stress in the shaft between *A* and *B* and between *B* and *C*.

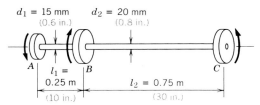

PROBLEM 7-14

7-15 A 1750-rpm, 150-kW (200-hp) motor drives, through a coupling at *A*, a straight shaft that runs from *A* to *B* to *C* to *D*. A gear is mounted on the shaft at *B*, 1.5 m (60 in.) from *A*. The gear at *B* drives a pump that requires 45 kW (60 hp). A gear at *C*, 2.5 m (100 in.) from *B*, drives a fan and imposes a torsional drag of 200 N · m (1800 lb · in.) on the shaft. The end of the shaft at *D*, 3.0 m (120 in.) from *C*, is connected to a generator that requires the remaining power. If the material in the shaft can safely carry a shear stress of 80 MPa (12 ksi), determine the minimum diameter of each portion of the shaft. Material: Steel.

7-16 Determine the angle of twist in degrees in the shaft in Prob. 7-9 if the shaft length is 10 m (33 ft).

7-17 If the shaft in Prob. 7-15 is 40 mm (1.58 in.) in diameter, determine the angle (degrees) by which the generator lags behind the motor.

7-18 For the shaft sizes required in Prob. 7-15, determine the angle (degrees) by which the generator lags behind the motor.

7-19 The propeller shaft of a ship turns at 250 rpm and transmits 1000 kW (1340 hp). The shaft is 8 m (26 ft) long and is made of steel having an allowable shear stress of 90 MPa (13 ksi).

a. Determine the minimum diameter of the shaft.
b. Determine the angle of twist in the shaft.

7-20 A circular shaft, 4 m (13 ft) long, is to be designed to carry a torque load of 200 N · m (150 lb · ft) with an allowable shear stress of 80 MPa (12 ksi).

a. Determine the required minimum diameter and the angle of twist.
b. If, when manufactured, the shaft is 1% undersize, determine the change (percent) in the stress and the angle of twist.

7-21 a. Determine the angle through which the arm in Prob. 7-1 rotates when the weight is suspended from the end of the arm.
b. Recalculate the torque load on the rod, taking into account the rotation of the arm.
c. Determine the percent error in torque if the rotation is neglected.

7-22 a. Calculate the angle through which the arm in Prob. 7-1 rotates and the downward motion of the load.
b. Repeat (a) using as the torsion member a steel tube having an outside diameter of 25 mm (1 in.) and a 1.5 mm (0.059 in.) wall thickness.
c. In a torsion-bar suspension system, would the solid rod or the tube give a better ride?

7-23 A coil spring is made of steel wire having a diameter of 6 mm (0.236 in.) wound to a coil diameter of 48 mm (1.89 in.). There are 5 active turns. Determine the stress and deflection when an axial load of 700 N (160 lb) is applied.

7-24 A coil spring is wound from hard-drawn steel wire having a 6-mm (0.236-in.) diameter. The mean diameter of the coils is 50 mm (2 in.) and there are 7 active turns.

a. If the allowable shear stress is 500 MPa (72 ksi), determine the safe load. (*Note:* The performance of springs is very dependent on working to high stresses. Consequently, it is profitable to pay a premium price for high-strength material. Design stresses of the given magnitude are not uncommon for spring materials.)

b. Calculate the deflection of the spring under the safe load.

c. What is the approximate percentage error in the stress in part (a) due to neglecting the direct shear?

7-25 A coil spring is to be made of wire that is 4.85 mm (0.191 in.) in diameter and can be safely stressed to 540 MPa (78 ksi) in shear. Specify the mean diameter and the number of turns if the spring is subjected to a load of 900 N (200 lb) and is required to deflect 30 mm (1.2 in.) under that load.

7-26 A motor-generator set is to be vibration-isolated using four springs, each located at the corner of the motor-generator support platform. If the motor-generator exerts a total load of 1600 N (360 lb) on the spring system, select a mean spring diameter if the wire size is 5 mm (0.2 in.) and the stress is limited to 450 MPa (65 ksi).

7-27 Determine the deflection of the motor-generator set in Prob. 7-26 if the spring material is phosphor bronze and the number of active coils is 6.

7-28 Determine the angle of twist in a rectangular bar 50 mm (2 in.) wide, 10 mm (0.4 in.) thick, and 1 m (40 in.) long due to a torsional load when the shear stress reaches a value of 90 MPa (13 ksi). The bar material is 6061-T6 aluminum.

7-29 A piece of steel is 2.5 m (100 in.) long and has a rectangular cross section 45 mm (1.8 in.) by 15 mm (0.6 in.). One end is fixed; a torque is applied to the other end causing the end to rotate through 12°. Determine:

a. The torque.

b. The maximum shear stress.

7-30 A steel member that is 3 m (10 ft) long must carry a torque load of 227 N·m (170 lb·ft). Seven different cross sections are being considered. Note that the amount of material is the same for each shaft. Determine the maximum stress and the angle of twist for each of these cases. Which shape is best for a torsion member?

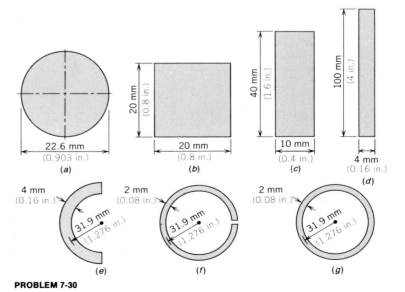

PROBLEM 7-30

7-31 A yellow brass shaft, with a 30-mm (1.2-in.) diameter, is designed to transmit torque at a constant speed. Determine the reduction in the torque carrying capacity of the shaft if one section of the shaft is changed to a square 26.6 mm (1.064 in.) on a side. The factor of safety is 2 based on the yield strength in shear. The yield stress in shear may be assumed to be 55% of the yield stress in tension.

7-32 A member that is to carry a torsional load of 400 N·m (3500 lb·in.) is to be made of material having an allowable shear stress of 80 MPa (12 ksi). The cross section is to be rectangular with one dimension 20 mm (0.8 in.). Determine the minimum value of the other dimension.

7-33 A member is 4 m (13 ft) long and carries a torque load of 2400 N·m (1800 lb·ft). If the member is an L200 × 100 × 20 (L8 × 4 × $\frac{3}{4}$), determine the maximum stress (neglecting stress concentration) and the angle of twist. How do these values compare with those of a solid circular rod having the same cross-sectional area?

7-34 A torque is applied to a 2.5-m-long (100-in.-long) channel, C180 × 22 (C7 × 15). One end is fixed and the other rotates through 10 degrees.

a. Determine the applied torque.
b. Determine the maximum shear stress (neglecting stress concentration).
c. If the allowable shear stress is 85 MPa (12 ksi), determine the safe torque load.

7-35 A torque load of 1000 N·m (738 lb·ft) is applied to a W460 × 52 (W18 × 35), 4 m (13 ft) long.

a. Calculate the maximum shear stress on the section, neglecting the stress concentration.
b. Select a pipe for the same application such that the shear stress does not exceed that found in part (a).
c. Compare the amount of steel in the pipe with that of the W-shape.

7-36 A W920 × 223 (W36 × 150) is subjected to a torque load. If the allowable shear stress is 95 MPa (14 ksi), neglecting stress concentration, determine the safe torque.

7-37 A hollow box girder is formed by welding four L200 × 100 × 20 (L8 × 4 × $\frac{3}{4}$) as shown.

PROBLEM 7-37

a. Determine the shear stress if the member is subjected to a torque of 100 kN·m (73,800 lb·ft).
   If the allowable shear stress is 90 MPa (13 ksi), determine the safe torque load for
b. The member as designed.
c. The member without a weld at A.

7-38 The cross section of a member in a deep-sea oil drilling platform has the shape shown. If the allowable shear stress is 85 MPa (12 ksi), determine the torque that can be safely applied to the member.

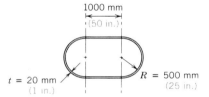

PROBLEM 7-38

7-39 By how much (percent) will the torque capacity of the member in Prob. 7-38 be increased if the same amount of 20-mm (1-in.) plate is rolled into a circular tube?

7-40 If the allowable shear stress in a standard 300-mm (12-in.) pipe is 100 MPa (14 ksi), determine the safe torque that can be applied to the pipe. Determine the safe torque if the bore of the pipe has an eccentricity of 5 mm (0.2 in.).

7-41 By how much (percent) will the torque capacity of the member in Prob. 7-38 be increased if the thickness of the side plates is doubled?

7-42 A hollow rectangular section is constructed from an aluminum channel having the dimensions of a C75 × 9 (C3 × 6) by welding a flat plate 76 mm (3.0 in.) wide and 5 mm (0.2 in.) thick to the open side of the channel. Determine the maximum torque that can be applied to the hollow section assuming the material to be 6061-T6 and a factor of safety of 2.5 based on the yield strength in shear. The yield stress in shear may be assumed to be 60% of the yield strength in tension.

7-43 A 24-mm-square (1-in.-square) steel shaft has a shear yield stress of 160 MPa (23 ksi). Calculate the torque load

    a. When the most highly stressed fibers are at the yield stress.
    b. When the cross section is fully plastic.
    c. What is the shape factor for the shaft in torsion?

7-44 Solve Prob. 7-43 using a 12 × 48 mm (0.5 × 2.0 in.) rectangular shaft.

7-45 Solve Prob. 7-43 changing the cross section to a circle with the same area as the square.

7-46 Solve Prob. 7-43 using a hollow-circular shaft having the same cross-sectional area as the square shaft and an outside to inside diameter ratio of 2.

7-47 Determine the shape factor for an L200 × 100 × 20 (L8 × 4 × $\frac{3}{4}$) in torsion (neglect stress concentrations).

Challenging Problems

7-A Modify the system given in Prob. 7-1 by making the steel rod 12 m (40 ft) long. If you are concerned about the rod bending under its own weight, assume that it is supported where required in frictionless bearings. Determine the maximum stress and the downward motion of the load. (*Note:* This is not an easy problem. Check the usual basic assumptions; they may not be valid.)

7-B A gear on the end of an AISI 1090 shaft is rotated in order to raise an arm on the other end of the shaft and thus lift a weight that hangs on the end of the arm. Determine the angle, from the given position, through which the arm rotates when the gear is turned through:

    a. 30°.   b. 60°.   c. 90°.
    d. 120°   e. 150°   f. 180°.

7-C Repeat Prob. 7-B for a shaft that is 20 m (65 ft) long.

7-D The torsional vibration characteristics of the propulsion system in Prob. 7-19 are found to be unacceptable. The shaft must be altered so that the angle of twist is halved, and this is to be accomplished by using a hollow shaft. If the allowable stress and all other parameters remain unaltered, determine the outside and inside diameters of the new shaft.

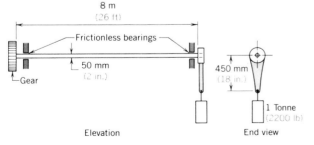

Elevation   End view

PROBLEM 7-B

8

BEAMS IN BENDING II

8-1 INTRODUCTION

We have seen that beams are important elements in buildings and other civil engineering structures. Our illustrations have been, and will continue to be, chosen from structures because of their familiarity and because many beams that are used in actual structures can be treated by the methods at our disposal. Beam theory is applicable in other fields, where the members are given different names but are subjected to bending and hence act as beams. For example, the frame of an automobile is a beam; the hull of a ship, when subjected to bending due to wave action, also behaves as a beam. In both cases the shapes are far too complex for us to analyze, but beam theory does form the basis of the analysis.

During the derivation of the beam formulas certain limitations were pointed out, such as the material being always stressed within the proportional range and the cross section having a certain symmetry. Failure to recognize the limitations can lead to faulty design. For example, if a beam has an unsymmetrical cross section, as in Fig. 8-1, the present bending formulas do not apply. Also, we would expect, from past experience, that the best place to apply the load would be directly over the shear-carrying part of the beam, that is, the web. This would also be wrong. In this chapter we will see how to deal with members that do not have symmetrical cross sections. It will be necessary to introduce some rather difficult concepts, but we have no choice because there is no easy way to solve nonsymmetrical beam problems.

At this stage we are familiar with the design of beams that are subjected to forces in the plane of symmetry of the cross section such as force, F, in Fig. 8-2a. We did not give much thought to the transverse location of the force and might have located it at F'. If we had stopped to question the proper force location to match our analysis and to ensure that our formulas were valid, we would have found the answer by examining the shear stress distribution on the cross section. A complete treatment of shear stress on the cross section will be given later in this chapter. Until then it is best to accept the shear stresses as they are shown in Fig. 8-2b. These stresses can be integrated over the various rectangular areas to give forces $F_1 \cdots F_5$ as depicted in Fig. 8-2c. It will be found that because of symmetry, F_1 equals F_2 and F_3 equals F_4. With these equalities it is readily seen that the resultant of the five forces acts downward in the plane of symmetry. Consequently, for the shear stress formula and the bending stress formula to be valid, the load must lie in the plane of symmetry. We could say the same thing by calling the centroid of the cross-sectional area the *shear center* and saying that the line of action of the load must pass through the shear center. The concept of shear center, which will be dealt with fully later, is somewhat superfluous at this stage, but it is advisable to become acquainted with it in this simple case before it appears again in more complex applications.

If the location of the load F' is unavoidable, the beam can be treated by a two-stage process as follows: First, replace the original force by an equal one, F in Fig. 8-2a, that passes through the shear center, and a torque equal to $F' \times e$. Second, solve the shear and bending stresses due to F and, as a separate problem, the shear stresses due to the torque.

FIG. 8-1

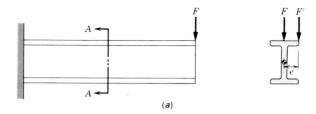

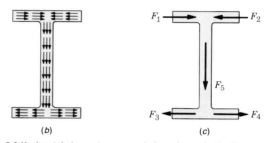

FIG. 8-2 Horizontal shear stresses and shear forces in the flanges of an *I*-section.
***(a)* Location of *F* with respect to the web of an *I*-section. *(b)* Shear stress on section *A-A*.**
***(c)* Shear forces on section *A-A*.**

The stress at any point can then be found by combining the stresses due to each load. You will appreciate that a considerable effort must be expended in finding the stresses and in combining shear stress due to torque with shear and bending stress due to *F*. Because of difficulties encountered in the analysis and the inherent torsional weakness of the usual beam cross sections, designers try to arrange members so that loads pass through the shear center. We will return later to the problem of determining the location of the shear center after we have considered some other beam characteristics; it is not always at the centroid.

8-2 INCLINED BENDING OF BEAMS HAVING SYMMETRICAL CROSS SECTIONS

The load *P* in Fig. 8-3 passes through the shear center but is inclined at an angle θ to the plane of symmetry. Since the load is in a plane that is not a plane of symmetry, we are facing our first *unsymmetrical bending* problem and must recognize that the bending formulas of Chapter 2 cannot be used directly. However, this problem can readily be decomposed into two simple problems by resolving *P* into a vertical and a horizontal component. Each component alone constitutes a load in a plane of symmetry and can be treated by existing formulas. Hence the problem can be solved as two separate problems and the results

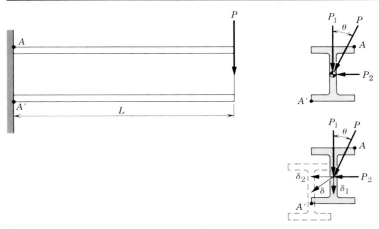

FIG. 8-3 Unsymmetrical bending of a symmetrical cross section, where the applied load passes through the shear center.

superimposed. In this case combining the bending stresses is quite simple because the points A and A' are subjected to the maximum stresses in each subproblem. Note that bending stresses at A would be tensile and at A' compressive. Consequently, the stress from (2-13) using load P_1 merely needs to be added to the stress from the same equation with load P_2 in order to determine the maximum bending stress. Since P_1 and P_2 cause bending about different axes of symmetry, appropriate I and c values with respect to these bending axes must be substituted into the bending stress formula.

The deflection can be obtained by solving the same two subproblems: P_1 causing deflection δ_1 in Fig. 8-3, and P_2 causing deflection δ_2. To get the actual deflection under load P, δ_1 and δ_2 need to be combined as vectors to give δ. It is interesting to note that *the direction of the displacement vector is not the same as the direction of the load vector*. This is a general characteristic of unsymmetrical bending.

Example 8-1

Determine the stresses at A and A' in the beam shown in Fig. 8-4a due to the load P applied at an angle of 30° to the y-axis.

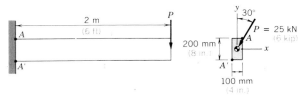

FIG. 8-4a Beam loading and cross section.

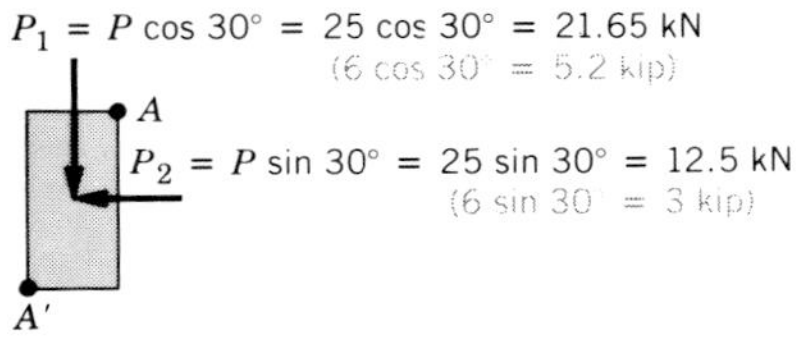

$P_1 = P \cos 30° = 25 \cos 30° = 21.65$ kN
(6 cos 30 = 5.2 kip)

A

$P_2 = P \sin 30° = 25 \sin 30° = 12.5$ kN
(6 sin 30 = 3 kip)

A'

FIG. 8-4b Vertical and horizontal components of P.

Solution. The load P can be resolved into two components, namely $P_1 = P \cos 30°$ and $P_2 = P \sin 30°$ acting as shown in Fig. 8-4b. P_1 causes bending about the x-axis and will give rise to bending stresses at A and A' of

$$\sigma_A = \frac{MC_y}{I_x} = + \frac{21.65 \times 10^3 \times 2000 \times 100}{(100 \times 200^3)/12} = 65.0 \ \frac{N}{mm^2} = 65 \ \text{MPa}$$

and

$$\sigma'_A = \frac{MC_y}{I_x} = - \frac{21.65 \times 10^3 \times 2000 \times 100}{(100 \times 200^3)/12} = -65.0 \ \frac{N}{mm^2} = -65 \ \text{MPa}$$

Similarly, the component P_2 causes bending stresses at A and A' of

$$\sigma_A = \frac{MC_x}{I_y} = + \frac{12.5 \times 10^3 \times 2000 \times 50}{(200 \times 100^3)/12} = 75 \ \frac{N}{mm^2} = 75 \ \text{MPa}$$

and

$$\sigma'_A = \frac{MC_x}{I_y} = - \frac{12.5 \times 10^3 \times 2000 \times 50}{(200 \times 100^3)/12} = -75 \ \frac{N}{mm^2} = -75 \ \text{MPa}$$

Combining the stresses due to the load components P_1 and P_2 yields the following stresses: at A,

$$\sigma = 65 + 75 = \mathbf{140 \ MPa}$$

and at A',

$$\sigma = -65 + (-75) = -\mathbf{140 \ MPa}$$

Example 8-1

Determine the stresses at A and A' in the beam shown in Fig. 8-4a, due to the load P applied at an angle of 30° to the y-axis.

Solution. The load P can be resolved into two components, namely $P_1 = P \cos 30°$ and $P_2 = P \sin 30°$ acting as shown in Fig. 8-4b. P_1 causes bending about the x-axis and will give rise to bending stresses at A and A' of

$$\sigma_A = \frac{MC_y}{I_x} = + \frac{5.2 \times 6 \times 12 \times 4}{4 \times 8^3/12} = 8.78 \ \text{kip/in.}^2$$

and

$$\sigma'_A = \frac{MC_y}{I_x} = - \frac{5.2 \times 6 \times 12 \times 4}{4 \times 8^3/12} = -8.78 \ \text{kip/in.}^2$$

Similarly, the component P_2 causes bending stresses at A and A' of

$$\sigma_A = \frac{MC_x}{I_y} = + \frac{3 \times 6 \times 12 \times 2}{8 \times 4^3/12} = 10.13 \ \text{kip/in.}^2$$

and

$$\sigma'_A = \frac{MC_x}{I_y} = -\frac{3 \times 6 \times 12 \times 2}{8 \times 4^3/12} = -10.13 \text{ kip/in.}^2$$

Combining the stresses due to the load components P_1 and P_2 yields the following stresses: at A,

$$\sigma = 8.78 + 10.13 = \textbf{18.91 kip/in.}^2$$

and at A',

$$\sigma = -8.78 + (-10.13) = -\textbf{18.91 kip/in.}^2$$

Problems 8-1 to 8-4

8-3 BEAMS HAVING NONSYMMETRICAL CROSS SECTIONS

In the last section we saw how to treat a cross section when it was subjected to bending about an axis that was not an axis of symmetry. The case that was treated was not perfectly general because the cross section had axes of symmetry and we still have not analyzed bending on sections that lack symmetry, such as Z-shapes and angles having unequal legs.

Turning now to the general case, we consider bending, as in Fig. 8-5a, in a member having a completely general cross-sectional shape. We will assume as before that plane sections remain plane under bending so that section A-B and section C-D remain plane and are displaced to A'-B' and C'-D' when the bending load is applied. We will take these planes as rotating relative to one another through an angle, α, about an x-axis through the centroid of the cross-sectional area. Notice that we have said nothing about the load but have specified a certain displacement. We will now establish the load that is necessary to impart that displacement and in so doing will discover a new property of area and some unexpected results.

For the given rotation a longitudinal element at distance y from the axis of rotation is subjected to a change in length of αy and hence a strain,

$$\varepsilon = \frac{\alpha y}{l}$$

and a stress,

$$\sigma = E\varepsilon = \frac{E\alpha y}{l} \tag{8-1}$$

The force required on the end of the element to give it the prescribed strain is

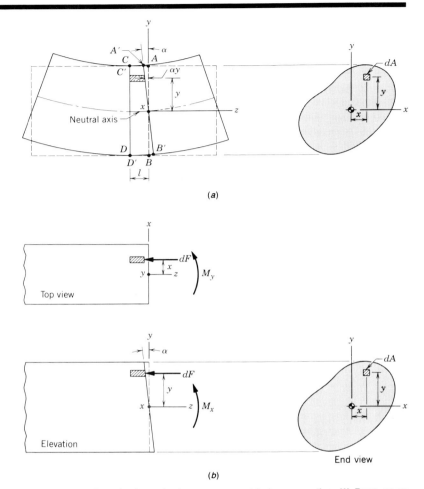

(a)

(b)

FIG. 8-5 (a) Bending of a beam having a nonsymmetrical cross section. (b) Force on an internal element as seen in a top view, elevation, and end view.

$$dF = \sigma \, dA$$

$$= \frac{E\alpha y}{l} \, dA \qquad \textbf{(8-2)}$$

This force is indicated in Fig. 8-5b, where the three views of the body are shown. The top view has been introduced for reasons that will soon become clear.

First let us check to see if an axial force is required on the member to produce the specified motion. The total axial force on the member will be the sum over the cross-sectional area of all the incremental forces; that is,

$$F = \int dF$$

$$= \int \frac{E\alpha}{l} y \, dA = \frac{E\alpha}{l} \int y \, dA$$

Since the x-axis passes through the centroid of the area, the moment of the area about the x-axis is zero. Or

$$\int y \, dA = 0$$

Hence

$$F = 0 \qquad\qquad \text{(8-3)}$$

This is consistent with our previous findings pertaining to bending and comes as no surprise. It indicates that for pure bending the neutral axis passes through the centroid, irrespective of symmetry or lack of symmetry.

Next let us see what moment is required about the x-axis. The moment required to exert force, dF, is

$$dM_x = y \, dF$$

Substituting from (8-2) yields

$$dM_x = \frac{E\alpha}{l} y^2 \, dA$$

Then the total moment is

$$M_x = \int \frac{E\alpha}{l} y^2 \, dA$$

$$= \frac{E\alpha}{l} \int y^2 \, dA$$

$$= \frac{E\alpha I_x}{l} \qquad\qquad \text{(8-4)}$$

This is exactly the same as (2-8), and hence the moment-rotation and moment-stress relationships that were established for symmetrical sections still hold. But we have one more check to make; that is, what moment about the y-axis, M_y, is required to hold the member in the given deformed state?

It can be seen in the top view, Fig. 8-5b, that the incremental force requires an incremental moment about the y-axis of

$$dM_y = x \, dF$$

Substituting from (8-2) gives

$$dM_y = x \frac{E\alpha}{l} y \, dA$$

$$= \frac{E\alpha}{l} xy \, dA$$

Then the total moment is

$$M_y = \frac{E\alpha}{l} \int xy \, dA \qquad \text{(8-5)}$$

We might hope that there would be some way to show that $\int xy \, dA$ is zero, but in general it is not, so we are forced to accept this unexpected moment as a part of the required loading system. The integral bears some resemblance to that which is defined as a moment of inertia, but instead of a squared variable there is a product term. The integral is therefore called the *product of inertia* and is represented by I_{xy}. Thus

$$\boxed{I_{xy} = \int xy \, dA} \qquad \text{(8-6)}$$

The moment equation then becomes

$$M_y = \frac{E\alpha}{l} I_{xy} \qquad \text{(8-7)}$$

Note that M_x and M_y are vector quantities that can be combined into a resultant bending moment and this moment will, in general, not be in the plane in which the deflection takes place. This is another manifestation of the principle noted earlier and seen in Fig. 8-3 where the displacement vector was found to have a direction that differs from the direction of the load vector. Before we carry out more work in the area of unsymmetrical bending, we will need to examine the new property, the product of inertia, more closely.

Problems 8-5 to 8-9

8-4 PROPERTIES OF CROSS SECTIONS: MOMENT OF INERTIA, PRODUCT OF INERTIA, AND MOHR'S CIRCLE OF INERTIA

We will now look at the properties of plane areas and pay particular attention to the unfamiliar property, the product of inertia, I_{xy}. Consider first an area such as that shown in Fig. 8-6a and assume that I_{xy} is required. The shape is quite general but does have one axis of symmetry, the y-axis. Before starting the laborious task of integrating (8-6), we may save some effort by an examination of the integration process.

Consider two equal elements of area dA_1 and dA_2; each makes a contribution toward I_{xy}. The elements have been chosen so that both have the same y-distance and the x-distances are numerically equal, but opposite in sign. Consequently, the contributions are the same in magnitude, but opposite in sign, so that together they contribute zero to the product of inertia. Similarly, all elements of area taken in symmetrical pairs cancel one another and, because of the symmetry, the whole area can be covered by symmetrical pairs of elements and the total I_{xy} will be zero.

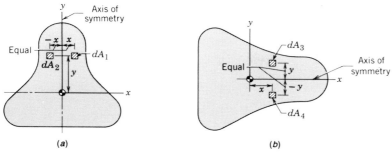

FIG. 8-6 Product of inertia for a cross section having one axis of symmetry.

If I_{xy} of the area of Fig. 8-6b is required, consider equal pairs of areas such as dA_3 and dA_4. Since the x's are identical for both areas and the y's are equal, but opposite in sign, the pair together contribute zero toward I_{xy}. By covering the whole area with similar pairs of areas, the total I_{xy} will be found to be zero. We can then conclude that *if either the x-axis or the y-axis is an axis of symmetry the product of inertia is zero*. This important conclusion can be used to avoid much unnecessary calculation. Areas that are not symmetrical may have a positive, negative, or zero product of inertia depending on the shape and dimensions.

When either axis is an axis of symmetry, I_{xy} is zero and (8-7) gives zero for M_y. In Chapter 2 the formulas were based on a cross section that was symmetrical about the y-axis, and this was a requirement of the cases solved by the bending formulas. We now see that the cross-sectional area may lack symmetry, and provided that I_{xy} is zero, the bending formulas of Chapter 2 are still applicable. Symmetry about the x-axis will make I_{xy} zero; hence the formulas in Chapter 2 could have been applied to bending of the channel in Fig. 8-7 or any other section that is symmetrical about the x-axis.

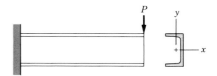

FIG. 8-7 The product of inertia, I_{xy}, is equal to zero for a channel section.

Problems 8-10 to 8-16

Let us now consider an area of general shape, such as that of Fig. 8-8, and assume that we know the values of I_x, I_y, and I_{xy}. Suppose that we want to know $I_{x'}$ and $I_{x'y'}$ with respect to a set of rotated axes o-x' and o-y'. The required properties could be obtained by integrating

$$I_{x'} = \int [y']^2 \, dA$$

and

$$I_{x'y'} = \int x'y' \, dA$$

In terms of the original coordinates

$$x' = x \cos \theta + y \sin \theta$$

and

$$y' = -x \sin \theta + y \cos \theta$$

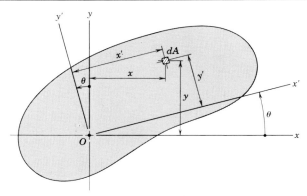

FIG. 8-8 Moment of inertia and product of inertia for a set of rotated axes.

Substituting these into the equation for the moment of inertia gives

$$I_{x'} = \int (-x \sin \theta + y \cos \theta)^2 \, dA$$
$$= \int (x^2 \sin^2 \theta - 2xy \sin \theta \cos \theta + y^2 \cos^2 \theta) \, dA$$
$$= \sin^2 \theta \int x^2 \, dA - 2 \sin \theta \cos \theta \int xy \, dA + \cos^2 \theta \int y^2 \, dA$$

But the integrals are all known properties with respect to the original set of axes. Replacing the integrals gives

$$I_{x'} = \sin^2 \theta \, I_y - 2 \sin \theta \cos \theta \, I_{xy} + \cos^2 \theta \, I_x$$
$$= \cos^2 \theta \, I_x + \sin^2 \theta \, I_y - 2 \sin \theta \cos \theta \, I_{xy} \qquad \textbf{(8-8)}$$

Substituting into the equation for the product of inertia gives

$$I_{x'y'} = \int (x \cos \theta + y \sin \theta)(-x \sin \theta + y \cos \theta) \, dA$$
$$= \int (-x^2 \sin \theta \cos \theta + xy \cos^2 \theta - xy \sin^2 \theta + y^2 \sin \theta \cos \theta) \, dA$$
$$= \int [\sin \theta \cos \theta (y^2 - x^2) + (\cos^2 \theta - \sin^2 \theta) xy] \, dA$$
$$= \sin \theta \cos \theta [\int y^2 \, dA - \int x^2 \, dA] + (\cos^2 \theta - \sin^2 \theta) \int xy \, dA$$
$$= \sin \theta \cos \theta (I_x - I_y) + (\cos^2 \theta - \sin^2 \theta) I_{xy} \qquad \textbf{(8-9)}$$

Equations (8-8) and (8-9) can be used to find the moment of inertia with respect to any inclined axis, and also the product of inertia, provided we have the three properties related to any set of axes. The formulas are useful in that way, but their real value lies in recognizing that they are identical in form to (5-1) and (5-2), with the following replacements

$$\sigma_x \rightarrow I_x$$
$$\sigma_y \rightarrow I_y$$
$$\tau_{xy} \rightarrow I_{xy}$$
$$\sigma_\theta \rightarrow I_{x'}$$
$$\tau_\theta \rightarrow I_{x'y'}$$

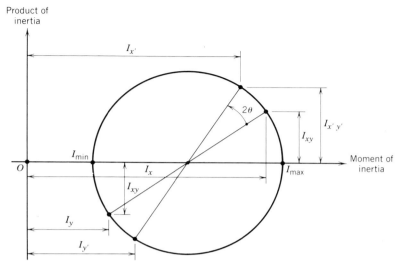

FIG. 8-9 Mohr's circle of inertia.

You will recall that we learned much about stress by picturing (5-1) and (5-2) as Mohr's circle of stress. Because of the similarity in the equations it is evident that (8-8) and (8-9) can also be represented by a circle, in this case *Mohr's circle of inertia*. A typical Mohr's circle of inertia is shown in Fig. 8-9. Mohr's circle of inertia can be used to determine quantities, but its greatest value lies in the insight that it can give us into the properties of areas as the following presentation will illustrate.

Suppose that we are designing a column that is to be made of angles and plates to give a cross section as in Fig. 8-10a. The properties with respect to the axes of symmetry can be calculated by well-known techniques. The I_x will obviously be less than the I_y. Therefore, it seems that I_x and its associated radius of gyration should be used in the buckling formula. But can we be sure that there is not another axis about which the moment of inertia is less than I_x and, hence, another mode of failure at a lower buckling load? To answer this question consider the Mohr's circle in Fig. 8-10b; drawing the circle is simplified by the fact that I_{xy} is known to be zero because of symmetry. It can be seen that no point on the circle has an abscissa less that I_x; hence there is no axis having a moment of inertia less than I_x and our original approach to the design does not need to be modified.

This illustration points out a useful principle:

The maximum and minimum moments of inertia, I_{max} and I_{min}, are associated with axes of symmetry or, in more general terms, *the maximum and minimum moments of inertia are associated with two orthogonal axes oriented so that the product of inertia with respect to these axes is zero.*

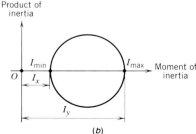

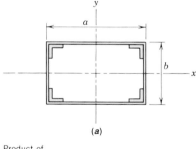

FIG. 8-10 (a) Column cross section with two axes of symmetry. (b) Mohr's circle of inertia for the column cross section.

The maximum and minimum moments of inertia are referred to as the *principal moments of inertia* and the associated axes as the *principal axes*. Columns can be expected to fail by bending about the principal axis having the smaller principal moment of inertia unless they are laterally restrained to prevent this from occurring.

It is interesting to consider a special case of the column in Fig. 8-10*a* that has *b* equal to *a*, and hence I_x equal to I_y. Mohr's circle then becomes a point, and the moment of inertia is the same for all inclinations of the axes. A column having such a cross section does not have a preferred direction for buckling; in this respect it behaves as a circular cross section in spite of lacking complete polar symmetry.

The parallel axis theorem was an invaluable aid in the computation of moment of inertia. The following derivation will lead to an equally useful formula in the product of inertia.

Consider an area such as that shown in Fig. 8-11 and say that we need to evaluate I_{xy}. We will treat as knowns the location of the centroid of the area and also the product of inertia with respect to the x''-axis and the y''-axis, which have their origin at the centroid of the area. The required quantity is

$$I_{xy} = \int xy \, dA$$

But

$$x = \bar{X} + x''$$

and

$$y = \bar{Y} + y''$$

Hence

$$I_{xy} = \int (\bar{X} + x'')(\bar{Y} + y'') \, dA$$
$$= \int \bar{X}\bar{Y} \, dA + \int x'' \bar{Y} \, dA + \int \bar{X}y'' \, dA + \int x''y'' \, dA$$
$$= \bar{X}\bar{Y}\int dA + \bar{Y}\int x'' \, dA + \bar{X}\int y'' \, dA + \int x''y'' \, dA$$

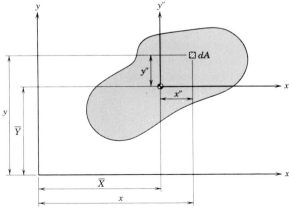

FIG. 8-11 Derivation of parallel axis theorem for product of inertia.

But

$$\bar{X}\bar{Y}\int dA = \bar{X}\bar{Y}\,\text{area}$$

$$\int x'' dA = 0$$

$$\int y'' dA = 0$$

and

$$\int x'' y'' dA = I_{x''y''}$$

Therefore,

$$I_{xy} = \bar{X}\bar{Y}\,\text{area} + I_{x''y''}$$

$$\boxed{I_{xy} = I_{x''y''} + \text{area}\,\bar{X}\bar{Y}}$$ (8-10)

Equation (8-10) expresses the Parallel Axis Theorem as applied to the product of inertia. Note the similarity to the Parallel Axis Theorem for the moment of inertia as expressed in (2-14).

Example 8-2

Determine the location of the principal axes as well as the principal moments of inertia for the area given in Fig. 8-12a. The origin of the axes is located at the centroid of the area.

Solution. With reference to Fig. 8-12b, the centroid is first determined as follows:

$$\text{area} = 10 \times 80 + 120 \times 10 = 800 + 1200 = 2000 \text{ mm}^2$$

$$A = \frac{5 \times 800 + 60 \times 1200}{2000} = 38.00 \text{ mm}$$

$$B = \frac{40 \times 800 + 85 \times 1200}{2000} = 67.00 \text{ mm}$$

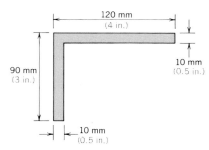

FIG. 8-12a

Properties relative to x- and y-axes are indicated in Fig. 8-12c.

$$I_x = \tfrac{1}{12} \times 10 \times 80^3 + 800 \times 27^2 + \tfrac{1}{12} \times 120 \times 10^3 + 1200 \times 18^2$$

$$= 1.409 \times 10^6 \text{ mm}^4$$

$$I_y = \tfrac{1}{12} \times 80 \times 10^3 + 800 \times 33^2 + \tfrac{1}{12} \times 10 \times 120^3 + 1200 \times 22^2$$

$$= 2.899 \times 10^6 \text{ mm}^4$$

Substituting into (8-10) for both areas gives

$$I_{xy} = 0 + 800 \times (-33) \times (-27) + 0 + 1200 \times 22 \times 18$$

$$= 1.188 \times 10^6 \text{ mm}^4$$

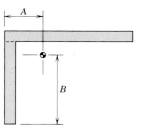

FIG. 8-12b

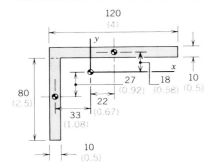

FIG. 8-12c

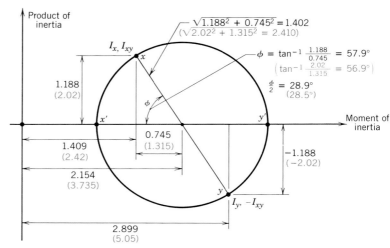

FIG. 8-12d

Mohr's circle of inertia, in 10^6 mm^4 units, can now be constructed as shown in Fig. 8-12d.

$$I_{x'} = (2.154 - 1.402) \times 10^6 \text{ mm}^4 = 0.752 \times 10^6 \text{ mm}^4$$

$$I_{y'} = (2.154 + 1.402) \times 10^6 \text{ mm}^4 = 3.556 \times 10^6 \text{ mm}^4$$

The principal moments of inertia are

$$I_{min} = \mathbf{0.752 \times 10^6 \text{ mm}^4}$$

$$I_{max} = \mathbf{3.556 \times 10^6 \text{ mm}^4}$$

The principal axes for the area are indicated in Fig. 8-12e.

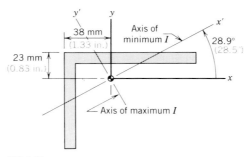

FIG. 8-12e

Example 8-2

Determine the location of the principal axes as well as the principal moments of inertia for the area given in Fig. 8-12a. The origin of the axes is located at the centroid of the area.

Solution. With reference to Fig. 8-12*b*, the centroid is first determined as follows:

$$\text{area} = 0.5 \times 2.5 + 4 \times 0.5 = 1.25 + 2.0 = 3.25 \text{ in.}^2$$

$$A = \frac{0.25 \times 1.25 + 2.0 \times 2.0}{3.25} = 1.33 \text{ in.}$$

$$B = \frac{1.25 \times 1.25 + 2.75 \times 2.0}{3.25} = 2.17 \text{ in.}$$

Properties relative to *x*- and *y*-axes are indicated in Fig. 8-12*c*.

$$I_x = \tfrac{1}{12} \times 0.5 \times 2.5^3 + 1.25 \times 0.92^2 + \tfrac{1}{12} \times 4 \times 0.5^3 + 2 \times 0.58^2$$

$$= 2.42 \text{ in.}^4$$

$$I_y = \tfrac{1}{12} \times 2.5 \times 0.5^3 + 1.25 \times 1.08^2 + \tfrac{1}{12} \times 0.5 \times 4^3 + 2 \times 0.67^2$$

$$= 5.05 \text{ in.}^4$$

Substituting into (8-10) for both areas gives

$$I_{xy} = 0 + 1.25 \times (-1.08) \times (-0.92) + 0 + 2 \times 0.67 \times 0.58$$

$$= 2.02 \text{ in.}^4$$

Mohr's circle of inertia, in in.^{4} units, can now be constructed as shown in Fig. 8-12*d*.

$$I_{x'} = (3.735 - 2.410) \text{ in.}^4 = 1.325 \text{ in.}^4$$

$$I_{y'} = (3.735 + 2.410) \text{ in.}^4 = 6.145 \text{ in.}^4$$

The principal moments of inertia are

$$I_{min} = \mathbf{1.325 \text{ in.}^4}$$

$$I_{max} = \mathbf{6.145 \text{ in.}^4}$$

The principal axes for the area are indicated in Fig. 8-12*e*.

Problems 8-17 to 8-25

We have seen that the formulas for rotation and stress from Chapter 2 are applicable only to cases of bending about a principal axis, and we have no convenient formulas for bending about other axes. However, such bending problems can be solved in a manner similar to that used to solve the case shown in Fig. 8-3. *In general, the approach is to resolve the bending moment on the section into components of moment about each of the principal axes.* By treating each moment separately and using the equations of Chapter 2, stresses and deflections can be found for each moment. Combining the stresses and deflections gives the solution to the original problem.

Example 8-3

Calculate the maximum bending stress in the cantilever beam of Fig. 8-13a.

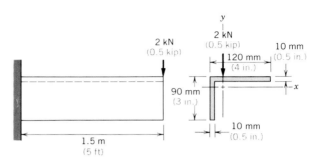

FIG. 8-13a

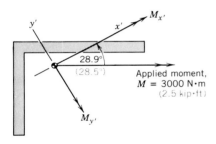

FIG. 8-13b

Solution. The maximum bending moment occurs at the left support and is

$$M = 2.0 \times 1.5 \text{ kN} \cdot \text{m}$$

$$= 3000 \text{ N} \cdot \text{m}$$

Since the cantilever section is the same as the section of Example 8-2, the properties of the section and the principal axes given in Fig. 8-13b are known. Resolving the maximum bending moment into components about the principal axes,

$$M_{x'} = M \cos 28.9° = 3000 \cos 28.9° = 2626 \text{ N} \cdot \text{m}$$

$$M_{y'} = M \sin 28.9° = 3000 \sin 28.9° = 1450 \text{ N} \cdot \text{m}$$

Referring to Fig. 8-13c, consider the load $M_{x'}$ only:

$$c_1 = 23 \cos 28.9° + 38 \sin 28.9°$$

$$= 20.14 + 18.36 = 38.50$$

$$c_2 = 67 \cos 28.9° - 28 \sin 28.9°$$

$$= 58.66 - 13.53 = 45.24$$

where c_1 and c_2 are the distances to A and B *normal* to the bending axis x'. Since $c_2 > c_1$, the stress at B will be greater than at A; hence

$$\sigma_B = \frac{M_{x'} c_2}{I_{x'}} = \frac{2626 \times 10^3 \times 45.24}{75.2 \times 10^4} \frac{\text{N} \cdot \text{mm} \cdot \text{mm}}{\text{mm}^4}$$

$$= 158.0 \frac{\text{N}}{\text{mm}^2} \quad \text{(compression)}$$

Considering the load $M_{y'}$ only as in Fig. 8-13d, we obtain

$$c_3 = 67 \sin 28.9° + 28 \cos 28.9° = 56.9 \text{ mm}$$

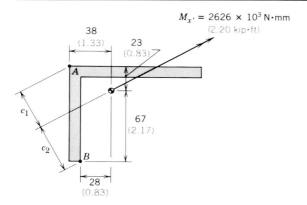

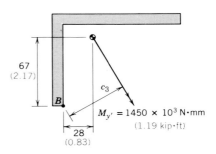

FIG. 8-13c

FIG. 8-13d

$$\sigma_B = \frac{M_{y'} c_3}{I_{y'}} = \frac{1450 \times 10^3 \times 56.9}{355.6 \times 10^4} \frac{\text{N} \cdot \text{mm} \cdot \text{mm}}{\text{mm}^4}$$

$$= 23.3 \frac{\text{N}}{\text{mm}^2} \qquad \text{(compression)}$$

and total stress at B

$$= 158.0 + 23.2 = 181.2 = 181.2 \frac{\text{N}}{\text{mm}^2} = \mathbf{181.2 \ MPa} \qquad \textbf{(compression)}$$

In this example the largest stress occurs at point B. However, no proof of this fact has been offered and there is no way to predict the location of the maximum stress point. All likely locations must be considered and the total stress calculated at each point so that the largest stress and its location can be identified.

Example 8-3

Calculate the maximum bending stress in the cantilever beam of Fig. 8-13a.

Solution. The maximum bending moment occurs at the left support and is

$$M = 0.5 \times 5 \text{ kip} \cdot \text{ft}$$

$$= 2.5 \text{ kip} \cdot \text{ft}$$

Since the cantilever section is the same as the section of Example 8-2, the properties of the section and the principal axes given in Fig. 8-13b are known. Resolving the maximum bending moment into components about the principal axes,

$$M_{x'} = M \cos 28.5 = 2.5 \cos 28.5° = 2.20 \text{ kip} \cdot \text{ft}$$

$$M_{y'} = M \sin 28.5° = 2.5 \sin 28.5° = 1.19 \text{ kip} \cdot \text{ft}$$

Referring to Fig. 8-13c, consider the load $M_{x'}$ only:

$$c_1 = 0.83 \cos 28.5° + 1.33 \sin 28.5°$$

$$= 0.73 + 0.63 = 1.36$$

$$c_2 = 2.17 \cos 28.5° - 0.83 \sin 28.5°$$

$$= 1.91 - 0.40 = 1.51$$

where c_1 and c_2 are the distances to A and B *normal* to the bending axis x'. Since $c_2 > c_1$, the stress at B will be greater than at A; hence

$$\sigma_B = \frac{M_{x'} c_2}{I_{x'}} = \frac{2.20 \times 12 \times 1.51}{1.325} \frac{\text{kip} \cdot \text{in.} \times \text{in.}}{\text{in.}^4}$$

$$= 30.09 \text{ kip/in.}^2 \quad \text{(compression)}$$

Considering the load $M_{y'}$ only as in Fig. 8-13d, we obtain

$$c_3 = 2.17 \sin 28.5° + 0.83 \cos 28.5° = 1.77 \text{ in.}$$

$$\sigma_B = \frac{M_{y'} c_3}{I_{y'}} = \frac{1.19 \times 12 \times 1.77}{6.145} \frac{\text{kip} \cdot \text{in.} \times \text{in.}}{\text{in.}^4}$$

$$= 4.11 \text{ kip/in.}^2 \quad \text{(compression)}$$

and total stress at B

$$= 30.09 + 4.11 = \mathbf{34.20 \text{ kip/in.}^2} \quad \textbf{(compression)}$$

In this example the largest stress occurs at point B. However, no proof of this fact has been offered and there is no way to predict the location of the maximum stress point. All likely locations must be considered and the total stress calculated at each point so that the largest stress and its location can be identified.

Problems 8-26 to 8-32

8-5 SHEAR CENTER

The shear center is a point in the plane of the cross section of a beam such that a load acting through the shear center will not subject the beam to any torsional load. The beam will, of course, be subjected to bending and shear. To find the shear center, a load is applied in a direction that will produce bending about a principal axis. The shear stresses on the cross section are determined on the assumption of zero torsion. The location of the resultant of the shear stresses gives the location of the line of action that the load must have in order to meet the assumed condition of zero torsion. Thus the resultant of the shear stresses must pass through the shear center. By taking a second direction for the load, at right angles to the first direction so that bending is about the other principal axis, a second resultant is obtained. The shear center is located at the intersection of the two resultants.

There is no general solution to the shear center location problem; each case must be worked separately. However, a few principles will emerge as we determine the shear centers for a variety of cross sections. These principles will be useful in reducing the effort required when the cross sections are more complex. Since the end conditions of the beam and the nature of the load have no bearing on the location of the shear center (it is a property of the cross section only), we will use the simplest of all beams in our determination of the shear center, that is, the cantilever beam with a concentrated load at the end. We will also restrict the cross sections to those having *thin* flanges and webs. This will enable us to assume that the stress is constant over the thickness of the part; otherwise the problem could not be solved by the methods at our disposal.

From the brief introduction to shear center at the beginning of this chapter and the illustration in Figure 8-2, it is evident that we need a good understanding of shear stress. Consequently, we will now review and extend our knowledge of shear by considering a flanged member, paying more attention to the shear in the flanges than was required previously. A typical flanged member is shown in Figure 8-14. The familiar bending stresses are shown on neighboring transverse planes, and the source of shearing stress, τ_L, is evident in the top view of the beam. The stress value is given by (2-24c) with the subscript L being introduced to remind us that the stress is on a *longitudinal* plane. Thus

$$\tau_L = \frac{VQ}{It}$$

We previously found in Section 2-6, Eq. (2-21) that shear stresses on planes at right angles are equal. Consequently, the stress on the

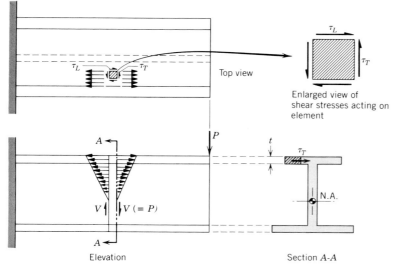

Enlarged view of shear stresses acting on element

Elevation

Section A-A

FIG. 8-14

transverse plane, τ_T, is equal to the stress on the longitudinal plane and is given by

$$\tau_T = \frac{VQ}{It} \tag{8-11}$$

where

$Q=$ the moment of the cross-hatched area about the neutral axis

A similar treatment of all flanges and the web will give the stress distribution pattern shown in Figure 8-15a. When calculations are required, it is simpler to deal with shear flow rather than stress. As in the treatment of closed sections in torsion, the shear flow is given by (7-21b) as

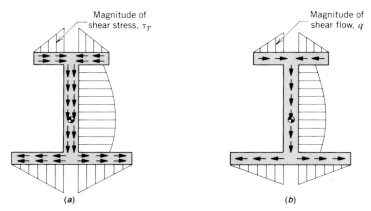

FIG. 8-15 (*a*) Shear stress distribution. (*b*) Shear flow distribution.

$$q = \tau t$$

or in the present application

$$q = \frac{VQ}{It} t$$

Hence

$$\boxed{q = \frac{VQ}{I}} \tag{8-12}$$

The force on any part of the cross section can be determined by integrating the shear flow over the length of that part. Thus forces such as those in Fig. 8-16 can be determined.

For the particular cross section in this illustration, the flanges are symmetrical about the vertical centerline. Without actually calculating quantities, it is evident that for corresponding points across the axis of symmetry, the shear stresses are equal but opposite in direction. The

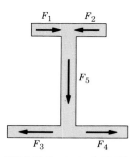

FIG. 8-16 Forces equivalent to shear stresses.

same equality follows for the shear flow and, when shear flow is integrated to give forces, F_1 will equal F_2 and F_3 will equal F_4. This means that the resultant of the five forces is in fact F_5 and coincides with the axis of symmetry. Then the shear center must be located at some point on the axis of symmetry.

From the foregoing we may conclude that *when there is an axis of symmetry in the plane of a cross section, the shear center lies on the axis of symmetry.*

If we extend this principle to beams such as S-shapes and W-shapes that have two axes of symmetry, the shear center must lie on both axes and hence is located at the intersection of the axes. This means that *when a cross section has two axes of symmetry, the shear center coincides with the centroid.*

Example 8-4

Find the location of the shear center of an L200 × 200 × 20.

Solution. We will use the angle as a cantilever and turn it so that bending is about one of the principal axes; otherwise the shear stress relations will not be valid. The cantilever is then as shown in Fig. 8-17a.

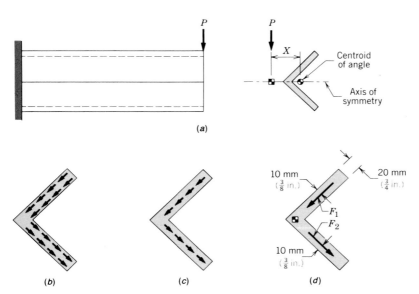

FIG. 8-17 (a) Cantilever beam with equal leg angle cross section. (b) Shear stress. (c) Shear flow. (d) Shear forces.

By the principle previously stated, the shear center is known to lie on the axis of symmetry. Its location will be determined when the dimension X has been evaluated. In most cases numerical values would be calculated for forces due to shear on the areas of the cross section. However, in the present case that is unnecessary; the solution can be found with a minimum of calculation. Consider the shear stress and shear flow on a typical cross section as shown in Figs. 8-17b and 8-17c. The distributed shear forces can be integrated to give F_1 and F_2. These forces act along the centerline of each leg, and the resultant passes through the point at which they converge. **The shear center is then located as shown in Fig. 8-17d.**

The same approach can be used for any angle; the shear center will always lie at the intersection of the centerlines of the legs.

Example 8-4

Find the location of the shear center of an L8 × 8 × $\frac{3}{4}$.

Solution. We will use the angle as a cantilever and turn it so that bending is about one of the principal axes; otherwise the shear stress relations will not be valid. The cantilever is then as shown in Fig. 8-17a.

By the principle previously stated, the shear center is known to lie on the axis of symmetry. Its location will be determined when the dimension X has been evaluated. In most cases numerical values would be calculated for forces due to shear on the areas of the cross section. However, in the present case that is unnecessary; the solution can be found with a minimum of calculation. Consider the shear stress and shear flow on a typical cross section as shown in Figs. 8-17b and 8-17c. The distributed shear forces can be integrated to give F_1 and F_2. These forces act along the centerline of each leg, and the resultant passes through the point at which they converge. **The shear center is then located as shown in Fig. 8-17d.**

The same approach can be used for any angle; the shear center will always lie at the intersection of the centerlines of the legs.

Next we will determine the shear center for a slightly more complex cross section, a channel section.

Example 8-5

Find the shear center for the channel section of Fig. 8-18a.

Solution. Since the shear center must lie on the axis of symmetry, only one locating dimension, X, is required as indicated in Fig. 8-18b.

We will approximate the true cross section by three rectangular areas as shown in Fig. 8-18c. Note that there is some overlapping of areas and a small area of the actual cross section has been omitted. This is part of the approximating process and introduces a small error. However, the areas are distorted in a region where the stress is poorly approximated by the stress formulas. Where thicknesses are small, the error is negligible. The subdivision used in this example is only one of many that could be devised; we might have elected to leave out the corner region entirely or to include it as part of both adjacent rectangles. The course chosen is midway between these extremes.

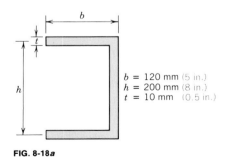

b = 120 mm (5 in.)
h = 200 mm (8 in.)
t = 10 mm (0.5 in.)

FIG. 8-18a

$$I_x = 2\left[\frac{1}{12}bt^3 + bt\left(\frac{h}{2}\right)^2\right] + \frac{1}{12}th^3 = 2\left[\frac{1}{12} \times 120 \times 10^3 + 120 \times 10 \times 100^2\right]$$

$$+ \frac{1}{12} \times 10 \times 200^3 \text{ mm}^4$$

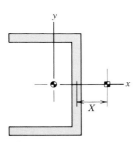

FIG. 8-18*b*

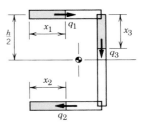

FIG. 8-18*c*

$$= 0.02 \times 10^6 + 24 \times 10^6 + 6.7 \times 10^6 \text{ mm}^4 = 30.7 \times 10^6 \text{ mm}^4$$

The shear flows, depicted in Fig. 8-18*d*, are calculated as

$$q = \frac{VQ}{I} \quad \text{(from 8-12)}$$

$$q_1 = \frac{V x_1 t (h/2)}{I} = \frac{Vth}{2I} x_1$$

$$q_2 = \frac{V x_2 t (h/2)}{I} = \frac{Vth}{2I} x_2$$

$$q_3 = \text{(not required)}$$

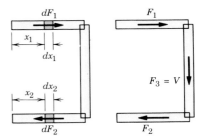

FIG. 8-18*d*

The shear forces shown in Fig. 8-18*e* are

$$dF_1 = q_1\, dx_1 = \frac{Vth}{2I} x_1\, dx_1 \qquad F_1 = \int_0^b dF_1 = \frac{Vth}{2I} \int_0^b x_1\, dx_1 = \frac{Vthb^2}{4I}$$

$$dF_2 = q_2\, dx_2 = \frac{Vth}{2I} x_2\, dx_2 \qquad F_2 = \int_0^b dF_2 = \frac{Vth}{2I} \int_0^b x_2\, dx_2 = \frac{Vthb^2}{4I}$$

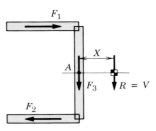

FIG. 8-18*e*

The resultant can now be located by determining the dimension X in Fig. 8-18*f*.
Equating the moment of the resultant R about A to moments of forces gives:

$$-RX = -F_1 \frac{h}{2} - F_2 \frac{h}{2}$$

$$X = \frac{(Vthb^2/4I)\,h}{V} = \frac{tb^2 h^2}{4I}$$

$$= \frac{10 \times 120^2 \times 200^2}{4 \times 30.7 \times 10^6} \frac{\text{mm} \cdot \text{mm}^2 \cdot \text{mm}^2}{\text{mm}^4} = 46.9 \text{ mm}$$

The shear center is located 46.9 mm to the right of the center of the web. It is
interesting to have a closer look at the general expression for X and in this way

FIG. 8-18*f*

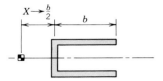

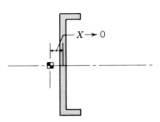

FIG. 8-18g

FIG. 8-18h

extend our knowledge of the location of the shear center for channel sections. We had

$$X = \frac{tb^2 h^2}{4I}$$

but from the beginning of the example problem

$$I \approx 2bt \left(\frac{h}{2}\right)^2 + \frac{th^3}{12}$$

Substituting for I in the X relation and simplifying, we obtain

$$X \approx \frac{1}{2 + (h/3b)} b$$

It can now be seen that when b is very large, X approaches $b/2$; alternatively, when h is very large, then X approaches zero. The two extreme cases for a channel section are shown in Fig. 8-18g and h.

Example 8-5

Find the shear center for the channel section of Fig. 8-18a.

Solution. Since the shear center must lie on the axis of symmetry, only one locating dimension, X, is required as indicated in Fig. 8-18b.

We will approximate the true cross section by three rectangular areas as shown in Fig. 8-18c. Note that there is some overlapping of areas and a small area of the actual cross section has been omitted. This is part of the approximating process and introduces a small error. However, the areas are distorted in a region where the stress is poorly approximated by the stress formulas. Where thicknesses are small, the error is negligible. The subdivision used in this example is only one of many that could be devised; we might have elected to leave out the corner region entirely or to include it as part of both adjacent rectangles. The course chosen is midway between these extremes.

$$I_x = 2 \left[\frac{1}{12} bt^3 + bt \left(\frac{h}{2}\right)^2 \right] + \frac{1}{12} th^3 = 2 \left[\frac{1}{12} \times 5 \times 0.5^3 + 5 \times 0.5 \times 4^2 \right]$$

$$+ \frac{1}{12} \times 0.5 \times 8^3 \text{ in.}^4$$

$$= 0.104 + 80.0 + 21.333 \text{ in.}^4 = 101.44 \text{ in.}^4$$

The shear flows, depicted in Fig. 8-18d, are calculated as

$$q = \frac{VQ}{I} \qquad \text{(from 8-12)}$$

$$q_1 = \frac{Vx_1 t(h/2)}{I} = \frac{Vth}{2I} x_1$$

$$q_2 = \frac{Vx_2 t(h/2)}{I} = \frac{Vth}{2I} x_2$$

$$q_3 = \text{(not required)}$$

The shear forces shown in Fig. 8-18e are

$$dF_1 = q_1\,dx_1 = \frac{Vth}{2I}\,x_1\,dx_1 \qquad F_1 = \int_0^b dF_1 = \frac{Vth}{2I}\int_0^b x_1\,dx_1 = \frac{Vthb^2}{4I}$$

$$dF_2 = q_2\,dx_2 = \frac{Vth}{2I}\,x_2\,dx_2 \qquad F_2 = \int_0^b dF_2 = \frac{Vth}{2I}\int_0^b x_2\,dx_2 = \frac{Vthb^2}{4I}$$

The resultant can now be located by determining the dimension X in Fig. 8-18f.
Equating the moment of the resultant R about A to moments of forces gives:

$$-RX = -F_1\frac{h}{2} - F_2\frac{h}{2}$$

$$X = \frac{(Vthb^2/4I)\,h}{V} = \frac{tb^2 h^2}{4I}$$

$$= \frac{0.5 \times 5^2 \times 8^2}{4 \times 101.44} = \frac{\text{in.}\cdot\text{in.}^2\cdot\text{in.}^2}{\text{in.}^4} = 1.97 \text{ in.}$$

The shear center is located 1.97 in. to the right of the center of the web. It is
interesting to have a closer look at the general expression for X and in this way
extend our knowledge of the location of the shear center for channel sections. We
had

$$X = \frac{tb^2 h^2}{4I}$$

but from the beginning of the example problem

$$I \simeq 2bt\left(\frac{h}{2}\right)^2 + \frac{th^3}{12}$$

Substituting for I in the X relation and simplifying, we obtain

$$X \simeq \frac{1}{2 + (h/3b)}\,b$$

It can now be seen that when b is very large, X approaches $b/2$; alternatively, when
h is very large, then X approaches zero. The two extreme cases for a channel
section are shown in Fig. 8-18g and h.

Problems 8-33 to 8-46

8-6 LOCAL BUCKLING

Column action may be present in local areas and may cause *local
buckling* failure when the part as a whole seems to be capable of carrying
its load. Although local buckling may appear in varied and often complex

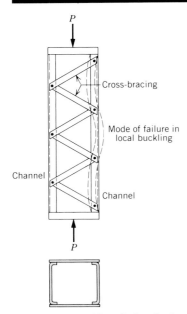

FIG. 8-19 Local buckling of a laced column.

forms, we will deal with only the most common and relatively simple cases. Local buckling is most readily understood through consideration of a column such as that shown in Fig. 8-19. The two channels constitute the main load-carrying elements, and the cross-bracing is provided to ensure that the channels act as integral parts of the column. The cross-bracing performs in a manner analogous to the web of a girder and its design, which is beyond the scope of this book, must receive careful attention. Returning to the channels, we see that they are constrained against lateral motion by the cross-bracing only at discrete points. Between these points the channels have an opportunity to fail by buckling, the mode of failure being illustrated by the broken lines in Fig. 8-19.

Less obvious forms of local buckling can occur in beams. In any beam subjected to bending, such as that of Fig. 8-20a, there is one flange in compression. In this case it is the upper flange; if this flange were viewed in isolation, it would appear as seen in Fig. 8-20b. Since a

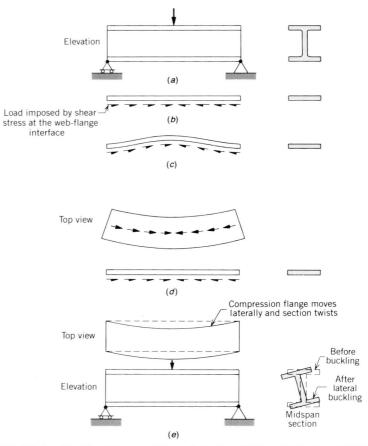

FIG. 8-20 Local buckling in beams. (a) Beam in bending. (b) Upper flange as a compression member. (c) Buckling about the axis having the least radius of gyration. (d) Alternate buckling mode. (e) Failure by lateral buckling.

compression member will tend to fail by buckling about the axis having the least radius of gyration, it might be expected to buckle as in Fig. 8-20c. However, in reality, the flange is attached to the web, which is capable of preventing the motion required for this mode of failure. Another possible mode is shown in Fig. 8-20d, where the upper flange moves laterally and buckling takes place by bending about a vertical axis. The web again tends to constrain the flange against this action, but it is neither well positioned nor proportioned to offer much resistance to the lateral motion. Failure of the beam by this type of local buckling is shown in Fig. 8-20e and is described as *lateral buckling of the compression flange*. You will appreciate that the loading on the upper flange is not as simple as it was in the case for columns. The complexity of the load, combined with the partial constraint offered by the web, make the determination of the local buckling load quite a complicated problem. When the design must conform to a code, an allowance is made for the tendency to buckle by reducing the allowable compressive stress according to some empirical formula based largely on test evidence. Note that if the beam is part of a system that prevents the compression flange from moving laterally, the buckling mode does not have to be considered. Beams in floor systems are frequently tied to other parts of the system that furnish lateral support. In some cases it becomes profitable to provide additional members for the sole purpose of giving lateral support to the compression flange of a beam.

The beam in bending can have yet another mode of flange failure. This is the result of ripples forming in the compression flange as shown in Fig. 8-21a. Since the flange has low torsional rigidity and the web offers little resistance to rotation, this type of failure, referred to as *crippling*, can take place. It is more likely to occur in flanges that are very thin relative to their width. Web stiffeners, as shown in Fig. 8-21b, in addition to supporting the web, may provide support for the flange and thus strengthen it against crippling.

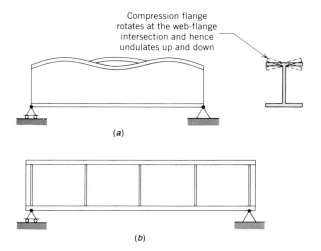

Compression flange rotates at the web-flange intersection and hence undulates up and down

(a)

(b)

FIG. 8-21 (a) Failure by local crippling. (b) Web stiffeners to strengthen not only the web, but also to prevent local buckling of the compression flange.

Problems 8-47 to 8-48

8-7 COMPOSITE BEAMS

In Chapter 1, we used our knowledge of stress–strain relations to solve for stresses in systems consisting of more than one material. The cases were all of such a nature that the materials were subjected to uniaxial stress and they were so connected that the strain was equal in both materials. Simple elementary methods were sufficient to solve some practical cases, including short reinforced concrete columns. When members composed of more than one material, that is, *composite members*, are subjected to bending, the analysis is considerably more difficult and a more systematic approach is required.

Consider a beam such as that shown in Fig. 8-22. It consists of a wooden core with steel plates fastened to the top and bottom surfaces. This is an arrangement that can result from the need to strengthen an existing timber building to accommodate an anticipated increase in load.

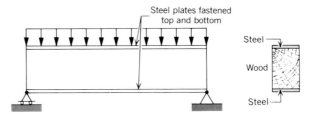

FIG. 8-22 Composite wood–steel beam.

For the case of composite members we will now derive bending equations similar to those of Chapter 2. For the sake of generality we will use a cross section that is not symmetrical about the neutral axis; it will, however, have a vertical axis of symmetry. Such a section is shown in Fig. 8-23a, where the areas are shown as rectangular, but will be treated as being unrestricted in shape except for the axis of symmetry already noted.

Again, plane sections will be assumed to remain plane and the strain distribution will be as in Fig. 8-23b. This gives a stress distribution as in Fig. 8-23c, where it is seen that two stress lines are required, one for each material, because of the two different values for Young's modulus. The stress formulas are

$$\sigma_1 = -\frac{E_1 \alpha y}{l} \tag{8-12a}$$

$$\sigma_2 = -\frac{E_2 \alpha y}{l} \tag{8-12b}$$

Since *y* must be measured from the neutral axis, the formulas cannot be further developed until the position of the neutral axis has been

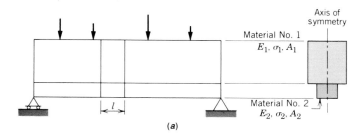

(a)

(b)

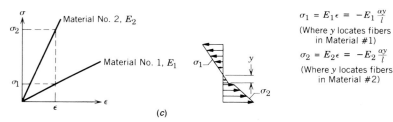

(c)

FIG. 8-23 Derivation of bending equations for a composite beam. (a) Composite beam. (b) Rotation and strain distribution. (c) Stress-strain diagram and stress distribution.

determined. To do this, we will once more use the fact that the axial load on the member is zero, which means that

$$\int_{A_1} \sigma_1 \, dA_1 + \int_{A_2} \sigma_2 \, dA_2 = 0$$

Substituting from (8-12) gives

$$\int_{A_1} \frac{E_1 \alpha y \, dA_1}{l} + \int_{A_2} \frac{E_2 \alpha y \, dA_2}{l} = 0$$

$$\frac{E_1 \alpha}{l} \int_{A_1} y \, dA_1 + \frac{E_2 \alpha}{l} \int_{A_2} y \, dA_2 = 0$$

The factor α/l can be removed from each term, but the Young's modulus terms cannot be eliminated. We can, however, simplify the equations by dividing through by E_1 and substituting

$$\boxed{n = \frac{E_2}{E_1}} \tag{8-13}$$

to get

$$\int_{A_1} y\, dA_1 + n \int_{A_2} y\, dA_2 = 0 \qquad (8\text{-}14)$$

Because n is the ratio between two moduli of elasticity, it is referred to as the *modular ratio*.

Each of the integrals is a moment of area; the equation is usually most useful in the form

$$\bar{Y}_1 A_1 + n\bar{Y}_2 A_2 = 0 \qquad (8\text{-}15)$$

In either form the relationship indicates that the neutral axis is located at a position that is determined by a calculation somewhat similar to the calculation performed in locating the centroid. The calculation can be made to be identical if in the second term of (8-15) we associate the n with A_2; thus

$$\bar{Y}_1 A_1 + \bar{Y}_2 (nA_2) = 0$$

and let

$$\tilde{A}_2 = nA_2 \qquad (8\text{-}16)$$

The equation then becomes

$$\bar{Y}_1 A_1 + \bar{Y}_2 \tilde{A}_2 = 0 \qquad (8\text{-}17)$$

The operation in (8-16) can be shown graphically on the cross section (Fig. 8-24a) by increasing the area A_2 by the factor n. The area would then be as in Fig. 8-24b and the cross section is referred to as the *transformed cross section*. The transformation was effected by expanding the dimension of the area *parallel* to the neutral axis. This was necessary so that the transformation would not disturb the distribution of strain in the fibers. Equation 8-17 indicates that the neutral axis passes through the centroid of the transformed cross section.

The moment required to cause the rotation α is given by

$$M = \int_{A_1} \sigma_1 y\, dA_1 + \int_{A_2} \sigma_2 y\, dA_2$$

Substituting for stresses gives

$$M = \int_{A_1} \frac{E_1 \alpha}{l} y^2\, dA_1 + \int_{A_2} \frac{E_2 \alpha}{l} y^2\, dA_2$$

$$= \frac{\alpha E_1}{l} \left[\int_{A_1} y_2\, dA_1 + \int_{A_2} y^2 n\, dA_2 \right]$$

The product n by dA_2 can be changed to $d\tilde{A}_2$, and we then have

$$M = \frac{\alpha E_1}{l} \left[\int_{A_1} y^2\, dA_1 + \int_{A_2} y^2\, d\tilde{A}_2 \right]$$

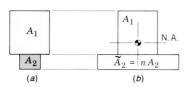

(a)   (b)

**FIG. 8-24 (a) Composite cross section.
(b) Transformed cross section.**

The integrals now give the moment of inertia of the transformed section, and the equation becomes

$$M = \frac{\alpha E_1}{l} \tilde{I} \tag{8-18}$$

where $\tilde{I}$ is the *moment of inertia of the transformed section* with respect to the axis through its centroid. From (8-18)

$$\alpha = \frac{Ml}{E_1 \tilde{I}} \tag{8-19}$$

which when substituted into (8-1) gives stress equations

$$\sigma_1 = - E_1 \frac{Ml}{E_1 \tilde{I}} \frac{y}{l}$$

$$\boxed{\sigma_1 = - \frac{My}{\tilde{I}}} \tag{8-20}$$

$$\sigma_2 = - \frac{E_2}{E_1} \frac{My}{\tilde{I}}$$

$$\boxed{\sigma_2 = - n \frac{My}{\tilde{I}}} \tag{8-21}$$

The equations can be seen to be similar to those of Chapter 2. To summarize, the composite beam can be treated by the following procedure:

1. Choose one of the materials as the base material; its Young's modulus is E_1. The other material will be referred to as the secondary material.
2. Determine the modular ratio, n, from (8-13) and transform the cross section of the two-material member to an equivalent one-material member by applying n to the area of the secondary material.
3. Perform whatever calculations are required (for stresses, deflections, etc.) as though the member were homogeneous and made of the primary material.
4. To determine stresses in the secondary material at a given location, calculate the stress in primary material at that location and then multiply by n to get the actual stress in the secondary material.

The most commonly used composite beams are made of concrete reinforced with steel, and the above procedure can be used to determine stresses in such a beam under some conditions. However, when concrete is under tension it cracks at a very low stress; hence the concrete is usually assumed as having no tensile strength. This means that in reinforced concrete beams, the reinforcing steel resists the entire tensile load while the concrete resists the compressive load. Because reinforced concrete is the most important composite structural material, it is treated in depth in specialized courses. We will restrict our study of the material to relatively simple flexural members.

Example 8-6

A Western hemlock timber is reinforced by fastening steel plates to the sides as shown in Fig. 8-25a. Calculate the maximum tensile and compressive stress in the steel and in the wood.

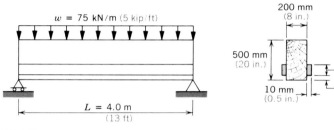

FIG. 8-25a

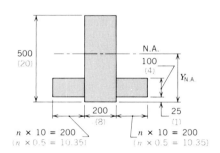

FIG. 8-25b

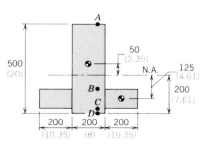

FIG. 8-25c Transformed all-wood cross section.

Solution

$$\text{maximum bending moment} = \tfrac{1}{8} wL^2$$

$$= \tfrac{1}{8} \times 75 \times 4.0^2 = 150 \text{ kN} \cdot \text{m}$$

$$= 150 \times 10^6 \text{ N} \cdot \text{mm}$$

From Table A,

$$\text{Young's modulus for Western hemlock, } E_1 = 10\ 000 \text{ MPa}$$

$$\text{Young's modulus for steel, } E_2 = 200\ 000 \text{ MPa}$$

$$\text{modular ratio, } n = \frac{E_2}{E_1} = \frac{200\ 000}{10\ 000} = 20$$

The transformed cross section (dimensions in millimetres) is shown in Fig. 8-25b and c.

$$Y_{NA} = \frac{500 \times 200 \times 250 + 2 \times 200 \times 100 \times 75}{500 \times 200 + 2 \times 200 \times 100} \quad \frac{\text{mm} \cdot \text{mm} \cdot \text{mm}}{\text{mm} \cdot \text{mm}}$$

$$= \frac{100\ 000 \times 250 + 40\ 000 \times 75}{100\ 000 + 40\ 000} = 200 \text{ mm}$$

$$\bar{I} = \tfrac{1}{12} \times 200 \times 500^3 + 100\ 000 \times 50^2 + \tfrac{1}{12} \times 400 \times 100^3 + 40\ 000 \times 125^2$$

$$= 2992 \times 10^6 \text{ mm}^4$$

The stresses in MPa are

$$\sigma = -\frac{My}{I} = -\frac{150 \times 10^6 y}{2992 \times 10^6} = -0.0501 y \frac{\text{N} \cdot \text{mm} \cdot \text{mm}}{\text{mm}^4}$$

and are listed in the following table.

Point	y(mm)	σ_{wood}(N/mm^2)	σ_{steel}(N/mm^2)
A	300	−15.04	
B	−75	+3.76	$20 \times 3.76 = +75.2$
C	−175	+8.77	$20 \times 8.77 = +175.5$
D	−200	+10.03	

maximum compression in wood = **15 MPa**
maximum tension in wood = **10 MPa**
maximum compression in steel = **None**
maximum tension in steel = **175 MPa**

Example 8-6

A Western hemlock timber is reinforced by fastening steel plates to the sides as shown in Fig. 8-25a. Calculate the maximum tensile and compressive stress in the steel and in the wood.

Solution

$$\text{maximum bending moment} = \tfrac{1}{8} wL^2$$

$$= \tfrac{1}{8} \times 5 \times 13^2 = 105.63 \text{ kip} \cdot \text{ft}$$

$$= 1.268 \times 10^3 \text{ kip} \cdot \text{in.}$$

From Table A,

Young's modulus for Western hemlock, $E_1 = 1400$ ksi

Young's modulus for steel, $E_2 = 29 \times 10^3$ ksi

$$\text{modular ratio, } n = \frac{E_2}{E_1} = \frac{29 \times 10^3}{1.4 \times 10^3} = 20.7$$

The transformed cross section (dimensions in inches) is shown in Fig. 8-25b and c.

$$Y_{NA} = \frac{20 \times 8 \times 10 + 2 \times 10.35 \times 4 \times 3}{20 \times 8 + 2 \times 10.35 \times 4} \quad \frac{\text{in.} \cdot \text{in.} \cdot \text{in.}}{\text{in.} \cdot \text{in.}}$$

$$= \frac{160 \times 10 + 82.8 \times 3}{160 + 82.8} = 7.61 \text{ in.}$$

$$\bar{I} = \tfrac{1}{12} \times 8 \times 20^3 + 160 \times 2.39^2 + \tfrac{1}{12} \times 2.07 \times 4^3 + 82.8 \times 4.61^2$$

$$= 8.02 \times 10^3 \text{ in.}^4$$

The stresses in ksi are

$$\sigma = -\frac{My}{\tilde{I}} = -\frac{1.268 \times 10^3\, y}{8.02 \times 10^3} = -0.158\, y\,\frac{\text{kip} \cdot \text{in.} \cdot \text{in.}}{\text{in.}^4}$$

and are listed in the following table.

Point	y(in.)	σ_{wood}(ksi)	σ_{steel}(ksi)
A	12.39	−1.96	
B	−2.61	+0.41	20.7 × 0.41 = +8.49
C	−6.61	+1.04	20.7 × 1.04 = +21.53
D	−7.61	+1.20	

maximum compression in wood = **1.96 kip/in.**^{2}
maximum tension in wood = **1.20 kip/in.**^{2}
maximum compression in steel = **None**
maximum tension in steel = **21.53 kip/in.**^{2}

Problems 8-49 to 8-54

8-8 NONLINEAR BENDING

The analysis of bending that has been presented so far is based on the assumption that stress is directly proportional to strain; that is, all fibers of the member in bending are represented by points on the straight-line portion of the stress–strain curve. We can state this condition by saying that the material obeys Hooke's law or that the proportional limit has not been exceeded. For metals there is always a straight-line portion in the stress–strain curve. However, for some metals, such as aluminum, the proportional limit occurs at such a low stress level that in actual structures some of the material may be stressed beyond the proportional limit. There are other materials, such as wood and concrete, that have practically no straight-line portion in the stress–strain curve. We will now consider the effect of nonlinear stress–strain on the distribution of stress. In all that follows, the curve for tensile stress and strain will be given and it will always be assumed that the same relationship applies for compressive stress and strain.

Consider a beam in bending, Fig. 8-26a. Transverse sections AB and CD, which were initially plane, will be assumed to remain plane after deformation. It follows from this assumption that strain varies linearly as in Fig. 8-26b. With the nonlinear stress–strain relationship in Fig. 8-26c, the stress is no longer linear and is curved as in Fig. 8-26d. For a nonsymmetrical cross section the neutral axis will not pass through the centroid.

In order to solve for stresses in a member, the equation of the stress–strain curve must be determined. Since there is no theory predicting the curve, the equation can only be obtained by fitting some assumed functions to the experimental points on the curve. This leads to

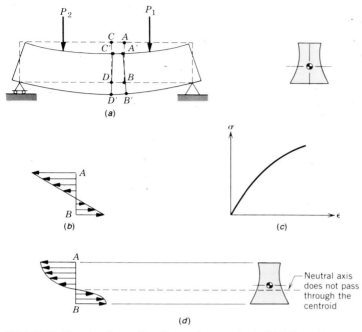

FIG. 8-26 Nonlinear bending and its effect on the neutral axis. (*a*) Deflected beam and cross section. (*b*) Strain distribution. (*c*) Stress-strain relationship. (*d*) Stress distribution.

very cumbersome solutions when the cross section is anything other than rectangular or when the stress–strain curve is more advanced than a second-order polynomial.

In practice, it is customary to approximate the stress–strain curve by a straight line such as that shown in Fig. 8-27*a*. This enables the designer to use the familiar linear formulas and to solve problems with a reasonable amount of effort. The solution is approximate, and the error becomes larger as the stress–strain curve departs more from the assumed straight line. For the case given, the largest stress, by the approximate method, is larger than the exact stress, which means that we have erred on the side of safety.

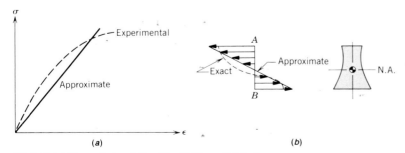

FIG. 8-27 (*a*) Stress-strain relationship. (*b*) Stress distribution.

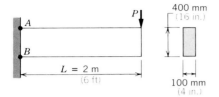

FIG. 8-28*a* Beam elevation and cross section.

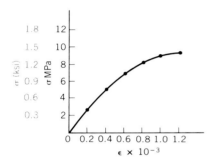

FIG. 8-28*b* Stress-strain diagram.

Example 8-7

The beam shown in Fig. 8-28*a* is made of a material having a stress–strain relationship that can be approximated by $\sigma = 15\,000\varepsilon - 6\,000\,000\varepsilon^2$. If the allowable bending stress is 6 MPa, determine the safe load P.

    Solution.   The given stress–strain formula when plotted gives the curve shown in Fig. 8-28*b*. The fibers subjected to the allowable stress, $\sigma_a = 6$ MPa, will be strained an amount given by the solution to

$$\sigma_a = 15\,000\varepsilon_a - 6\,000\,000\varepsilon_a^2 = 6$$

There are two solutions to this quadratic: $\varepsilon_a = 0.0005$ and $\varepsilon_a = 0.0020$. The latter value has no real meaning and will be disregarded. When the beam is carrying the allowable load, the strains and stresses on the cross section *A-B* will be as shown in Fig. 8-28*c*, *d*, and *e*.

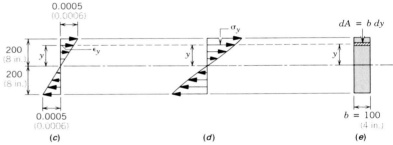

FIG. 8-28 (*c*) Strains. (*d*) Stresses. (*e*) Cross section.

At a typical location given by *y*, from similar triangles,

$$\frac{\varepsilon_y}{0.0005} = \frac{y}{200}$$

Hence

$$\varepsilon_y = \frac{0.0005}{200}\,y = 2.5 \times 10^{-6}y$$

Substituting into the given stress–strain relationship gives

$$\sigma_y = 15\,000(2.5 \times 10^{-6}y) - 6\,000\,000(2.5 \times 10^{-6}y)^2$$

$$= 0.0375y - 37.5 \times 10^{-6}y^2$$

The allowable bending moment is

$$M_a = 2\int\limits_{0}^{200} \sigma_y y\,dA\,\frac{\text{N}}{\text{mm}^2}\cdot\text{mm}\cdot\text{mm}^2$$

$$= 2 \int_0^{200} (0.0375y - 37.5 \times 10^{-6}y^2) y100 \, dy$$

$$= 200 \int_0^{200} (0.0375y^2 - 37.5 \times 10^{-6}y^3) \, dy$$

$$= 200 \left[0.0125y^3 - 9.375 \times 10^{-6}y^4 \right]_0^{200}$$

$$= 17.0 \times 10^6 \, \text{N} \cdot \text{mm} = 17.0 \, \text{kN} \cdot \text{m}$$

The allowable load is

$$P = \frac{M_a}{L} = \frac{17.0}{2} \frac{\text{kN} \cdot \text{m}}{\text{m}} = \textbf{8.5 kN}$$

An interesting comparison can be made if we disregard the nonlinear effects and solve the problem by the usual equations. From (2-13)

$$M_a = \frac{\sigma_a I}{c}$$

$$= \frac{6 \times \frac{1}{12} \times 100 \times 400^3}{200} \frac{\text{N}}{\text{mm}^2} \cdot \frac{\text{mm}^4}{\text{mm}}$$

$$= 16.0 \times 10^6 \, \text{N} \cdot \text{mm}$$

$$= 16.0 \, \text{kN} \cdot \text{m}$$

Hence

$$P = \frac{M_a}{L} = \frac{16.0}{2} \frac{\text{kN} \cdot \text{m}}{\text{m}} = \textbf{8.0 kN}$$

The linear solution is obviously much shorter and gives, in this case, a reasonably accurate answer that errs on the safe side. If the allowable stress in the linear solution had been taken as 6.375 MPa, the answers would have been identical. In practice, an increased allowable stress is often used in the linear equations when materials are known to have a nonlinear stress–strain relationship. The shape of the stress–strain curve and the strength of the material are taken into account in arriving at the increased allowable stress, which is referred to as the *rupture modulus*. The allowable stresses for wood, given in Table A, show two values; one for compression and another for bending. The value for bending has been raised to take into account the nonlinearity as well as some other effects that are peculiar to the wood.

Example 8-7

The beam shown in Fig. 8-28a is made of a material having a stress–strain relationship that can be approximated by $\sigma = 2180\varepsilon - 870,000\varepsilon^2$. If the allowable bending stress is 1 ksi, determine the safe load P.

Solution. The given stress–strain formula when plotted gives the curve shown in Fig. 8-28b. The fibers subjected to the allowable stress, $\sigma_a = 1$ kip/in.^{2}, will be strained an amount given by the solution to

$$\sigma_a = 2180\varepsilon_a - 870{,}000\varepsilon_a^2 = 1$$

There are two solutions to this quadratic: $\varepsilon_a = 0.0006$ and $\varepsilon_a = 0.0019$. The latter value has no real meaning and will be disregarded. When the beam is carrying the allowable load, the strains and stresses on the cross section A-B will be as shown in Fig. 8-28c, d, and e.

At a typical location given by y, from similar triangles,

$$\frac{\varepsilon_y}{0.0006} = \frac{y}{8}$$

Hence

$$\varepsilon_y = \frac{0.0006}{8} y = 75 \times 10^{-6} y$$

Substituting into the given stress–strain relationship gives

$$\sigma_y = 2180(75 \times 10^{-6} y) - 870{,}000(75 \times 10^{-6} y)^2$$

$$= 0.1635 y - 4.89 \times 10^{-3} y^2$$

The allowable bending moment is

$$M_a = 2 \int_0^8 \sigma_y y \, dA \frac{\text{kip}}{\text{in.}^2} \cdot \text{in.} \cdot \text{in.}^2$$

$$= 2 \int_0^8 (0.1635 y - 4.89 \times 10^{-3} y^2) y 4 \, dy$$

$$= 8 \int_0^8 (0.1635 y^2 - 4.89 \times 10^{-3} y^3) \, dy$$

$$= 8 \left[0.0545 y^3 - 1.22 \times 10^{-3} y^4 \right]_0^8$$

$$= 183.25 \text{ kip} \cdot \text{in.} = 15.27 \text{ kip} \cdot \text{ft}$$

The allowable load is

$$P = \frac{M_a}{L} = \frac{15.27}{6} \frac{\text{kip} \cdot \text{ft}}{\text{ft}} = \textbf{2.54 kip}$$

An interesting comparison can be made if we disregard the non-linear effects and solve the problem by the usual equations. From (2-13)

$$M_a = \frac{\sigma_a I}{c}$$

$$= \frac{1 \times \frac{1}{12} \times 4 \times 16^3}{8} \frac{\text{kip}}{\text{in.}^2} \cdot \frac{\text{in.}^4}{\text{in.}}$$

$$= 170.7 \text{ kip} \cdot \text{in.}$$

$$= 14.2 \text{ kip} \cdot \text{ft}$$

Hence

$$P = \frac{M_a}{L} = \frac{14.2 \text{ kip} \cdot \text{ft}}{6} = 2.37 \text{ kip}$$

The linear solution is obviously much shorter and gives, in this case, a reasonably accurate answer that errs on the safe side. If the allowable stress in the linear solution had been taken as 1.07 ksi, the answers would have been identical. In practice, an increased allowable stress is often used in the linear equations when materials are known to have a nonlinear stress–strain relationship. The shape of the stress–strain curve and the strength of the material are taken into account in arriving at the increased allowable stress, which is referred to as the *rupture modulus*. The allowable stresses for wood, given in Table A, show two values; one for compression and another for bending. The value for bending has been raised to take into account the nonlinearity as well as some other effects that are peculiar to the wood.

Problems 8-55 to 8-56

8-9 BENDING INTO THE INELASTIC RANGE

The nonlinear cases treated in the previous section were considered to have all the strains within the elastic limit. Consequently, the members would return to their original shape when the load was removed. Another type of nonlinear problem arises when a low-carbon steel member carries a bending moment of such magnitude that some of the fibers are stressed beyond the proportional limit. This poses a special nonlinear problem in that the stress–strain curve departs suddenly from a straight line and is of a shape that is not readily approximated by mathematical functions. Since the point of departure from the straight line and the point where plastic flow begins are practically coincident, we describe this case by saying that *the beam has been bent into the inelastic range*. Some permanent deformation will take place and the beam will not return to its original shape upon removal of the load.

A mathematical approximation to the stress–strain curve for low-carbon steel requires many terms if it is to give reasonably good accuracy. The function is then too difficult to use in practice. However, a fairly good approximation can be obtained by substituting straight lines for the curve as shown in Fig. 8-29. The lines define a region where the material is treated as elastic and another region where the material is treated as though plastic deformation takes place without change in stress. This is described as an *idealized material* and is said to be *elastic-perfectly plastic*. Even with this simplification, the general formulas for stress and deflection become unmanageable and it is best to solve each case from first principles. It will, however, be profitable to look at the behavior and arrive at some basic conclusions.

Consider that the member shown in Fig. 8-30 is curved by applying a rotation, α, and that we want to know the required bending moment. We will continue to assume that plane sections remain plane under load. This means that the strain will vary linearly with distance from the neutral

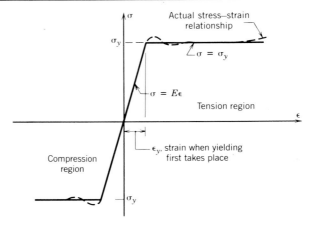

FIG. 8-29 Stress-strain relationship for idealized material.

axis and that stress will also vary linearly from the neutral axis in all fibers that have not yielded. For the case shown in Fig. 8-30, some of the fibers are strained into the plastic range and all of these, regardless of the amount of strain, will be stressed to the yield stress. The stress distribution will then be as shown, with a straight-line variation from the neutral axis for fibers within the elastic range and a constant stress in the plastic fibers. In the elastic analysis, where the stress curve is linear, it was shown that the neutral axis passes through the centroid of the cross section. The departure from the straight line, when plastic flow takes place, invalidates that analysis and the neutral axis does not necessarily pass through the centroid.

The neutral axis can be located, as it was in the elastic analysis, by equating the total axial force to zero. It is not practical to write a general formula for its location; hence each case must be considered separately. The moment-stress relations of Chapter 2 are still applicable in the elastic region provided the moment is taken as that part of the total applied moment that is carried by the elastic fibers. Problems that have a

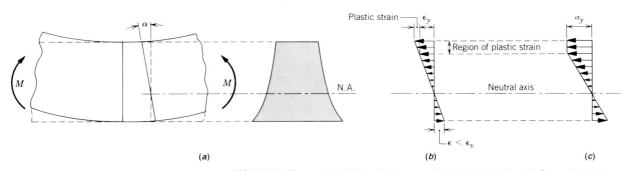

FIG. 8-30 (a) Beam strained into plastic range. (b) Strain distribution. (c) Stress distribution.

given α, or given strains, are readily solved, but the more practical problems, in which the total bending moment is given, are very difficult except for the simplest cross sections.

Example 8-8

An S380 × 64 is made of AISI 1020. Gages at A and B on the beam shown in Fig. 8-31*a* indicate axial strains of 0.002 m/m, tensile at B and compressive at A. Calculate the load P.

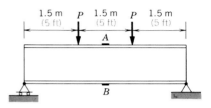

FIG. 8-31*a*

Solution.   For the given steel,

$$E = 200\ 000\ \text{MPa} \qquad \text{(from Table A)}$$

$$\sigma_y = 250\ \text{MPa} \qquad \text{(from Table A)}$$

$$\varepsilon_y = \frac{\sigma_y}{E} = \frac{250}{200\ 000} = 0.00125$$

The distance, Y, to the beginning of the plastic region as shown in Fig. 8-31*b* is found as follows:
From similar triangles

$$\frac{Y}{\varepsilon_y} = \frac{c}{\varepsilon_{\text{max}}}$$

$$\frac{Y}{0.00125} = \frac{\frac{381}{2}}{0.002}$$

$$Y = \frac{0.00125 \times 381}{0.002 \times 2} = 119\ \text{mm}$$

The stress distribution and the resulting forces are shown in Fig. 8-31*c* and *d*.

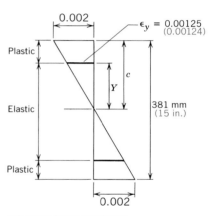

FIG. 8-31*b* Strain distribution.

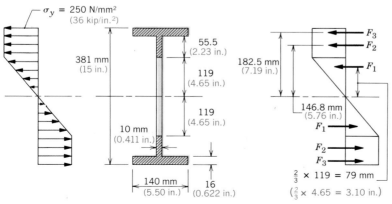

FIG. 8-31*c* Stress distribution.   **FIG. 8-31*d* Forces.**

$$F_1 = \tfrac{250}{2} \times 119 \times 10 = 149\ 000\ N = 149\ kN$$

$$F_2 = 250 \times 55.5 \times 10 = 139\ 000\ N = 139\ kN$$

$$F_3 = 250 \times 140 \times 16 = 560\ 000\ N = 560\ kN$$

$$\text{bending moment} = 2(149 \times 79 + 139 \times 146.8 + 560 \times 182.5)\ kN \cdot mm$$

$$= 269 \times 10^3\ kN \cdot mm = 269\ kN \cdot m$$

$$P = \frac{269}{1.5}\ \frac{kN \cdot m}{m} = \textbf{179 kN}$$

Example 8-8

An S15 × 43 is made of AISI 1020. Gages at *A* and *B* on the beam shown in Fig. 8-31*a* indicate axial strains of 0.002 in./in., tensile at *B* and compressive at *A*. Calculate the load *P*.

Solution. For the given steel,

$$E = 29 \times 10^3\ ksi \qquad \text{(from Table A)}$$

$$\sigma_y = 36\ ksi \qquad \text{(from Table A)}$$

$$\varepsilon_y = \frac{\sigma_y}{E} = \frac{36}{29 \times 10^3} = 0.00124$$

The distance, *Y*, to the beginning of the plastic region as shown in Fig. 8-31*b* is found as follows:
From similar triangles

$$\frac{Y}{\varepsilon_y} = \frac{c}{\varepsilon_{max}}$$

$$\frac{Y}{0.00124} = \frac{\tfrac{15}{2}}{0.002}$$

$$Y = \frac{0.00124 \times 15}{0.002 \times 2} = 4.65\ in.$$

The stress distribution and the resulting forces are shown in Fig. 8-31*c* and *d*.

$$F_1 = \tfrac{36}{2} \times 5.76 \times 0.411 = 42.64\ kip$$

$$F_2 = 36 \times 2.23 \times 0.411 = 33.0\ kip$$

$$F_3 = 36 \times 5.50 \times 0.622 = 123.16\ kip$$

$$\text{bending moment} = 2(42.64 \times 3.10 + 33.0 \times 5.76 + 123.16 \times 7.19)\ kip \cdot in.$$

$$= 2415.6\ kip \cdot in. = 201.3\ kip \cdot ft$$

$$P = \frac{201.3}{5}\ \frac{kip \cdot ft}{ft} = \textbf{40.3 kip}$$

The problem in Example 8-8 is indeed a very simple problem of its type. Consider the difficulty that would have been encountered if the question had been to find the stress distribution for a given P, or the added complications if the cross section had not been symmetrical about the neutral axis.

If a larger moment had been applied in the example, the measured strains would have been larger and the plastic zone would have penetrated more deeply into the beam. Further increase in moment would cause an increase in the depth of the plastic zone. However, it would be found that successive increments in the plastic flow would bring diminishing increments in moment. A theoretical limit is reached when the elastic core is reduced to an insignificant size. The limiting *moment for such a fully plastic section*, M_p, is of interest to designers and is much easier to calculate than the moment in Example 8-8. This plastic moment is also quite readily determined for nonsymmetrical sections. Since large values of rotation, α, are required for deep penetration of the plastic zones, this state is described as a *plastic hinge*.

Example 8-9

Calculate the bending moment that would be required to develop a plastic hinge in the T-section shown in Fig. 8-32a. The material is AISI 1020 steel.

Solution. The distribution of stress in the T-section with the entire section in the plastic state is shown in Fig. 8-32b.

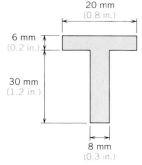

FIG. 8-32a

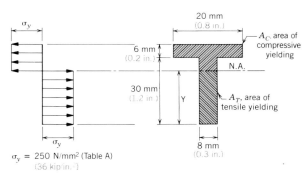

$\sigma_y = 250$ N/mm² (Table A)
(36 kip/in.²)

FIG. 8-32b

For equilibrium of forces, the net axial load on the cross section must be equal to zero. Hence

$$A_T\sigma_y - A_C\sigma_y = 0$$

$$A_T = A_C$$

$$A_T = 8 \times Y$$

$$A_C = 6 \times 20 + 8(30 - Y) = 120 + 240 - 8Y = 360 - 8Y$$

Equating areas, we obtain

$$8Y = 360 - 8Y$$

$$16Y = 360 \quad \text{or} \quad Y = \tfrac{360}{16} = 22.5 \text{ mm}$$

$$A_T = 8 \times 22.5 = 180 \text{ mm}^2$$

The forces indicated in Fig. 8-32c can now be calculated.

$$F_1 = 180 \times 250 \text{ mm}^2 \cdot \frac{N}{mm^2} = 45\,000 \text{ N}$$

$$F_2 = 7.5 \times 8 \times 250 = 15\,000 \text{ N}$$

$$F_3 = 6 \times 20 \times 250 = 30\,000 \text{ N}$$

Check: $\quad \sum F_x = 45\,000 - 30\,000 - 15\,000$

$$= 0$$

Taking moments about the base and using forces shown in Fig. 8-32c gives

$$M_p = 30\,000 \times 33 + 15\,000 \times 26.25 - 45\,000 \times 11.25$$

$$= 877\,500 \text{ N} \cdot \text{mm}$$

$$= 877 \text{ N} \cdot \text{m}$$

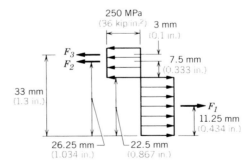

FIG. 8-32c

The bending moment required to develop the plastic hinge is **877 N·m.**

Example 8-9

Calculate the bending moment that would be required to develop a plastic hinge in the T-section shown in Fig. 8-32a. The material is AISI 1020 steel.

Solution.   The distribution of stress in the T-section with the entire section in the plastic state is shown in Fig. 8-32b.

For equilibrium of forces, the net axial load on the cross section must be equal to zero. Hence

$$A_T \sigma_y - A_C \sigma_y = 0$$

$$A_T = A_C$$

$$A_T = 0.3 \times Y$$

$$A_C = 0.2 \times 0.8 + 0.3(1.2 - Y) = 0.16 + 0.36 - 0.3Y = 0.52 - 0.3Y$$

Equating areas, we obtain

$$0.3Y = 0.52 - 0.3Y$$

$$0.6Y = 0.52 \quad \text{or} \quad Y = \frac{0.52}{0.6} = 0.867 \text{ in.}$$

$$A_T = 0.3 \times 0.867 = 0.26 \text{ in.}^2$$

The forces indicated in Fig. 8-32c can now be calculated.

$$F_1 = 0.3 \times 0.867 \times 36 \text{ in.}^2 \cdot \frac{\text{kip}}{\text{in.}^2} = 9.36 \text{ kip}$$

$$F_2 = 0.333 \times 0.3 \times 36 = 3.60 \text{ kip}$$

$$F_3 = 0.2 \times 0.8 \times 36 = 5.76 \text{ kip}$$

Check: $\sum F_x = 9.36 - 5.76 - 3.6$

$$= 0$$

Taking moments about the base and using forces shown in Fig. 8-32c gives

$$M_p = 5.76 \times 1.3 + 3.60 \times 1.034 - 9.36 \times 0.434$$

$$= 7.15 \text{ kip} \cdot \text{in.}$$

$$= 0.60 \text{ kip} \cdot \text{ft}$$

The bending moment required to develop the plastic hinge is **0.60 kip·ft.**

The solution to Example 8-9 was based on the discovery that the tensile area is equal to the compressive area. The areas involved are all rectangles in this case, but that did not influence the conclusion: The same area relationship would have resulted regardless of the shape of the section. This important conclusion may be stated:

In a plastic hinge, when the material is elastic-perfectly plastic and the yield stresses in compression and tension are equal, the neutral axis divides the cross section into two equal areas.

Except for certain regular shapes, the location of the neutral axis under fully plastic conditions does not correspond to the location under elastic or elastic-partially plastic conditions.

8-10 PLASTIC DESIGN

For a small bending moment the stresses in a section are entirely within the elastic range. If the moment is gradually increased, the stresses rise and a point is reached where some of the fibers are just on the verge of yielding. This moment will be referred to as the *yield moment*, M_y. Since the bulk of the material is not yet stressed to the limit, the section is capable of carrying a greater moment. When the moment is further increased, more of the fibers will become plastic. The deflection associated with the additional moment is greater than for an equal increment of moment in the elastic range. The limit to moment, the *plastic moment*, M_p, is reached when the section is fully plastic and we have a plastic hinge. The ratio of the plastic moment to the moment at first yielding is a significant property of the cross section. It is referred to as the shape factor, Z, and is given by

$$Z = \frac{M_p}{M_y} \tag{8-22}$$

The plastic hinge cannot actually exist, since it would require infinite rotation of neighboring planes relative to one another. However, the concepts of a plastic hinge and the moment, M_p, are still useful in that they give the absolute maximum loads that a structure can carry when it is in a state of collapse. Design based on plastic hinges is referred to as *plastic design*. The design process consists of analyzing a structure to determine the ultimate load under fully plastic conditions and then applying the safety factor to the ultimate load to get the load that can be safely carried by the structure under service conditions. This will give a structure that differs from that obtained when the same safety factor is applied to stress determined by an elastic analysis. For indeterminate structures the analysis for plastic design is simpler than for elastic design. The following example will illustrate these points.

Example 8-10

For a safety factor of 2, find the safe load that the AISI 1020 beam shown in Fig. 8-33a can carry (1) by elastic analysis, with the factor of safety based on yield; and (2) by plastic design.

Solution

1. *Elastic Design.* The maximum elastic bending moment, M_{max}, is not easily determined but can be calculated according to the methods of Section 3-3 for statically indeterminate beams. Its magnitude is

$$M_{max} = \frac{PL}{8}$$

$$= \frac{P \times 2000}{8} = 250P \qquad \text{(N·mm if } P \text{ is in N)}$$

$$\sigma_{max} = \frac{Mc}{I} = \frac{250P \times 90}{\frac{1}{12}50 \times 180^3} \frac{\text{N·mm·mm}}{\text{mm}^4}$$

$$= 0.926 \times 10^{-3} P \frac{\text{N}}{\text{mm}^2}$$

$$\sigma_y = 250 \text{ MPa} \qquad \text{(Table A)}$$

$$\sigma_a = \frac{\sigma_y}{N} = \frac{250}{2} = 125 \frac{\text{N}}{\text{mm}^2}$$

$$0.926 \times 10^{-3} P_{safe} = \sigma_{max} = \sigma_a = 125$$

$$P_{safe} = \frac{125}{0.926 \times 10^{-3}} \, N = 135\,000 \, N = \textbf{135 kN}$$

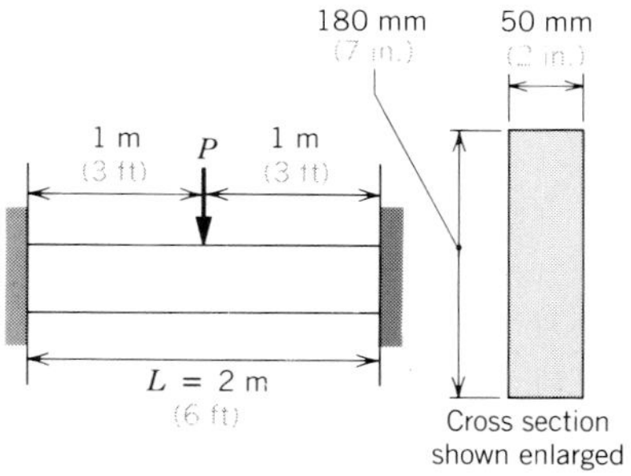

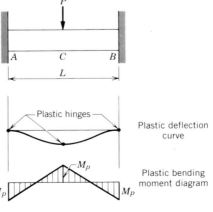

FIG. 8-33a

2. *Plastic Design.* In order to establish where along the beam plastic hinges will form, it is necessary to consider first the elastic case shown in Fig. 8-33b. Under load, *P*, we can visualize the elastic deflection curve with its points of contraflexure and sketch the general shape of the bending moment diagram with its peak moments occurring at the two supports and at midspan. As the load, *P*, is increased beyond the elastic range, the most highly stressed beam sections at *A*, *B*, and *C* experience plastic deformations; when *P* reaches the ultimate load under plastic conditions, P_p, plastic hinges will have formed at sections *A*, *B*, and *C* and the corresponding moments at the sections will be equal to M_p as indicated in Fig. 8-33c.

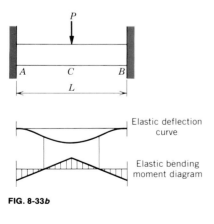

FIG. 8-33b

FIG. 8-33c

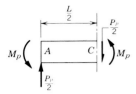

FIG. 8-33d

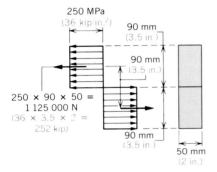

FIG. 8-33e

The formation of the three hinges collapses the beam so that no further load can be sustained and we speak of a *collapse mechanism* under the ultimate load, P_p. The ultimate load in terms of M_p can be readily found from the free-body diagram of half the beam shown in Fig. 8-33d.

From $\sum M = 0$

$$M_p + M_p - \frac{P_p}{2} \times \frac{L}{2} = 0$$

$$P_p = \frac{8M_p}{L}$$

By considering the fully plastic stresses acting on the beam cross section as indicated in Fig. 8-33e, M_p can now be determined.

$$M_p = 1\,125\,000 \times 90 \text{ N} \cdot \text{mm}$$

$$= 101\,250\,000 \text{ N} \cdot \text{mm}$$

$$P_p = \frac{8 \times M_p}{L} = \frac{8 \times 101\,250\,000}{2000} \frac{\text{N} \cdot \text{mm}}{\text{mm}}$$

$$= 405\,000 \text{ N}$$

$$P_{\text{safe}} = \frac{P_p}{N} = \frac{405\,000}{2} = 202\,500 \text{ N}$$

$$= \mathbf{202 \text{ kN}}$$

Example 8-10

For a safety factor of 2, find the safe load that the AISI 1020 beam shown in Fig. 8-33a can carry (1) by elastic analysis, with the factor of safety based on yield; and (2) by plastic design.

Solution

1. *Elastic Design.* The maximum elastic bending moment, M_{max}, is not easily determined but can be calculated according to the methods of Section 3-3 for statically indeterminate beams. Its magnitude is

$$M_{\text{max}} = \frac{PL}{8}$$

$$= \frac{P \times 6 \times 12}{8} = 9P \quad \text{(kip} \cdot \text{in. if } P \text{ is in kip)}$$

$$\sigma_{\text{max}} = \frac{Mc}{I} = \frac{9P \times 3.5}{\frac{1}{12}2 \times 7^3} \frac{\text{kip} \cdot \text{in.} \cdot \text{in.}}{\text{in.}^4}$$

$$= 0.551 \, P \text{ kip/in.}^2$$

$$\sigma_y = 36 \text{ kip/in.}^2 \quad \text{(Table A)}$$

$$\sigma_a = \frac{\sigma_y}{N} = \frac{36}{2} = 18 \text{ kip/in.}^2$$

$$0.551 P_{\text{safe}} = \sigma_{\max} = \sigma_a = 18$$

$$P_{\text{safe}} = \textbf{32.67 kip}$$

2. *Plastic Design.*   In order to establish where along the beam plastic hinges will form, it is necessary to consider first the elastic case shown in Fig. 8-33b. Under load, P, we can visualize the elastic deflection curve with its points of contraflexure and sketch the general shape of the bending moment diagram with its peak moments occurring at the two supports and at midspan. As the load, P, is increased beyond the elastic range, the most highly stressed beam sections at A, B, and C experience plastic deformations; when P reaches the ultimate load under plastic conditions, P_p, plastic hinges will have formed at sections A, B, and C and the corresponding moments at the sections will be equal to M_p as indicated in Fig. 8-33c.

The formation of the three hinges collapses the beam so that no further load can be sustained and we speak of a *collapse mechanism* under the ultimate load, P_p. The ultimate load in terms of M_p can be readily found from the free-body diagram of half the beam shown in Fig. 8-33d.

From $\sum M = 0$

$$M_p + M_p - \frac{P_p}{2} \times \frac{L}{2} = 0$$

$$P_p = \frac{8 M_p}{L}$$

By considering the fully plastic stresses acting on the beam cross section as indicated in Fig. 8-33e, M_p can now be determined.

$$M_p = 252 \times 3.5 \text{ kip} \cdot \text{in.}$$

$$= 882 \text{ kip} \cdot \text{in.}$$

$$P_p = \frac{8 \times M_p}{L} = \frac{8 \times 882}{6 \times 12} \frac{\text{kip} \cdot \text{in.}}{\text{in.}}$$

$$= 98 \text{ kip}$$

$$P_{\text{safe}} = \frac{P_p}{N} = \frac{98}{2} = 49 \text{ kip}$$

This example shows the difference in results for the two cases. In the elastic analysis no advantage is taken of the fact that part of the cross section is not loaded to capacity and constitutes a reserve that is ready to take additional load when called upon. It is interesting to note that if the two approaches are used in the design of a simple member in tension, the results will be identical.

In Example 8-10, if the load had not been at midspan, constructing the elastic bending moment diagram would have required a considerable effort. The corresponding diagram in the plastic design can be readily drawn once the designer becomes acquainted with some simple principles. Plastic design is of relatively recent origin, but promises to be adopted more widely in the future.

Problems 8-57 to 8-65

PROBLEMS

8-1 Calculate the maximum bending stress due to the weight of the beam.

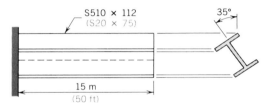

PROBLEM 8-1

8-2 Determine the load, P, that can be safely carried by the Eastern spruce beam.

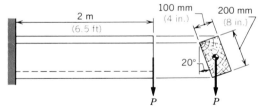

PROBLEM 8-2

8-3 A horizontal cantilever beam consists of an S310 × 47 (S12 × 32). The beam projects 8 m (26 ft) from a vertical supporting wall. The member is tilted so that the web makes an angle of 45° with a vertical plane. Determine the vertical and horizontal components of deflection at the free end due to the force of gravity acting on the member. Draw the deflections to scale on a sketch showing an end view of the beam.

8-4 A crane boom, S310 × 74 (S12 × 50), has a cantilever length of 3 m (10 ft). If the boom is horizontal, determine the maximum vertical load that can be applied to the end of the boom if the factor of safety based on yield is 3.5. The material is medium-strength structural steel. If the load is applied at an angle of 30° to the vertical, determine the load reduction to maintain the factor of safety of 3.5.

8-5 to 8-7 Working from first principles, that is, starting with equation (8-6), for the given sections determine the product of inertia with respect to the x- and y-axes shown in Table B-4 and B-5.

8-5 L125 × 90 × 16 (L5 × $3\frac{1}{2}$ × $\frac{5}{8}$).

8-6 L100 × 90 × 13 (L4 × $3\frac{1}{2}$ × $\frac{1}{2}$).

8-7 L100 × 100 × 16 (L4 × 4 × $\frac{5}{8}$).

8-8 Determine the moment of inertia about A-A for the given thin sections. All thicknesses = 5 mm (0.2 in.).

8-9 Calculate the moment with respect to the x-axis and the moment with respect to the y-axis required to rotate the end of the aluminum beam about the x-axis through 0.004 radians as shown.

Note: This somewhat academic problem gives you a good opportunity to become acquainted with the determination of the product of inertia. If you find the moment of inertia by double integration from first principles, you will be well prepared for determining the product of inertia of other cross sections.

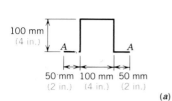

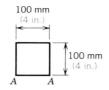

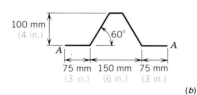

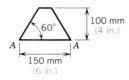

(a)

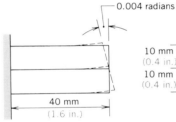

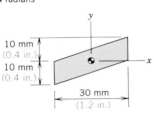

(b)

PROBLEM 8-8

0.004 radians

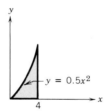

PROBLEM 8-9

8-10 to 8-13 Working from first principles, that is, starting with equation (8-6), determine the product of inertia of the area with respect to the given axes.

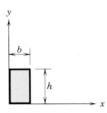

PROBLEM 8-10

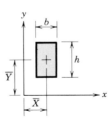

PROBLEM 8-11

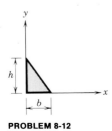

PROBLEM 8-12

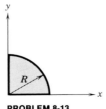

PROBLEM 8-13

8-14 to 8-15 Determine the moment of inertia and product of inertia for the areas with respect to the given axes.

$y = 0.5x^2$

PROBLEM 8-14

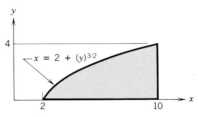

$x = 2 + (y)^{3/2}$

PROBLEM 8-15

8-16 Calculate the product of inertia of the Z-shape.

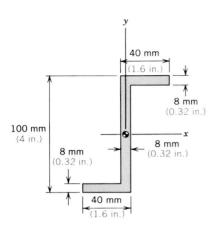

PROBLEM 8-16

8-17 to 8-19 Determine the moment of inertia of the given areas with respect to the x'-axis and the y'-axis.

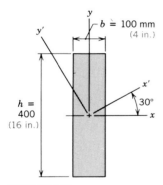

PROBLEM 8-17

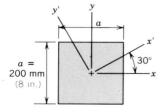

PROBLEM 8-18

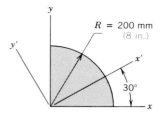

PROBLEM 8-19

8-20 to 8-22 Determine the location of the principal axes, having their origin at the centroid, and the value of the principal moments of inertia for the given sections.

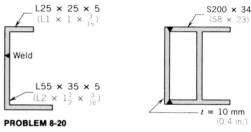

PROBLEM 8-20

PROBLEM 8-21

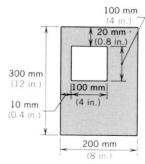

PROBLEM 8-22

8-23 Determine $I_{x'}$, $I_{y'}$ and $I_{x'y'}$ for an S610 × 149 (S24 × 100) with respect to axes through the centroid that have been rotated through 20°.

8-24 An L200 × 100 × 20 (L8 × 4 × $\frac{3}{4}$) is used for a compression member that is 3 m (10 ft) long and has pivoted ends. Determine the Euler buckling load for the member.

8-25 A steel member, 5 m (16 ft) long, has pivoted ends and the given cross section. Calculate the Euler buckling load.

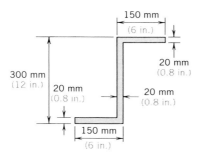

PROBLEM 8-25

8-26 a. For the beam shown, determine the vertical and horizontal components of deflection at the end of the beam.

    b. Determine the bending stress at A and B.

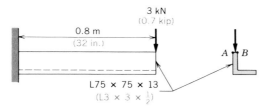

PROBLEM 8-26

8-27 Determine the vertical and horizontal components of displacement at the end of the steel beam.

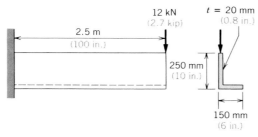

PROBLEM 8-27

8-28 Determine the horizontal component of displacement at the end of the steel member due to the given load.

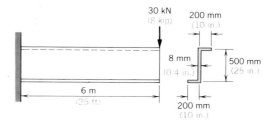

PROBLEM 8-28

8-29 The given beam deflects horizontally as well as vertically due to the force of gravity. Determine the angle through which the beam should be rotated so that there will be no horizontal deflection. Draw a sketch showing the rotated beam.

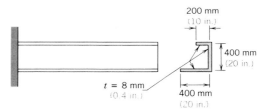

PROBLEM 8-29

8-30 to 8-31 Determine the maximum stress in the simply supported beams loaded as shown.

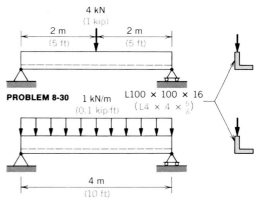

PROBLEM 8-30

PROBLEM 8-31

8-32 A roof has trusses spaced at 5 m (16 ft), and purlins, simply supported at the ends, span the distance between adjacent trusses. The purlins support the roof sheathing and waterproofing and are subjected to a distributed downward load of 18 kN/m (1.2 kip/ft).

Different cross-sectional shapes, all having the same area, are being considered. Determine the maximum bending stress for each given shape.

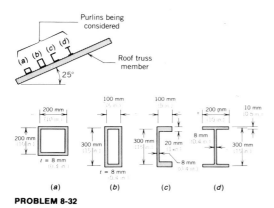

PROBLEM 8-32

8-33 to 8-36 Determine the distance between the centroid and the shear center of the given cross sections. All the thicknesses = 10 mm (0.4 in.).

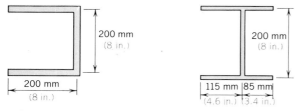

PROBLEM 8-33

PROBLEM 8-34

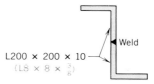

PROBLEM 8-35

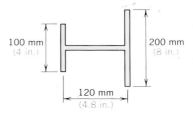

PROBLEM 8-36

8-37 to 8-39 Determine the centroid and the location of the shear center for the given sections.

L55 × 35 × 5
(L2 × 1½ × $\frac{3}{16}$)

PROBLEM 8-37

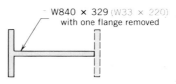

W840 × 329 (W33 × 220)
with one flange removed

PROBLEM 8-38

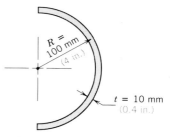

Weld

L200 × 200 × 10
(L8 × 8 × $\frac{3}{8}$)

PROBLEM 8-39

8-40 to 8-41 Determine the distance between the center of the circular arc and the shear center for the given sections.

$R = 100$ mm (4 in.)

$t = 10$ mm (0.4 in.)

PROBLEM 8-40

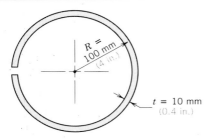

PROBLEM 8-41

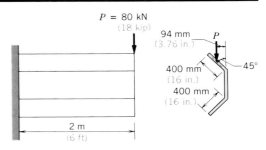

PROBLEM 8-45

8-42 A member is formed by bending a long strip of metal that is 300 mm (12 in.) wide and 5 mm (0.2 in.) thick to form the equilateral triangular cross section shown. Determine the location of the shear center for the member

a. As shown.   b. If the weld is omitted.

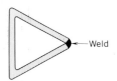

PROBLEM 8-42

8-43 to 8-45 The load on the given cantilever is applied at the location shown in the end view. Calculate the torsional shear stress neglecting stress concentration. All thicknesses = 10 mm (0.4 in.).

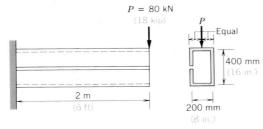

PROBLEM 8-43

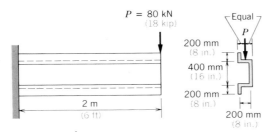

PROBLEM 8-44

8-46 A cantilever beam has the cross section given in Prob. 8-41. A downward load of 30 kN (7 kip) is applied through the centroid of the cross section. Calculate the maximum torsional shear stress.

8-47 to 8-48 For the beam shown:

a. Determine the load P at which the compression flange will fail by local buckling. Make assumptions that you find necessary for a solution.
b. Determine the safe load if a safety factor of 2 is required on Euler buckling.
c. Calculate the bending stress when the safe load is applied and compare it with the allowable stress for structural steel.

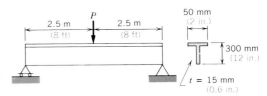

PROBLEM 8-47

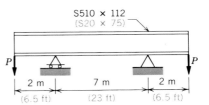

PROBLEM 8-48

8-49 to 8-50 Calculate the maximum stress:

a. In the wood if there is no steel reinforcement.
b. In the wood if the timber is reinforced as shown.
c. In the steel for the beam shown.

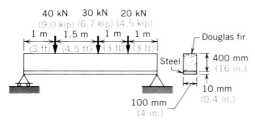

PROBLEM 8-49

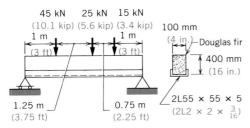

PROBLEM 8-50

8-51 A 300-mm-square (12-in.-square) timber is reinforced by having 300 × 10 mm (12 × 0.4 in.) steel plates securely attached to two opposite surfaces. The allowable stress in the timber is 13 MPa (1.9 ksi) and in the steel is 140 MPa (20 ksi). Determine the allowable bending moment when the beam is placed:

a. With the plates on the sides of the timber.
b. With the plates on the top and the bottom of the timber.

8-52 Calculate the maximum deflection in the reinforced timber beam due to the given uniformly distributed load.

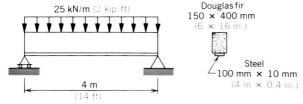

PROBLEM 8-52

8-53 Calculate the maximum bending stress in the concrete and in the steel of the given beam.

Note: In this type of composite beam, shear connectors are welded to the top flange of the steel member before the concrete is poured. The connectors are designed to ensure that both parts bend together as a single unit. The connector design is not a part of this problem.

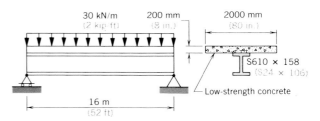

PROBLEM 8-53

8-54 The allowable stresses in the reinforced concrete beam are

| concrete | 9.0 MPa (1.3 ksi) | (compression) |
| steel | 150 MPa (22 ksi) | |

Regions of the concrete that tend to be stressed in tension may be assumed to crack, and hence to not carry any tensile stress. The design of the beam is to be optimum in that the maximum stresses in the materials are to be equal to the allowable. Determine the size of the bars required and the safe load for the configuration shown.

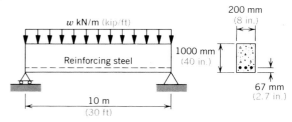

PROBLEM 8-54

8-55 to 8-56 The beam shown in Fig. 8-28a is made of material having stress and strain related by

$$\sigma(\text{MPa}) = 15\,000\varepsilon - 6\,000\,000\varepsilon^2$$

$$(\sigma(\text{ksi}) = 2175\varepsilon - 870{,}000\varepsilon^2)$$

If the fibers at A and B are strained to +0.0010 and −0.0010, determine the load P assuming the given straight line approximation to the stress–strain curves.

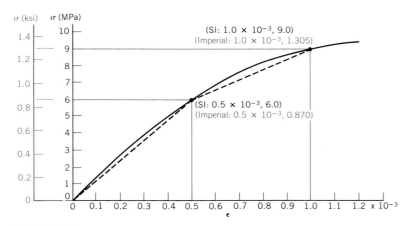

PROBLEM 8-55

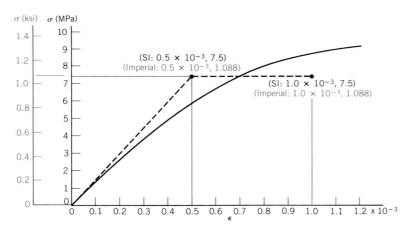

PROBLEM 8-56

8-57 Calculate the moment required to produce a plastic hinge in a low-strength steel beam that is formed by welding a 140-mm-wide (5.5-in.-wide) by 15-mm-thick (0.6-in.-thick) plate to the top flange of an S380 × 64 (S15 × 43).

8-58 A beam having the given cross section and a yield stress of 250 MPa (36 ksi) is subjected to bending about a horizontal axis. Determine M_y, M_p, and Z.

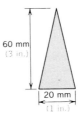

PROBLEM 8-58

8-59 Determine the shape factor for

    a. A rectangular section.

    b. A W310 × 79 (W12 × 53).

    c. An S310 × 47 (S12 × 32).

8-60 to 8-63 Determine M_y, M_p, and Z for the given cross section in bending about a horizontal axis. Note that all sections have the same area.

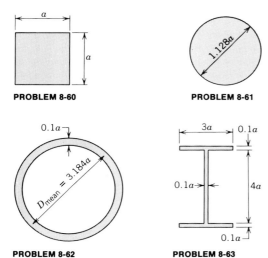

PROBLEM 8-60

PROBLEM 8-61

PROBLEM 8-62

PROBLEM 8-63

8-64 Calculate the ultimate load under fully plastic conditions for the given low-strength steel beam.

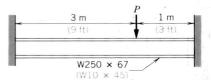

PROBLEM 8-64

8-65 Calculate M_p for bending about a horizontal axis for the section shown.

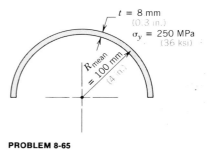

PROBLEM 8-65

Challenging Problems

8-A A prestressed concrete member is made by the following procedure: (1) Hydraulic jacks are arranged to apply an initial tension to the special high-strength rods; in this case they apply a force of 840 kN (190 kip) to each rod. (2) With this pretension applied, high-strength concrete is poured and allowed to set. (3) The rods are secured to the end plates, the jacks are released, and the ends of the rods are cut off. The member is now a *prestressed* beam.

    a. Calculate the maximum stress in the concrete and in the steel after step 3 above has been completed.

    b. If the prestressed beam is simply supported at the ends and is subjected to a load of 40 kN/m (2.75 kip/ft), determine the maximum stress in the steel and draw a profile of the stress in the concrete.

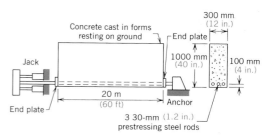

PROBLEM 8-A

8-B When the jacks are released in Prob. 8-A, the beam will arch upward. If the ends are held at a constant elevation, by how much will the midsection rise when the jacks are released?

8-C A strip of AISI 1045 steel is 40 mm (1.6 in.) wide and 8 mm (0.32 in.) thick. Initially, it is curved on a radius of 1.5 m (60 in.) as shown and in this state is stress-free. The strip will be more useful as a spring leaf if it is prestressed by placing it between the platens of a press, pressing it flat and then releasing. The steel will now have a larger radius of curvature and will have certain residual stresses. Determine

a. The new radius of curvature.
b. The profile of the residual stresses.

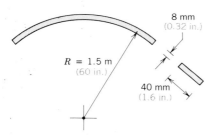

PROBLEM 8-C

8-D Specify the force capacity of press that you would recommend for flattening the bar in Prob. 8-C.

8-E Solve Prob. 8-32 for purlins having cross sections as shown.

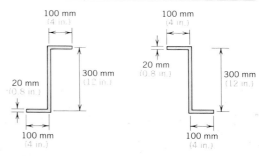

PROBLEM 8-E

8-F In Example 8-8, P is 150 kN (34 kip). Determine the strain at A and draw a stress distribution profile similar to Fig. 8-31c.

8-G Gages indicate a strain of -0.0045 at A and 0.0045 at B. Determine the load, w, when the beam is made of material that

a. Is elastic-perfectly plastic.
b. Strain hardens.

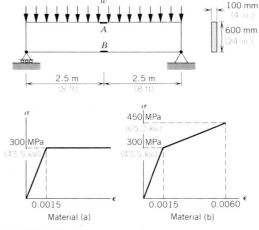

PROBLEM 8-G

9

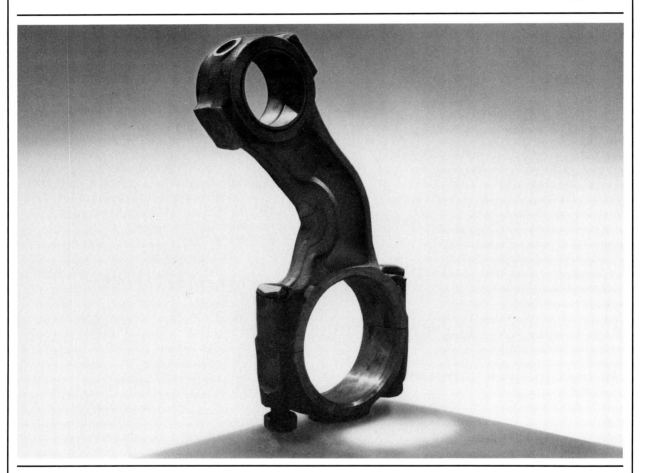

COLUMNS II

9-1 INTRODUCTION

Columns were introduced in Chapter 4 and Euler's formula, which is basic to all column analysis, was derived. Practical design formulas, partly empirical and partly based on Euler, were given and the effect of certain end constraints was studied. It was seen that an apparently subtle change in end constraint can make a large difference in the load-carrying capacity of a column. This sensitivity, combined with the tendency for overloaded columns to collapse without warning, causes engineers to approach column design with great caution. Hence a generous safety factor is incorporated into column design and assumptions usually err on the side of safety. We will now look more closely at the effects of end constraints, of eccentric loads, and of superimposed lateral loads.

9-2 PARTIAL END CONSTRAINTS

The basic formula in buckling is the Euler formula (4-8), which gives the critical load as

$$P_{cr} = \frac{\pi^2}{L^2} EI$$

This is the load required to make a slender column become unstable and to fail by deflecting laterally. It is based on the assumptions that

1. The column is initially straight.
2. The load is concentric.
3. The ends are not constrained against rotation.
4. The material is not stressed beyond the proportional limit.

A column that meets these conditions is described as an *ideal column* or as a *Euler column*. We will refer to the load required to buckle such a column as the Euler load and designate it as P_e. Thus

$$P_e = \frac{\pi^2}{L^2} EI \qquad (9\text{-}1)$$

We have seen that some nonideal end conditions can be taken into account by using an effective length L' in the buckling formula. The critical load is then

$$P_{cr} = \frac{\pi^2 EI}{(L')^2} \qquad (9\text{-}2)$$

We found it convenient to express L' as the product of the *effective length factor* by the actual length. Thus

$$\boxed{L' = KL} \qquad (9\text{-}3)$$

This gives a critical load formula

$$P_{cr} = \frac{\pi^2 EI}{K^2 L^2} \qquad (9\text{-}4)$$

It is useful to express this in terms of the Euler load for an ideal column by substituting from (9-1) to get

$$P_{cr} = \frac{P_e}{K^2}$$ (9-5)

To find the buckling load of any column by (9-5), the Euler load for an ideal column is calculated by (9-1) and then corrected by applying the appropriate value of K for the end conditions. For certain end conditions K values were given in Section 4-6; the remainder of this section will be devoted to the determination of K for other end constraints.

We will first consider a column that is pivoted at one end and has the other end *partially constrained* against rotation. This arrangement is shown in Fig. 9-1a, where the spiral spring offers some resistance to rotation about the pivot at the right end of the column. The spiral spring can be taken as a device that applies, to the end of the column, a moment that starts at zero and increases linearly as the end of the column rotates about the right support. In Fig. 9-1b the forces that act on the column are shown. The moment equilibrium equation for the free body in Fig. 9-1c gives the formula for the bending moment on a typical section,

$$M = -P_{cr}y + \frac{k\theta}{L}x$$

From (3-8)

$$\frac{d^2y}{dx^2} = \frac{M}{EI} = -\frac{P_{cr}}{EI}y + \frac{k\theta}{EIL}x$$

$$\frac{d^2y}{dx^2} + \frac{P_{cr}}{EI}y = \frac{k\theta}{EIL}x$$ (9-6)

This is a commonly occurring form of differential equation that has a solution consisting of a complementary function and a particular integral. In this case the solution is

$$y = C_1 \sin\sqrt{\frac{P_{cr}}{EI}}x + C_2\cos\sqrt{\frac{P_{cr}}{EI}}x + \frac{k\theta}{P_{cr}L}x$$ (9-7)

If you are not thoroughly familiar with the process that leads from (9-6) to (9-7), you should at least substitute the solution given in (9-7) into (9-6) to assure yourself that the solution does in fact satisfy the differential equation.

Rearranging (9-4) gives

$$\frac{P_{cr}}{EI} = \frac{\pi^2}{K^2L^2}$$ (9-8a)

or

$$\sqrt{\frac{P_{cr}}{EI}} = \sqrt{\frac{\pi^2}{K^2L^2}} = \frac{\pi}{KL}$$ (9-8b)

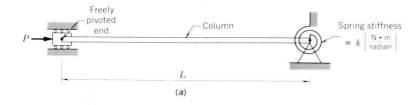

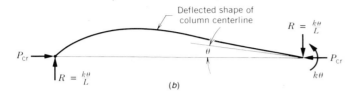

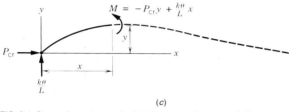

FIG. 9-1 Case of a column pivoted at one end and partially constrained at the other end. (a) Column with one end partially constrained. (b) Reactions on column. (c) Bending moment on typical section.

Substituting into (9-7), we obtain

$$y = C_1 \sin \frac{\pi}{KL} x + C_2 \cos \frac{\pi}{KL} x + \left[\frac{k\theta}{\pi^2 EI/(K^2 L^2)L} \right] x$$

There are advantages in arranging the parameters into dimensionless groups,

$$y = C_1 \sin \left(\frac{\pi x}{KL} \right) + C_2 \cos \left(\frac{\pi x}{KL} \right) + \left[\frac{k}{EI/L} \frac{\theta}{(\pi/K)^2} \right] x \qquad \textbf{(9-9a)}$$

Notice that k, the rotational stiffness of the end support, is associated with EI/L, which is also a rotational stiffness; it is the stiffness of a cantilever, having the same dimensions as the given column, and subjected to a moment at the free end (see Case 7, Table G).

Whether the physical significance is recognized or not, it is convenient to relate the two rotational stiffnesses and to let

$$k' = \frac{k}{EI/L} \qquad \textbf{(9-10)}$$

Then (9-9a) becomes

$$y = C_1 \sin \left(\frac{\pi x}{KL} \right) + C_2 \cos \left(\frac{\pi x}{KL} \right) + \frac{k'\theta x}{(\pi/K)^2} \qquad \textbf{(9-9b)}$$

There are three known end conditions and three quantities (C_1, C_2, and K) to be determined. From

$$y]_{x=0} = 0$$

$$C_1 \sin 0 + C_2 \cos 0 + 0 = 0$$

Therefore,

$$C_2 = 0$$

Then

$$y = C_1 \sin\left(\frac{\pi x}{KL}\right) + \frac{k'\theta}{(\pi/K)^2} x$$

From

$$y]_{x=L} = 0$$

we obtain

$$C_1 \sin\left(\frac{\pi}{K}\right) + \frac{k'\theta}{(\pi/K)^2} L = 0$$

Therefore,

$$C_1 = -\frac{k'\theta L}{(\pi/K)^2 \sin(\pi/K)}$$

and

$$y = \frac{k'\theta}{(\pi/K)^2}\left[\frac{-L}{\sin(\pi/K)} \sin\left(\frac{\pi x}{KL}\right) + x\right]$$

Differentiating gives

$$\frac{dy}{dx} = \frac{k'\theta}{(\pi/K)^2}\left[\frac{-(\pi/K)}{\sin(\pi/K)} \cos\left(\frac{\pi x}{KL}\right) + 1\right]$$

From

$$\frac{dy}{dx}\bigg]_{x=L} = -\theta$$

we obtain

$$\frac{k'\theta}{(\pi/K)^2}\left[\frac{-(\pi/K)}{\sin(\pi/K)} \cos\left(\frac{\pi}{K}\right) + 1\right] = -\theta$$

Rearranging and simplifying yields

$$k' = \frac{\pi/K}{\cot(\pi/K) - 1/(\pi/K)} \qquad \textbf{(9-11)}$$

In most applications k' is known and K must be evaluated. However, it is impossible to rearrange (9-11) to give K in terms of k' so that practically a solution must be obtained by selecting a trial value of K and finding the corresponding value of k'. The calculated value of k' is unlikely to match the given k', so a new K is selected and the process is repeated until a satisfactory value of K is found. To assist in this trial-and-error process, it is common practice to plot the calculated k' values against the selected values of K. For a particular problem the parameters would be plotted over a small range near the solution. Curve (a) in Fig. 9-2 shows the relationship defined by (9-11) over a range that covers nearly all values of practical interest. The curve can be used to solve problems directly or to indicate the approximate answer for use as a starting point in a more accurate trial-and-error solution. Note that for small values of k, that is, for very weak rotational constraint, the value of K approaches 1.0. This, in the limit, is in agreement with case (c) in Fig. 4-9.

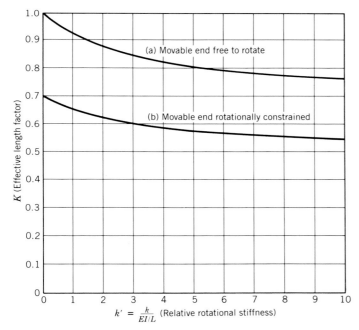

FIG. 9-2 Effective length factor for a column with elastic rotational constraint at one end.

At the other end of the spectrum, when k becomes very large, K approaches 0.70 as an asymptote. An infinitely stiff spring will impose complete constraint against rotation at the end. The conditions are then as in case (b) of Fig. 4-9 and for the second time the value of K obtained by (9-11) is verified.

Figure 9-2 also gives, in curve (b), the effective length factor that applies when the left end of the column in Fig. 9-1 is constrained against all rotation. The governing equation is given later as (9-13) and the derivation is explained.

Example 9-1

For the structure of Fig. 9-3a, determine the load P that will cause the column to buckle.

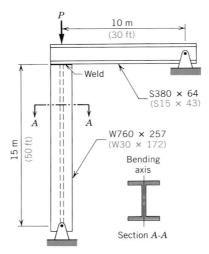

FIG. 9-3a

Solution.   Before proceeding with detailed calculations, it is important to visualize likely buckling configurations for the structure. Since the column of the given structure is loaded concentrically at the top and is concentrically pin-supported at the bottom, the column can buckle either to the right or to the left as shown in Fig. 9-3b. Both buckling modes will give the same solution. Because the horizontal beam is rigidly connected to the column at its top, a right angle must be maintained at the connection during buckling as indicated in Fig. 9-3b. Having established the possible buckling shapes, we can now proceed with the buckling calculations.

We will first concern ourselves with the stiffness of the beam. The beam is pin-ended at the right support and prevented from moving vertically by the column at the left support as shown in Fig. 9-3c. The rotational constraint can be determined from Case 8 of Table G. By letting the subscripts B and C refer to the beam and column, respectively, we obtain

$$\theta = \frac{1}{3}\frac{ML_B}{EI_B} \quad \text{(Case 8, Table G)}$$

$$k = \frac{M}{\theta} = \frac{3EI_B}{L_B}$$

$$k' = \frac{k}{EI_C/L_C} = \frac{3EI_B}{L_B}\frac{L_C}{EI_C} = 3\frac{L_C}{L_B}\frac{I_B}{I_C}$$

$$= 3 \times \frac{15}{10} \times \frac{187 \times 10^6}{250 \times 10^6}\frac{\text{m}\cdot\text{mm}^4}{\text{m}\cdot\text{mm}^4} = 3.37$$

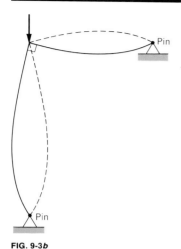

FIG. 9-3b

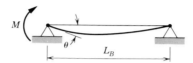

FIG. 9-3c

From Fig. 9-2 the effective length factor is

$$K = 0.84$$

$L' = 0.84 \times 15\,000 = 12\,600$ mm
$r = 87$ mm (from Table B1)

The slenderness ratio is

$$\frac{L'}{r} = \frac{12\,600}{87} = 145$$

Since $145 > 110$, Euler's formula applies:

$$P_{cr} = \frac{P_e}{K^2} = \frac{\pi^2 EI}{(L')^2} = \frac{\pi^2 \times 200\,000 \times 250 \times 10^6}{12\,600^2} \frac{\text{N/mm}^2 \cdot \text{mm}^4}{\text{mm}^2}$$

$$= 3.11 \times 10^6 \text{ N} = \mathbf{3110 \text{ kN}}$$

Example 9-1

For the structure of Fig. 9-3a, determine the load P that will cause the column to buckle.

Solution.   Before proceeding with detailed calculations, it is important to visualize likely buckling configurations for the structure. Since the column of the given structure is loaded concentrically at the top and is concentrically pin-supported at the bottom, the column can buckle either to the right or to the left as shown in Fig. 9-3b. Both buckling modes will give the same solution. Because the horizontal beam is rigidly connected to the column at its top, a right angle must be maintained at the connection during buckling as indicated in Fig. 9-3b. Having established the possible buckling shapes, we can now proceed with the buckling calculations.

We will first concern ourselves with the stiffness of the beam. The beam is pin-ended at the right support and prevented from moving vertically by the column at the left support as shown in Fig. 9-3c. The rotational constraint can be determined from Case 8 of Table G. By letting the subscripts B and C refer to the beam and column, respectively, we obtain

$$\theta = \frac{1}{3} \frac{ML_B}{EI_B} \qquad \text{(Case 8, Table G)}$$

$$k = \frac{M}{\theta} = \frac{3EI_B}{L_B}$$

$$k' = \frac{k}{EI_C/L_C} = 3EI_B/L_B \frac{L_C}{EI_C} = 3 \frac{L_C}{L_B} \frac{I_B}{I_C}$$

$$= 3 \times \frac{50}{30} \times \frac{448}{598} \frac{\text{ft}}{\text{ft}} \cdot \frac{\text{in.}^4}{\text{in.}^4} = 3.75$$

From Fig. 9-2 the effective length factor is

$$K = 0.83$$

$$L' = 0.83 \times 50 \times 12 = 498 \text{ in.}$$
$$r = 3.43 \text{ in. (from Table B1)}$$

The slenderness ratio is

$$\frac{L'}{r} = \frac{498}{3.43} = 145$$

Since $145 > 110$, Euler's formula applies

$$P_{\text{cr}} = \frac{P_e}{K^2} = \frac{\pi^2 EI}{(L')^2} = \frac{\pi^2 \times 29 \times 10^3 \times 598}{498^2} \frac{\text{kip/in.}^2 \cdot \text{in.}^4}{\text{in.}^2}$$

$$= \textbf{690 kip}$$

Problems 9-1 to 9-9

When some constraint is imposed on the rotation at both ends, the effective length factor cannot be determined by the previously developed formula. By extending the methods just used, an equation to take the place of (9-11) will be generated. It will be appreciated that the manipulations are much more complicated and that there are *two rotational stiffness factors*, k'_1 and k'_2, instead of the one, k'. This equation works out to be

$$\left[\left(\frac{\pi}{K} \cos \frac{\pi}{K} - \sin \frac{\pi}{K} \right) k'_1 - \left(\frac{\pi}{K} \right)^2 \sin \frac{\pi}{K} \right] \left[\left(\frac{\pi}{K} \cos \frac{\pi}{K} - \sin \frac{\pi}{K} \right) k'_2 \right.$$

$$\left. - \left(\frac{\pi}{K} \right)^2 \sin \frac{\pi}{K} \right] - \left(\frac{\pi}{K} - \sin \frac{\pi}{K} \right)^2 k'_1 k'_2 = 0 \qquad \textbf{(9-12)}$$

Again we would like to express K in terms of the other variables, but this is obviously impossible, so we must resort to trial-and-error methods or to graphical solutions. The curves in Fig. 9-4 give the solution to K for values of k_1' and k_2'.

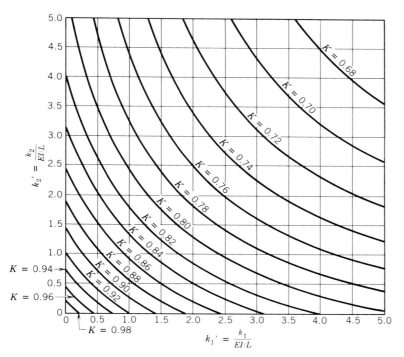

FIG. 9-4 Effective length factor for a column with elastic rotational constraint at both ends.

Equation (9-12) is more general than (9-11), and it can be seen that the latter could have been obtained by making k_2' equal to zero in (9-12). The case of a column with complete constraint at one end and partial constraint at the other can be obtained from (9-12) by letting one k approach infinity. The other k then becomes

$$k' = \frac{(\pi/K)^2((\pi/K)\cot(\pi/K) - 1)}{((\pi/K)\cot(\pi/K) - 1)^2 - [(\pi/K)\,\mathrm{cosec}\,(\pi/K) - 1]^2} \qquad \textbf{(9-13)}$$

This cannot be solved for K; therefore, for a given k', K must be evaluated by trial and error or by using curve (b) in Fig. 9-2.

Example 9-2

For the structure of Fig. 9-5a, determine the load P that will cause the column to buckle.

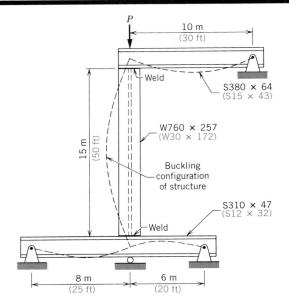

FIG. 9-5a

Solution.   The upper beam and column arrangement is identical to Example 9-1; hence from the previous example we know that $k'_1 = 3.37$. For the lower beam depicted in Fig. 9-5b we can write

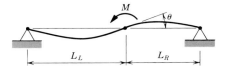

FIG. 9-5b

$$M = \frac{3EI_B\theta}{L_L} + \frac{3EI_B\theta}{L_R} \qquad \text{(from Case 8, Table G)}$$

$$k_2 = \frac{M}{\theta} = 3EI_B\left(\frac{1}{L_L} + \frac{1}{L_R}\right)$$

$$k'_2 = \frac{k_2}{EI_C/L_C} = \frac{3EI_BL_C}{EI_C}\left(\frac{1}{L_L} + \frac{1}{L_R}\right)$$

$$= 3 \times \frac{90.7 \times 10^6 \times 15}{250 \times 10^6}\left(\frac{1}{8} + \frac{1}{6}\right) \frac{\text{mm}^4 \cdot \text{mm}}{\text{mm}^4 \cdot \text{mm}}$$

$$= 4.76$$

For $k'_1 = 3.35$ and $k'_2 = 4.76$ Fig. 9-4 gives

$$K = 0.686$$

The slenderness ratio is

$$\frac{L'}{r} = \frac{0.686 \times 15\,000}{87} \frac{\text{mm}}{\text{mm}} = 118$$

Since $118 > 110$, Euler's formula is valid and

$$P_{cr} = \frac{P_e}{K^2} = \frac{\pi^2 EI}{K^2 L^2} = \frac{\pi^2 \times 200\,000 \times 250 \times 10^6}{0.686^2 \times 15\,000^2} \frac{\text{N/mm}^2 \cdot \text{mm}^4}{\text{mm}^2}$$

$$= 4.66 \times 10^6 \text{ N} = \textbf{4660 kN}$$

Example 9-2

For the structure of Fig. 9-5a, determine the load P that will cause the column to buckle.

Solution. The upper beam and column arrangement is identical to Example 9-1; hence from the previous example we know that $k'_1 = 3.75$. For the lower beam depicted in Fig. 9-5b we can write

$$M = \frac{3EI_B \theta}{L_L} + \frac{3EI_B \theta}{L_R} \qquad \text{(from Case 8, Table G)}$$

$$k_2 = \frac{M}{\theta} = 3EI_B\left(\frac{1}{L_L} + \frac{1}{L_R}\right)$$

$$k'_2 = \frac{k_2}{EI_C/L_C} = \frac{3EI_B L_C}{EI_C}\left(\frac{1}{L_L} + \frac{1}{L_R}\right)$$

$$= 3 \times \frac{219 \times 50}{598}\left(\frac{1}{25} + \frac{1}{20}\right) \frac{\text{in.}^4 \cdot \text{ft}}{\text{in.}^4 \cdot \text{ft}}$$

$$= 4.94$$

For $k'_1 = 3.77$ and $k'_2 = 4.94$ Fig. 9-4 gives

$$K = 0.678$$

The slenderness ratio is

$$\frac{L'}{r} = \frac{0.678 \times 50 \times 12}{3.43} \frac{\text{in.}}{\text{in.}} = 118$$

Since $118 > 110$, Euler's formula is valid and

$$P_{cr} = \frac{P_e}{K^2} = \frac{\pi^2 EI}{K^2 L^2} = \frac{\pi^2 \times 29 \times 10^3 \times 598}{0.678^2 \times (50 \times 12)^2} \frac{\text{kip/in.}^2 \cdot \text{in.}^4}{\text{in.}^2}$$

$$= \textbf{1034 kip}$$

In comparing the results from this example with Example 9-1, we see that the partial constraint at the bottom of the column leads to a reduced slenderness ratio and hence to a significant increase in buckling capacity. If we were to provide even more constraint at the bottom of the column, the limit being complete end fixity, the column would be able to withstand still higher loads before buckling.

Problems 9-10 to 9-17

9-3 ECCENTRICALLY LOADED COLUMNS

The load on an ideal column is applied at the centroid of the cross section. In reality such a condition rarely exists, there is usually some eccentricity in the application of the load. When the eccentricity can be assumed to be negligible, the formulas developed up to this point may be used. The formulas that we will now derive will enable us to allow for eccentricity in the design of compression members.

Consider a column loaded with a force that has an eccentricity, e, as shown in Fig. 9-6a. The centerline will curve under load as in Fig. 9-6b.

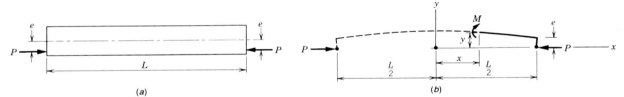

FIG. 9-6 (a) **Eccentrically loaded column.** (b) **Deflected shape of the column centerline.**

The bending moment at a typical cross section will be

$$M = -Py$$

From (3-8)

$$\frac{d^2y}{dx^2} = \frac{M}{EI} = \frac{-P}{EI}y$$

To determine y, we must solve the differential equation

$$\frac{d^2y}{dx^2} + \frac{P}{EI}y = 0$$

The solution is known to be

$$y = C_1 \sin\sqrt{\frac{P}{EI}}x + C_2\cos\sqrt{\frac{P}{EI}}x$$

or

$$y = C_1 \sin\lambda x + C_2\cos\lambda x$$

where

$$\lambda = \sqrt{\frac{P}{EI}} \qquad \textbf{(9-14)}$$

The symbol λ is introduced merely to simplify the writing of equations. It is interesting to note that it has dimension 1/length so that when it is multiplied by a length the product is dimensionless.

From known y at the left end

$$y]_{x=(-L/2)} = e$$

$$C_1 \sin\left(-\lambda\frac{L}{2}\right) + C_2\cos\left(-\lambda\frac{L}{2}\right) = e$$

$$-\sin\left(\lambda\frac{L}{2}\right)C_1 + \cos\left(\lambda\frac{L}{2}\right)C_2 = e \tag{1}$$

From known y at the right end

$$y]_{x=(L/2)} = e$$

$$\sin\left(\lambda\frac{L}{2}\right)C_1 + \cos\left(\lambda\frac{L}{2}\right)C_2 = e \tag{2}$$

Solving these simultaneous equations for the constants gives

$$C_1 = 0 \quad\text{and}\quad C_2 = \frac{e}{\cos[\lambda(L/2)]}$$

Substituting into the deflection equation gives

$$y = \frac{e}{\cos[\lambda(L/2)]}\cos(\lambda x) = e\sec\left(\lambda\frac{L}{2}\right)\cos(\lambda x)$$

$$y_{max} = y]_{x=0} = e\sec\left(\lambda\frac{L}{2}\right) \tag{9-15}$$

The greatest stress will occur on the section having the largest bending moment, that is at midlength in the column, where

$$\text{direct stress} = \frac{P}{A}$$

and

$$\text{maximum bending stress} = \frac{M_{max}c}{I} = \frac{Py_{max}c}{I}$$

$$= \frac{Pe\sec[\lambda(L/2)]c}{I}$$

The total stress, which is the maximum compressive stress, is then

$$\sigma_{max} = \frac{P}{A} + \frac{Pe\sec[\lambda(L/2)]c}{I} = \frac{P}{A}\left[1 + \frac{ec}{r^2}\sec\left(\lambda\frac{L}{2}\right)\right] \tag{9-16}$$

It is interesting to note that when there is eccentricity, any small load will cause some deflection and the amplitude of this deflection, the quantity C_2 or Y_{max} in (9-15), can be determined. This means that under an increasing load, the column begins to bend from the start and will give a visual indication that it is being loaded. Such behavior is quite different from that of an ideal column, which remains straight until it suddenly collapses when the load reaches its critical level.

The Euler buckling load does not appear in (9-16) and it may seem that P_e is no longer significant. However, P_e will appear when we change the form of the equation by altering the argument of the secant. Recall that λ was defined in (9-14) as

$$\lambda = \sqrt{\frac{P}{EI}}$$

Then

$$\lambda \frac{L}{2} = \sqrt{\frac{PP_e}{EIP_e}}\,\frac{L}{2} = \sqrt{\frac{P}{EIP_e}\,\frac{\pi^2 EI}{L^2}}\,\frac{L}{2} = \frac{\pi}{2}\sqrt{\frac{P}{P_e}} \tag{9-17}$$

Equation (9-16) can then be written as

$$\sigma_{max} = \frac{P}{A}\left[1 + \frac{e}{r}\frac{c}{r}\sec\left(\frac{\pi}{2}\sqrt{\frac{P}{P_e}}\right)\right]$$

Because it contains a secant term, the equation frequently is referred to as the *secant formula*. Since P/A is the average stress on the column for a concentric load, it can be represented by σ_{ave}, and putting the equation completely into dimensionless form, we obtain

$$\frac{\sigma_{max}}{\sigma_{ave}} = 1 + \frac{e}{r}\frac{c}{r}\sec\left(\frac{\pi}{2}\sqrt{\frac{P}{P_e}}\right) \tag{9-18}$$

The expression on the right side of the equation is then seen to be an *amplification factor* that must be applied to the average stress in order to obtain the maximum stress. The effect that eccentricity has on stress in columns can be seen by studying the curves in Fig. 9-7, which are plots of the amplification factor in (9-18). It can also be seen that although the column bends in a manner that is quite different from Euler buckling, the load relative to the Euler load is significant. We see that as the Euler load is approached the stress amplification factor increases without limit. At the other end of the scale, for small P relative to P_e, the curves give the stress that would be determined by the methods of Section 2-5.

Example 9-3

A Douglas fir column is loaded as shown in Fig. 9-8. Determine the maximum stress when the length is

(a) 2 m.
(b) 15 m.

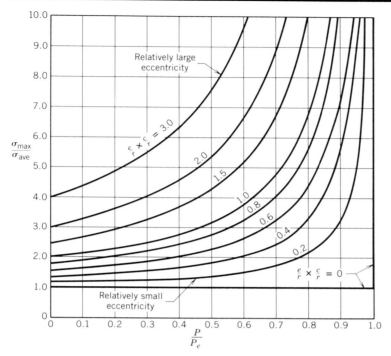

FIG. 9-7 Stress in a column due to load having a uniform eccentricity.

Solution

$$r = \frac{h}{\sqrt{12}} \quad \text{(Case 4, Table H)}$$

$$= \frac{300}{\sqrt{12}} = 86.6 \text{ mm}$$

$$e = 50 \text{ mm} \quad \text{(given)}$$

$$c = \frac{h}{2} = \frac{300}{2} = 150 \text{ mm}$$

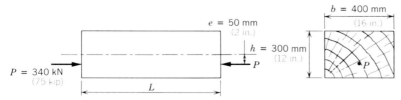

FIG. 9-8

$$\frac{e}{r} \times \frac{c}{r} = \frac{50}{86.6} \times \frac{150}{86.6} \frac{mm \cdot mm}{mm \cdot mm} = 1.000$$

$$\sigma_{ave} = \frac{340\ 000}{300 \times 400} \frac{N}{mm \cdot mm} = 2.833 \frac{N}{mm^2}$$

$$P_e = \frac{\pi^2 EI}{L^2}$$

$$= \frac{\pi^2 \times 12\ 000 \times \frac{1}{12} \times 400 \times 300^3}{L^2} \frac{N/mm^2 \cdot mm^4}{mm^2}$$

$$= \frac{106.6 \times 10^{12}}{L^2} N$$

Case a

$$L = 2\ m = 2000\ mm$$

$$P_e = \frac{106.6 \times 10^{12}}{2000^2} = 26\ 650 \times 10^3\ N$$

$$\frac{P}{P_e} = \frac{340 \times 10^3}{26\ 650 \times 10^3} = 0.0128$$

$$\frac{\sigma_{max}}{\sigma_{ave}} = 1 + \frac{e}{r} \frac{c}{r} \sec \frac{\pi}{2} \sqrt{\frac{P}{P_e}} \qquad (9\text{-}18)$$

$$= 1 + 1.000 \sec \frac{\pi}{2} \sqrt{0.0128}$$

$$= 2.016 \qquad \text{(could have been obtained from Fig. 9-7)}$$

$$\sigma_{max} = 2.016 \times \sigma_{ave}$$

$$= 2.016 \times 2.833 \frac{N}{mm^2} = \textbf{5.71 MPa}$$

Case b

$$L = 15\ m = 15\ 000\ mm$$

$$P_e = \frac{106.6 \times 10^{12}}{15\ 000^2} = 474 \times 10^3\ N$$

$$\frac{P}{P_e} = \frac{340 \times 10^3}{474 \times 10^3} = 0.718$$

$$\frac{\sigma_{max}}{\sigma_{ave}} = 1 + 1.000 \sec \frac{\pi}{2} \sqrt{0.718}$$

$$= 5.211 \qquad \text{(could have been obtained from Fig. 9-7)}$$

$$\sigma_{max} = 5.211 \times \sigma_{ave}$$

$$= 5.211 \times 2.833 \frac{N}{mm^2} = \textbf{14.76 MPa}$$

Note that the stress in the short member does not exceed the allowable stress of 6 MPa (Table A), but the longer member is overstressed. This means that if the dimensions were as in (b) and the column effect was not taken into account, the calculations would indicate a safe member although in reality the column would be overloaded.

Example 9-3

A Douglas fir column is loaded as shown in Fig. 9-8. Determine the maximum stress when the length is

(a) 6 ft.
(b) 50 ft.

Solution

$$r = \frac{h}{\sqrt{12}} \quad \text{(Case 4, Table H)}$$

$$= \frac{12}{\sqrt{12}} = 3.46 \text{ in.}$$

$$e = 2 \text{ in.} \quad \text{(given)}$$

$$c = \frac{h}{2} = \frac{12}{2} = 6 \text{ in.}$$

$$\frac{e}{r} \times \frac{c}{r} = \frac{2}{3.46} \times \frac{6}{3.46} = 1.00$$

$$\sigma_{ave} = \frac{75}{12 \times 16} \frac{\text{kip}}{\text{in.} \cdot \text{in.}} = 0.391 \text{ kip/in.}^2$$

$$P_e = \frac{\pi^2 EI}{L^2}$$

$$= \frac{\pi^2 \times 1700 \times \frac{1}{12} \times 16 \times 12^3}{L^2} \frac{\text{kip/in.}^2 \cdot \text{in.} \cdot \text{in.}^3}{\text{in.}^2}$$

$$= \frac{38.66 \times 10^6}{L^2} \text{kip}$$

Case a

$$L = 6 \text{ ft} = 72 \text{ in.}$$

$$P_e = \frac{38.66 \times 10^6}{72^2} = 7.46 \times 10^3 \text{ kip}$$

$$\frac{P}{P_e} = \frac{75}{7.46 \times 10^3} = 0.0101$$

$$\frac{\sigma_{max}}{\sigma_{ave}} = 1 + \frac{ec}{rr} \sec \frac{\pi}{2} \sqrt{\frac{P}{P_e}} \qquad \textbf{(9-18)}$$

$$= 1 + 1.000 \sec \frac{\pi}{2} \sqrt{0.0101}$$

$$= 2.013 \qquad \text{(could have been obtained from Fig. 9-7)}$$

$$\sigma_{max} = 2.013 \times \sigma_{ave}$$

$$= 2.013 \times 0.391 \frac{\text{kip}}{\text{in.}^2} = \textbf{0.787 kip/in.}^2$$

Case b

$$L = 50 \text{ ft} = 600 \text{ in.}$$

$$P_e = \frac{38.66 \times 10^6}{600^2} = 107.4 \text{ kip}$$

$$\frac{P}{P_e} = \frac{75}{107.4} = 0.698$$

$$\frac{\sigma_{max}}{\sigma_{ave}} = 1 + 1.000 \sec \frac{\pi}{2} \sqrt{0.698}$$

$$= 4.913 \qquad \text{(could have been obtained from Fig. 9-7)}$$

$$\sigma_{max} = 4.913 \times \sigma_{ave}$$

$$= 4.913 \times 0.391 \frac{\text{kip}}{\text{in.}^2} = \textbf{1.92 kip/in.}^2$$

Note that the stress in the short member does not exceed the allowable stress of 0.9 ksi (Table A), but the longer member is overstressed. This means that if the dimensions were as in (b) and the column effect was not taken into account, the calculations would indicate a safe member although in reality the column would be overloaded.

Problems 9-18 to 9-20

In practice it frequently happens that the eccentricity is not the same at both ends of the column. The deflection curve is then as shown in Fig. 9-9. The differential equation is the same as before but the boundary conditions are different. This seemingly minor change makes the solution much more complicated; the final results are presented here without proof. Because of the transcendental nature of the functions, it is impossible to write stress in terms of load, eccentricity, and so on, and we must again resort to the use of a set of curves.

The curves in Fig. 9-10 will enable us to solve the problem of eccentrically loaded columns, where the eccentricity at one end may be different from the eccentricity at the other end in both magnitude and direction. To use the curves, we take e_1 as the larger end eccentricity and take its direction as positive. The other eccentricity will be e_0 and the ratio of e_0 to e_1 may range between -1 and $+1$; negative values indicate

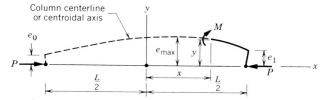

FIG. 9-9 Case of a column with two different end eccentricities.

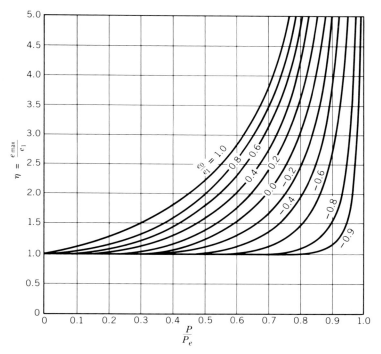

FIG. 9-10 Stress in a column due to load having a nonuniform eccentricity.

that the forces act on the *opposite* sides of the axis of the column. The ratio of e_0 to e_1, and the ratio of the load to the Euler load are required as input. The curves in Fig. 9-10 will then give a corresponding value for η, which is the ratio of the maximum eccentricity to e_1. That is

$$\eta = \frac{e_{max}}{e_1}$$

Note that the eccentricity e_{max} refers to the maximum distance between the centerline of the column section and the line of action of the load.

The maximum bending moment (without regard to sign) will be

$$M_{max} = Pe_{max} = Pe_1\eta$$

This will cause a compressive stress in the extreme fibers that is given by

$$\frac{M_{max}c}{I} = \frac{Pe_1 \eta c}{Ar^2}$$

There is also a direct stress

$$\frac{P}{A}$$

Hence a portion of the member is subjected to a stress

$$\sigma_{max} = \frac{P}{A} + \frac{Pe_1 c}{Arr} \eta$$

$$= \frac{P}{A}\left(1 + \frac{e_1}{r}\frac{c}{r}\eta\right)$$

But P/A is the average stress on the cross section and can be represented by σ_{ave}. Then

$$\boxed{\frac{\sigma_{max}}{\sigma_{ave}} = \left(1 + \frac{e_1}{r}\frac{c}{r}\eta\right)} \qquad \textbf{(9-19)}$$

This is similar to the formula for equal eccentricities (9-18) with η taking the place of the secant term.

Example 9-4

Calculate the maximum stress in the column of Fig. 9-11.

Solution

$$P_e = \frac{\pi^2 EI}{L^2} = \frac{\pi^2 \times 200\,000 \times 64 \times 10^6}{(10\,000)^2} \frac{N/mm^2 \cdot mm^4}{mm^2}$$

$$= 1.263 \times 10^6 \, N = 1263 \, kN$$

$$\frac{P}{P_e} = \frac{815}{1263} = 0.65$$

Taking e_1 as the larger eccentricity and letting its direction be positive, we have

$$e_1 = 40 \, mm \qquad \text{and} \qquad e_0 = -8 \, mm$$

Note that e_0 is negative because it acts on the side of the column centerline opposite to e_1.

$$\frac{e_0}{e_1} = \frac{-8}{40} = -0.2$$

With the values of P/P_e and e_0/e_1, we use Fig. 9-10 and determine

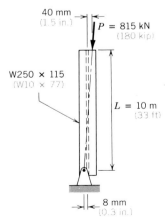

40 mm
(1.5 in.)

$P = 815 \, kN$
(180 kip)

W250 × 115
(W10 × 77)

$L = 10 \, m$
(33 ft)

8 mm
(0.3 in.)

FIG. 9-11

$$\eta = 1.45$$

From Table B1 we obtain

$$r = 66 \text{ mm}$$

$$c = \text{flange width}/2 = 259/2 = 129.5 \text{ mm}$$

$$A = 14\ 600 \text{ mm}^2$$

Substituting into (9-19) gives

$$\frac{\sigma_{max}}{\sigma_{ave}} = 1 + \frac{40}{66} \times \frac{129.5}{66} \times 1.45 = 2.72$$

$$\text{average stress} = \frac{P}{A}$$

$$= \frac{815\ 000}{14\ 600} \frac{\text{N}}{\text{mm}^2}$$

$$= 55.8 \frac{\text{N}}{\text{mm}^2}$$

$$\text{maximum stress} = 2.72 \times 55.8 \frac{\text{N}}{\text{mm}^2}$$

$$= 152 \frac{\text{N}}{\text{mm}^2} = \textbf{152 MPa}$$

We can see that for the particular column of this example, the eccentricities at each end lead to bending stresses that approximately triple the average compressive stress.

Example 9-4

Calculate the maximum stress in the column of Fig. 9-11.

Solution

$$P_e = \frac{\pi^2 EI}{L^2} = \frac{\pi^2 \times 29 \times 10^3 \times 154}{(33 \times 12)^2} \frac{\text{kip/in.}^2 \cdot \text{in.}^4}{\text{in.}^2}$$

$$= 281 \text{ kip}$$

$$\frac{P}{P_e} = \frac{180}{281} = 0.64$$

Taking e_1 as the larger eccentricity and letting its direction be positive, we get

$$e_1 = 1.5 \text{ in.} \quad \text{and} \quad e_0 = -0.3 \text{ in.}$$

Note that e_0 is negative because it acts on the side of the column centerline opposite to e_1.

$$\frac{e_0}{e_1} = \frac{-0.3}{1.5} = -0.2$$

With the values of P/P_e and e_0/e_1, we use Fig. 9-10 and determine

$$\eta = 1.45$$

From Table B1 we obtain

$$r = 2.61 \text{ in.}$$

$$c = \text{flange width}/2 = 10.2/2 = 5.1 \text{ in.}$$

$$A = 22.6 \text{ in.}^2$$

Substituting into (9-19) gives

$$\frac{\sigma_{max}}{\sigma_{ave}} = 1 + \frac{1.5}{2.61} \times \frac{5.1}{2.61} \times 1.45 = 2.63$$

$$\text{average stress} = \frac{P}{A}$$

$$= \frac{180}{22.6} \frac{\text{kip}}{\text{in.}^2}$$

$$= 7.96 \frac{\text{kip}}{\text{in.}^2}$$

$$\text{maximum stress} = 2.63 \times 7.96 \frac{\text{kip}}{\text{in.}^2}$$

$$= 20.9 \frac{\text{kip}}{\text{in.}^2}$$

We can see that for the particular column of this example, the eccentricities at each end lead to bending stresses that approximately triple the average compressive stress.

Problems 9-21 to 9-25

9-4 BEAM-COLUMNS

Many members in real structures are simultaneously subjected to axial and lateral loads. When the axial load is compressive, there is column action as well as beam bending and the member is referred to as a *beam-column*. The two actions cannot be treated separately and the result superimposed because the deflection due to bending has an influence on the column action that would be ignored in the solution by superposition. This is unfortunate, since the solution in one step requires such complicated manipulation that only cases having the simplest lateral load can be solved analytically.

A beam-column loaded by a uniformly distributed load, as shown in Fig. 9-12 is subjected to a bending moment on a typical section of

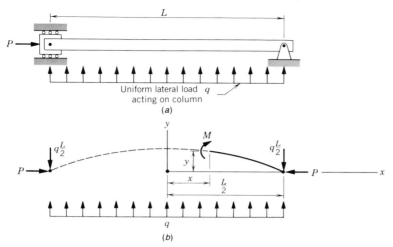

FIG. 9-12 Beam-column under the combined action of axial and uniform lateral load. (a) Beam column (b) Deflected shape of the beam-column centerline.

$$M = -Py + q\left(-\frac{L^2}{8} + \frac{x^2}{2}\right) \qquad \textbf{(9-20)}$$

From (3-8)

$$\frac{d^2y}{dx^2} = \frac{M}{EI} = -\frac{P}{EI}y + \frac{q}{EI}\left(-\frac{L^2}{8} + \frac{x^2}{2}\right)$$

$$\frac{d^2y}{dx^2} + \frac{P}{EI}y = \frac{q}{EI}\left(-\frac{L^2}{8} + \frac{x^2}{2}\right) \qquad \textbf{(9-21)}$$

The solution to this differential equation, which can be checked by substitution, is

$$y = C_1 \sin\sqrt{\frac{P}{EI}}x + C_2 \cos\sqrt{\frac{P}{EI}}x + \frac{q}{2P}x^2 - q\frac{EI}{P}\left(\frac{L^2}{8EI} + \frac{1}{P}\right)$$

which simplifies to

$$y = C_1 \sin\lambda x + C_2 \cos\lambda x + \frac{q}{2P}x^2 - q\frac{EI}{P}\left(\frac{L^2}{8EI} + \frac{1}{P}\right)$$

where

$$\lambda = \sqrt{\frac{P}{EI}}$$

Then

$$\frac{dy}{dx} = C_1\lambda\cos\lambda x - C_2\lambda\sin\lambda x + \frac{q}{P}x$$

From symmetry

$$\frac{dy}{dx}\bigg]_{x=0} = 0$$

$$C_1\lambda - 0C_2 + 0 = 0$$

Therefore,

$$C_1 = 0$$

and

$$y = C_2\cos\lambda x + \frac{q}{2P}x^2 - q\frac{EI}{P}\left(\frac{L^2}{8EI} + \frac{1}{P}\right)$$

At the right end there is no deflection and

$$y]_{x=(L/2)} = 0$$

$$C_2\cos\lambda\frac{L}{2} + \frac{q}{2P}\frac{L^2}{4} - q\frac{EI}{P}\left(\frac{L^2}{8EI} + \frac{1}{P}\right) = 0$$

$$C_2 = \frac{qEI}{P^2\cos\lambda(L/2)}$$

and

$$y = \frac{qEI}{P^2}\frac{\cos\lambda x}{\cos\lambda(L/2)} + \frac{q}{2P}x^2 - q\frac{EI}{P}\left(\frac{L^2}{8EI} + \frac{1}{P}\right) \qquad \textbf{(9-22)}$$

Substituting into (9-20) gives the bending moment

$$M = -\frac{qEI}{P}\frac{\cos\lambda x}{\cos\lambda(L/2)} - \frac{q}{2}x^2 + qEI\left(\frac{L^2}{8EI} + \frac{1}{P}\right) + q\left(-\frac{L^2}{8} + \frac{x^2}{2}\right)$$

$$= -\frac{qEI}{P}\frac{\cos\lambda x}{\cos\lambda(L/2)} + \frac{qEI}{P}$$

$$= \frac{qEI}{P}\left(1 - \frac{\cos\lambda x}{\cos\lambda(L/2)}\right)$$

$$M_{max} = M]_{x=0} = \frac{qEI}{P}\left(1 - \frac{1}{\cos\lambda(L/2)}\right) = \frac{qEI}{P}\left(1 - \sec\lambda\frac{L}{2}\right)$$

The maximum bending compressive stress is

$$\frac{-M_{max}c}{I} = \frac{qEc}{P}\left(\sec\lambda\frac{L}{2} - 1\right)$$

and the maximum direct compressive stress, $\dfrac{P}{A}$.

The maximum compressive stress is

$$\sigma_{max} = \frac{P}{A} + \frac{qEc}{P}\left(\sec\lambda\frac{L}{2} - 1\right)$$

$$= \frac{P}{A}\left[1 + \frac{qEcA}{P^2}\left(\sec \lambda \frac{L}{2} - 1\right)\right]$$

Substituting $\frac{\pi}{2}\sqrt{\frac{P}{P_e}}$ for $\lambda \frac{L}{2}$, from (9-17), gives

$$\sigma_{\max} = \frac{P}{A}\left[1 + \frac{qEcA}{P^2}\left(\sec \frac{\pi}{2}\sqrt{\frac{P}{P_e}} - 1\right)\right]$$

The factor $\dfrac{qEcA}{P^2}$ can be altered thus

$$\frac{qEcA}{P^2} = \frac{qEcA}{P(P/P_e)P_e} = \frac{qEcA}{P(P/P_e)(\pi^2 EI/L^2)}$$

$$= \frac{q\not{E}cAL^2}{P(P/P_e)\pi^2 \not{E}Ar^2} = \frac{qLcL}{Prr}\frac{1}{4(\pi/2\sqrt{P/P_e})^2}$$

The stress equation can then be written

$$\sigma_{\max} = \frac{P}{A}\left\{1 + \frac{qL}{P}\frac{c}{r}\frac{L}{r}\left[\frac{\sec(\pi/2\sqrt{P/P_e}) - 1}{4(\pi/2\sqrt{P/P_e})^2}\right]\right\}$$

$$\frac{\sigma_{\max}}{\sigma_{\text{ave}}} = 1 + \frac{qL}{P}\frac{c}{r}\frac{L}{r}\left[\frac{\sec(\pi/2\sqrt{P/P_e}) - 1}{4(\pi/2\sqrt{P/P_e})^2}\right] \tag{9-23}$$

where $\sigma_{\text{ave}} = P/A$.

The relationships when plotted as in Fig. 9-13 are more useful, but, because of the complex nature of the beam-column, practical problems which require the selection of steel sections are still very lengthy.

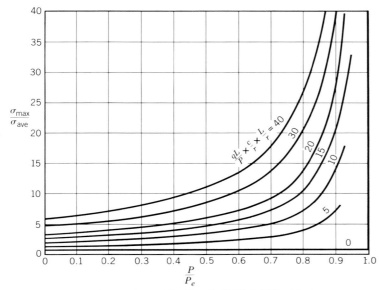

FIG. 9-13 Stress in a beam-column with uniformly distributed lateral load.

Example 9-5

A horizontal member in a pin-jointed structure is 7 m long and carries a compressive load of 70 kN. A 88-mm low-strength steel pipe is being considered for the member. Calculate the maximum stress in the pipe due to the axial load and gravity.

Solution. The properties of the pipe (see Table C) are

$$D \quad = 101.6 \text{ mm}$$

$$c \quad = \frac{D}{2} = \frac{101.6}{2} = 50.8 \text{ mm}$$

$$A \quad = 1730 \text{ mm}^2$$

$$I \quad = 1.99 \times 10^6 \text{ mm}^4$$

$$r \quad = 34.0 \text{ mm}$$

$$\text{mass} = 13.6 \frac{\text{kg}}{\text{m}}$$

$$\text{weight} = 9.81 \times 13.6 = 133.4 \frac{\text{N}}{\text{m}}$$

The loads are acting as shown in Fig. 9-14.

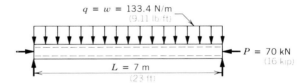

FIG. 9-14

$$P_e = \frac{\pi^2 EI}{L^2}$$

$$= \pi^2 \frac{200\,000 \times 1.99 \times 10^6}{(7000)^2} \frac{\text{N/mm}^2 \cdot \text{mm}^4}{\text{mm}^2}$$

$$= 80\,200 \text{ N} = 80.2 \text{ kN}$$

$$\frac{P}{P_e} = \frac{70}{80.2} \frac{\text{kN}}{\text{kN}} = 0.873$$

$$\frac{\pi}{2} \sqrt{\frac{P}{P_e}} = \frac{\pi}{2} \sqrt{0.873} = 1.467$$

$$\frac{qL}{P} \frac{c}{r} \frac{L}{r} = \frac{133.4 \times 7}{70\,000} \frac{50.8}{34.0} \frac{7000}{34.0} \frac{\text{N/m} \cdot \text{m}}{\text{N}} \cdot \frac{\text{mm}}{\text{mm}} \cdot \frac{\text{mm}}{\text{mm}} = 4.10$$

$$\frac{\sigma_{max}}{\sigma_{ave}} = 1 + \frac{qL}{P} \frac{c}{r} \frac{L}{r} \left\{ \frac{\sec\left[(\pi/2)\sqrt{(P/P_e)}\right] - 1}{4[(\pi/2)\sqrt{(P/P_e)}]^2} \right\} \tag{9-23}$$

$$= 1 + 4.10 \left[\frac{\sec 1.467 - 1}{4 \times 1.467^2} \right]$$

$$= 5.12$$

$$\sigma_{ave} = \frac{P}{A} = \frac{70\ 000}{1730}\ \frac{N}{mm^2} = 40.5\ \frac{N}{mm^2}$$

$$\sigma_{max} = 5.12 \times 40.5\ \frac{N}{mm^2}$$

$$= 207\ \frac{N}{mm^2} = \textbf{207 MPa}$$

It is interesting to note in this problem the effect of combined compressive and lateral loads. For the axial load alone the stress is 40.5 MPa, and for the lateral load alone the stress is 20.9 MPa (calculations not shown). We might then expect the two loads acting simultaneously to give a stress equal to the sum of the stress due to each load, that is, 61.4 MPa, when in fact the combined loads give a stress of 207 MPa. This illustrates a misuse of superposition. The error comes about because the deflection due to the lateral load causes the axial load to have an eccentricity that is not taken into account when the two loads are considered separately.

Example 9-5

A horizontal member in a pin-jointed structure is 23 ft long and carries a compressive load of 16 kip. A $3\frac{1}{2}$-in. low-strength steel pipe is being considered for the member. Calculate the maximum stress in the pipe due to the axial load and gravity.

Solution. The properties of the pipe (see Table C) are

D = 4.00 in.
c = $D/2 = 4.00/2 = 2.00$ in.
A = 2.68 in.^{2}
I = 4.79 in.^{4}
r = 1.34 in.
mass = 9.11 lb/ft
weight = 32.2/g × 9.11 = 9.11 lb/ft

The loads are acting as shown in Fig. 9-14.

$$P_e = \frac{\pi^2 EI}{L^2}$$

$$= \pi^2 \frac{29 \times 10^3 \times 4.79}{(23 \times 12)^2}\ \frac{kip/in.^2 \cdot in.^4}{in.^2}$$

$$= 18.0\ kip$$

$$\frac{P}{P_e} = \frac{16.0}{18.0} = 0.889$$

$$\frac{\pi}{2}\sqrt{\frac{P}{P_e}} = \frac{\pi}{2}\sqrt{0.889} = 1.481$$

$$\frac{qL}{P}\frac{c}{r}\frac{L}{r} = \frac{9.1 \times 23}{16,000} \times \frac{2.00}{1.34} \times \frac{23 \times 12}{1.34} \frac{\text{lb/ft} \cdot \text{ft}}{\text{lb}} \times \frac{\text{in.} \cdot \text{in.}}{\text{in.} \cdot \text{in.}}$$

$$= 4.02$$

$$\frac{\sigma_{max}}{\sigma_{ave}} = 1 + \frac{qL}{P}\frac{c}{r}\frac{L}{r}\left\{\frac{\sec\left[(\pi/2)\sqrt{(P/P_e)}\right] - 1}{4[(\pi/2)\sqrt{(P/P_e)}]^2}\right\} \qquad (9\text{-}23)$$

$$= 1 + 4.02\left[\frac{\sec 1.481 - 1}{4 \times 1.481^2}\right]$$

$$= 5.66$$

$$\sigma_{ave} = \frac{P}{A} = \frac{16}{2.68}\frac{\text{kip}}{\text{in.}^2} = 5.97 \text{ kip/in.}^2$$

$$\sigma_{max} = 5.66 \times 5.97 \text{ kip/in.}^2$$

$$= \textbf{33.8 ksi}$$

It is interesting to note in this problem the effect of combined compressive and lateral loads. For the axial load alone the stress is 5.97 ksi, and for the lateral load alone the stress is 3.02 ksi (calculations not shown). We might then expect the two loads acting simultaneously to give a stress equal to the sum of the stress due to each load, that is, 8.99 ksi, when in fact the combined loads give a stress of 33.8 ksi. This illustrates a misuse of superposition. The error comes about because the deflection due to the lateral load causes the axial load to have an eccentricity that is not taken into account when the two loads are considered separately.

Problems 9-26 to 9-30

The column that is subjected to a uniformly distributed lateral load has obvious application in horizontal compression members, where the lateral load comes from the force of gravity. We will now consider a column with a concentrated lateral load at the center; this load condition, in the beginning, may seem unrealistic but will later be seen to be of great practical interest. The column is shown in Fig. 9-15. At a typical cross section, the bending moment is given by

$$M = -Py - \frac{F}{2}\left(\frac{L}{2} - x\right) \qquad \text{for } 0 \leqslant x \leqslant \frac{L}{2}$$

From (3-8)

$$\frac{d^2y}{dx^2} = \frac{M}{EI} = -\frac{P}{EI}y - \frac{F}{2EI}\left(\frac{L}{2} - x\right)$$

$$\frac{d^2y}{dx^2} + \frac{P}{EI}y = -\frac{F}{2EI}\left(\frac{L}{2} - x\right) \qquad (9\text{-}24)$$

The solution to this differential equation, which you should check by substitution, is

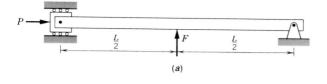

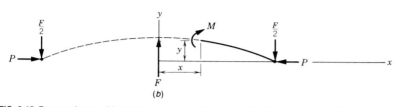

FIG. 9-15 Beam-column subjected to concentrated lateral load. (*a*) Beam-column. (*b*) Deflected shape of beam-column centerline.

$$y = C_1 \sin \lambda x + C_2 \cos \lambda x - \frac{F}{2P}\left(\frac{L}{2} - x\right)$$

where $\lambda = \sqrt{P/EI}$. Then

$$\frac{dy}{dx} = C_1 \lambda \cos \lambda x - C_2 \lambda \sin \lambda x + \frac{F}{2P}$$

But from symmetry

$$\frac{dy}{dx}\bigg]_{x=0} = 0$$

$$C_1 \lambda + \frac{F}{2P} = 0$$

Therefore,

$$C_1 = -\frac{F}{2P\lambda}$$

and

$$y = -\frac{F}{2P\lambda} \sin \lambda x + C_2 \cos \lambda x - \frac{F}{2P}\left(\frac{L}{2} - x\right)$$

At the right end the deflection is zero, then

$$y]_{x=(L/2)} = 0$$

$$-\frac{F}{2P\lambda} \sin \lambda \frac{L}{2} + C_2 \cos \lambda \frac{L}{2} - 0 = 0$$

$$C_2 = \frac{F}{2P\lambda} \frac{\sin \lambda(L/2)}{\cos \lambda(L/2)} = \frac{F}{2P\lambda} \tan \lambda \frac{L}{2}$$

and

$$y = -\frac{F}{2P\lambda} \sin \lambda x + \frac{F}{2P\lambda} \tan \lambda \frac{L}{2} \cos \lambda x - \frac{F}{2P}\left(\frac{L}{2} - x\right)$$

$$y = \frac{F}{2P}\left(-\frac{1}{\lambda} \sin \lambda x + \frac{1}{\lambda} \tan \lambda \frac{L}{2} \cos \lambda x - \frac{L}{2} + x\right) \qquad \textbf{(9-25)}$$

$$\frac{d^2 y}{dx^2} = \frac{F}{2P}\left(\lambda \sin \lambda x - \lambda \tan \lambda \frac{L}{2} \cos \lambda x\right)$$

The bending moment is

$$M = EI\frac{d^2 y}{dx^2} = \frac{EIF}{2P}\left(\lambda \sin \lambda x - \lambda \tan \lambda \frac{L}{2} \cos \lambda x\right)$$

$$= \frac{F}{2\lambda}\left(\sin \lambda x - \tan \lambda \frac{L}{2} \cos \lambda x\right)$$

and the maximum bending moment is

$$M_{\text{max}} = M]_{x=0}$$

$$= \frac{F}{2\lambda}\left(-\tan \lambda \frac{L}{2}\right)$$

The maximum compressive bending stress is

$$\frac{-M_{\text{max}}c}{I} = \frac{Fc}{2I}\frac{\tan \lambda(L/2)}{\lambda} = \frac{Fc}{2I}\frac{\tan \lambda(L/2)}{\lambda}$$

$$= \frac{Fc}{2I}\frac{L}{2}\frac{\tan \lambda(L/2)}{\lambda(L/2)}$$

$$= \frac{FcL}{4I}\frac{\tan (\pi/2\sqrt{P/P_e})}{(\pi/2\sqrt{P/P_e})} \qquad \text{[from (9-17)]}$$

Adding the direct stress gives the total maximum stress,

$$\sigma_{\text{max}} = \frac{P}{A} + \frac{FcL}{4I}\frac{\tan (\pi/2\sqrt{P/P_e})}{(\pi/2\sqrt{P/P_e})}$$

$$= \frac{P}{A} + \frac{FcL}{4Ar^2}\frac{\tan (\pi/2\sqrt{P/P_e})}{(\pi/2\sqrt{P/P_e})}$$

$$= \frac{P}{A}\left[1 + \frac{1}{4}\frac{F}{P}\frac{c}{r}\frac{L}{r}\frac{\tan (\pi/2\sqrt{P/P_e})}{(\pi/2\sqrt{P/P_e})}\right]$$

But P/A is the average stress, σ_{ave}, and the stress can be expressed in dimensionless form as

$$\boxed{\frac{\sigma_{\text{max}}}{\sigma_{\text{ave}}} = 1 + \frac{F}{P}\frac{c}{r}\frac{L}{r}\frac{\tan (\pi/2\sqrt{P/P_e})}{4(\pi/2\sqrt{P/P_e})}} \qquad \textbf{(9-26)}$$

The relationships are again complicated and design must be done by rather lengthy manipulations. The curves in Fig. 9-16 are useful in such a design and also in giving us some insight into the relative importance of the factors that influence stress.

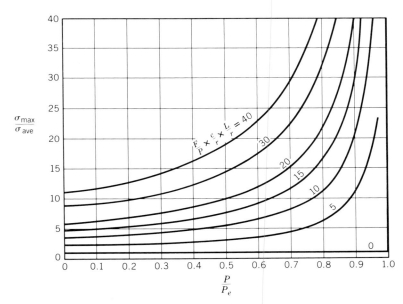

FIG. 9-16 Stress in a beam-column with a concentrated lateral load at midlength.

A practical problem that is related to the above analysis arises in the design of columns in an unbraced frame. Such a structure is shown in Fig. 9-17a. There is no diagonal bracing to carry the lateral load, so it is imperative that the column ends be secured against rotation. This poses no problem at the foundation but requires that the floor system be made more rigid than usual. The rigidity is commonly provided by a spandrel beam, which usually is incorporated into the exterior walls. With this arrangement the lateral load causes the columns to flex and the floor to move as shown in Fig. 9-17b. The lateral deflection of a reinforced concrete structure in a model test is illustrated in Fig. 9-17c.

The total lateral load is taken as being carried equally by those columns that are capable of resisting lateral deflection. With this assumption the load on a particular column can be determined. A typical column with its lateral load, Q, is shown in Fig. 9-18b. The moment and deflection variations in the upper half of this column are similar to those of the lower half of the column in Fig. 9-18a, which is the column that has just been analyzed. In Fig. 9-18c the upper half of the column in Fig. 9-18b is reproduced as well as its mirror image. From this it can be seen that the column can be designed as a beam column with a lateral load at the center equal to twice the lateral load imposed on the actual column.

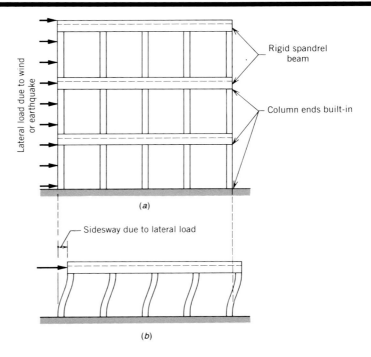

Lateral load due to wind or earthquake

Rigid spandrel beam

Column ends built-in

(a)

Sidesway due to lateral load

(b)

(c)

FIG. 9-17 Column action in a building subjected to lateral load. (*a*) Building elevation. (*b*) Sidesway of a storey high section. (*c*) Lateral deflection of a reinforced concrete structure in a model test.

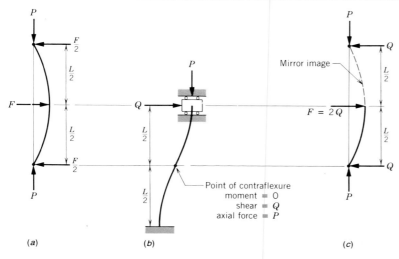

FIG. 9-18 (a) Basic beam-column. (b) Structural column with sidesway. (c) Equivalent basic beam-column.

Example 9-6

A column line such as that of Fig. 9-17b carries vertical and horizontal loads. All columns are W250 × 89 and are 7 m long. The columns are oriented in the structure so that bending takes place about the axis having the minimum radius of gyration. Determine the maximum stress in a typical column if it carries an axial load of 400 kN and a lateral load of 25 kN.

Solution. The column under the actual loads is shown in Fig. 9-19a and the loading of the equivalent beam column is indicated in Fig. 9-19b.

$L = 7000$ mm
$c = \frac{256}{2} = 128$ mm
$I = 48 \times 10^6$ mm^4
$r = 65$ mm
$A = 11\,400$ mm^2
$F = 2Q = 2 \times 25 = 50$ kN

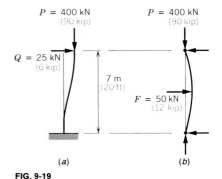

FIG. 9-19

$$P_e = \frac{\pi^2 EI}{L^2} = \frac{\pi^2 \times 200\,000 \times 48 \times 10^6}{(7000)^2} \frac{\text{N/mm}^2 \cdot \text{mm}^4}{\text{mm}^2}$$

$$= 1.934 \times 10^6 \text{ N} = 1934 \text{ kN}$$

$$\frac{P}{P_e} = \frac{400}{1934} \frac{\text{kN}}{\text{kN}} = 0.207$$

$$\frac{F}{P} \times \frac{c}{r} \times \frac{L}{r} = \frac{50}{400} \times \frac{128}{65} \times \frac{7000}{65} \frac{\text{kN}}{\text{kN}} \cdot \frac{\text{mm}}{\text{mm}} \cdot \frac{\text{mm}}{\text{mm}} = 26.5$$

$$\frac{\sigma_{max}}{\sigma_{ave}} = 1 + 26.5 \frac{\tan(\pi/2\sqrt{0.207})}{4(\pi/2\sqrt{0.207})} \qquad \text{from (9-26)}$$

$$= 9.04$$

$$\sigma_{ave} = \frac{P}{A} = \frac{400\,000}{11\,400} = 35.1 \frac{N}{mm^2}$$

$$\sigma_{max} = 9.04 \times \sigma_{ave} = 9.04 \times 35.1 = 317 \frac{N}{mm^2} = \textbf{317 MPa}$$

The analysis in this example will be invalid if the yield stress of the material is less than 317 MPa, since it was assumed that stress and strain are linearly related. The analysis will be correct, although the structure may not have an adequate safety factor if the column is made of high-strength structural steel.

Example 9-6

A column line such as that of Fig. 9-17b carries vertical and horizontal loads. All columns are W10 × 60 and are 20 ft long. The columns are oriented in the structure so that bending takes place about the axis having the minimum radius of gyration. Determine the maximum stress in a typical column if it carries an axial load of 90 kip and a lateral load of 6 kip.

Solution.   The column under the actual loads is shown in Fig. 9-19a and the loading of the equivalent beam column is indicated in Fig. 9-19b.

$L = 20$ ft $= 240$ in.

$c = \dfrac{10.08}{2} = 5.04$ in.

$I = 116$ in.^{4}

$r = 2.57$ in.

$A = 17.6$ in.^{2}

$F = 2Q = 2 \times 6 = 12$ kip

$$P_e = \frac{\pi^2 EI}{L^2} = \frac{\pi^2 \times 29 \times 10^3 \times 116}{240^2} \frac{kip/in.^2 \cdot in.^4}{in.^2}$$

$$= 576 \text{ kip}$$

$$\frac{P}{P_e} = \frac{90}{576} = 0.156$$

$$\frac{F}{P} \times \frac{c}{r} \times \frac{L}{r} = \frac{12}{90} \times \frac{5}{2.57} \times \frac{240}{2.57} \frac{kip}{kip} \cdot \frac{in.}{in.} \cdot \frac{in.}{in.} = 24.2$$

$$\frac{\sigma_{max}}{\sigma_{ave}} = 1 + 24.2 \frac{\tan(\pi/2\sqrt{0.156})}{4(\pi/2\sqrt{0.156})} \qquad \text{from (9-26)}$$

$$= 7.97$$

$$\sigma_{ave} = \frac{P}{A} = \frac{90}{17.6} = 5.1 \frac{kip}{in.^2}$$

$$\sigma_{max} = 7.97 \times \sigma_{ave} = 7.97 \times 5.1 = \mathbf{40.6} \; \frac{\mathbf{kip}}{\mathbf{in.}^2}$$

The analysis in this example will be invalid if the yield stress of the material is less than 40.6 ksi, since it was assumed that stress and strain are linearly related. The analysis will be correct, although the structure may not have an adequate safety factor if the column is made of high-strength structural steel.

Problems 9-31 to 9-34

9-5 COMPLEX BUCKLING MODES

The aspects of buckling that have been treated up to this point are the principal effects that influence the design of compression members. In all cases a simple differential equation was established and its solution was used to determine a required relationship. All solutions were based on the assumption that the mode of failure would be a general or overall buckling by bending in a certain plane and for that type of failure the formulas are applicable. There are, however, other modes of failure that are not so obvious and if ignored can lead to failure that is no less disastrous.

The built-up member shown in Fig. 9-20 might be adequately designed for general buckling but might fail by local buckling of an angle because it is supported by the *lattice bracing* at points that are too far apart.

Lattice bracing has, in recent years, been replaced by *cover plates* such as in Fig. 9-21. It is conceivable that the weld could be intermittent and so widely spaced that the angles would be in danger of buckling between welds in the same manner as the lattice-braced member. However, a continuous weld is more likely to be specified and local buckling in angles would no longer be possible. If the cover plates are thin, they may buckle locally by rippling as shown in Fig. 9-21. The rippled plate is no longer capable of carrying its full share of the load, and additional load would be transferred to the angles that might then be overloaded.

We know that the W-shape is well proportioned to give a high moment of inertia about both principal axes and is consequently an efficient member for a column. We also know from our studies of torsion that open cross sections are not well shaped for carrying torsion and are relatively flexible in torsion. Because of its torsional flexibility it is possible for a W-shape under compression to fail by buckling in the torsional mode shown in Fig. 9-22. Even with the ends constrained against torsional rotation, the central region of the column may rotate, as shown, and lead to collapse. If cover plates are provided to run from flange to flange on both sides of the W-shape, the cross section becomes a closed section and the torsional rigidity and torsional buckling strength are greatly increased.

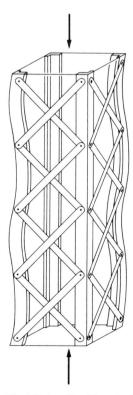

FIG. 9-20 Local buckling of column angle sections due to lattice bracing spaced too far apart.

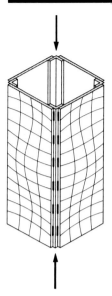

FIG. 9-21 Local buckling of cover plates and angles if welds are spaced too far apart.

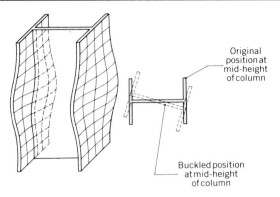

FIG. 9-22 Buckling by torsional rotation of an open section such as a W-shape.

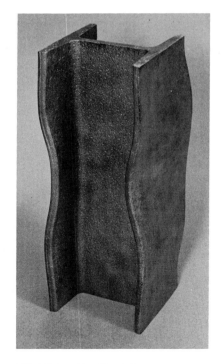

FIG. 9-23 Local buckling in flange.

The flange of a W-shape can fail by local buckling, usually referred to as *crippling*, in the same manner as the compression flange of a beam in bending. This mode is shown in Fig. 9-23.

In view of the difficulties encountered in the solution to the general buckling problem, you can appreciate that the more complex buckling modes are not amenable to analysis. Experimental evidence and experience gained through actual structures have provided the engineering profession with an adequate set of constraints on the proportions of standard columns to guard against failure in these complex modes. However, when a design requires an entirely new cross-sectional shape, all modes of buckling should be explored and, if doubt exists, a model of the part should be tested to prove its adequacy.

Problems 9-35 to 9-36

9-6 COLUMN DESIGN

The impression may have been created that a column can be designed in isolation with either pivoted ends, ends constrained to some degree, eccentric load, or lateral load. In actual structures, columns usually act in conjunction with a number of beams and other columns. To determine the loads on a column in a structure, the interaction of all the beams and columns must be taken into account. This is done by a structural analysis, which gives the axial force and end moments that act on any particular column. At each end the moment and axial force can be combined to give a resultant eccentric compressive load, a load state for which we have a theoretical analysis. Since the system is frequently statically indeterminate, a column section must be chosen before the analysis can be performed. After the analysis has been carried out, all structural elements are checked to see if they meet minimum requirements. If any element is not adequate and consequently is changed, the analysis must be repeated and all elements rechecked because a change in one element will redistribute the loads.

To determine whether a given column is acceptable or not requires that general and local buckling, as well as stress, be taken into consideration. Because this analysis is extremely complex in most cases, it is usually done by making certain that conditions expressed by empirical formulas are satisfied. For buildings, in the interest of safety, local authorities have established codes that state the condition that must be satisfied. The design of columns is one of many areas in engineering where designers require a long period of practice under experienced supervision before they are competent enough to undertake independent design.

PROBLEMS

9-1 to 9-5 Determine the load P that will cause the column to buckle.

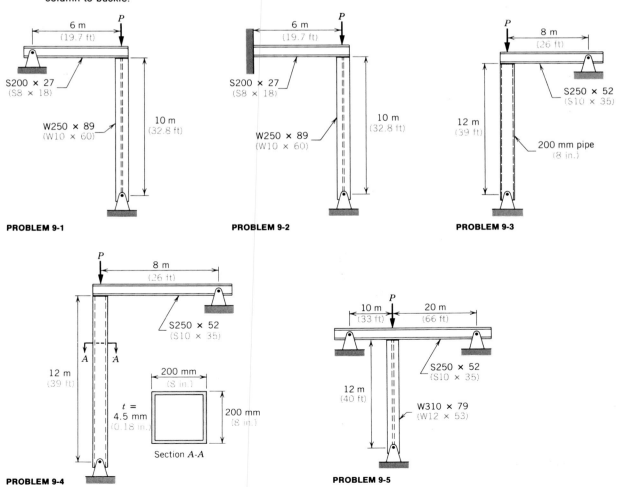

PROBLEM 9-1

PROBLEM 9-2

PROBLEM 9-3

PROBLEM 9-4

PROBLEM 9-5

9-6 Determine the safe load P if the structure is to have a safety factor of 2 on buckling.

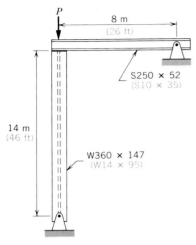

PROBLEM 9-6

9-7 Determine the safe load in Prob. 9-6 if the right end of the beam is built-in.

9-8 Determine the safe load P if the structure is to have a safety factor of 2 on buckling.

a. If the column is securely welded to the beam.
b. If the column is pin-connected to the beam.

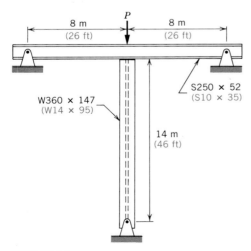

PROBLEM 9-8

9-9 Determine the buckling load if the load is applied:

a. At A.
b. At B.

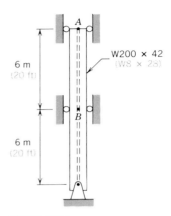

PROBLEM 9-9

9-10 Determine the load P that will cause the column to buckle.

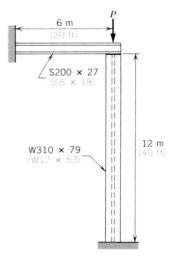

PROBLEM 9-10

9-11 Determine the safe load in Prob. 9-6 if the foot of the column is built-in and the column length is 20 m (65 ft).

9-12 Determine the load *P* required to buckle the column.

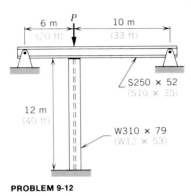

PROBLEM 9-12

9-13 Determine the load *P* required to buckle the column.

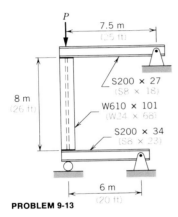

PROBLEM 9-13

9-14 Determine the buckling load in Prob. 9-13 when the ends of the beams are built-in rather than pinned.

9-15 Determine the load *P* required to buckle the column.

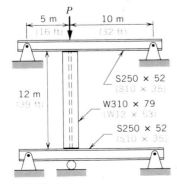

PROBLEM 9-15

9-16 The given column is constrained in rotation at the ends by welded-on S-shapes that extend 4 m (13 ft) out of the plane of the figure and have built-in ends. Determine the buckling load of the column.

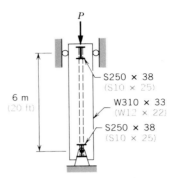

PROBLEM 9-16

9-17 Determine the buckling load of the column in Prob. 9-16 when the S-shapes are replaced by 125-mm (5-in.) steel pipes.

9-18 Calculate the maximum stress in the eccentrically loaded column.

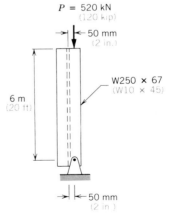

PROBLEM 9-18

9-19 Determine the maximum lateral deflection when the load is applied to the column in Prob. 9-18.

9-20 Determine the safe load on the column in Prob. 9-18 if the allowable stress is 140 MPa (20 ksi).

9-21 Calculate the maximum stress.

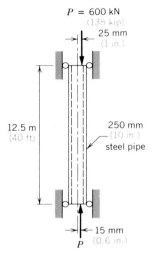

$P = 600$ kN
(135 kip)

25 mm
(1 in.)

12.5 m
(40 ft)

250 mm
(10 in.)
steel pipe

15 mm
(0.6 in.)
P

PROBLEM 9-21

9-22 Calculate the maximum stress.

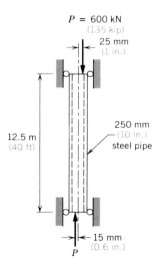

$P = 600$ kN
(135 kip)

25 mm
(1 in.)

12.5 m
(40 ft)

250 mm
(10 in.)
steel pipe

15 mm
(0.6 in.)
P

PROBLEM 9-22

9-23 If the maximum allowable compressive stress is 140 MPa (20 ksi), calculate the safe P on the column given in Prob. 9-21.

9-24 An axial load of 1600 kN (360 kip) is carried by a 12 m (40 ft) long W360 × 196 (W14 × 136). At one end the load may be off the centroid by as much as 80 mm

(3.2 in.) while at the other end it may be off by 60 mm (2.4 in.). Be as pessimistic as possible with respect to the location of the loads and calculate the maximum stress in the column.

9-25 The column is subjected to an axial load and to the end moments as shown. Calculate the maximum stress in the column.

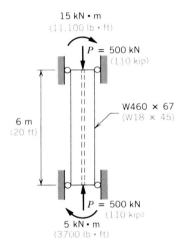

15 kN • m
(11,100 lb • ft)

$P = 500$ kN
(110 kip)

6 m
(20 ft)

W460 × 67
(W18 × 45)

$P = 500$ kN
(110 kip)

5 kN • m
(3700 lb • ft)

PROBLEM 9-25

9-26 A horizontal member of a pin-jointed structure is 25 m (80 ft) long and is made of four L200 × 200 × 20 (L8 × 8 × $\frac{3}{4}$) welded together to form a box section. The member carries a distributed, downward load of 1.0 kN/m (70 lb/ft) in addition to its own weight and an axial compressive load of 1250 kN (280 kip).

Calculate the maximum stress in the member due to

a. The lateral load only.
b. The axial load only.
c. Both loads acting simultaneously.

9-27 If the stress in the member described in Prob. 9-26 is limited to 140 MPa (20 ksi) and the axial load is as given in the problem, determine the allowable lateral load.

9-28 If the stress in the member described in Prob. 9-26 is limited to 140 MPa (20 ksi) and the lateral loads are as given in the problem, determine the allowable axial load.

9-29 A horizontal member of a pin-jointed truss is 4.0 m (13 ft) long. The member consists of a C200 × 21 (C8 × 14) turned so that the web is horizontal and the flanges are upward. It is subjected to an axial compressive load of 40 kN (9 kip) and must carry its own weight. Determine the maximum stress.

9-30 A compression member in a truss consists of two L150 × 150 × 20 (L6 × 6 × $\frac{3}{4}$) placed back to back and welded together as shown. The material is high-strength structural steel.

   a. Determine the maximum stress when P is 650 kN (145 kip) and w is 4 kN/m (275 lb/ft).
   b. For P of 650 kN (145 kip), determine the upper limit on w if the maximum stress is not to exceed the allowable as given in Table A.
   c. For w of 4 kN/m (275 lb/ft), determine the upper limit on P.

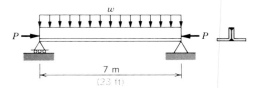

PROBLEM 9-30

9-31 A truss member consists of two L200 × 200 × 25 (L8 × 8 × 1) welded together as shown. The material is medium-strength structural steel. Neglect the force of gravity.

   a. Determine the maximum stress when P is 1000 kN (225 kip) and W is 30 kN (6.7 kip).
   b. If P is 1000 kN (225 kip), determine the safe limit on W.
   c. If W is 30 kN (6.7 kip), determine the safe limit on P.

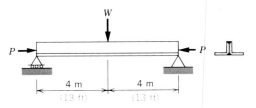

PROBLEM 9-31

9-32 A wind load on a building causes the columns to be subjected to a lateral loading in addition to the vertical loading. If the column is subjected to bending about its axis of minimum radius of gyration, determine the maximum stress in the column. The column is constructed from a W310 × 79 (W12 × 53), is 3 m (10 ft) long, and is subjected to an axial load of 500 kN (112 kip) and a lateral load, applied at the upper end of the column, of 20 kN (4.5 kip). No end rotation of the column is permitted.

9-33 If the allowable stress is 140 MPa (20 ksi), select the lightest, safe W-shape for the column described in Example 9-6.

9-34 A utility pole consists of a vertical 150-mm (6-in.) pipe that is securely attached to a foundation at ground level. The pole supports a concentrically placed 5300-kg (12000-lb) transformer at a point 7.5 m (25 ft) above ground level. If a horizontal force of 800 N (180 lb) acts on the transformer, calculate the maximum stress in the pole.

9-35 For the given steel column, find the Euler buckling load considering

   a. General buckling.
   b. Local buckling.

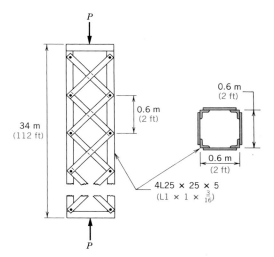

PROBLEM 9-35

9-36 For the given steel column find the Euler buckling load considering

    a. General buckling.
    b. Local buckling.

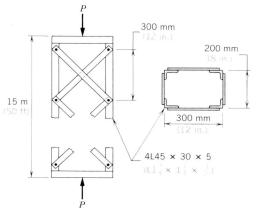

PROBLEM 9-36

Challenging Problems

9-A For a safety factor of 2 on collapse, calculate the safe load

    a. If the structure is as given.
    b. If cross-bracing is provided so that there is no sidesway.

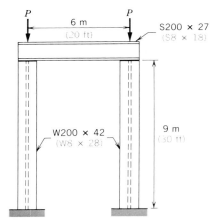

PROBLEM 9-A

9-B A standard 200-mm (8-in.) steel pipe is used as a utility pole to support two transformers. Assume that there are no wind loads or loads from attached wires. Calculate the maximum stress when:

    a. The utility pole is constructed as shown.
    b. When only the lighter transformer is in place.

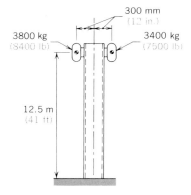

PROBLEM 9-B

9-C A rectangular high-strength concrete beam 10 m (32 ft) long, 600 mm (24 in.) deep, and 250 mm (10 in.) wide has been cast on the ground and is to be lifted into place. For each of the lifting methods indicated, determine the stress at midspan in the upper fibers and in the lower fibers.

    a. The beam is lifted by two cranes, one at each end.
    b. The beam is lifted by a single crane and cable as shown in the figure.
    c. Discuss practical methods by which you could overcome the tensile stress problems encountered in your solutions to parts (a) and (b).

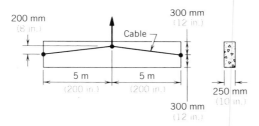

PROBLEM 9-C(b)

9-D A 20-mm (0.8-in.) horizontal rod moves a certain flight-control surface in an aircraft. The rod is 1200 mm (48 in.) long and simply supported at the ends, and when in use carries an axial compressive load of 10 kN (2250 lb). The material is AISI 4340. Determine the maximum stress in the rod:

a. Due to the axial load only.

b. Due to the weight of the rod only.

c. Due to both loads acting simultaneously.

d. Due to inertia and the axial load acting simultaneously when the aircraft is in a maneuver that subjects it to an upward acceleration of 7 g.

e. If the safety factor is 1.2 based on yield, is the aircraft safe?

9-E Would the aircraft in Prob. 9-D be safe if the rod was replaced by a tube having an outside diameter of 25 mm (1 in.) and an inside diameter of 20 mm (0.8 in.). Note that the tube contains much less material than the rod.

9-F Calculate the sidesway (see Fig. 9-17) in the building described in Prob. 9-32.

9-G Calculate the horizontal motion at the top of the pole in Prob. 9-B due to the horizontal force:

a. Before the second transformer is in place.

b. When both transformers are in place.

10

ENERGY METHODS

10-1   INTRODUCTION

Several types of problems that we are able to solve with great difficulty could be more readily treated by energy methods. Also, other types that cannot be solved by present methods lend themselves to solution by energy methods. We will now make use of some well-known principles of physics, such as conservation of energy and minimum potential energy, in order to develop new concepts in the form of theorems that are useful in solving certain types of engineering problems.

Statically indeterminate problems such as those found in Section 1-13 were solved by taking displacement into account. We will shortly find that most problems involving displacement are best approached by energy methods whether the displacement is an end in itself or whether it is the means by which stresses in a statically indeterminate structure are determined.

10-2   STRAIN ENERGY IN A STRETCHED SPRING

When a load is applied to a spring, the spring deforms or strains and a certain amount of work is done on the spring. In its strained state the spring contains an amount of strain energy which can be calculated by formulas that we will now derive.

Consider a spring that has a free length of L_0 as shown in Fig. 10-1a. The stiffness of the spring is given by the quantity k, which is related to the force necessary to stretch the spring an amount, x, by the linear formula

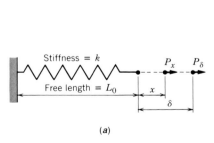

FIG. 10-1 Strain energy stored in a stretched spring.

$$P_x = kx \qquad \textbf{(10-1a)}$$

Hence, for a stretch of δ, the force required is

$$P_\delta = k\delta \qquad \textbf{(10-1b)}$$

Let us say that our objective is to determine the energy that is stored in the spring when the spring has been stretched an amount δ. In other words, we are about to determine the strain energy in a spring that has one end displaced an amount δ and, as a consequence, has an axial force P_δ in the spring.

Starting with the undeformed, and hence unloaded, spring we will gradually displace one end of the spring and apply the required force. At a typical displacement, x, where the force is P_x, an increment of displacement, dx, is applied. During the incremental displacement the force remains substantially constant and does an amount of work

$$dU = P_x\, dx$$

It is interesting to note that the cross-hatched area in Fig. 10-1b is a measure of the work.

Substituting from (10-1a) we obtain

$$dU = kx\, dx \tag{10-2}$$

Since no energy is lost, the work done goes into strain energy stored in the deformed spring.

To obtain the *total strain energy*, U, stored in the spring when it has been stretched by an amount, δ, we need to integrate (10-2). Thus

$$U = \int_0^\delta kx\, dx = k\int_0^\delta x\, dx = \frac{k}{2}x^2 \bigg]_0^\delta$$

$$U = \frac{1}{2}k\delta^2 \tag{10-3}$$

This can be written as

$$U = \tfrac{1}{2}(k\delta)\delta$$

and substituting from (10-1b) gives

$$U = \tfrac{1}{2}P_\delta\delta$$

Since the force varies linearly from zero to P_δ, the term $\tfrac{1}{2}P_\delta$ can be taken as the *average* force and the strain energy can be expressed as

$$U = P_{\text{ave}}\delta \tag{10-4}$$

Taking this process one step further, (10-1b) can be used to eliminate δ from (10-4), giving

$$U = P_{\text{ave}}\frac{P_\delta}{k} = \frac{P_\delta}{2}\frac{P_\delta}{k}$$

$$\boxed{U = \frac{1}{2k}P_\delta^2} \tag{10-5}$$

10-3 STRAIN ENERGY DENSITY

When a load is applied to any elastic member, there is a stress imposed on the material and the dimensions of the member change. Energy is stored in the fibers of the member by virtue of the strain that occurs and the associated stress. It is not possible to develop an expression for strain energy when the stresses are perfectly general. However, it will be sufficient for most purposes if we determine the energy for a number of very simple stress states.

Consider an element cut from a stressed member having a *uniaxial* stress σ_x. The element, when unstressed, has dimensions dx, dy, and dz as shown in Fig. 10-2.

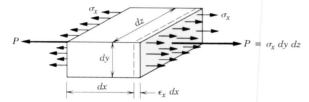

FIG. 10-2 Strain energy in an element stressed uniaxially.

When the load is gradually applied, the stress rises to σ_x and the total force on the element rises *gradually* to $\sigma_x \, dy \, dz$ while the element dimension dx changes by $\varepsilon_x \, dx$. The force in the x-direction then moves through the distance $\varepsilon_x \, dx$ and does work equal to the average force multiplied by the distance through which it acts [see (10-4)]. The work done, or strain energy, is then

$$U = (\tfrac{1}{2}\sigma_x \, dy \, dz)(\varepsilon_x \, dx)$$

$$U = \tfrac{1}{2}\sigma_x \varepsilon_x \, dx \, dy \, dz \qquad \textbf{(10-6)}$$

It is often convenient to deal with the energy per unit volume, rather than the total strain energy, and this is called the *strain energy density*, U_0. By definition

$$U_0 = \frac{U}{\text{vol}} \qquad \textbf{(10-7)}$$

In the present case

$$U_0 = \frac{U}{dx \, dy \, dz} = \frac{\tfrac{1}{2}\sigma_x \varepsilon_x \, dx \, dy \, dz}{dx \, dy \, dz}$$

$$U_0 = \tfrac{1}{2}\sigma_x \varepsilon_x \qquad \textbf{(10-8a)}$$

As long as the proportional limit has not been exceeded, $\sigma_x = E\varepsilon_x$. Substituting into (10-8a) gives

$$\boxed{U_0 = \frac{E}{2} \varepsilon_x^2} \qquad \textbf{(10-8b)}$$

The stress-strain relation can be expressed as $\varepsilon_x = (1/E)\sigma_x$ and substituted into (10-8a) giving

$$U_0 = \frac{1}{2E}\sigma_x^2 \qquad \text{(10-8c)}$$

These are equations that have an immediate usefulness, but before applying them, let us find the strain energy density when the stress is shear only.

Consider an element cut from a material that is subjected to shear stress only. The element, when unstressed, has dimensions dx, dy, and dz, as shown in Fig. 10-3, and all angles are right angles.

The loads on the stressed surfaces and the distortions are shown in Fig. 10-4.

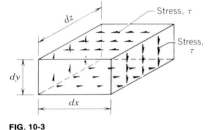

FIG. 10-3

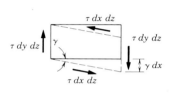

FIG. 10-4

The only force to move through a distance, and hence to do work, is that which acts on the surface at the right. That force increases, as the material is gradually stressed, to $\tau\,dy\,dz$ and moves through a distance $\gamma\,dx$. The work done, and hence the strain energy, is then

$$U = \tfrac{1}{2}\tau\,dy\,dz\,\gamma\,dx$$

$$U = \tfrac{1}{2}\tau\,\gamma\,dx\,dy\,dz \qquad \text{(10-9)}$$

and the strain energy density is

$$U_0 = \tfrac{1}{2}\tau\,\gamma \qquad \text{(10-10a)}$$

As long as the proportional limit has not been exceeded, $\tau = G\gamma$, which, when substituted into (10-10a), gives

$$U_0 = \frac{G}{2}\gamma^2 \qquad \text{(10-10b)}$$

The stress-strain relationship can also be expressed as $\gamma = (1/G)\tau$ and substituted into (10-10a), giving

$$U_0 = \frac{1}{2G}\tau^2 \qquad \text{(10-10c)}$$

It is useful to note the similarity between (10-8) and (10-10) for uniaxial and shear loading, respectively; E compares with G, σ_x with τ, and ε_x with γ.

10-4 STRAIN ENERGY IN PRISMATIC MEMBERS

To calculate the strain energy in a member, the strain energy density is determined in a typical elemental volume. This is substituted into (10-7), rearranged into the form

$$U = U_0 \text{ vol} \tag{10-11}$$

When the stress is uniform over the member, the strain energy density is uniform and merely needs to be multiplied by the volume of the member to get the total strain energy. In cases where the stress varies from point to point, the strain energy density also varies. This requires that an incremental volume be chosen and an integration be performed to determine the total strain energy.

Consider a member, as in Fig. 10-5, subjected to an axial load. The strain energy density from (10-8c) is

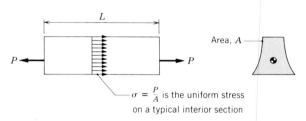

$\sigma = \dfrac{P}{A}$ is the uniform stress
on a typical interior section

FIG. 10-5 Consideration of strain energy in a prismatic member loaded axially.

$$U_0 = \frac{1}{2E}\sigma_x^2 = \frac{1}{2E}\left(\frac{P}{A}\right)^2$$

Since the stress is *uniform* in all particles of the member, (10-11) gives

$$U = \frac{1}{2E}\left(\frac{P}{A}\right)^2 LA$$

or

$$U = \frac{1}{2}\frac{P^2 L}{EA} \tag{10-12}$$

Instead of a load being imposed on the member, a displacement may be applied changing the length by ΔL as shown in Fig. 10-6. We will disregard the axial force associated with the change in length and direct our attention to the strains. The strain in all fibers is $\varepsilon_x = \Delta L/L$ and by (10-8b)

$$U_0 = \frac{E}{2}\varepsilon_x^2 = \frac{E}{2}\left(\frac{\Delta L}{L}\right)^2$$

FIG. 10-6 Displacement ΔL applied to a prismatic member.

The strain, and hence the strain energy density, is constant over the whole member. Equation (10-11) can then be used to give

$$U = \frac{E}{2}\left(\frac{\Delta L}{L}\right)^2 LA$$

$$U = \frac{1}{2}\frac{EA\,\Delta L^2}{L} \tag{10-13}$$

Energy is also stored in a member when it is subjected to *bending* as in Fig. 10-7.

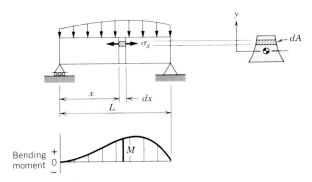

FIG. 10-7 Consideration of strain energy due to bending.

Choosing a typical elemental volume, the stress, from (2-11) gives

$$\sigma_x = -\frac{My}{I}$$

and the strain energy density, from (10-8c), is

$$U_0 = \frac{1}{2E}\left(-\frac{My}{I}\right)^2$$

By (10-11) the strain energy in the element is

$$dU = U_0\,d\,\mathrm{vol}$$

$$= \frac{1}{2E}\left(\frac{My}{I}\right)^2 dA\,dx = \frac{1}{2E}\frac{M^2 y^2}{I^2}\,dA\,dx$$

In the whole member this total strain energy is

$$U = \frac{1}{2E} \int\limits_{length} \frac{M^2}{I^2} \int\limits_{area} y^2 \, dA \, dx$$

The integral over the area is readily recognized as the moment of inertia I which cancels one of the I's already in the equation, giving

$$U = \frac{1}{2E} \int\limits_{length} \frac{M^2}{I} \, dx$$

For a *prismatic* member the moment of inertia is constant and the equation can be simplified to

$$U = \frac{1}{2EI} \int\limits_{length} M^2 \, dx \qquad \text{(10-14)}$$

This strain energy relation due to *bending* is one of the most frequently used energy equations. For the case of constant bending moment, which rarely occurs in real problems, the equation becomes

$$U = \frac{1}{2} \frac{M^2 L}{EI} \qquad \text{(10-15)}$$

When a member is subjected to a bending moment, it is also, except in the rare case of constant moment, subjected to shear forces that induce shear stress and strain. Consequently, for beams the *shear strain energy* must also be taken into account if the total strain energy is required.

Taking a typical element from the beam subjected to shear, as in Fig. 10-8, we obtain the strain energy density from (10-10c):

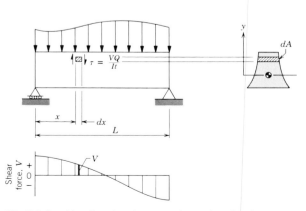

FIG. 10-8 Consideration of strain energy due to shear in a beam.

$$U_0 = \frac{1}{2G}\left(\frac{VQ}{It}\right)^2$$

The shear strain energy in the whole member is

$$U = \frac{1}{2G} \int_{\text{length}} \int_{\text{area}} \left(\frac{VQ}{It}\right)^2 dA\,dx \qquad \text{(10-16)}$$

This equation does not lend itself to further development in the general case but can be used in its present form for particular cases as they arise. The equation is not difficult to use for beams having a rectangular cross section. However, for more complex sections, such as standard structural shapes, determining the strain energy is a lengthy process.

Shear strain energy is also significant in *torsional* problems. However, for all but the simplest of shapes the shear stress is not known and consequently we cannot calculate the shear strain energy. The case of a shaft having a circular cross section is often encountered in practice and will be the only torsional case for which we will develop an equation.

Consider a cylinder subjected to a torsional load as in Fig. 10-9.

FIG. 10-9 Consideration of shear strain energy in a shaft subjected to torsion.

On the annular face of the ring element the shear stress can be obtained by substituting τ_{max} from (7-9) into (7-6) to get

$$\tau = \frac{Tr}{J}$$

The strain energy density from (10-10c) is

$$U_0 = \frac{1}{2G}\left(\frac{Tr}{J}\right)^2$$

and the strain energy in the ring element from (10-11) is

$$dU = \frac{1}{2G}\left(\frac{Tr}{J}\right)^2 dA\,dx$$

The total shear strain energy then becomes

$$U = \frac{1}{2G}\frac{T^2}{J^2} \int_{\text{length}} \int_{\text{area}} r^2\,dA\,dx$$

The area integral is recognized as the polar moment of inertia, J, which cancels one of the other J's and gives

$$U = \frac{1}{2} \frac{T^2 L}{GJ}$$

(10-17)

It is interesting to note the similarity between the strain energy equations for the three cases of uniformly stressed prismatic bars. Those formulas, appearing in (10-12), (10-15), and (10-17), are repeated for ease of comparison.

$$U = \frac{1}{2} \frac{P^2 L}{EA}$$

(10-12)

$$U = \frac{1}{2} \frac{M^2 L}{EI}$$

(10-15)

$$U = \frac{1}{2} \frac{T^2 L}{GJ}$$

(10-17)

All formulas have $\frac{1}{2}$ and L as common factors. In each case the "load," whether it be force, moment, or torque, enters as a squared term. We find in the denominator the elastic constant that indicates the stiffness of the material with respect to the type of stress. Also, in the denominator there is the property of the area that appears in the stress formula for the type of load being applied.

Problems 10-1 to 10-12

10-5 THE DIRECT ENERGY METHOD

When forces are gradually applied to a conservative system, no work is lost to friction or plastic deformation and there is no significant amount of kinetic energy. Consequently, the work done by the forces that are applied to the system is, by the Law of Conservation of Energy, equal to the strain energy in the system. This principle is true regardless of the number of forces that act on the system, but we will find the principle most useful when only *one* force is applied and it is required to find the displacement *in the direction* of the force at its point of application. We will call this the *direct energy method*. When the load is applied gradually, it increases from zero to its final value, P, while the point of application displaces linearly to a final value, δ. The work done by the load is then

$$\tfrac{1}{2} P \delta$$

which can be equated to the strain energy, giving

$$\tfrac{1}{2} P \delta = U$$

Since the displacement is usually the unknown quantity, the equation is more useful in the form

$$\delta = \frac{2U}{P}$$

(10-18)

The following example illustrates the application of the direct energy method to a familiar beam deflection problem.

Example 10-1

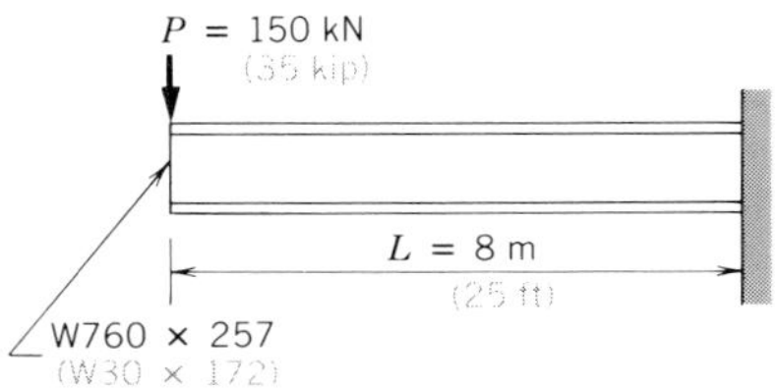

Determine the deflection at the end of the cantilever beam given in Fig. 10-10a.

Solution. Moment equilibrium applied to the free body of Fig. 10-10b shows that $M = -Px$. From (10-14) the *bending* strain energy in the member is

$$U = \frac{1}{2EI} \int_0^L (M)^2 \, dx = \frac{1}{2EI} \int_0^L (-Px)^2 \, dx$$

$$= \frac{1}{2EI} P^2 \int_0^L x^2 \, dx$$

$$= \frac{P^2}{6EI} x^3 \Big]_0^L = \frac{P^2 L^3}{6EI}$$

We will treat this as the total strain energy, although there is in fact shear stress and, hence, shear strain energy that has not been taken into account. Substituting into (10-18) gives

$$\delta = \frac{2U}{P} = \frac{2}{P} \frac{P^2 L^3}{6EI} = \frac{1}{3} \frac{PL^3}{EI}$$

for the given beam,

$$I = 3420 \times 10^6 \text{ mm}^4 \qquad \text{(Table B-1)}$$
$$E = 200\,000 \, \frac{N}{mm^2} \qquad \text{(Table A)}$$

Substituting these and other known quantities gives

$$\delta = \frac{1}{3} \frac{150\,000 \times 8000^3}{200\,000 \times 3420 \times 10^6} \frac{N \cdot mm^3}{N/mm^2 \cdot mm^4}$$

$$= 37.4 \text{ mm}$$

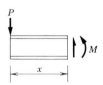

$P = 150$ kN
(35 kip)

$L = 8$ m
(25 ft)

W760 × 257
(W30 × 172)

FIG. 10-10a

P

M

x

FIG. 10-10b

Example 10-1

Determine the deflection at the end of the cantilever beam given in Fig. 10-10a.

Solution. Moment equilibrium applied to the free body of Fig. 10-10b shows that $M = -Px$. From (10-14) the *bending* strain energy in the member is

$$U = \frac{1}{2EI} \int_0^L (M)^2 \, dx = \frac{1}{2EI} \int_0^L (-Px)^2 \, dx$$

$$= \frac{1}{2EI} P^2 \int_0^L x^2 \, dx$$

$$= \frac{P^2}{6EI} x^3 \Big]_0^L = \frac{P^2 L^3}{6EI}$$

We will treat this as the total strain energy, although there is in fact shear stress and, hence, shear strain energy that has not been taken into account. Substituting into (10-18) gives

$$\delta = \frac{2U}{P} = \frac{2}{P} \frac{P^2 L^3}{6EI} = \frac{1}{3} \frac{PL^3}{EI}$$

For the given beam,

I = 8200 in.^{4} (Table B-1)
E = 29,000 ksi   (Table A)

Substituting these and other known quantities gives

$$\delta = \frac{1}{3} \frac{35 \times (25 \times 12)^3}{29,000 \times 8200} \frac{\text{kip} \cdot \text{in.}^3}{\text{kip/in.}^2 \cdot \text{in.}^4}$$

$$= \textbf{1.33 in.}$$

The deflection formula in this example is the same as that which would be obtained by the application of the integration methods given in Chapter 3. It will be noted that the energy method is considerably shorter; however, it gives the displacement at *one* point only, whereas the integration method gives the deflection at *all* points along the beam in return for the additional effort.

It may seem strange that the two methods should be in exact agreement when shear strain energy has been omitted from the energy solution. The agreement results from the shear deformation being left out of *both* methods, which introduces exactly the same error into each solution. It would be more correct to say that we have determined the *bending* deflection. This would imply that deflection due to shear has been omitted. Later we will see that for beams having the proportions usually found in real structures, the error due to neglecting shear is quite insignificant.

When the structure consists of several parts, it becomes necessary to calculate the energy in each part separately and to sum the energy in all parts. In order to keep the required effort from becoming excessive, and to facilitate checking, the calculations are usually put into tabular form. When solving a truss structure, the total energy is the sum of the strain energy due to axial forces in all members. If the forces in the members are F_1, F_2, F_3, ..., F_M, the total strain energy by (10-12) is

$$U = \sum_{i=1}^{M} \frac{1}{2} \frac{F_i^2 L_i}{E_i A_i} \qquad \textbf{(10-19)}$$

where M is the number of members.

When the same material is used throughout the structure, Young's modulus is the same for all members and the constants can be placed in front of the summation symbol giving

$$U = \frac{1}{2E} \sum_{i=1}^{M} \frac{F_i^2 L_i}{A_i} \qquad \textbf{(10-20)}$$

The following example illustrates a convenient tabular format and shows that the direct energy method can be used to solve a problem that is unmanageable by any other method now known to us.

Example 10-2

Calculate the vertical deflection at the end of the steel cantilever truss given in Fig. 10-11a due to the load, P.

Solution. To identify the members, we will arbitrarily assign numbers as shown in Fig. 10-11b. The forces in the members can be readily determined by the methods of Chapter 1. The forces in kN are indicated in Fig. 10-11c.

The above values can now be entered into a table and the energy calculated by (10-20).

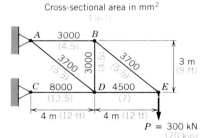

Cross-sectional area in mm² (in.²)

FIG. 10-11a

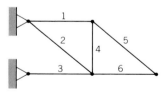

FIG. 10-11b

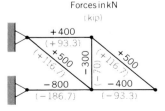

Forces in kN (kip)

FIG. 10-11c

i	F_i (N)	L_i (mm)	A_i (mm²)	$F_i^2 L_i / A_i$ (N²/mm)
1	$+400 \times 10^3$	4×10^3	3.0×10^3	213×10^9
2	$+500$	5	3.7	338
3	-800	4	8.0	320
4	-300	3	3.0	90
5	$+500$	5	3.7	338
6	-400	4	4.5	142

$$\sum \frac{F_i^2 L_i}{A_i} = 1441 \times 10^9 \frac{N^2}{mm}$$

$$U = \frac{1}{2E} \sum \frac{F_i^2 L_i}{A_i} \qquad \text{(10-20)}$$

$$= \frac{1441 \times 10^9 \ N^2/mm}{2 \times 200\,000 \ N/mm^2} = 3.60 \times 10^6 \ N \cdot mm$$

From (10-18)

$$\delta = \frac{2U}{P} = \frac{2 \times 3.60 \times 10^6}{300\ 000} \frac{N \cdot mm}{N}$$

$$= \textbf{24.0 mm}$$

Example 10-2

Calculate the vertical deflection at the end of the steel cantilever truss given in Fig. 10-11a due to the load, P.

Solution.   To identify the members, we will arbitrarily assign numbers as shown in Fig. 10-11b. The forces in the members can be readily determined by the methods of Chapter 1. The forces in kip are indicated in Fig. 10-11c.

The above values can now be entered into a table and the energy calculated by (10-20).

i	F_i (kip)	L_i (in.)	A_i (in.^{2})	$F_i^2 L_i / A_i$ (kip^2/in.)
1	+ 93.3	144	4.5	278.6×10^3
2	+ 116.7	180	5.5	445.7
3	− 186.7	144	12.5	401.5
4	− 70.0	108	4.5	117.6
5	+ 116.7	180	5.5	445.7
6	− 93.3	144	7.0	179.1

$$\sum \frac{F_i^2 L_i}{A_i} = 1868.2 \times 10^3 \text{ kip}^2/\text{in.}$$

$$U = \frac{1}{2E} \sum \frac{F_i^2 L_i}{A_i} \qquad (10\text{-}20)$$

$$= \frac{1868.2 \times 10^3}{2 \times 29 \times 10^3} \frac{\text{kip}^2/\text{in.}}{\text{kip/in.}^2} = 32.21 \text{ kip} \cdot \text{in.}$$

From (10-18)

$$\delta = \frac{2U}{P} = \frac{2 \times 32.21}{70} \frac{\text{kip} \cdot \text{in.}}{\text{kip}}$$

$$= \textbf{0.92 in.}$$

The next example will show that for the commonly used length-to-depth ratio, the deformation due to *shear strain* in a rectangular beam is negligible.

Example 10-3

For the beam shown in Fig. 10-12a, determine the deflection under the load when the shear strain energy is (a) neglected and (b) included; and (c) determine the length-to-depth ratio at which the error caused by omitting the shear strain is equal to 5%.

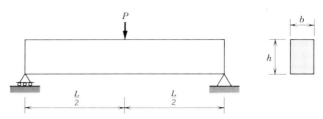

FIG. 10-12a

Solution. Figure 10-12b shows the beam together with the moment and shear equations.

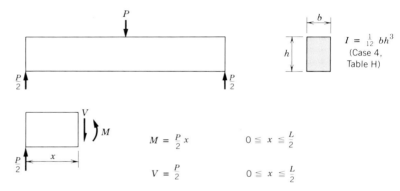

$$M = \frac{P}{2}x \qquad 0 \leq x \leq \frac{L}{2}$$

$$V = \frac{P}{2} \qquad 0 \leq x \leq \frac{L}{2}$$

FIG. 10-12b

a. The strain energy due to bending is given by

$$U_B = \frac{1}{2EI} \int M^2\, dx \qquad (10\text{-}14)$$

$$= \frac{1}{2EI} 2 \int_0^{L/2} \left(\frac{P}{2}x\right)^2 dx$$

$$= \frac{1}{EI}\frac{P^2}{4} \int_0^{L/2} x^2\, dx = \frac{P^2}{4EI}\frac{1}{3}x^3 \Big]_0^{L/2}$$

$$= \frac{P^2}{12EI}\left(\frac{L}{2}\right)^3 = \frac{P^2 L^3}{96EI}$$

The deflection under the load due to bending only is

$$\delta_B = \frac{2U_B}{P} \qquad (10\text{-}18)$$

$$\delta_B = \frac{2P^2L^3}{P96EI} = \frac{PL^3}{48EI} = \frac{\mathbf{PL^3}}{\mathbf{4Ebh^3}}$$

b.   To determine the shear strain energy we select a typical element as in Fig. 10-12c and integrate (10-16). Thus

$$U_S = \frac{1}{2G} 2 \int_0^{L/2} 2 \int_0^{h/2} \left(\frac{VQ}{It}\right)^2 b \, dy \, dx$$

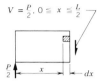

FIG. 10-12c

In this case

$$V = \frac{P}{2}$$

$$Q = b\left(\frac{h}{2} - y\right)\left[\frac{(h/2) + y}{2}\right] = \frac{b}{2}\left(\frac{h^2}{4} - y^2\right)$$

$$I = \frac{1}{12} bh^3$$

$$t = b$$

and

$$\frac{VQ}{It} = \frac{(P/2)(b/2)[(h^2/4) - y^2]}{\frac{1}{12}bh^3 b} = \frac{3P}{bh^3}\left(\frac{h^2}{4} - y^2\right)$$

Hence

$$U_S = \frac{2}{G} \int_0^{L/2} \int_0^{h/2} \frac{9P^2}{b^2 h^6}\left(\frac{h^2}{4} - y^2\right)^2 b \, dy \, dx$$

$$= \frac{18P^2}{Gbh^6} \int_0^{L/2} \int_0^{h/2} \left(\frac{h^4}{16} - \frac{h^2}{2}y^2 + y^4\right) dy \, dx$$

$$= \frac{18P^2}{Gbh^6} \int_0^{L/2} \left[\frac{h^4}{16}y - \frac{h^2}{6}y^3 + \frac{1}{5}y^5\right]_0^{h/2} dx$$

$$= \frac{18P^2}{Gbh^6}\left(\frac{h^5}{32} - \frac{h^5}{48} + \frac{h^5}{160}\right) \int_0^{L/2} dx$$

$$= \frac{18P^2}{Gbh^6} \frac{h^5}{60} \frac{L}{2}$$

$$= \frac{3}{20} \frac{P^2 L}{Gbh}$$

The total strain energy is

$$U = U_B + U_S$$

$$= \frac{1}{96} \frac{P^2 L^3}{EI} + \frac{3}{20} \frac{P^2 L}{Gbh}$$

$$= \frac{1}{8} \frac{P^2 L^3}{Ebh^3} + \frac{3}{20} \frac{P^2 L}{E/[2(1 + v)]bh}$$

$$= \frac{1}{8} \frac{P^2 L^3}{Ebh^3} + \frac{3(1 + v)}{10} \frac{P^2 L}{Ebh}$$

$$= \frac{1}{8} \frac{P^2 L^3}{Ebh^3} \left[1 + \frac{12}{5}(1 + v)\frac{h^2}{L^2} \right]$$

Deflection under the load due to bending and shear,

$$\delta = \delta_B + \delta_S = \frac{2U}{P} \qquad (10\text{-}18)$$

$$\delta = \frac{1}{4} \frac{PL^3}{Ebh^3} \left[1 + \frac{12}{5}(1 + v)\left(\frac{h}{L}\right)^2 \right]$$

We see for the first time a deflection that depends upon Poisson's ratio. Since that elastic constant is approximately 0.3 for all metals, we will substitute that value into the displacement equation to obtain

$$\delta = \frac{1}{4} \frac{PL^3}{Ebh^3} \left[1 + 3.120\left(\frac{h}{L}\right)^2 \right]$$

c. Since the first term in the bracket came from bending strain energy and the second from shear strain energy, the relative magnitudes of the terms gives a measure of the relative contributions to displacement. Thus, for shear to contribute 5% to deformation, the second term must be equal to 0.05 or

$$3.120\left(\frac{h}{L}\right)^2 = 0.05$$

Solving, we obtain

$$\frac{h}{L} = \sqrt{\frac{0.05}{3.120}} = 0.127 \approx \frac{1}{8}$$

or

$$\frac{L}{h} = 8$$

Based on these calculations, we can say that the error caused by neglecting the shear deformation in a beam having a rectangular cross section will be less than 5% **when the length of the beam is more than eight times the depth.**

The foregoing example can be used to draw some important conclusions. It shows that much more effort is required to determine shear strain energy than bending energy. The disparity would be much greater if the member did not have such a simple cross section or if the loading had been of a different form. In fact, the effort required to repeat the problem with a W-shape would be prohibitive if exact methods were used. It would be necessary to determine the shear strain energy in the flanges as well as in the web. A small error will be introduced if the shear strain energy in the flanges is not taken into account. Omitting the shear strain energy of flanges and taking the stress in the web as determined by (2-24c) will give reasonably good accuracy. The accuracy will still be acceptable in most cases if the shear stress in the web is assumed to be constant.

The conclusion, reached in the example, that less than 5% error (which in most engineering calculations may be treated as insignificant) will be caused by neglecting the shear strain when the length is more than eight times the depth, is applicable only to a centrally loaded beam with a rectangular cross section. In practice, the length-to-depth ratio is usually in the 12 to 20 range; and engineers usually ignore shear deformation and accept the resulting small error. However, when a beam is unusually short, relative to its depth, shear deformation is taken into account.

Example 10-4

A steel rod, 40 mm in diameter, is shaped as shown in Fig. 10-13a. Determine the downward deflection at the end when the load is applied.

Solution. The properties of the cross section are

diameter, $D = 40$ mm   (given)
$\qquad A = 1260$ mm^2 (Table D)
$\qquad I = 126\,000$ mm^4 (Table D)

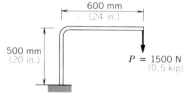

FIG. 10-13a

Let

$\qquad U_1$ = bending strain energy in horizontal member
$\qquad U_2$ = shear strain energy in horizontal member
$\qquad U_3$ = bending strain energy in vertical member
$\qquad U_4$ = axial strain energy in vertical member.

For the horizontal member, Fig. 10-13b shows the free-body diagram. From (10-14)

$$U_1 = \frac{1}{2EI} \int_0^{L_1} (-Px)^2 \, dx$$

$$= \frac{P^2}{2EI} \int_0^{L_1} x^2 \, dx = \frac{P^2}{2EI} \frac{x^3}{3} \Big]_0^{L_1}$$

$$= \frac{P^2 L_1^3}{6EI} = \frac{1500^2 \times 600^3}{6 \times 200\,000 \times 126\,000} \frac{\text{N}^2 \cdot \text{mm}^3}{\text{N/mm}^2 \cdot \text{mm}^4}$$

$$= 3214 \text{ N} \cdot \text{mm}$$

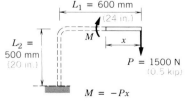

FIG. 10-13b

Since the length-to-depth ratio is $(\frac{600}{40}=)$ 15, we will assume that the shear strain energy is negligible, that is,

$$U_2 = 0$$

In the vertical member the bending moment is PL_1 and is constant; hence (10-15) applies. Substituting into (10-15) gives

$$U_3 = \frac{1}{2} \frac{(PL_1)^2 L_2}{EI}$$

$$= \frac{1}{2} \frac{(1500 \times 600)^2 500}{200\,000 \times 126\,000} \frac{N^2 \cdot mm^2 \cdot mm}{N/mm^2 \cdot mm^4}$$

$$= 8036 \; N \cdot mm$$

Substituting into (10-12) will give the axial strain energy in the vertical member

$$U_4 = \frac{1}{2} \frac{(-P)^2 L_2}{EA}$$

$$= \frac{(-1500)^2 \times 500}{2 \times 200\,000 \times 1260} \frac{N^2 \cdot mm}{N/mm^2 \cdot mm^2}$$

$$= 2.23 \; N \cdot mm$$

The total strain energy is

$$U = U_1 + U_2 + U_3 + U_4$$

$$= 3214 + 0 + 8036 + 2$$

$$= 11252 \; N \cdot mm$$

Substituting into (10-18) gives the vertical displacement at the end,

$$\delta = \frac{2U}{P} = \frac{2 \times 11\,252}{1500} \frac{N \cdot mm}{N}$$

$$= \textbf{15.0 mm}$$

Example 10-4

A steel rod, 1.5 in. in diameter, is shaped as shown in Fig. 10-13a. Determine the downward deflection at the end when the load is applied.

Solution. The properties of the cross section are

diameter, $D = 1.5$ in.   (given)
$A = 1.77$ in.^{2} (Table D)
$I = 0.248$ in.^{4} (Table D)

Let

U_1 = bending strain energy in horizontal member
U_2 = shear strain energy in horizontal member

$U_3 =$ bending strain energy in vertical member
$U_4 =$ axial strain energy in vertical member.

For the horizontal member, Fig. 10-13b shows the free-body diagram. From (10-15)

$$U_1 = \frac{1}{2EI} \int_0^{L_1} (-Px)^2 \, dx$$

$$= \frac{P^2}{2EI} \int_0^{L_1} x^2 \, dx = \frac{P^2}{2EI} \frac{x^3}{3} \Big]_0^{L_1}$$

$$= \frac{P^2 L_1^3}{6EI} = \frac{0.5^2 \times 24^3}{6 \times 29{,}000 \times 0.248} \frac{kip^2 \cdot in.^3}{kip/in.^2 \cdot in.^4}$$

$$= 0.08 \; kip \cdot in.$$

Since the length-to-depth ratio is ($\frac{24}{1.5} =$) 16, we will assume that the shear strain energy is negligible, that is,

$$U_2 = 0$$

In the vertical member the bending moment is PL_1 and is constant; hence (10-15) applies. Substituting into (10-15) gives

$$U_3 = \frac{1}{2} \frac{(PL_1)^2 L_2}{EI}$$

$$= \frac{1}{2} \frac{(0.5 \times 24)^2 20}{29{,}000 \times 0.248} \frac{kip^2 \cdot in.^2 \cdot in.}{kip/in.^2 \cdot in.^4}$$

$$= 0.20 \; kip \cdot in.$$

Substituting into (10-12) will give the axial strain energy in the vertical member

$$U_4 = \frac{1}{2} \frac{(-P)^2 L_2}{EA}$$

$$= \frac{(-0.5)^2 \times 20}{2 \times 29{,}000 \times 1.77} \frac{kip^2 \cdot in.}{kip/in.^2 \cdot in.^2}$$

$$= 48.7 \times 10^{-6} \; kip \cdot in.$$

The total strain energy is

$$U = U_1 + U_2 + U_3 + U_4$$

$$= 0.08 + 0 + 0.20 + 48.7 \times 10^{-6}$$

$$= 0.2800 \; kip \cdot in.$$

Substituting into (10-18) gives the vertical displacement at the end,

$$\delta = \frac{2U}{P} = \frac{2 \times 0.2800}{0.5} \frac{kip \cdot in.}{kip}$$

$$= \mathbf{1.12 \; in.}$$

Note that in this example the contribution to the total energy, and hence to the displacement, made by the axial strain energy, is less than 0.02%. It could have been neglected without appreciable error.

Where a structure consists of members subjected to a combination of bending, torsion, and axial loads, it is common practice to ignore the axial strains in determining the deformation. However, if the proportions are unusual, all strains should be taken into consideration.

Problems 10-13 to 10-28

We have seen that the direct energy method is very powerful when certain displacement problems are encountered. However, the method has some severe limitations. For example, when more than one force acts on the structure each force does work as the structure deflects and equating the total work to the strain energy gives a true equation. However, the equation contains, as unknowns, all the displacements associated with the forces and the single equation is insufficient for the solution of the two or more unknowns. Also, if the displacement is required *at a point other than the point of application of the load*, the direct energy method fails. While the direct energy method has severely limited application, it has served us well in that it enabled us to get acquainted with strain energy in the context of a very simple method. We will now proceed to more advanced methods, which do not suffer from the same limitations.

10-6   MINIMUM POTENTIAL ENERGY

A basic law of physics states that any system will come to rest in an equilibrium position such that the potential energy of the system is a minimum. This is illustrated by a ball that is free to roll under the influence of gravity on a vertical curved path as in Fig. 10-14a. The potential energy is defined as the work required to move the system from some arbitrary reference configuration to the displaced configuration. In Fig. 10-14a the ball at A could be chosen as the reference configuration and B would be a typical displaced configuration. The potential energy at B would then be the work required to move the ball from A to B. If the shape of the profile is known, the potential energy, PE, of the ball when it has been moved a distance D can be written in terms of D. That is,

$$PE = f(D)$$

For the system shown in Fig. 10-14a, the PE function will have the general shape of the curve shown in Fig. 10-14b. The potential energy will be a minimum when

$$\frac{d\text{PE}}{dD} = 0 \qquad\qquad (10\text{-}21)$$

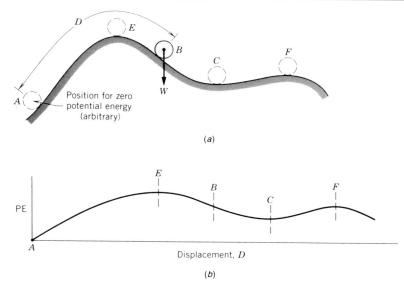

FIG. 10-14 Potential energy concepts. (*a*) Ball and profile of path. (*b*) Potential energy function for ball on path.

By performing the differentiation and solving for *D*, we will have found the location of point *C* where the ball will be in equilibrium. Solving (10-21) in this case would also give the location of *E* and *F*, which are also equilibrium positions. It would be more accurate to say that (10-21) will give all locations where there is stationary potential energy and that these are equilibrium positions, but some, such as *C*, are *stable* equilibrium positions while others, such as *E* and *F*, may be *unstable*. A simple test can be applied to the potential energy function to determine stability. If the second derivative of the PE function is positive at the equilibrium point, the system is stable. A negative second derivative indicates instability. The test is not usually required, since most practical structures have only one equilibrium position and that is a stable position.

We really do not need the principle of stationary potential energy to solve the above problem because it is obvious from intuition where the ball will come to rest. However, if the system is made slightly more complicated, such as by the addition of a spring as in Fig. 10-15, intuition will no longer be adequate. To use the method of minimum potential energy, we must first select a reference position. This is normally arbitrary, but we will make it a rule to start wherever possible with the unstrained configuration, that is, the position in which the spring elements are unstretched so that there is no initial strain energy. This rule would lead us to choose position *A* in Fig. 10-15 such that the distance, *A-G*, is equal to the free length of the spring. The following example is very simple but adequately illustrates the principle by solving a problem that would be very difficult to solve by other methods.

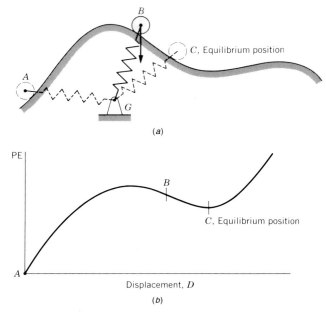

FIG. 10-15 Potential energy concepts including a spring. (a) Ball, spring and profile of path. (b) Potential energy function for ball-spring systems.

Example 10-5

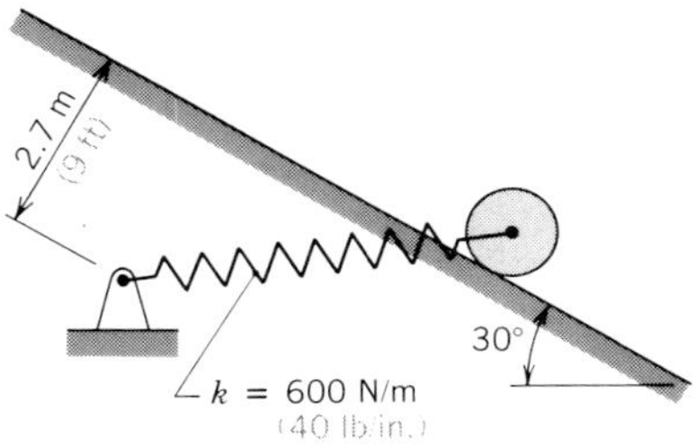

A cylinder has a diameter of 0.6 m and a mass of 200 kg. It rests on an inclined plane and is constrained by a spring as shown in Fig. 10-16a. The spring has a stiffness, k, of 600 N/m and a free length of 3.0 m. Determine the equilibrium position.

Solution. We will take the configuration shown in Fig. 10-16b, where the spring is unstretched, as the reference position. Then for a typical displacement, D, the configuration is shown in Fig. 10-16c.

$$L = \sqrt{3^2 + D^2} \qquad \Delta L = L - 3 = \sqrt{3^2 + D^2} - 3$$

The strain energy in the spring is given by

$$U = \tfrac{1}{2} k \delta^2 \qquad (10\text{-}3)$$

$$= \tfrac{1}{2} k \, \Delta L^2$$

$$\mathrm{PE} = \tfrac{1}{2} k \, \Delta L^2 - WD \sin 30°$$

$$= \tfrac{1}{2} k [(3^2 + D^2)^{1/2} - 3]^2 - W \sin 30° \, D$$

$$= \tfrac{1}{2} k [3^2 + D^2 - 6(3^2 + D^2)^{1/2} + 9] - W \sin 30° \, D$$

For equilibrium, from (10-21):

$$\frac{d\mathrm{PE}}{dD} = 0$$

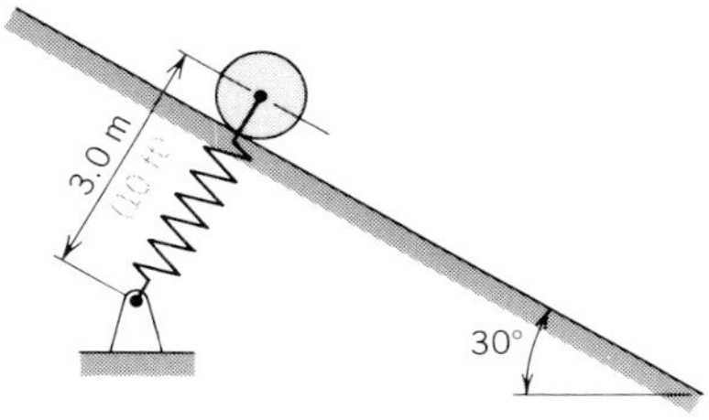

FIG. 10-16a

k = 600 N/m
(40 lb in.)

2.7 m
(9 ft)

30°

FIG. 10-16b

3.0 m
(10 ft)

30°

$$\tfrac{1}{2}k[2D - 6 \times \tfrac{1}{2} \times (3^2 + D^2)^{-1/2}2D] - W \sin 30° = 0$$

$$D\left[2 - \frac{6}{\sqrt{3^2 + D^2}}\right] = \frac{W \sin 30°}{\tfrac{1}{2}k}$$

$$= \frac{1962 \times 0.500}{\tfrac{1}{2}(600)} \ \frac{N}{N/m}$$

$$= 3.27 \ m$$

Solving by trial and error, we obtain

$$D = 4.04 \ m$$

The equilibrium position for the cylinder is then **4.04 m down the incline from the reference position.**

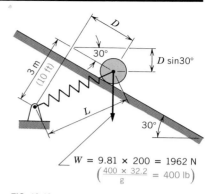

FIG. 10-16c

$$W = 9.81 \times 200 = 1962 \ N$$
$$\left(\frac{400 \times 32.2}{g} = 400 \ lb\right)$$

Example 10-5

A cylinder has a diameter of 2 ft and a mass of 400 lb. It rests on an inclined plane and is constrained by a spring as shown in Fig. 10-16a. The spring has a stiffness, k, of 40 lb/ft and a free length of 10 ft. Determine the equilibrium position.

Solution. We will take the configuration shown in Fig. 10-16b, where the spring is unstretched, as the reference position. Then for a typical displacement, D, the configuration is shown in Fig. 10-16c.

$$L = \sqrt{10^2 + D^2} \qquad \Delta L = L - 10 = \sqrt{10^2 + D^2} - 10$$

The strain energy in the spring is given by

$$U = \tfrac{1}{2}k\,\delta^2 \qquad (10\text{-}3)$$

$$= \tfrac{1}{2}k\,\Delta L^2$$

$$PE = \tfrac{1}{2}k\,\Delta L^2 - WD \sin 30°$$

$$= \tfrac{1}{2}k[(10^2 + D^2)^{1/2} - 10]^2 - W \sin 30° \ D$$

$$= \tfrac{1}{2}k[10^2 + D^2 - 20(10^2 + D^2)^{1/2} + 100] - W \sin 30° \ D$$

For equilibrium, from (10-21):

$$\frac{d\text{PE}}{dD} = 0$$

$$\tfrac{1}{2}k[2D - 20 \times \tfrac{1}{2} \times (10^2 + D^2)^{-1/2}2D] - W \sin 30° = 0$$

$$D\left[2 - \frac{20}{\sqrt{10^2 + D^2}}\right] = \frac{W \sin 30°}{\tfrac{1}{2}k}$$

$$= \frac{400 \times 0.500}{\tfrac{1}{2}(40)} \ \frac{lb}{lb/ft}$$

$$= 10 \ ft$$

Solving by trial and error, we obtain

$$D = 12.9 \text{ ft}$$

The equilibrium position for the cylinder is then **12.9 ft down the incline from the reference position.**

We have developed a principle of minimum potential energy that is adequate for solving one-dimensional problems. It can only be applied to systems in which the displaced configuration can be described by a single dimension. Such systems are said to have *one degree of freedom*. We can extend the principle to two dimensions by considering a ball that is seeking its equilibrium position in the bottom of a depression such as in Fig. 10-17. For a typical location of the ball, given by D_x and D_y, the potential energy can be written in terms of D_x and D_y. The equilibrium position will correspond to the position of stationary potential energy with respect to the two variables. This requires that

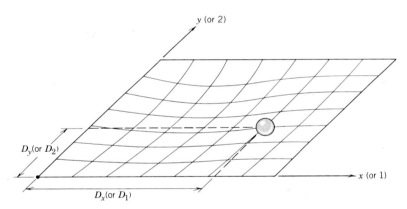

FIG. 10-17 Illustration of a two-degrees-of-freedom system by a ball on a plane.

$$\frac{\partial PE}{\partial D_x} = 0 \quad \text{and} \quad \frac{\partial PE}{\partial D_y} = 0 \tag{10-22}$$

These equations are sufficient to give the values of the two displacements that define the equilibrium position. Looking ahead to the solution of structural problems, we will need to deal with a large number of displacements; that is, with systems having many degrees of freedom. In such cases, it is more convenient to have displacement subscripts that are numbers, as indicated in Fig. 10-17. Then (10-22) becomes

$$\frac{\partial PE}{\partial D_1} = 0 \quad \text{and} \quad \frac{\partial PE}{\partial D_2} = 0$$

Extending this to N displacement components, or N degrees of freedom, we have

$$\frac{\partial PE}{\partial D_1} = 0$$

$$\frac{\partial PE}{\partial D_2} = 0$$

$$\frac{\partial PE}{\partial D_3} = 0 \qquad\qquad \textbf{(10-23)}$$

$$\vdots$$

$$\frac{\partial PE}{\partial D_N} = 0$$

These relations will always give N simultaneous equations in N unknown displacements. Solving the equations will give the displacements that can then be used to solve for stresses. The fact that a structure may be indeterminate does not alter the method, or add to the difficulty, of solution.

Example 10-6

Determine the displacements at A and the stress in member AB due to the load P acting on the steel structure shown in Fig. 10-18a.

Solution. Identification numbers and the calculated length of members are indicated in Fig. 10-18b. Assume that point A moves to the right a distance, D_1, and downward a distance, D_2. Taking components of these displacements along the members in turn gives changes in length:

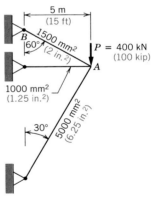

FIG. 10-18a

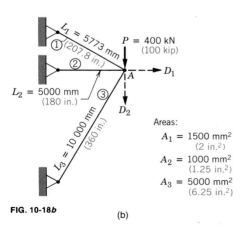

Areas:
$$A_1 = 1500 \text{ mm}^2$$
$$(2 \text{ in.}^2)$$
$$A_2 = 1000 \text{ mm}^2$$
$$(1.25 \text{ in.}^2)$$
$$A_3 = 5000 \text{ mm}^2$$
$$(6.25 \text{ in.}^2)$$

FIG. 10-18b

(b)

$$\Delta L_1 = \cos 30° \, D_1 + \cos 60° \, D_2 = 0.866 D_1 + 0.5 D_2$$

$$\Delta L_2 = D_1$$

$$\Delta L_3 = \cos 60° \, D_1 - \cos 30° \, D_2 = 0.5 D_1 - 0.866 D_2$$

The strain energy in each member is given by (10-13):

$$U = \frac{1}{2}\frac{EA\,\Delta L^2}{L}$$

For the dimensions and displacements of the structure:

$$U_1 = \frac{1}{2}\frac{1\,200\,000 \times 1500(0.866D_1 + 0.5D_2)^2}{5773}$$

$$= 19\,500D_1^2 + 22\,500D_1D_2 + 6500D_2^2$$

$$U_2 = \frac{1}{2}\frac{1\,200\,000 \times 1000D_1^2}{5000}$$

$$= 20\,000D_1^2$$

$$U_3 = \frac{1}{2}\frac{1\,200\,000 \times 5000(0.5D_1 - 0.866D_2)^2}{10\,000}$$

$$= 12\,500D_1^2 - 43\,300D_1D_2 + 37\,500D_2^2$$

$$U = U_1 + U_2 + U_3$$

$$= 52\,000D_1^2 - 20\,800D_1D_2 + 44\,000D_2^2$$

In moving to the displaced position, the vertical force moves through distance D_2 and does work on the system. Hence the potential energy is

$$PE = U - PD_2$$

$$= 52\,000D_1^2 - 20\,800D_1D_2 + 44\,000D_2^2 - 400\,000D_2$$

For equilibrium the displacements must satisfy (10-23), that is,

$$\frac{\partial PE}{\partial D_1} = 0 \qquad \text{or} \qquad 104\,000D_1 - 20\,800D_2 = 0 \tag{1}$$

and

$$\frac{\partial PE}{\partial D_2} = 0 \qquad \text{or} \qquad -20\,800D_1 + 88\,000D_2 - 400\,000 = 0 \tag{2}$$

$$\textbf{(1)} + \frac{20\,800}{88\,000} \times \textbf{(2)} \text{ gives } 99\,080D_1 - 94\,500 = 0$$

$$D_1 = 0.954 \text{ mm}$$

Substituting into **(1)** gives

$$D_2 = \frac{104\,000 \times 0.954}{20\,800} = \textbf{4.771 mm}$$

Therefore, **the point A moves downward 4.771 mm and to the right 0.954 mm.** The strain in the member connecting A to B is

$$\varepsilon_1 = \frac{\Delta L_1}{L_1} = \frac{0.866 D_1 + 0.5 D_2}{5773}$$

$$= \frac{0.866 \times 0.954 \times 0.5 + 4.771}{5773} = 0.000556$$

The stress is

$$\sigma = E\varepsilon_1$$

$$= 200\,000 \times 0.000556 = \mathbf{111}\ \frac{\mathbf{N}}{\mathbf{mm}^2}$$

Example 10-6

Determine the displacements at A and the stress in member AB due to the load P acting on the steel structure shown in Fig. 10-18a.

Solution.   Identification numbers and the calculated length of members are indicated in Fig. 10-18b. Assume that point A moves to the right a distance, D_1, and downward a distance, D_2. Taking components of these displacements along the members in turn gives changes in length:

$$\Delta L_1 = \cos 30° D_1 + \cos 60° D_2 = 0.866 D_1 + 0.5 D_2$$

$$\Delta L_2 = D_1$$

$$\Delta L_3 = \cos 60° D_1 - \cos 30° D_2 = 0.5 D_1 - 0.866 D_2$$

The strain energy in each member is given by (10-13):

$$U = \frac{1}{2} \frac{EA\,\Delta L^2}{L}$$

For the dimensions and displacements of the structure:

$$U_1 = \frac{1}{2} \frac{29{,}000 \times 2(0.866 D_1 + 0.5 D_2)^2}{207.8}$$

$$= 104.66 D_1^2 + 120.86 D_1 D_2 + 34.89 D_2^2$$

$$U_2 = \frac{1}{2} \frac{29{,}000 \times 1.25 D_1^2}{180}$$

$$= 100.69 D_1^2$$

$$U_3 = \frac{1}{2} \frac{29{,}000 \times 6.25(0.5 D_1 - 0.866 D_2)^2}{360}$$

$$= 62.93 D_1^2 - 218.0 D_1 D_2 + 188.79 D_2^2$$

$$U = U_1 + U_2 + U_3$$

$$= 268.28 D_1^2 - 97.14 D_1 D_2 + 223.68 D_2^2$$

In moving to the displaced position, the vertical force moves through distance D_2 and does work on the system. Hence the potential energy is

$$PE = U - PD_2$$

$$= 268.28D_1^2 - 97.14D_1 D_2 + 223.68D_2^2 - 100D_2$$

For equilibrium the displacements must satisfy (10-23), that is,

$$\frac{\partial PE}{\partial D_1} = 0 \quad \text{or} \quad 536.56D_1 - 97.14D_2 = 0 \tag{1}$$

and

$$\frac{\partial PE}{\partial D_2} = 0 \quad \text{or} \quad -97.14D_1 + 447.36D_2 - 100 = 0 \tag{2}$$

(1) $+ \dfrac{97.14}{447.36} \times$ **(2)** gives $515.467D_1 - 21.714 = 0$

$$D_1 = 0.042 \text{ in.}$$

Substituting into **(1)** gives

$$D_2 = \frac{536.56 \times 0.042}{97.14} = 0.23 \text{ in.}$$

Therefore, **the point A moves downward 0.23 in. and to the right 0.042 in.** The strain in the member connecting A to B is

$$\varepsilon_1 = \frac{\Delta L_1}{L_1} = \frac{0.866D_1 + 0.5D_2}{207.8}$$

$$= \frac{0.866 \times 0.042 + 0.5 \times 0.23}{207.8} = 0.000728$$

The stress is

$$\sigma = E\varepsilon_1$$

$$= 29,000 \times 0.000728 = 21.1 \text{ kip/in.}^2$$

This example shows that in energy methods, although the principles are quite simple, the applications are rather lengthy and great care must be exercised to avoid error. It also shows that no special treatment was required for this statically indeterminate problem. If one member had been removed to make it statically determinate, the same principles would have been employed to determine the displacements; the strain energy in two members would have been calculated instead of three, thus slightly reducing the amount of effort. On the other hand, if another redundant member had been added the method would be the same, and there would be only a slight increase in the effort required to solve the problem.

Problems 10-29 to 10-32

10-7 THE RAYLEIGH-RITZ METHOD

An approximate method, which solves with a reasonable amount of effort some otherwise difficult problems, was described in 1877 by Lord Rayleigh and was put onto a sound mathematical basis by Walter Ritz in 1908. Today this method, known as the *Rayleigh-Ritz* method, is used in determining the deflection and stress in beams, plates, and shells. We will restrict our applications to beams in bending.

     Consider a beam such as that in Fig. 10-19. If we tried to obtain the deflection of the beam by the principle of minimum potential energy, the expression for strain energy would be quite complicated and the method could become impractical. To solve by the Rayleigh-Ritz method, we assume the form of the displacement function for the member. The displacement function is usually taken to consist of either polynomial terms or trigonometric terms. The function must be chosen carefully and must not violate any of the constraints that are imposed on the member. Since the amplitude of the deflection is unknown, the displacement function must contain multipliers that allow for the unknown amplitudes of the displacement functions. For example, there may be a problem in which we assumed the displacement to be

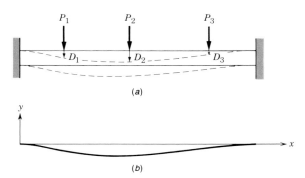

(a)

(b)

FIG. 10-19 Rayleigh-Ritz method for beams in bending. (*a*) Loaded beam. (*b*) Deflection curve.

$$y = a_1 + a_2 x + a_3 x^2 + a_4 x^3$$

or

$$y = a_1 + a_2 x^2 + a_3 x^4$$

or

$$y = a_1 + a_2 \cos \frac{\pi x}{L} + a_3 \cos \frac{2\pi x}{L}$$

or

$$y = a_1 \sin \frac{\pi x}{L} + a_2 \sin \frac{3\pi x}{L}$$

or

$$y = a_1 + a_2 \sin cx + a_3 \cos cx$$

In using the Rayleigh-Ritz method, skill is required in choosing the displacement function and experience is invaluable.

Once the displacement function has been assumed, the strain energy can be written in terms of a_1, a_2, ..., a_n. Also, the displacements at the points of load application can be evaluated in terms of a_1, a_2, ..., a_n. Thus the potential energy of the system can be written and will contain n unknown a's. But the a's are a measure of displacement and can be treated as unknown displacements in (10-23), and hence the equations can be rewritten:

$$\frac{\partial \text{PE}}{\partial a_1} = 0$$

$$\frac{\partial \text{PE}}{\partial a_2} = 0 \qquad\qquad \textbf{(10-24)}$$

$$\vdots$$

$$\frac{\partial \text{PE}}{\partial a_n} = 0$$

These will give n simultaneous equations in unknowns a_1, a_2, ..., a_n. Solving for the a's will give the displacement function for the member, which can then be used to solve for displacements, moments, or stresses as required.

In general, accuracy is improved by adding more terms to the function, but each extra term means another unknown and another simultaneous equation. The effort required to solve simultaneous equations increases rapidly with the number of unknowns, and it is impractical to solve for more than about three or four unless a computer is used. Thus there is always a compromise between the accuracy that we are prepared to accept and the effort that we are willing to expend.

In applying the Rayleigh-Ritz method, it is convenient to have the strain energy of a deflected beam expressed in terms of the deflection curve. In terms of deflection y, from (3-8), the bending moment is given by

$$M = EI \frac{d^2 y}{dx^2}$$

Substituting into (10-14) gives the strain energy in a deflected beam as

$$U = \frac{1}{2EI} \int\limits_{\text{length}} \left(EI \frac{d^2 y}{dx^2} \right)^2 dx$$

$$U = \frac{EI}{2} \int_{length} \left(\frac{d^2y}{dx^2}\right)^2 dx \qquad \textbf{(10-25)}$$

Although this equation is valid for all deflected beams, its principal application is in the Rayleigh-Ritz method.

The following examples illustrate some of the techniques required in using the Rayleigh-Ritz method.

Example 10-7

The cantilever beam of Fig. 10-20a supports an upward load at its end. Using the Rayleigh-Ritz method, and assuming a polynominal displacement function, determine the deflection at the end. Compare the answer with the exact solution.

Solution.   Assume a displacement function

$$y = a_1 x^2 + a_2 x^3$$

See Figure 10-20b.

Note that this function satisfies the end conditions of Fig. 10-20a at the left; that is, the function ensures that there will be zero displacement and zero rotation at the left end. The end constraints would be violated if the function contained a constant or a linear term. Differentiating the assumed function gives

$$\frac{d^2y}{dx^2} = 2a_1 + 6a_2 x$$

Substituting into (10-25) gives

$$U = \frac{EI}{2} \int_0^L (2a_1 + 6a_2 x)^2 \, dx$$

$$= \frac{EI}{2} \int_0^L (4a_1^2 + 24a_1 a_2 x + 36a_2^2 x^2) \, dx$$

$$= \frac{EI}{2} \left[4a_1^2 x + 12a_1 a_2 x^2 + 12a_2^2 x^3 \right]_0^L$$

$$= \frac{EI}{2} (4La_1^2 + 12L^2 a_1 a_2 + 12L^3 a_2^2)$$

$$PE = U - Py]_{x=L}$$

$$= \frac{EI}{2} (4La_1^2 + 12L^2 a_1 a_2 + 12L^3 a_2^2) - P(L^2 a_1 + L^3 a_2)$$

From (10-24)

$$\frac{\partial PE}{\partial a_1} = 0$$

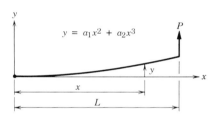

FIG. 10-20a Cantilever loaded at its end.

$$y = a_1 x^2 + a_2 x^3$$

FIG. 10-20b Assumed displacement function for the Rayleigh-Ritz method.

$$EI[4La_1 + 6L^2a_2] - PL^2 = 0$$

$$4La_1 + 6L^2a_2 = \frac{L^2}{EI}P \qquad \text{(1)}$$

From (10-24)

$$\frac{\partial PE}{\partial a_2} = 0$$

$$EI[6L^2a_1 + 12L^3a_2] - PL^3 = 0$$

$$6La_1 + 12L^2a_2 = \frac{L^2}{EI}P \qquad \text{(2)}$$

(2) $-2 \times$ **(1)** gives $-2La_1 + 0a_2 = -\dfrac{L^2}{EI}P$

$$a_1 = \frac{1}{2}\frac{L}{EI}P$$

Substituting into **(1)** gives

$$2\frac{L^2}{EI}P + 6L^2a_2 = \frac{L^2}{EI}P$$

$$6L^2a_2 = -\frac{L^2}{EI}P$$

$$a_2 = -\frac{1}{6}\frac{P}{EI}$$

$$y = \frac{1}{2}\frac{L}{EI}Px^2 - \frac{1}{6}\frac{P}{EI}x^3$$

$$\delta = y]_{x=L} = \left(\frac{1}{2} - \frac{1}{6}\right)\frac{PL^3}{EI} = \frac{1}{3}\frac{PL^3}{EI}$$

Note that this is, in fact, the exact solution. The Rayleigh–Ritz method is normally approximate, but in this case the assumed function happens to be exact, and hence we obtained the exact solution. If we had included another term, say a_3x^4, in the displacement function in order to improve accuracy, the solution would have given $a_3 = 0$.

Example 10-8

Solve the cantilever beam of Example 10-7 using a trigonometric displacement function.

Solution. Assume a displacement function

$$y = a_1\left(1 - \cos\frac{\pi x}{2L}\right)$$

Note in Fig. 10-21 that the function satisfies the known displacements at the support.

$$\frac{d^2y}{dx^2} = a_1 \left(\frac{\pi}{2L}\right)^2 \cos\frac{\pi x}{2L}$$

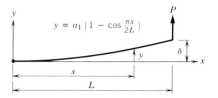

FIG. 10-21 Assumed trigonometric displacement function for the Rayleigh-Ritz method.

Substituting into (10-25) gives

$$U = \frac{EI}{2}\int_0^L \left[a_1 \left(\frac{\pi}{2L}\right)^2 \cos\frac{\pi x}{2L}\right]^2 dx$$

$$= \frac{EIa_1^2}{2}\left(\frac{\pi}{2L}\right)^4 \int_0^L \cos^2\frac{\pi x}{2L}\, dx$$

$$= \frac{EIa_1^2}{2}\left(\frac{\pi}{2L}\right)^4 \left[\frac{x}{2} + \frac{\sin(\pi/L)x}{2\pi/L}\right]_0^L$$

$$= \frac{ELa_1^2}{2}\left(\frac{\pi}{2L}\right)^4 \frac{L}{2} = \frac{EI\pi^4}{64L^3}a_1^2$$

$$PE = U - Py]_{x=L}$$

$$= \frac{EI\pi^4}{64L^3}a_1^2 - Pa_1$$

From (10-24) $\dfrac{\partial PE}{\partial a_1} = 0$

$$\frac{EI\pi^4}{32L^3}a_1 - P = 0$$

$$a_1 = \frac{32}{\pi^4}\frac{L^3}{EI}P$$

$$y = \frac{32L^3P}{\pi^4EI}\left(1 - \cos\frac{\pi x}{2L}\right)$$

$$\delta = y]_{x=L} = a_1 = \frac{32}{\pi^4}\frac{L^3}{EI}P = 0.3285\frac{PL^3}{EI}$$

Since the exact solution has a coefficient of 0.333, we see that this solution is only 1.5% in error. We can extend this problem to determine moments at critical sections by substituting a_1 into the bending moment equation (3-8) to give

$$M = EI\frac{32}{\pi^4}\frac{L^3}{EI}P\left(\frac{\pi}{2L}\right)^2 \cos\frac{\pi x}{2L}$$

$$= \frac{8}{\pi^2}PL \cos\frac{\pi x}{2L}$$

In this case we know the bending moment from elementary theory and so can check on the accuracy of the Rayleigh–Ritz method. For instance, at the left end we obtain, by the Rayleigh-Ritz solution,

$$M = 0.811PL \qquad (\text{error} = 19\%)$$

and at midlength

$$M = 0.573PL \qquad (\text{error} = 15\%)$$

In this example we see a characteristic of the Rayleigh-Ritz method: it gives good accuracy for displacements but may be quite inaccurate for moments, and hence stresses.

The Rayleigh-Ritz method gives approximate solutions of varying degrees of accuracy. When the assumed function coincides with the actual deflection, the solution is exact. This rarely happens; in most cases the deflection, although not exact, is quite accurate, but the stresses are commonly found to be in error by as much as 30%. Introducing more terms into the assumed displacement function will give better accuracy but will require more effort. By a prudent choice of the displacement functions, accuracy may often be improved without an increase in the difficulty of the solution. Thus there is an art to the choice of the functions that improves with experience. Although the Rayleigh-Ritz method is very powerful and is widely used, at this stage in your education it is sufficient for you to have a general knowledge of the method; the required skills and experience can be acquired later if you elect to specialize in stress analysis.

Problems 10-33 to 10-34

10-8 CASTIGLIANO'S FIRST THEOREM

We will now reexamine the principle of minimum potential energy and develop a method that is slightly easier to apply than the method of minimum PE.

When we used equation (10-23),

$$\frac{\partial \text{PE}}{\partial D_1} = 0$$

$$\frac{\partial \text{PE}}{\partial D_2} = 0 \qquad (10\text{-}23)$$

$$\vdots$$

$$\frac{\partial \text{PE}}{\partial D_N} = 0$$

the potential energy was always in the form

$$\text{PE} = U - [P_1 D_1 + P_2 D_2 + \ldots + P_N D_N] \qquad \mathbf{(10\text{-}26)}$$

In (10-26) we have made provision for a force component occurring with each displacement component, but some forces are zero and the corresponding terms will disappear. For completeness we will retain all terms.

Substituting the formula for PE given in (10-26) into the first equation (10-23), we get

$$\frac{\partial}{\partial D_1}[U - (P_1 D_1 + P_2 D_2 + P_3 D_3 + \ldots + P_N D_N)] = 0$$

which simplifies to

$$\frac{\partial U}{\partial D_1} - \frac{\partial}{\partial D_1}(P_1 D_1 + P_2 D_2 + P_3 D_3 + \ldots + P_N D_N) = 0$$

or

$$\frac{\partial U}{\partial D_1} - P_1 = 0$$

or

$$\frac{\partial U}{\partial D_1} = P_1$$

Treating the second equation of (10-23) in this way gives

$$\frac{\partial U}{\partial D_2} = P_2$$

A similar equation can be found for all N equations in (10-23). Rewriting then, we have

$$
\begin{aligned}
\frac{\partial U}{\partial D_1} &= P_1 \\
\frac{\partial U}{\partial D_2} &= P_2 \\
&\vdots \\
\frac{\partial U}{\partial D_N} &= P_N
\end{aligned}
\tag{10-27}
$$

These equations can be expressed by saying: *The partial derivative of the strain energy* (when expressed in terms of displacements), *with respect to any displacement component, is equal to the corresponding force component.* This is known as *Castigliano's first theorem.*

The theorem is useful for solving the same problems as the minimum potential energy method. The principles behind the theorem are more obscure, but most engineers, once they are familiar with it, prefer the theorem because it is slightly easier to use.

Example 10-9

Determine the displacements of the panel points in the steel structure shown in Fig. 10-22a, due to the given load.

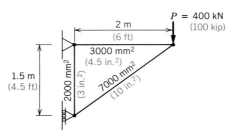

FIG. 10-22a

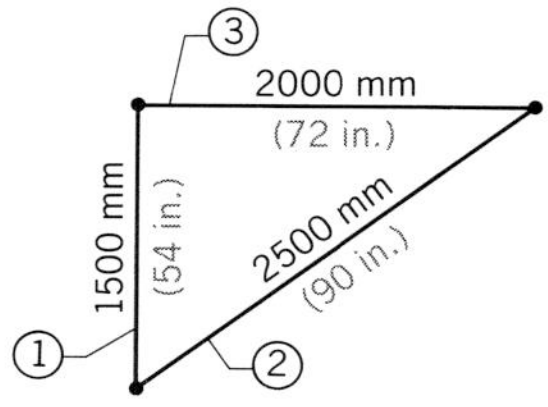

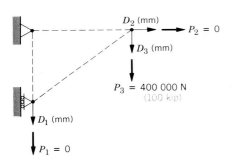

FIG. 10-22b Member numbers and lengths.

Solution.   Member identification numbers and displacements are shown in Fig. 10-22b and c. Changes in member lengths are shown in Fig. 10-22d. The strain energy in each member is given by

$$U = \frac{1}{2}\frac{EA\,\Delta L^2}{L} \qquad (10\text{-}13)$$

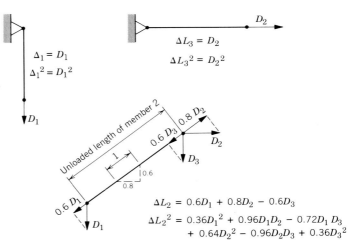

FIG. 10-22c Displacements (assumed directions) and loads.

FIG. 10-22d Changes in member lengths.

When the same Young's modulus applies to all members, the total strain energy can be determined by

$$U = \frac{1}{2}E\sum_{i=1}^{M}\frac{A_i\Delta L_i^2}{L_i} \qquad \textbf{(10-28)}$$

The calculation can best be done in tabular form.

i	A_i (mm^2)	ΔL_i^2 (mm^2)	L_i (mm)	$\dfrac{A_i \Delta L_i^2}{L_i}$ (mm^3)
1	2000	D_1^2	1500	$1.333 D_1^2$
2	7000	$0.36 D_1^2 + 0.96 D_1 D_2$	2500	$1.008 D_1^2 + 2.688 D_1 D_2$
		$-0.72 D_1 D_3 + 0.64 D_2^2$		$-2.016 D_1 D_3 + 1.792 D_2^2$
		$-0.96 D_2 D_3 + 0.36 D_3^2$		$-2.688 D_2 D_3 + 1.008 D_3^2$
3	3000	D_2^2	2000	$1.500 D_2^2$

$$\sum \frac{A_i \Delta L_i^2}{L_i} = \quad 2.341 D_1^2 + 2.688 D_1 D_2$$
$$-2.016 D_1 D_3 + 3.292 D_2^2$$
$$-2.688 D_2 D_3 + 1.008 D_3^2$$

$$U = \frac{1}{2} E \sum \frac{A_i \Delta L_i^2}{L_i} = \frac{200\,000}{2} \sum \frac{A_i \Delta L_i^2}{L_i}$$

$$= 234\,100 D_1^2 + 268\,800 D_1 D_2 - 201\,600 D_1 D_3 + 329\,200 D_2^2 - 268\,800 D_2 D_3$$

$$+ 100\,800 D_3^2$$

By Castigliano's first theorem

$$\frac{\partial U}{\partial D_1} = P_1 = 0$$

$$468\,200 D_1 + 268\,800 D_2 - 201\,600 D_3 = 0 \qquad \textbf{(1)}$$

$$\frac{\partial U}{\partial D_2} = P_2 = 0$$

$$268\,800 D_1 + 658\,400 D_2 - 268\,800 D_3 = 0 \qquad \textbf{(2)}$$

$$\frac{\partial U}{\partial D_3} = P_3 = 400\,000$$

$$-201\,600 D_1 - 268\,800 D_2 + 201\,600 D_3 = 400\,000 \qquad \textbf{(3)}$$

These equations would normally be solved by inverting the matrix consisting of the coefficients of the D's. However, in the present case the coefficients are of such a nature that the equations can be solved without great difficulty.

(1) + (3)   gives $266\,600 D_1 + 0 D_2 + 0 D_3 = 400\,000$

$$D_1 = \textbf{1.500 mm}$$

$$\frac{201\,600}{268\,800} \times \textbf{(2)} + \textbf{(3)} \quad \text{gives} \quad 0 D_1 + 225\,000 D_2 + 0 D_3 = 400\,000$$

$$D_2 = \textbf{1.778 mm}$$

Substituting into **(1)**   gives

$$D_3 = \frac{468\,200 \times 1.500 + 268\,800 \times 1.778}{201\,600} = \textbf{5.854 mm}$$

Example 10-9

Determine the displacements of the panel points in the steel structure shown in Fig. 10-22a, due to the given load.

Solution. Member identification numbers and displacements are shown in Fig. 10-22b and c. Changes in member lengths are shown in Fig. 10-22d.

The strain energy in each member is given by

$$U = \frac{1}{2} \frac{EA \, \Delta L^2}{L} \qquad (10\text{-}13)$$

When the same Young's modulus applies to all members, the total strain energy can be determined by

$$U = \frac{1}{2} E \sum_{i=1}^{M} \frac{A_i \Delta L_i^2}{L_i} \qquad \textbf{(10-28)}$$

The calculation can best be done in tabular form.

i	A_i (in.²)	ΔL_i^2 (in.²)	L_i (in.)	$\dfrac{A_i \Delta L_i^2}{L_i}$ (in.³)
1	3	D_1^2	54	$0.0556 D_1^2$
2	10	$0.36 D_1^2 + 0.96 D_1 D_2$ $- 0.72 D_1 D_3 + 0.64 D_2^2$ $- 0.96 D_2 D_3 + 0.36 D_3^2$	90	$0.0400 D_1^2 + 0.1067 D_1 D_2$ $- 0.0800 D_1 D_3 + 0.0711 D_2^2$ $- 0.1067 D_2 D_3 + 0.0400 D_3^2$
3	4.5	D_2^2	72	$0.0625 D_2^2$

$$\sum \frac{A_i \Delta L_i^2}{L_i} = 0.0956 D_1^2 + 0.1067 D_1 D_2$$
$$- 0.0800 D_1 D_3 + 0.1336 D_2^2$$
$$- 0.1067 D_2 D_3 + 0.0400 D_3^2$$

$$U = \frac{1}{2} E \sum \frac{A_i \Delta L_i^2}{L_i} = \frac{29{,}000}{2} \sum \frac{A_n \Delta L_i^2}{L_n}$$

$$= 1386.2 D_1^2 + 1547.2 D_1 D_2 - 1160.0 D_1 D_3 + 1937.2 D_2^2 - 1547.2 D_2 D_3$$

$$+ 580.0 D_3^2$$

By Castigliano's first theorem

$$\frac{\partial U}{\partial D_1} = P_1 = 0$$

$$2772.4 D_1 + 1547.2 D_2 - 1160.0 D_3 = 0 \qquad \textbf{(1)}$$

$$\frac{\partial U}{\partial D_2} = P_2 = 0$$

$$1547.2 D_1 + 3874.4 D_2 - 1547.2 D_3 = 0 \qquad \textbf{(2)}$$

$$\frac{\partial U}{\partial D_3} = P_3 = 100$$

$$-1160.0D_1 - 1547.2D_2 + 1160.0D_3 = 100 \qquad \textbf{(3)}$$

These equations would normally be solved by inverting the matrix consisting of the coefficients of the D's. However, in the present case the coefficients are of such a nature that the equations can be solved without great difficulty.

$$\textbf{(1)} + \textbf{(3)} \quad \text{gives} \quad 1612.4D_1 + 0D_2 + 0D_3 = 100$$

$$D_1 = \textbf{0.062 in.}$$

$$\frac{1160.0}{1547.2} \times \textbf{(2)} + \textbf{(3)} \quad \text{gives} \quad 0D_1 + 1357.6D_2 + 0D_3 = 100$$

$$D_2 = \textbf{0.074 in.}$$

Substituting into **(1)** gives

$$D_3 = \frac{2772.4 \times 0.062 + 1547.2 \times 0.074}{1160.0} = \textbf{0.247 in.}$$

Problems 10-35 to 10-39

When any truss structure, such as that shown in Fig. 10-23, is displaced the strain energy can be expressed in terms of dimensions of the structure, Young's modulus and the displacements.

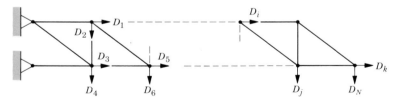

FIG. 10-23

The energy function will contain terms such as D_1^2, D_1D_2, D_1D_5, D_3D_6, ..., D_iD_j. The terms are all of second order. A great variety of subscripts will occur together but not necessarily all possible combinations, because these combinations will depend upon which points are connected by the truss members. Each term will be multiplied by a constant, and the energy function can be written in the form:

$$U = a_{11}D_1^2 + a_{12}D_1D_2 + a_{15}D_1D_5 \dots a_{ij}D_iD_j + \dots a_{NN}D_N^2$$

Using the subscript scheme shown allows us to readily relate the constants to the displacements.

Applying Castigliano's first theorem to the energy function in this form gives

$$\frac{\partial U}{\partial D_1} = 2a_{11}D_1 + a_{12}D_2 + a_{13}D_3 \ldots a_{1j}D_j + \ldots a_{1N}D_N = P_1$$

$$\frac{\partial U}{\partial D_2} = a_{12}D_1 + 2a_{22}D_2 + a_{23}D_3 + \ldots a_{2j}D_j + \ldots a_{2N}D_N = P_2$$

$$\vdots$$

$$\frac{\partial U}{\partial D_i} = a_{1i}D_1 + a_{2i}D_2 + \ldots 2a_{ii}D_i + \ldots a_{ij}D_j + \ldots a_{iN}D_N = P_i$$

$$\vdots$$

$$\frac{\partial U}{\partial D_N} = a_{1N}D_1 + a_{2N}D_2 + \ldots 2a_{NN}D_N = P_N \tag{10-29a}$$

These equations can be written in matrix form as:

$$
\begin{bmatrix}
2a_{11} & a_{12} & a_{13} & a_{14} & \ldots & a_{1j} & \ldots & a_{1N} \\
a_{12} & 2a_{22} & a_{23} & a_{24} & \ldots & a_{2j} & \ldots & a_{2N} \\
\vdots & & & & & & & \\
a_{1i} & a_{2i} & a_{3i} & \ldots & 2a_{ii} & a_{ij} & \ldots & a_{1N} \\
\vdots & & & & & & & \\
a_{1N} & & & & & & & 2a_{NN}
\end{bmatrix}
\times
\begin{Bmatrix}
D_1 \\ D_2 \\ \vdots \\ D_j \\ \vdots \\ D_N
\end{Bmatrix}
=
\begin{Bmatrix}
P_1 \\ P_2 \\ \vdots \\ P_i \\ \vdots \\ P_N
\end{Bmatrix}
$$

$$\tag{10-29b}$$

or as

$$[a]\{D\} = \{P\} \tag{10-29c}$$

The elements of the a-matrix can be determined directly from the strain energy function, U, and we will write the matrix for the case given in Example 10-9. But first we need to make one more observation. Consider a typical term in the U function, say

$$a_{35}D_3D_5$$

When the third equation is being formed by (10-29a)

$$\frac{\partial U}{\partial D_3} = \ldots a_{35}D_5 \ldots$$

Thus a_{35} goes into the third row and fifth column of the a-matrix.

The fifth equation will also contain a contribution from the term under consideration. That is,

$$\frac{\partial U}{\partial D_5} = \ldots a_{35}D_3 \ldots$$

so that a_{35} will be entered in the fifth row and the third column of a. For the general term $a_{ij}D_iD_j$ the value of a_{ij} is entered in row i, column j, and also in row j, column i. This means that the a-matrix is always symmetrical with respect to the main diagonal.

Using the strain energy function in Example 10-9, we could write the a-matrix directly as:

$$[a] = \begin{bmatrix} 468\ 200 & 268\ 800 & -201\ 600 \\ 268\ 800 & 658\ 400 & -268\ 800 \\ -201\ 600 & -268\ 800 & 201\ 600 \end{bmatrix}$$

$$[a] = \begin{bmatrix} 2772.4 & 1547.2 & -1160.0 \\ 1547.2 & 3874.4 & -1547.2 \\ -1160.0 & -1547.2 & 1160.0 \end{bmatrix}$$

When a spring is stretched by giving one end an axial displacement, D, the force, P, required to displace the end is given by

$$kD = P$$

where k is the spring stiffness.

This same format exists in Eq. (10-29c), where the a's are multiplied by displacements to give forces. Consequently, the a-matrix is referred to as the *stiffness matrix* of the structure. The elements of the stiffness matrix can be filled in directly once the energy function has been written. The usual practical problem requires that the displacements be determined for a set of known loads, P. In that case (10-29) represents N simultaneous equations in N unknown displacements. The best systematic way to solve the equations is by inverting the a-matrix and calculating the displacements by:

$$\{D\} = [a^{-1}]\{P\} \tag{10-30}$$

Except for the simplest of structures it is not practical to invert the a-matrix by hand. In practice a computer is used to solve (10-30). Usually, the computer is used for the whole process of filling in the a-matrix and solving for the displacements. Special techniques have been developed for determining the elements of the a-matrix. These techniques are readily programmable, and the method, referred to as the *stiffness matrix method*, is used almost exclusively for the analysis of large structures.

Example 10-10

Use the stiffness method to determine the displacements in the steel structure illustrated in Fig. 10-24a and the stress in member B-F.

Solution. Displacements and members identification numbers are shown in Fig. 10-24b. The change in the length of a typical inclined member is shown in Fig. 10-24c.

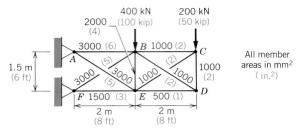

FIG. 10-24a

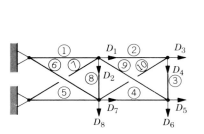

FIG. 10-24b

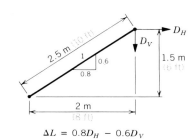

$$\Delta L = 0.8D_H - 0.6D_V$$

FIG. 10-24c

From Eq. (10-13) the energy is given by

$$U_i = \frac{1}{2} \frac{EA_i \Delta L_i^2}{L_i}$$

$$= \frac{200\,000}{2} \frac{A_i \Delta L_i^2}{L_i} = 100 \times 10^3 \frac{A_i \Delta L_i^2}{L_i}$$

which may be calculated in tabular form as follows:

	A_i (mm²)	L_i (mm)	ΔL_i (mm)	$\dfrac{A_i \Delta L_i^2}{L_i}$ (mm³)
1	3000	2000	D_1	$1.50(D_1^2)$
2	1000	2000	$D_3 - D_1$	$0.50(D_3^2 - 2D_1 D_3 + D_1^2)$
3	1000	1500	$D_6 - D_4$	$0.667(D_6^2 - 2D_4 D_6 + D_4^2)$
4	500	2000	$D_5 - D_7$	$0.25(D_5^2 - 2D_5 D_7 + D_7^2)$
5	1500	2000	D_7	$0.75(D_7^2)$
6	3000	2500	$0.8D_7 + 0.6D_8$	$1.20(0.64D_7^2 + 0.96D_7 D_8 + 0.36D_8^2)$
7	3000	2500	$0.8D_1 - 0.6D_2$	$1.20(0.64D_1^2 - 0.96D_1 D_2 + 0.36D_2^2)$
8	2000	1500	$D_8 - D_2$	$1.33(D_8^2 - 2D_2 D_8 + D_2^2)$
9	1000	2500	$0.8D_5 - 0.8D_1 + 0.6D_6 - 0.6D_2$	$0.40(0.64D_5^2 - 1.28D_1 D_5 + 0.96D_5 D_6 - 0.96D_2 D_5 + 0.64D_1^2$ $- 0.96D_1 D_6 + 0.96D_1 D_2 - 0.72D_2 D_6 + 0.36D_6^2 + 0.36D_2^2)$
10	1000	2500	$0.8D_3 - 0.8D_7 + 0.6D_8 - 0.6D_4$	$0.40(0.64D_3^2 - 1.28D_7 D_3 + 0.96D_3 D_8 - 0.96D_4 D_3 + 0.64D_7^2$ $- 0.96D_7 D_8 + 0.96D_7 D_4 - 0.72D_4 D_8 + 0.36D_8^2 + 0.36D_4^2)$

From (10-28)

$$U = \frac{1}{2}E\sum \frac{A_i \Delta L_i^2}{L_i} = \frac{200\,000}{2}\sum \frac{A_i \Delta L_i^2}{L_i}$$

$$U = 10^3 \times \left\{
\begin{aligned}
&(150 + 50 + 120 \times 0.64 + 40 \times 0.64)D_1^2 + (-120 \times 0.96 + 40 \times 0.96)D_1 D_2 \\
&\quad + (-50 \times 2)D_1 D_3 + (-40 \times 1.28)D_1 D_5 + (-40 \times 0.96)D_1 D_6 \\
&\quad + (120 \times 0.36 + 133 \times 1 + 40 \times 0.36)D_2^2 + (-40 \times 0.96)D_2 D_5 \\
&\quad + (-40 \times 0.72)D_2 D_6 + (-133 \times 2)D_2 D_8 \\
&\quad + (50 + 40 \times 0.64)D_3^2 + (-40 \times 0.96)D_3 D_4 + (-40 \times 1.28)D_3 D_7 \\
&\quad + (40 \times 0.96)D_3 D_8 \\
&\quad + (66.7 + 40 \times 0.36)D_4^2 + (-66.7 \times 2)D_4 D_6 + (40 \times 0.96)D_4 D_7 \\
&\quad + (-40 \times 0.72)D_4 D_8 \\
&\quad + (25 + 40 \times 0.64)D_5^2 + (40 \times 0.96)D_5 D_6 + (-25 \times 2)D_5 D_7 \\
&\quad + (66.7 + 40 \times 0.36)D_6^2 \\
&\quad + (25 + 75 + 120 \times 0.64 + 40 \times 0.64)D_7^2 \\
&\quad + (120 \times 0.96 - 40 \times 0.96)D_7 D_8 \\
&\quad + (120 \times 0.36 + 133 + 40 \times 0.36)D_8^2
\end{aligned}
\right.$$

$$U = 10^3 \times \left\{
\begin{aligned}
&302.4D_1^2 - 76.8D_1 D_2 - 100D_1 D_3 - 51.2D_1 D_5 - 38.4D_1 D_6 \\
&190.6D_2^2 - 38.4D_2 D_5 - 28.8D_2 D_6 - 266D_2 D_8 \\
&75.6D_3^2 - 38.4D_3 D_4 - 51.2D_3 D_7 + 38.4D_3 D_8 \\
&81.1D_4^2 - 133.4D_4 D_6 + 38.4D_4 D_7 - 28.8D_4 D_8 \\
&50.6D_5^2 + 38.4D_5 D_6 - 50D_5 D_7 \\
&81.1D_6^2 \\
&202.4D_7^2 + 76.8D_7 D_8 \\
&190.9D_8^2
\end{aligned}
\right.$$

$$[a] = 10^3 \times \begin{bmatrix}
604.8 & -76.8 & -100.0 & 0 & -51.2 & -38.4 & 0 & 0 \\
-76.8 & 381.2 & 0 & 0 & -38.4 & -28.8 & 0 & -266.0 \\
-100.0 & 0 & 151.2 & -38.4 & 0 & 0 & -51.2 & 38.4 \\
0 & 0 & -38.4 & 162.2 & 0 & -133.4 & 38.4 & -28.8 \\
-51.2 & -38.4 & 0 & 0 & 101.2 & 38.4 & -50.0 & 0 \\
-38.4 & -28.8 & 0 & -133.4 & 38.4 & 162.2 & 0 & 0 \\
0 & 0 & -51.2 & 38.4 & -50.0 & 0 & 404.8 & 76.8 \\
0 & -266.0 & 38.4 & -28.8 & 0 & 0 & 76.8 & 381.8
\end{bmatrix}$$

Inverting the a-matrix by computer and substituting into (10-30) along with the known components of the force vector gives the displacements.

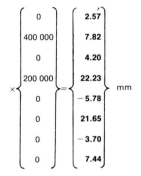

$$\{D\} = 10^6 \times \begin{bmatrix} 2.887 & 2.687 & 2.925 & 7.491 & -0.805 & 7.512 & -0.880 & 2.320 \\ 2.687 & 10.58 & 2.835 & 17.92 & -3.170 & 18.00 & -3.446 & 9.135 \\ 2.925 & 2.835 & 11.63 & 15.35 & -3.390 & 14.62 & -0.806 & 2.125 \\ 7.491 & 17.92 & 15.35 & 75.31 & -22.54 & 72.23 & -11.58 & 18.95 \\ -0.805 & -3.170 & -3.390 & -22.54 & 19.86 & -23.99 & 5.032 & -4.580 \\ 7.512 & 18.00 & 14.62 & 72.23 & -23.99 & 76.23 & -11.54 & 18.84 \\ -0.880 & -3.466 & -0.806 & -11.58 & 5.032 & -11.54 & 4.884 & -4.190 \\ 2.320 & 9.135 & 2.125 & 18.95 & -4.580 & 18.84 & -4.190 & 11.04 \end{bmatrix}$$

$$\times \begin{Bmatrix} 0 \\ 400\,000 \\ 0 \\ 200\,000 \\ 0 \\ 0 \\ 0 \\ 0 \end{Bmatrix} = \begin{Bmatrix} 2.57 \\ 7.82 \\ 4.20 \\ 22.23 \\ -5.78 \\ 21.65 \\ -3.70 \\ 7.44 \end{Bmatrix} \text{ mm}$$

To find the stress in member $B\text{-}F$, or member number 7, we substitute D_1 and D_2 into

$$\Delta L_7 = 0.8D_1 - 0.6D_2 = 0.8 \times 2.57 - 0.6 \times 7.82 = -2.64 \text{ mm}$$

The strain in the member is

$$\frac{\Delta L}{L} = \frac{-2.64}{2500} = -0.001054$$

and the stress

$$\sigma = E\varepsilon = 200\,000 \times (-0.001054) = \mathbf{-210} \ \frac{\mathbf{N}}{\mathbf{mm}^2}$$

Example 10-10

Use the stiffness method to determine the displacements in the steel structure illustrated in Fig. 10-24*a* and the stress in member *B-F*.

Solution. Displacements and members identification numbers are shown in Fig. 10-24*b*. The change in the length of a typical inclined member is shown in Fig. 10-24*c*.

From Eq. (10-13) the energy is given by

$$U_i = \frac{1}{2}\frac{EA_i \Delta L_i^2}{L_i}$$

$$= \frac{29{,}000}{2}\frac{A_i \Delta L_i^2}{L_i} = 14{,}500\,\frac{A_i \Delta L_i^2}{L_i}$$

which may be calculated in tabular form as follows:

i	A_i (in.^{2})	L_i (in.)	ΔL_i (in.)	$\dfrac{A_i \Delta L_i^2}{L_i}$ (in.^{3})
1	6	96	D_1	$0.0625(D_1^2)$
2	2	96	$D_3 - D_1$	$0.0208(D_3^2 - 2D_1 D_3 + D_1^2)$
3	2	72	$D_6 - D_4$	$0.0278(D_6^2 - 2D_4 D_6 + D_4^2)$
4	1	96	$D_5 - D_7$	$0.0104(D_5^2 - 2D_5 D_7 + D_7^2)$
5	3	96	D_7	$0.0313(D_7^2)$
6	5	120	$0.8D_7 + 0.6D_8$	$0.0417(0.64D_7^2 + 0.96D_7 D_8 + 0.36D_8^2)$
7	5	120	$0.8D_1 - 0.6D_2$	$0.0417(0.64D_1^2 - 0.96D_1 D_2 + 0.36D_2^2)$
8	4	72	$D_8 - D_2$	$0.0556(D_8^2 - 2D_2 D_8 + D_2^2)$
9	2	120	$0.8D_5 - 0.8D_1 + 0.6D_6 - 0.6D_2$	$0.0167(0.64D_5^2 - 1.28D_1 D_5 + 0.96D_5 D_6 - 0.96D_2 D_5 + 0.64D_1^2$ $- 0.96D_1 D_6 + 0.96D_1 D_2 - 0.72D_2 D_6 + 0.36D_6^2 + 0.36D_2^2)$
10	2	120	$0.8D_3 - 0.8D_7 + 0.6D_8 - 0.6D_4$	$0.0167(0.64D_3^2 - 1.28D_7 D_3 + 0.96D_3 D_8 - 0.96D_4 D_3 + 0.64D_7^2$ $- 0.96D_7 D_8 + 0.96D_7 D_4 - 0.72D_4 D_8 + 0.36D_8^2 + 0.36D_4^2)$

From (10-28)

$$U = \frac{1}{2}E\sum \frac{A_i \Delta L_i^2}{L_i} = \frac{29{,}000}{2}\sum \frac{A_i \Delta L_i^2}{L_i}$$

$$U = 14.5 \times \left\{ \begin{array}{l}
(6.25 + 20.8 + 41.7 \times 0.64 + 16.7 \times 0.64)D_1^2 + (-41.7 \times 0.96 + 16.7 \times 0.96)D_1 D_2 \\[4pt]
+ (-20.8 \times 2)D_1 D_3 + (-16.7 \times 1.28)D_1 D_5 + (-16.7 \times 0.96)D_1 D_6 \\[4pt]
+ (41.7 \times 0.36 + 55.6 \times 1 + 16.7 \times 0.36)D_2^2 + (-16.7 \times 0.96)D_2 D_5 \\[4pt]
+ (-16.7 \times 0.72)D_2 D_6 + (-55.6 \times 2)D_2 D_8 \\[4pt]
+ (20.8 + 16.7 \times 0.64)D_3^2 + (-16.7 \times 0.96)D_3 D_4 + (-16.7 \times 1.28)D_3 D_7 \\[4pt]
+ (16.7 \times 0.96)D_3 D_8 \\[4pt]
+ (27.8 + 16.7 \times 0.36)D_4^2 + (-27.8 \times 2)D_4 D_6 + (16.7 \times 0.96)D_4 D_7 \\[4pt]
+ (-16.7 \times 0.72)D_4 D_8 \\[4pt]
+ (10.4 + 16.7 \times 0.64)D_5^2 + (16.7 \times 0.96)D_5 D_6 + (-10.4 \times 2)D_5 D_7 \\[4pt]
+ (27.8 + 16.7 \times 0.36)D_6^2 \\[4pt]
+ (10.4 + 31.3 + 41.7 \times 0.64 + 16.7 \times 0.64)D_7^2 \\[4pt]
+ (41.7 \times 0.96 - 16.7 \times 0.96)D_7 D_8 \\[4pt]
+ (41.7 \times 0.36 + 55.6 + 16.7 \times 0.36)D_8^2
\end{array} \right.$$

$$U = 14.5 \times \begin{Bmatrix} 120.7D_1^2 - 24.0D_1D_2 - 41.6D_1D_3 - 21.4D_1D_5 - 16.0D_1D_6 \\ 76.6D_2^2 - 16.0D_2D_5 - 12.0D_2D_6 - 111.2D_2D_8 \\ 31.5D_3^2 - 16.0D_3D_4 - 21.4D_3D_7 + 16.0D_3D_8 \\ 33.8D_4^2 - 55.6D_4D_6 + 16.0D_4D_7 - 12.0D_4D_8 \\ 21.1D_5^2 + 16.0D_5D_6 - 20.8D_5D_7 \\ 33.8D_6^2 \\ 79.1D_7^2 + 24.0D_7D_8 \\ 76.6D_8^2 \end{Bmatrix}$$

$$[a] = 10^3 \times \begin{bmatrix}
3.500 & -0.348 & -0.603 & 0 & -0.310 & -0.232 & 0 & 0 \\
-0.348 & 2.221 & 0 & 0 & -0.232 & -0.174 & 0 & -1.612 \\
-0.603 & 0 & 0.932 & -0.232 & 0 & 0 & -0.310 & 0.232 \\
0 & 0 & -0.232 & 0.980 & 0 & -0.806 & 0.232 & -0.174 \\
-0.310 & -0.232 & 0 & 0 & 0.612 & 0.232 & -0.302 & 0 \\
-0.234 & -0.174 & 0 & -0.806 & 0.232 & 0.980 & 0 & 0 \\
0 & 0 & -0.310 & 0.232 & -0.302 & 0 & 2.294 & 0.348 \\
0 & -1.612 & 0.232 & -0.174 & 0 & 0 & 0.348 & 2.221
\end{bmatrix}$$

Inverting the *a*-matrix by computer and substituting into (10-30) along with the known components of the force vector gives the displacements.

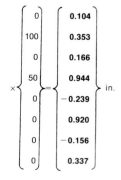

$$\{D\} = 10^{-3} \times \begin{bmatrix}
0.480 & 0.434 & 0.472 & 1.203 & -0.115 & 1.207 & -0.131 & 0.380 \\
0.434 & 1.952 & 0.445 & 3.159 & -0.534 & 3.173 & -0.589 & 1.710 \\
0.472 & 0.445 & 1.862 & 2.434 & -0.528 & 2.318 & -0.115 & 0.338 \\
1.203 & 3.159 & 2.434 & 12.56 & -3.720 & 12.06 & -1.936 & 3.326 \\
-0.115 & -0.534 & -0.528 & -3.720 & 3.297 & -3.962 & 0.854 & -0.758 \\
1.207 & 3.173 & 2.318 & 12.06 & -3.962 & 12.73 & -1.930 & 3.308 \\
-0.131 & -0.589 & -0.115 & -1.936 & 0.854 & -1.930 & 0.834 & -0.698 \\
0.380 & 1.710 & 0.338 & 3.326 & -0.758 & 3.308 & -0.698 & 2.206
\end{bmatrix} \times \begin{Bmatrix} 0 \\ 100 \\ 0 \\ 50 \\ 0 \\ 0 \\ 0 \\ 0 \end{Bmatrix} = \begin{Bmatrix} 0.104 \\ 0.353 \\ 0.166 \\ 0.944 \\ -0.239 \\ 0.920 \\ -0.156 \\ 0.337 \end{Bmatrix} \text{ in.}$$

To find the stress in member *B-F*, or member number 7, we substitute D_1 and D_2 into

$$\Delta L_7 = 0.8D_1 - 0.6D_2 = 0.8 \times 0.104 - 0.6 \times 0.353 = -0.129$$

The strain in the member is

$$\frac{\Delta L}{L} = \frac{-0.129}{120} = -0.001075$$

and the stress

$$\sigma = E\varepsilon = 29,000 \times (-0.001075) = -\mathbf{31.2 \ kip/in.}^2$$

It is seen from this example that even a very simple problem can be impractical to solve without the aid of a computer. Since a computer will be used for the inversion process, it seems a waste of engineering manpower to do the preliminary work by hand. Consequently, most problems of this type are solved by programs that read in the basic data describing the structure, do all the calculations, and print out the stresses in the members.

Problems 10-40 to 10-47

10-9 CASTIGLIANO'S SECOND THEOREM

The energy methods that we have encountered up to this point have required that we express energy in terms of displacements. They have been most useful in solving problems in which displacement was required. The fact that a structure was statically indeterminate did not alter the approach. In fact, making a structure indeterminate often reduces the number of displacements and, contrary to our intuitive expectation, makes the solution easier. However, it is sometimes inconvenient to deal with displacements, especially in problems where force, reaction, and stress are of the greatest interest. For problems of this nature *Castigliano's second theorem* is more useful.

To derive the second theorem, we will consider a system such as that of Fig. 10-25*a*. We know that if the forces are applied gradually, the work that they do will be equal to the total strain energy in all the members. We have calculated the strain energy in terms of displacements in many cases, but it is evident that in the structure of Fig. 10-25, it would be easier to determine the strain energy in terms of the forces with the aid of (10-20). We can then quite easily write the strain energy in the structure in terms of loads. Hence the strain energy is expressed as a function of the *n* loads acting on the structure.

$$U = U(P_1, P_2, \ldots, P_n)$$

Before applying the loads $P_1 \ldots P_n$, assume that we had first applied a small load ΔP_i as shown in Fig. 10-25*b*. Let us take ΔP_i to be so small

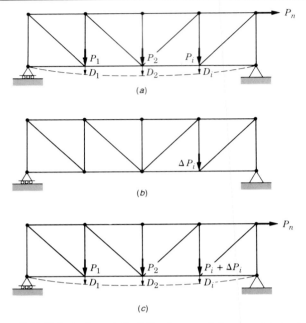

FIG. 10-25 Derivation of Castigliano's second theorem. (*a*) Loads and displacements. (*b*) Small initial load. (*c*) Total load system.

that the stresses and displacements that it causes are negligible, and hence the strain energy at this stage is negligible. Let us now apply the loads $P_1 \dots P_n$ and reconsider the strain energy. The energy will be substantially as it was for the forces in Fig. 10-25a but will be increased by a small amount due to the work done by ΔP_i. Equating the additional strain energy to the added work input, we have

$$\Delta U = \Delta P_i D_i$$

or   **(10-31)**

$$\frac{\Delta U}{\Delta P_i} = D_i$$

The assumption that the initial strain energy is negligible becomes more accurate as ΔP_i is made smaller or when ΔP_i becomes dP_i. Then

$$\frac{dU}{dP_i} = D_i$$

but since U is a variable containing P_1, P_2, $\dots P_1$, $\dots$, P_n, we must use partial derivatives; then

$$\boxed{\frac{\partial U}{\partial P_i} = D_i}$$ **(10-32)**

This is *Castigliano's second theorem: The partial derivative of the strain energy* (when expressed in terms of loads), *with respect to any load component, is equal to the corresponding displacement.*

The relationships expressed by the second theorem are even more obscure than those of the first theorem and are harder to visualize in engineering terms. However it can be used to solve many categories of difficult problems with a reasonable amount of effort. The range of problems that can be solved and the required techniques are demonstrated in the following examples.

Example 10-11

Calculate the vertical deflection at the end of the steel truss of Fig. 10-26a due to the 300 kN load.

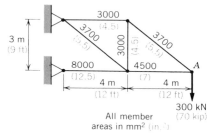

Solution. Let us call the 300 kN load P_1. Since we want to find the vertical deflection at A, that is, the deflection which is in the *same direction* as P_1 and *at the point* where P_1 is applied, the vertical deflection D_1 can be obtained by means of

$$\frac{\partial U}{\partial P_1} = D_1$$

FIG. 10-26a Example of a single load acting on a truss.

To carry out this operation, we first express all member forces in terms of P_1 and then find the total strain energy in the truss as a function of P_1.

The forces in members are shown in Fig. 10-26b.

The strain energy in the structure by (10-20) is

$$U = \frac{1}{2E} \sum_{i=1}^{6} \frac{F_i^2 L_i}{A_i}$$

$$= 2.5(10^{-6}) \sum_{i=1}^{6} \frac{F_i^2 L_i}{A_i} \quad \text{(for steel)}$$

We assign an identification number to each member as in Fig. 10-26c and tabulate the strain energy calculation as shown:

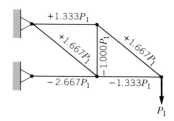

FIG. 10-26b

i	F_i^2 (N^2)	L_i (mm)	A_i (mm^2)	$(F_i^2 L_i/A_i)$ N^2/mm
1	$(1.333P_1)^2$	4000	3000	$2.369P_1^2$
2	$(1.667P_1)^2$	5000	3700	$3.755P_1^2$
3	$(-2.667P_1)^2$	4000	8000	$3.556P_1^2$
4	$(-1.000P_1)^2$	3000	3000	$1.000P_1^2$
5	$(1.667P_1)^2$	5000	3700	$3.755P_1^2$
6	$(-1.333P_1)^2$	4000	4500	$1.579P_1^2$

$$\sum \frac{F_i^2 L_i}{A_i} = 16.014P_1^2$$

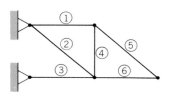

FIG. 10-26c

$$U = \frac{1}{2E} \sum \frac{F_i^2 L_i}{A_i}$$

$$= \frac{1}{2 \times 200\,000} \times 16.014P_1^2 \frac{\text{N}^2/\text{mm}}{\text{N/mm}^2} = 40.03 \times 10^{-6} P_1^2 \text{ (N} \cdot \text{mm)}$$

By Castigliano's second theorem:

$$D_1 = \frac{\partial U}{\partial P_1} = \frac{\partial}{\partial P_1} [40.03(10^{-6}) P_1^2]$$

$$= 80.06(10^{-6}) P_1$$

$$= 80.06(10^{-6})300\ 000 = 24.0 \text{ mm}$$

The vertical deflection at the end of the structure is **24.0 mm**. Note that the substitution of 300 kN for P_1 is only done at the end of the problem *after the differentiation has been carried out.*

Example 10-11

Calculate the vertical deflection at the end of the steel truss of Fig. 10-26a due to the 70 kip load.

Solution.   Let us call the 70 kip load P_1. Since we want to find the vertical deflection at A, that is, the deflection which is in the *same direction* as P_1 and *at the point* where P_1 is applied, the vertical deflection D_1 can be obtained by means of

$$\frac{\partial U}{\partial P_1} = D_1$$

To carry out this operation, we first express all member forces in terms of P_1 and then find the total strain energy in the truss as a function of P_1.

The forces in members are shown in Fig. 10-26b.

The strain energy in the structure by (10-20) is

$$U = \frac{1}{2E} \sum_{i=1}^{6} \frac{F_i^2 L_i}{A_i}$$

$$= 17.24(10^{-6}) \sum_{i=1}^{6} \frac{F_i^2 L_i}{A_i} \qquad \text{(for steel)}$$

We assign an identification number to each member as in Fig. 10-26c and tabulate the strain energy calculation as shown:

i	F_i^2 (kip^2)	L_i (in.)	A_i (in.^{2})	$\dfrac{F_i^2 L_i}{A_i} \dfrac{\text{kip}^2}{\text{in.}}$
1	$(1.333 P_1)^2$	144	4.5	$56.86 P_1^2$
2	$(1.667 P_1)^2$	180	5.5	$90.95 P_1^2$
3	$(-2.667 P_1)^2$	144	12.5	$81.94 P_1^2$
4	$(-1.000 P_1)^2$	108	4.5	$24.00 P_1^2$
5	$(1.667 P_1)^2$	180	5.5	$90.95 P_1^2$
6	$(-1.333 P_1)^2$	144	7	$36.55 P_1^2$

$$\sum \frac{F_i^2 L_i}{A_i} = 381.25 P_1^2$$

$$U = \frac{1}{2E} \sum \frac{F_i^2 L_i}{A_i}$$

$$= \frac{1}{2 \times 29,000} \times 381.25 P_1^2 \frac{kip^2/in.}{kip/in.^2} = 6.573 \times 10^{-3} P_1^2 \, (kip \cdot in.)$$

By Castigliano's second theorem:

$$D_1 = \frac{\partial U}{\partial P_1} = \frac{\partial}{\partial P_1} [6.573(10^{-3}) P_1^2]$$

$$= 13.146(10^{-3}) P_1$$

$$= 13.146(10^{-3})70 = 0.92 \, in.$$

The vertical deflection at the end of the structure is **0.92 in.** Note that the substitution of 70 kip for P_1 is only done at the end of the problem *after the differentiation has been carried out*.

This is the same problem that was solved by the direct method as Example 10-2; a comparison shows that the two methods require about equal effort. The superiority of the present method is illustrated in the next example, which cannot be solved by the direct method.

Example 10-12

Calculate the deflection at A and B, due to the loads at A and B, for the steel truss illustrated in Fig. 10-27a.

Solution. Letting the loads at A and B be represented by P_1 and P_2, we can determine the member forces shown in Fig. 10-27b. Using the same numbering scheme as in Example 10-11 and using a similar tabular form for the calculations, we obtain the strain energy function.

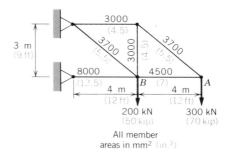

All member areas in mm² (in.²)

FIG. 10-27a Example of two loads on a truss.

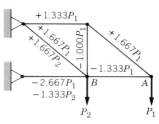

FIG. 10-27b

i	$F_i^2 \, (N^2)$	$L_i \, (mm)$	$A_i \, (mm^2)$	$(F_i^2 L_i / A_i) \, N^2/mm$
1	$(1.333P_1)^2$	4000	3000	$2.369P_1^2$
2	$(1.667P_1 + 1.667P_2)^2$	5000	3700	$3.755P_1^2 + 7.511P_1P_2$
	$= 2.779P_1^2 + 5.558P_1P_2$			$+ 3.755P_2^2$
	$+ 2.779P_2^2$			
3	$(-2.667P_1 - 1.333P_2)^2$	4000	8000	$3.557P_1^2 + 3.555P_1P_2$
	$= 7.113P_1^2 - 7.110P_1P_2$			$+ 0.888P_2^2$
	$1.777P_2^2$			
4	$(-1.000P_1)^2$	3000	3000	$1.000P_1^2$
5	$(1.667P_1)^2$	5000	3700	$3.755P_1^2$
6	$(-1.333P_1)^2$	4000	4500	$1.579P_1^2$

$$\frac{\sum F_i^2 \Delta L_i}{A_i} = 16.015P_1^2 + 11.066P_1P_2$$

$$+ 4.643P_2^2$$

$$U = \frac{1}{2E} \sum \frac{F_i^2 L_i}{A_i} = \frac{1}{2 \times 200\,000}(16.015P_2^2 + 11.066P_1 P_2 + 4.643P_2^2)$$

$$= (40.04P_1^2 + 27.67P_1 P_2 + 11.61P_2^2) \times 10^{-6}\,(\text{N} \cdot \text{mm})$$

By Castigliano's second theorem:

$$D_1 = \frac{\partial U}{\partial P_1} = (80.08P_1 + 27.67P_2)(10^{-6})$$

$$= [80.08 \times 0.3 + 27.67 \times 0.2](10^6)(10^{-6})$$

$$= 24.02 + 5.53 = \textbf{29.6 mm}$$

and

$$D_2 = \frac{\partial U}{\partial P_2} = (27.67P_1 + 23.22P_2)(10^{-6})$$

$$= [27.67 \times 0.3 + 23.22 \times 0.2](10^6)(10^{-6})$$

$$= 8.30 + 4.64 = \textbf{12.9 mm}$$

The vertical deflections *in the direction of the loads* applied at A and B are, therefore, both downward and equal to 29.6 mm and 12.9 mm, respectively. We see that we were able to find these deflections quite simply by first expressing the total strain energy in the truss as a function of the two loads P_1 and P_2 and then differentiating the energy with respect to P_1 and P_2.

Example 10-12

Calculate the deflection at A and B, due to the loads at A and B, for the steel truss illustrated in Fig. 10-27a.

Solution.   Letting the loads at A and B be represented by P_1 and P_2, we can determine the member forces shown in Fig. 10-27b. Using the same numbering scheme as in Example 10-11 and using a similar tabular form for the calculations, we obtain the strain energy function.

i	F_i^2 (kip^2)	L_i (in.)	A_i (in.^{2})	$\dfrac{F_i^2 L_i}{A_i} \dfrac{\text{kip}^2}{\text{in.}}$
1	$(1.333P_1)^2$	144	4.5	$56.86P_1^2$
2	$(1.667P_1 + 1.667P_2)^2$ $= 2.779P_1^2 + 5.558P_1 P_2$ $+ 2.779P_2^2$	180	5.5	$90.95P_1^2 + 181.89P_1 P_2$ $+ 90.95P_2^2$
3	$(-2.667P_1 - 1.333P_2)^2$ $= 7.113P_1^2 - 7.110P_1 P_2$ $1.777P_2^2$	144	12.5	$81.95P_1^2 + 81.91P_1 P_2$ $+ 20.47P_2^2$
4	$(-1.000P_1)^2$	108	4.5	$24.0P_1^2$
5	$(1.667P_1)^2$	180	5.5	$90.95P_1^2$
6	$(-1.333P_1)^2$	144	7	$27.42P_1^2$

$$\sum \frac{F_i^2 \Delta L_i}{A_i} = 372.13P_1^2 + 263.80P_1 P_2 + 111.42P_2^2$$

$$U = \frac{1}{2E} \sum \frac{F_i^2 L_i}{A_i} = \frac{1}{2 \times 29,000} (372.13P_2^2 + 263.80P_1P_2 + 111.42P_2^2)$$

$$= (6.416P_1^2 + 4.548P_1P_2 + 1.921P_2^2) \times 10^{-3} \text{ (kip·in.)}$$

By Castigliano's second theorem:

$$D_1 = \frac{\partial U}{\partial P_1} = (12.832P_1 + 4.548P_2)(10^{-3})$$

$$= [12.832 \times 70 + 4.548 \times 50](10^{-3})$$

$$= 0.898 + 0.227 = \textbf{1.125 in.}$$

and

$$D_2 = \frac{\partial U}{\partial P_2} = (4.548P_1 + 3.842P_2)(10^{-3})$$

$$= [4.548 \times 70 + 3.842 \times 50](10^{-3})$$

$$= 0.318 + 0.192 = \textbf{0.510 in.}$$

The vertical deflections *in the direction of the loads* applied at *A* and *B* are, therefore, both downward and equal to 1.125 in. and 0.510 in., respectively. We see that we were able to find these deflections quite simply by first expressing the total strain energy in the truss as a function of the two loads P_1 and P_2 and then differentiating the energy with respect to P_1 and P_2.

At first glance you may think that the theorem might not be applicable when a displacement is required that does not correspond to a given load. However, this problem is handled simply by creating a *fictitious* or *dummy load* to match the required displacement and solving in the usual manner. At the time the final substitution of forces is to be made, zero is entered for the dummy load. This approach is illustrated in the next example.

Example 10-13

Determine the horizontal displacement at point *A* in the structure given in Example 10-10.

Solution. Applying both the real load P_1 and the dummy load P_2 (note that P_2 is in the direction of the required horizontal displacement), we determine the member forces as given in Fig. 10-28.

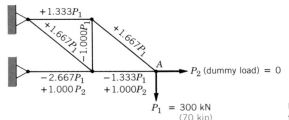

FIG. 10-28 Example of the dummy load method to find a horizontal truss displacement.

Again using a table, we determine the strain energy in terms of the loads making no special concession, at this stage, to the dummy load.

i	F_i^2 (N^2)	L_i (mm)	A_i (mm^2)	$(F_i^2 L_i/A_i)$ N^2/mm
1	$(1.333P_1)^2$	4000	3000	$2.369P_1^2$
2	$(1.667P_1)^2$	5000	3700	$3.755P_1^2$
3	$(-2.667P_1 + 1.000P_2)^2$ $= 7.113P_1^2 - 5.334P_1P_2$ $+ 1.000P_2^2$	4000	8000	$3.556P_1^2 - 2.667P_1P_2$ $+ 0.500P_2^2$
4	$(-1.000P_1)^2$	3000	3000	$1.000P_1^2$
5	$(1.667P_1)^2$	5000	3700	$3.755P_1^2$
6	$(-1.333P_1 + 1.000P_2)^2$ $= 1.777P_1^2 - 2.667P_1P_2$ $+ 1.000P_2^2$	4000	4500	$1.580P_1^2 - 2.371P_1P_2$ $+ 0.889P_2^2$

$$\sum \frac{F_i^2 L_i}{A_i} = 16.015P_1^2 - 5.038P_1P_2 + 1.389P_2^2$$

$$U = \frac{1}{2E} \sum \frac{F_i^2 L_i}{A_i} = \frac{1}{2 \times 200\,000}(16.015P_1^2 - 5.038P_1P_2 + 1.389P_2^2)$$

$$= (40.04P_1^2 - 12.59P_1P_2 + 3.47P_2^2) \times 10^{-6} \, (\text{N} \cdot \text{mm})$$

By Castigliano's second theorem:

$$D_2 = \frac{\partial U}{\partial P_2} = (-12.59P_1 + 6.94P_2)(10^{-6})$$

Substituting the known loads, the dummy load P_2 being zero, we obtain

$$D_2 = (-12.59 \times 300\,000 + 6.94 \times 0) \times 10^{-6}$$

$$= -\textbf{3.78 mm}$$

The fact that the horizontal deflection, D_2, turns out to be a negative quantity means that the displacement takes place in the direction opposite to the dummy load. Note that the fictitious load was set equal to zero *after* differentiation with respect to P_2 had been carried out. If P_2 had been made equal to zero at any earlier stage, no solution could be obtained and all our effort would have been wasted.

Example 10-13

Determine the horizontal displacement at point A in the structure given in Example 10-10.

Solution. Applying both the real load P_1 and the dummy load P_2 (note that P_2 is in the direction of the required horizontal displacement), we determine the member forces as given in Fig. 10-28.

Again using a table, we determine the strain energy in terms of the loads making no special concession, at this stage, to the dummy load.

i	F_i^2 (kip^2)	L_i (in.)	A_i (in.^{2})	$\dfrac{F_i^2 L_i}{A_i} \dfrac{\text{kip}^2}{\text{in.}^2}$
1	$(1.333 P_1)^2$	144	4.5	$58.86 P_1^2$
2	$(1.667 P_1)^2$	180	5.5	$90.95 P_1^2$
3	$(-2.667 P_1 + 1.000 P_2)^2$ $= 7.113 P_1^2 - 5.334 P_1 P_2$ $+ 1.000 P_2^2$	144	12.5	$81.95 P_1^2 - 61.45 P_1 P_2$ $+ 11.52 P_2^2$
4	$(-1.000 P_1)^2$	108	4.5	$24.00 P_1^2$
5	$(1.667 P_1)^2$	180	5.5	$90.95 P_1^2$
6	$(-1.333 P_1 + 1.000 P_2)^2$ $= 1.777 P_1^2 - 2.667 P_1 P_2$ $+ 1.000 P_2^2$	144	7	$36.56 P_1^2 - 54.86 P_1 P_2$ $+ 20.57 P_2^2$

$$\sum \frac{F_i^2 L_i}{A_i} = 381.27 P_1^2 - 116.31 P_1 P_2 + 32.09 P_2^2$$

$$U = \frac{1}{2E} \sum \frac{F_i^2 L_i}{A_i} = \frac{1}{2 \times 29,000} (381.27 P_1^2 - 116.31 P_1 P_2 + 32.09 P_2^2)$$

$$= (6.574 P_1^2 - 2.005 P_1 P_2 + 0.553 P_2^2) \times 10^{-3} \,(\text{kip} \cdot \text{in.})$$

By Castigliano's second theorem:

$$D_2 = \frac{\partial U}{\partial P_2} = (-2.005 P_1 + 1.106 P_2)(10^{-3})$$

Substituting the known loads, the dummy load P_2 being zero, we obtain

$$D_2 = (-2.005 \times 70 + 1.106 \times 0) \times 10^{-3}$$

$$= -0.140 \text{ in.}$$

The fact that the horizontal deflection, D_2, turns out to be a negative quantity means that the displacement takes place in the direction opposite to the dummy load. Note that the fictitious load was set equal to zero *after* differentiation with respect to P_2 had been carried out. If P_2 had been made equal to zero at any earlier stage, no solution could be obtained and all our effort would have been wasted.

Example 10-14

Determine the deflection at point A and at point B due to the load P applied at point A in the beam shown in Fig. 10-29a.

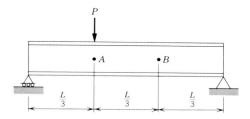

FIG. 10-29a Dummy load method applied to the determination of beam deflection.

Solution. The reactions with a dummy load applied at *B* are found by equilibrium and are shown in Fig. 10-29*b*. The corresponding bending moment diagram is given in Fig. 10-29*c*. The bending moment equations are determined by the method of sections and equilibrium.

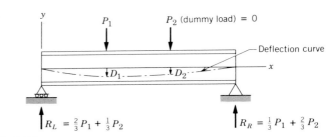

FIG. 10-29*b*

FIG. 10-29*c*

$$M_1 = \left(\frac{2}{3}P_1 + \frac{1}{3}P_2\right)x \qquad\qquad 0 \leqslant x \leqslant \frac{L}{3}$$

$$M_2 = \left(\frac{2}{3}P_1 + \frac{1}{3}P_2\right)x - P_1\left(x - \frac{L}{3}\right) \qquad \frac{L}{3} \leqslant x \leqslant \frac{2L}{3}$$

$$M_3 = \left(\frac{1}{3}P_1 + \frac{2}{3}P_2\right)(L - x) \qquad\qquad \frac{2L}{3} \leqslant x \leqslant L$$

From Eq. (10-14)

$$U = \frac{1}{2EI}\left[\int_0^{L/3} M_1^2\, dx + \int_{L/3}^{2L/3} M_2^2\, dx + \int_{2L/3}^{L} M_3^2\, dx\right]$$

$$\int_0^{L/3} M_1^2\, dx = \left(\frac{2}{3}P_1 + \frac{1}{3}P_2\right)^2 \int x^2\, dx = \left(\frac{2}{3}P_1 + \frac{1}{3}P_2\right)^2 \frac{1}{3}x^3\Big]_0^{L/3}$$

$$= \frac{1}{81}L^3\left(\frac{2}{3}P_1 + \frac{1}{3}P_2\right)^2$$

$$\int_{L/3}^{2L/3} M_2^2\, dx = \int_{L/3}^{2L/3}\left[\left(\frac{2}{3}P_1 + \frac{1}{3}P_2\right)x - P_1\left(x - \frac{L}{3}\right)\right]^2 dx$$

$$= \int_{L/3}^{2L/3} \left[\left(-\frac{1}{3}P_1 + \frac{1}{3}P_2 \right)x + \frac{P_1}{3}L \right]^2 dx$$

$$= \int_{L/3}^{2L/3} \left[\left(-\frac{1}{3}P_1 + \frac{1}{3}P_2 \right)^2 x^2 + \frac{2P_1L}{3}\left(-\frac{1}{3}P_1 + \frac{1}{3}P_2 \right)x + \left(\frac{P_1L}{3} \right)^2 \right] dx$$

$$= \left[\left(-\frac{1}{3}P_1 + \frac{1}{3}P_2 \right)^2 \frac{1}{3}x^3 + \frac{2P_1L}{3}\left(-\frac{1}{3}P_1 + \frac{1}{3}P_2 \right)\frac{1}{2}x^2 + \left(\frac{P_1L}{3} \right)^2 x \right]_{(1/3)L}^{2L/3}$$

$$= \frac{7}{81}L^3\left(-\frac{1}{3}P_1 + \frac{1}{3}P_2 \right)^2 + \frac{1}{9}P_1L^3\left(-\frac{1}{3}P_1 + \frac{1}{3}P_2 \right) + \frac{1}{27}P_1^2L^3$$

$$\int_{2L/3}^{L} M_3^2 \, dx = \int_{2L/3}^{L} \left(\frac{1}{3}P_1 + \frac{2}{3}P_2 \right)^2 (L-x)^2 \, dx$$

$$= \left[\left(\frac{1}{3}P_1 + \frac{2}{3}P_2 \right)^2 \left(-\frac{1}{3} \right)(L-x)^3 \right]_{2L/3}^{L}$$

$$= \frac{1}{81}L^3\left(\frac{1}{3}P_1 + \frac{2}{3}P_2 \right)^2$$

$$U = \frac{1}{2EI}\frac{L^3}{81}\left[\left(\frac{2}{3}P_1 + \frac{1}{3}P_2 \right)^2 + 7\left(-\frac{1}{3}P_1 + \frac{1}{3}P_2 \right)^2 \right.$$
$$\left. + 9\left(-\frac{1}{3}P_1^2 + \frac{1}{3}P_1P_2 \right) + 3P_1^2 + \left(\frac{1}{3}P_1 + \frac{2}{3}P_2 \right)^2 \right]$$

By Castigliano's second theorem:

$$D_1 = \frac{\partial U}{\partial P_1} = \frac{L^3}{162EI}\left[2\left(\frac{2}{3}P_1 + \frac{1}{3}P_2 \right)\frac{2}{3} + 7 \times 2\left(-\frac{1}{3}P_1 + \frac{1}{3}P_2 \right)\left(-\frac{1}{3} \right) \right.$$
$$\left. + 9\left(-\frac{1}{3}2P_1 + \frac{1}{3}P_2 \right) + 6P_1 + 2\left(\frac{1}{3}P_1 + \frac{2}{3}P_2 \right)\frac{1}{3} \right]$$

$$D_2 = \frac{\partial U}{\partial P_2} = \frac{L^3}{162EI}\left[2\left(\frac{2}{3}P_1 + \frac{1}{3}P_2 \right)\frac{1}{3} + 7 \times 2\left(-\frac{1}{3}P_1 + \frac{1}{3}P_2 \right)\frac{1}{3} + 9\left(\frac{1}{3}P_1 \right) \right.$$
$$\left. + 2\left(\frac{1}{3}P_1 + \frac{2}{3}P_2 \right)\frac{2}{3} \right]$$

The displacement at A is

$$D_1]_{\substack{P_1=P \\ P_2=0}} = \frac{L^3}{162EI}\left(\frac{8}{9}P + \frac{14}{9}P - 6P + 6P + \frac{2}{9}P \right)$$

$$= \frac{L^3}{162EI}\left(\frac{8}{3}P \right) = \frac{4}{243}\frac{PL^3}{EI}$$

The displacement at B is

$$D_2]_{\substack{P_1 = P \\ P_2 = 0}} = \frac{L^3}{162EI}\left(\frac{4}{9}P - \frac{14}{9}P + 3P + \frac{4}{9}P\right)$$

$$= \frac{L^3}{162EI}\left(\frac{7}{3}P\right) = \frac{7}{486}\frac{PL^3}{EI}$$

Problems 10-48 to 10-60

Up to this point Castigliano's second theorem has been applied only to statically determinate structures. When an *indeterminate structure* is to be solved, the internal forces in the structure, and hence the strain energy, cannot be written in terms of the loads. This obstruction is overcome by releasing a sufficient number of constraints to make the structure determinate and applying an unknown force (actually, the reaction) to match each released constraint. This introduces an unknown (the reaction) for each release, but since the corresponding displacement is known to be zero, we form one equation for each release. This technique is demonstrated in the following example.

Example 10-15

For the statically indeterminate beam given in Fig. 10-30*a*, determine the vertical reactions and the maximum bending moment.

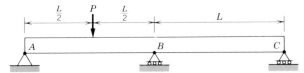

FIG. 10-30*a* Castigliano's second theorem applied to a statically indeterminate beam problem.

Solution. We recognize that the beam is statically indeterminate and, therefore, that we want to find one additional *independent* equation (additional to those available from statics) by means of a displacement condition. We can obtain a solution as follows:

If, as in Fig. 10-30*b*, we release the constraint at A and let the vertical force R represent the reaction at A, we can find the reactions at B and C as a function of R and the applied load, P. Next we can obtain the total strain energy in the beam as a function of R and P and finally differentiate U with respect to R to find D_R, the vertical displacement at A. The fact that in reality $D_R = 0$ gives us the one additional independent equation to solve the indeterminate problem as the following calculations will demonstrate.

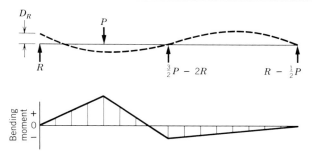

FIG. 10-30b Deflection curve and bending moment diagram.

The moment equations for the three beam regions are

$$M = Rx \qquad\qquad 0 \leqslant x \leqslant \frac{L}{2}$$

$$M = Rx - P\left(x - \frac{L}{2}\right) \qquad \frac{L}{2} \leqslant x \leqslant L$$

$$= (R - P)x + P\frac{L}{2}$$

$$M = \left(R - \frac{1}{2}P\right)(2L - x) \qquad L \leqslant x \leqslant 2L$$

$$U = \frac{1}{2EI} \int_{\text{length}} M^2 \, dx \qquad (10\text{-}14)$$

$$= \frac{1}{2EI}\left\{ \int_0^{L/2} R^2 x^2 \, dx + \int_{L/2}^L \left[(R-P)^2 x^2 + PL(R-P)x + \frac{1}{4}P^2L^2 \right] dx \right.$$

$$\left. + \int_L^{2L} \left[\left(R - \frac{1}{2}P\right)^2 (2L-x)^2 \right] dx \right\}$$

$$= \frac{1}{2EI}\left\{ \left[\frac{1}{3}R^2 x^3 \right]_0^{L/2} + \left[\frac{1}{3}(R-P)^2 x^3 + \frac{1}{2}PL(R-P)x^2 + \frac{1}{4}P^2L^2 x \right]_{L/2}^L \right.$$

$$\left. + \left[-\frac{1}{3}\left(R - \frac{1}{2}P\right)^2 (2L-x)^3 \right]_L^{2L} \right\}$$

$$= \frac{1}{2EI}\left\{ \frac{1}{24}R^2 L^3 + \frac{1}{3}(R-P)^2\left[L^3 - \left(\frac{L}{2}\right)^3 \right] + \frac{1}{2}PL(R-P)\left[L^2 - \left(\frac{L}{2}\right)^2 \right] \right.$$

$$\left. + \frac{1}{4}P^2L^2\left(L - \frac{L}{2}\right) - \frac{1}{3}\left(R - \frac{1}{2}P\right)^2 (-L^3) \right\}$$

We know that $D_R = 0$ or

$$\frac{\partial U}{\partial R} = 0$$

$$\frac{1}{12} RL^3 + \frac{2}{3}(R - P)\frac{7}{8}L^3 + \frac{1}{2} PL\frac{3}{4}L^2 + \frac{2}{3}\left(R - \frac{1}{2}P\right)L^3 = 0$$

$$\frac{1}{12} RL^3 + \frac{7}{12}(R - P)L^3 + \frac{3}{8}PL^3 + \frac{2}{3}RL^3 - \frac{1}{3}PL^3 = 0$$

$$\left(\frac{1}{12} + \frac{7}{12} + \frac{2}{3}\right)RL^3 + \left(-\frac{7}{12} + \frac{3}{8} - \frac{1}{3}\right)PL^3 = 0$$

$$\frac{4}{3} RL^3 - \frac{13}{24}PL^3 = 0$$

$$R = \frac{3}{4}\left(\frac{13}{24}P\right) = \frac{13}{32}P$$

$$M_{max} = \frac{13}{32}P\left(\frac{L}{2}\right) = \frac{13}{64}PL$$

Since we used R at the left support as the redundant force in this example, you might ask if a solution could also have been obtained if the reaction at B or C had been utilized instead. The answer is yes, but a word of caution might be in order. *It is important to choose the redundant force and the origin of x in as simple and straightforward a manner as possible;* otherwise the moment functions can become very cumbersome. In the present case, if the reaction at C had been chosen as the redundant force, then the origin of x should have been at the right end of the beam.

The expression in the above example became much simpler after the operation $\partial/\partial R$ was performed, and it seems likely that many manipulations could have been avoided if we had done the differentiation earlier, that is, before the integration.

By Castigliano's second theorem

$$\frac{\partial U}{\partial R} = 0$$

Substituting for U from (10-13) gives

$$\frac{\partial}{\partial R}\frac{1}{2EI}\int M^2\, dx = 0$$

which can be arranged as

$$\frac{1}{2EI}\int \frac{\partial (M^2)}{\partial R}\, dx = 0$$

or

$$\int_{length} \frac{\partial M^2}{\partial R}\, dx = 0$$

$$\int_{length} 2M\frac{\partial M}{\partial R}\, dx = 0$$

When we now substitute for the three moment regions, we get

$$\int_0^{L/2} [2Rx^2]\,dx + \int_{L/2}^{L} [2(R-P)x^2 + PLx]\,dx + \int_L^{2L} \left[2\left(R - \frac{1}{2}P\right)(2L - x)^2 \right] dx = 0$$

$$\left[\frac{2}{3} Rx^3 \right]_0^{L/2} + \left[\frac{2}{3}(R-P)x^3 + \frac{1}{2}PLx^2 \right]_{L/2}^{L} + \left[-\frac{2}{3}\left(R - \frac{1}{2}P\right)(2L - x)^3 \right]_L^{2L} = 0$$

$$\frac{1}{12}RL^3 + \frac{2}{3}(R-P)\left[L^3 - \left(\frac{L}{2}\right)^3\right] + \frac{1}{2}PL\left[L^2 - \left(\frac{L}{2}\right)^2\right] - \frac{2}{3}\left(R - \frac{1}{2}P\right)(-L)^3 = 0$$

$$\frac{1}{12}RL^3 + \frac{7}{12}(R-P)L^3 + \frac{3}{8}PL^3 + \frac{2}{3}\left(R - \frac{1}{2}P\right)L^3 = 0$$

$$\left(\frac{1}{12} + \frac{7}{12} + \frac{2}{3}\right)RL^3 + \left(-\frac{7}{12} + \frac{3}{8} - \frac{1}{3}\right)PL^3 = 0$$

$$R = \frac{13}{32}P$$

We see that this agrees with the first solution but is the preferred method because it offers fewer opportunities to make an error in the manipulations.

When a truss is statically indeterminate because of a redundant member, the technique differs slightly from that demonstrated above for a beam. The method of solving such a problem is shown in the next example.

Example 10-16

Determine the stress in member AB due to the load acting on the steel structure shown in Fig. 10-31a.

Solution. As in the case of the redundant support of Example 10-15, we remove the redundancy, in this case by cutting any one of the members and substituting a force. We will cut the horizontal member near point A and assume that the tension in the member is T, as illustrated in Fig. 10-31b.

We will consider that the displacement D_T corresponding to T is the relative motion of the two cross-sectional surfaces toward one another. We know that in reality this displacement is zero so that by Castigliano's second theorem

$$\frac{\partial U}{\partial T} = D_T = 0$$

Now that the structure is no longer indeterminate, we can write all member forces, and hence the strain energy, in terms of T and P. Forces in members as found by equilibrium are given in Fig. 10-31c.

From (10-20), we obtain

$$U = \frac{1}{2E} \sum_{i=1}^{3} \frac{F_i^2 L_i}{A_i}$$

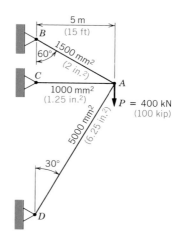

FIG. 10-31a Indeterminate truss problem solved by Castigliano's second theorem.

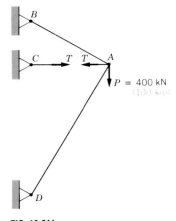

FIG. 10-31b

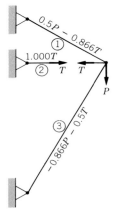

FIG. 10-31c

i	F_i^2 (N^2)	L_i (mm)	A_i (mm^2)	$F_i^2 L_i / A_i$ (N^2/mm)
1	$0.25P^2 - 0.866PT + 0.75T^2$	5773	1500	$0.962P^2 - 3.333PT + 2.887T^2$
2	$1.00T^2$	5000	1000	$+5.000T^2$
3	$0.75P^2 + 0.866PT + 0.25T^2$	10 000	5000	$1.500P^2 + 1.732PT + 0.500T^2$

$$\sum \frac{F_i^2 L_i}{A_i} = (2.462P^2 - 1.601PT + 8.387T^2)$$

$$U = \frac{1}{2 \times 0.2 \times 10^6} \sum \frac{F_i^2 L_i}{A_i} = (6.155P^2 - 4.003PT + 20.968T^2)10^{-6}$$

$$\frac{\partial U}{\partial T} = (-4.003P + 41.935T)10^{-6} = 0$$

$$T = 0.0955P$$

$$= 0.0955 \times 400\ 000 = 38\ 200 \text{ N}$$

tension in $AB = 0.5P - 0.866T$

$$= 0.5 \times 400\ 000 - 0.866 \times 38\ 200 = 167\ 000 \text{ N}$$

stress in $AB = \frac{167\ 000}{1500}\ \frac{\text{N}}{\text{mm}^2} = \textbf{111 MPa}$

Example 10-16

Determine the stress in member AB due to the load acting on the steel structure shown in Fig. 10-31a.

Solution. As in the case of the redundant support of Example 10-15, we remove the redundancy, in this case by cutting any one of the members and substituting a force. We will cut the horizontal member near point A and assume that the tension in the member is T, as illustrated in Fig. 10-31b.

We will consider that the displacement D_T corresponding to T is the relative motion of the two cross-sectional surfaces toward one another. We know that in reality this displacement is zero so that by Castigliano's second theorem

$$\frac{\partial U}{\partial T} = D_T = 0$$

Now that the structure is no longer indeterminate, we can write all member forces, and hence the strain energy, in terms of T and P. Forces in members as found by equilibrium are given in Fig. 10-31c.

From (10-20), we obtain

$$U = \frac{1}{2E} \sum_{i=1}^{3} \frac{F_i^2 L_i}{A_i}$$

i	F_i^2 (kip^2)	L_i (in.)	A_i (in.^{2})	$F_i^2 L_i / A_i$ (kip^2/in.)
1	$0.25P^2 - 0.866PT + 0.75T^2$	270.8	2	$33.85P^2 - 117.26PT + 101.55T^2$
2	$1.00T^2$	180	1.25	$+ 144.00T^2$
3	$0.75P^2 + 0.866PT + 0.25T^2$	360	6.25	$43.20P^2 + 49.88PT + 14.40T^2$

$$\sum \frac{F_i^2 L_i}{A_i} = (77.05P^2 - 67.38PT + 259.95T^2)$$

$$U = \frac{1}{2 \times 29,000} \sum \frac{F_i^2 L_i}{A_i} = (1.328P^2 - 1.162PT + 4.482T^2) 10^{-3}$$

$$\frac{\partial U}{\partial T} = (-1.162P + 8.964T) 10^{-3} = 0$$

$$T = 0.129P$$

$$= 0.129 \times 100 = 12.9 \text{ kip}$$

$$\text{tension in } AB = 0.5P - 0.866T$$

$$= 0.5 \times 100 - 0.866 \times 12.9 = 38.8 \text{ kip}$$

$$\text{stress in } AB = \frac{38.8}{2} = \textbf{19.4 kip/in.}^2$$

Problems 10-61 to 10-69

In all the foregoing developments of energy methods, the loads and their associated displacements were always combined as a product to give work. Up to this point the load has been force, P_i, and the displacement, D_i, has been motion in the *same direction* as the force. This has given us useful equations and theorems such as Castigliano's second theorem, equation (10-32)

$$\frac{\partial U}{\partial P_i} = D_i$$

The developments would have followed the same line and reached the same conclusion if we had been less definite about the load, merely calling it F and the associated displacement, δ, without specifying the nature of the displacement. We would impose the constraint that *the product of F by δ must be work*. With this approach, Castigliano's second theorem, for example, would have evolved as

$$\frac{\partial U}{\partial F} = \delta \qquad \textbf{(10-33)}$$

When using (10-33) to solve truss problems, for example, load F is force and δ is translation; the equation takes the form of (10-32). However, if we encounter problems involving loads that are couples, C_i, and corresponding rotations, θ_i, the product is known to be work, provided that θ_i is in radians. Then (10-33) becomes

$$\frac{\partial U}{\partial C_i} = \theta_i \qquad \textbf{(10-34)}$$

Similarly for torque loads, T_i, and angles of twist, ϕ_i, the product again meets the work stipulation. Equation (10-33) then becomes

$$\frac{\partial U}{\partial T_i} = \phi_i \qquad \textbf{(10-35)}$$

The above illustrations show combinations of forces and displacements in which both force and displacement have some well-known form. In more advanced work it is often necessary to deal with forces that are so obscure that we cannot possibly imagine them, for example, Force × length^2. The displacements then become equally obscure, in this case, length^{-1}, but, provided that the forces and displacements meet the work rule, they can be used in relationships such as Castigliano's second theorem.

To generalize all of the important conclusions of this chapter, we could rewrite the equations as we have done for Castigliano's second theorem. This would produce an inordinate number of equations. Instead we will retain the original equations and merely alter our perception of P_i and D_i. We will treat P_i as any load and D_i as the corresponding displacement such that the product of load by displacement is work.

Example 10-17

Determine the vertical displacement and the rotation at the end of the cantilever beam shown in Fig. 10-32a due to the 40 kN downward load and the 60 kN·m couple.

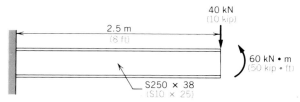

FIG. 10-32a

Solution. Referring to the force and couple at the free end of the cantilever as P_1 and P_2 as indicated in Fig. 10-32b, we want to find $\partial U/\partial P_1$ and $\partial U/\partial P_2$ to solve the problem. From (10-14) the strain energy is

$$U = \frac{1}{2EI} \int_0^L [P_2 - P_1(L - x)]^2 \, dx$$

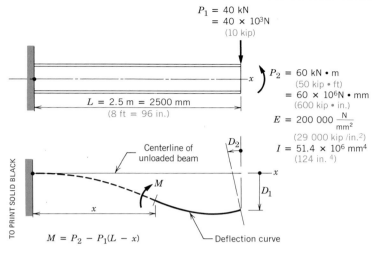

$$M = P_2 - P_1(L - x)$$

FIG. 10-32b

$$= \frac{1}{2EI} \int_0^L [P_2^2 - 2P_1P_2(L-x) + P_1^2(L-x)^2]\, dx$$

$$= \frac{1}{2EI}\left[P_2^2 x + P_1P_2(L-x)^2 - \frac{P_1^2}{3}(L-x)^3 \right]_0^L$$

$$= \frac{1}{2EI}\left[(P_2^2 L + 0 + 0) - \left(P_1P_2 L^2 - \frac{1}{3}P_1^2 L^3 \right) \right]$$

$$= \frac{1}{2EI}\left(P_2^2 L - P_1P_2 L^2 + \frac{1}{3}P_1^2 L^3 \right)$$

$$D_1 = \frac{\partial U}{\partial P_1} = \frac{1}{2EI}\left(-P_2 L^2 + \frac{2}{3}P_1 L^3 \right)$$

$$= \frac{1}{2 \times 200\,000 \times 51.4 \times 10^6}\left(-60 \times 10^6 \times 2500^2 + \frac{2}{3} \times 40 \right.$$

$$\left. \times 10^3 \times 2500^3 \right) \frac{N \cdot mm^3}{N/mm^2 \cdot mm^4}$$

$$= \frac{1}{20.56 \times 10^{12}}(-375 \times 10^{12} + 416.7 \times 10^{12})\ mm$$

$$= \mathbf{2.03\ mm}$$

$$D_2 = \frac{\partial U}{\partial P_2} = \frac{1}{2EI}(2P_2 L - P_1 L^2)$$

$$= \frac{1}{20.56 \times 10^{12}} (2 \times 60 \times 10^6 \times 2500 - 40 \times 10^3 \times 2500^2) \frac{N \cdot mm^2}{N/mm^2 \cdot mm^4}$$

$$= \frac{1}{20.56 \times 10^{12}} (300 \times 10^9 - 250 \times 10^9) \text{ radians}$$

$$= 0.00243 \text{ radians}$$

Example 10-17

Determine the vertical displacement and the rotation at the end of the cantilever beam shown in Fig. 10-32a due to the 10 kip downward load and the 50 kip·ft couple.

Solution. Referring to the force and couple at the free end of the cantilever as P_1 and P_2 as indicated in Fig. 10-31b, we want to find $\partial U/\partial P_1$ and $\partial U/\partial P_2$ to solve the problem. From (10-14) the strain energy is

$$U = \frac{1}{2EI} \int_0^L [P_2 - P_1(L - x)]^2 \, dx$$

$$= \frac{1}{2EI} \int_0^L [P_2^2 - 2P_1 P_2(L - x) + P_1^2(L - x)^2] \, dx$$

$$= \frac{1}{2EI} \left[P_2^2 x + P_1 P_2(L - x)^2 - \frac{P_1^2}{3}(L - x)^3 \right]_0^L$$

$$= \frac{1}{2EI} \left[(P_2^2 L + 0 + 0) - \left(P_1 P_2 L^2 - \frac{1}{3} P_1^2 L^3 \right) \right]$$

$$= \frac{1}{2EI} \left(P_2^2 L - P_1 P_2 L^2 + \frac{1}{3} P_1^2 L^3 \right)$$

$$D_1 = \frac{\partial U}{\partial P_1} = \frac{1}{2EI} \left(-P_2 L^2 + \frac{2}{3} P_1 L^3 \right)$$

$$= \frac{1}{2 \times 29{,}000 \times 124} \left(-600 \times 96^2 + \frac{2}{3} \times 10 \times 96^3 \right) \frac{kip \cdot in.^3}{kip/in.^2 \cdot in.^4}$$

$$= \frac{1}{7.192 \times 10^6} (-5.530 \times 10^6 + 5.898 \times 10^6) \text{ in.}$$

$$= 0.051 \text{ in.}$$

$$D_2 = \frac{\partial U}{\partial P_2} = \frac{1}{2EI} (2P_2 L - P_1 L^2)$$

$$= \frac{1}{7.192 \times 10^6} (2 \times 600 \times 96 - 10 \times 96^2) \frac{kip \cdot in.^2}{kip/in.^2 \cdot in.^4}$$

$$= \frac{1}{7.192 \times 10^6} (115.20 \times 10^3 - 92.16 \times 10^3) \text{ radians}$$

$$= 0.0032 \text{ radians}$$

Problems 10-70 to 10-75

PROBLEMS

10-1 Calculate the strain energy in the structure when the given load is applied.

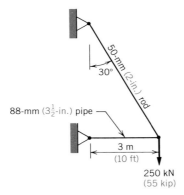

PROBLEM 10-1

10-2 Calculate the strain energy in the structure when displacements $D_1 = 3$ mm (0.12 in.) and $D_2 = 4$ mm (0.16 in.) are imposed.

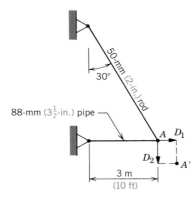

PROBLEM 10-2

10-3 to 10-6 Calculate the bending strain energy in the given cantilever. In all cases the beam is an S510 × 143 (S20 × 96).

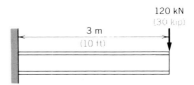

PROBLEM 10-3

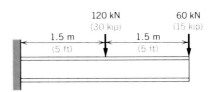

PROBLEM 10-4

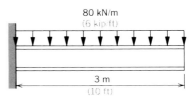

PROBLEM 10-5

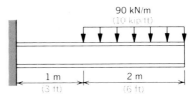

PROBLEM 10-6

10-7 to 10-8 Determine the bending strain energy in terms of E and the given load and dimensions.

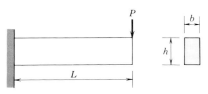

PROBLEM 10-7

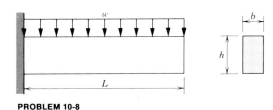

PROBLEM 10-8

10-9 Determine the shear strain energy in terms of G, the load and the dimensions

   a. Given in Prob. 10-7.
   b. Given in Prob. 10-8.

10-10 a. Calculate the total strain energy in the beam given in Prob. 10-7 if the material is steel and $P = 50$ kN (12 kip), $L = 1.2$ m (4 ft), $b = 60$ mm (2.4 in.), and $h = 200$ mm (8 in.).
   b. Calculate the percentage error if the shear strain energy is not included.

10-11 a. Calculate the total strain energy in the beam given in Prob. 10-8 if the material is steel and $w = 80$ kN/m (6 kip/ft), $L = 1.2$ m (4 ft), $b = 60$ mm (2.4 in.), and $h = 200$ mm (8 in.).
   b. Calculate the percentage error if the shear strain energy is not included.

10-12 Calculate the strain energy in a steel member 4 m (12 ft) long that is subjected to a torque of 45 kN·m (33 ft·kip) when the cross section has the given shape. (Note that the maximum shear stress is the same for all four cross sections.)

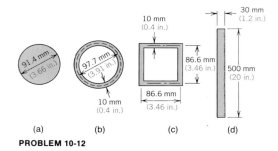

PROBLEM 10-12

10-13 to 10-14 Determine the vertical deflection under the load.

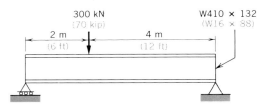

PROBLEM 10-13

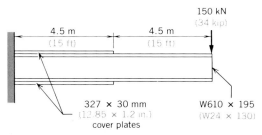

PROBLEM 10-14

10-15 For the truss shown determine by energy considerations the deflection at A under the vertical load. Compare your answer with the deflection found by elementary methods.

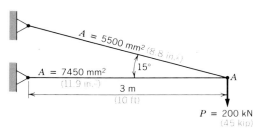

PROBLEM 10-15

10-16 Determine the deflection at A, by energy methods, due to the vertical load P applied at A.

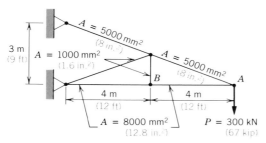

PROBLEM 10-16

10-17 The members of the truss shown below are constructed of channels and cover plates having the given cross-sectional areas. Determine the vertical deflection under a moving vertical load of 400 kN (90 kip) when the load is at point A.

Cross-sectional area in mm² (in.²)

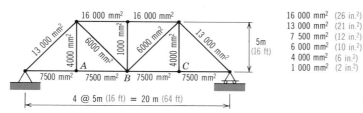

16 000 mm² (26 in.²)
13 000 mm² (21 in.²)
7 500 mm² (12 in.²)
6 000 mm² (10 in.²)
4 000 mm² (6 in.²)
1 000 mm² (2 in.²)

PROBLEM 10-17

10-18 Calculate the deflection under the load in Prob. 10-17 when the load is at point B.

10-19 to 10-26 Determine the deflection, in the direction of the load, at the point of load application.

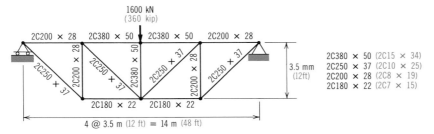

2C380 × 50 (2C15 × 34)
2C250 × 37 (2C10 × 25)
2C200 × 28 (2C8 × 19)
2C180 × 22 (2C7 × 15)

PROBLEM 10-19

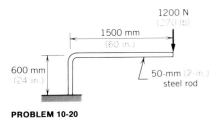

PROBLEM 10-20

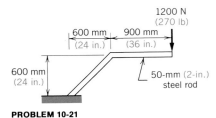

PROBLEM 10-21

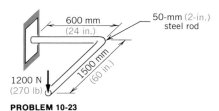

PROBLEM 10-22

PROBLEM 10-23

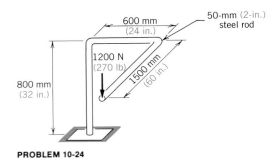

PROBLEM 10-24

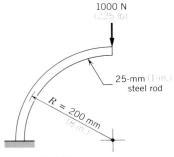

PROBLEM 10-25

Load, 1000 N (225 lb), normal
to plane of figure

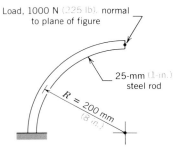

PROBLEM 10-26

10-27 A coil spring has N_c active coils of diameter $\bar{D}$. The spring is formed from wire having a diameter D. Using the direct energy method, derive a formula for the deflection of the spring, when an axial load P is applied. Compare your answer with equation (7-15).

10-28 Calculate the deflection under the load taking into account

a. Bending strain energy only.

b. Both bending and shear strain energy. Assume that there is no shear stress in the flanges and that the shear stress in the web is constant over the web cross section.

c. Determine the error (%) in the deflection caused by omitting the shear strain energy.

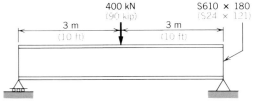

PROBLEM 10-28

10-29 Rework Prob. 10-15 using the minimum potential energy approach.

10-30 The members of the structure are steel. Determine the horizontal displacement at A when the load is applied.

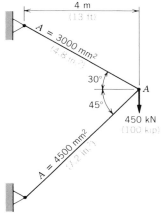

PROBLEM 10-30

10-31 Determine the stresses in the members of the given structure.

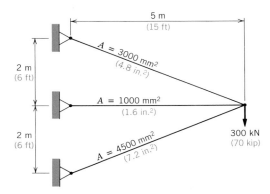

PROBLEM 10-31

10-32 All members of the truss have cross-sectional areas of 1800 mm² (3 in.²). Determine the stresses in the members.

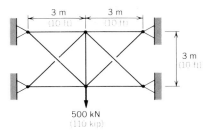

PROBLEM 10-32

10-33 A simply supported beam carries a concentrated load, P, at a point one-third of the length from the right end.

    a. Determine the deflection at midspan by the Rayleigh-Ritz method. Use a displacement function consisting of two trigonometric terms.

    b. By an exact analysis the center deflection is $0.01775 \, PL^3/EI$. What is the error (%) in your Rayleigh-Ritz solution?

    c. Draw a bending moment diagram for the exact moment and that given by the Rayleigh–Ritz method.

10-34 Repeat Prob. 10-33 using a displacement function consisting of three trigonometric functions.

10-35 to 10-38 Rework Probs. 10-29 to 10-32 using Castigliano's first theorem.

10-39 Use Castigliano's first theorem to determine

    a. The displacements at point A when load P is applied.

    b. The stress in the vertical member.

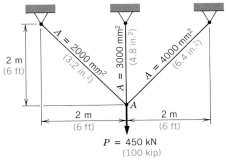

PROBLEM 10-39

10-40 Use the stiffness method to determine the displacements at *A*, due to the given loads, and the stresses in the members.

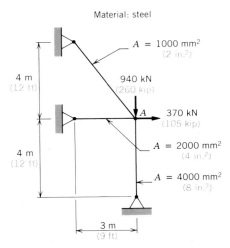

Material: steel

PROBLEM 10-40

10-41 to 10-44 Use the stiffness method to determine the stresses in the members of the steel structures.

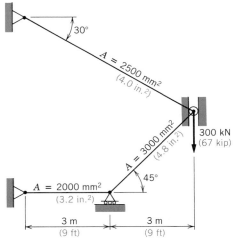

PROBLEM 10-41

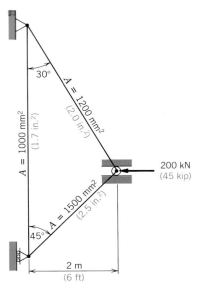

PROBLEM 10-42

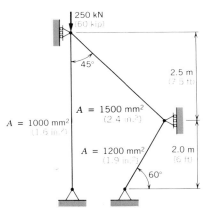

PROBLEM 10-43

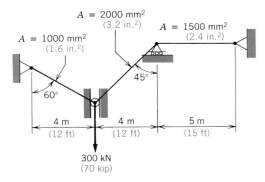

PROBLEM 10-44

10-45 Use the stiffness method to determine the displacements in the given steel structure.

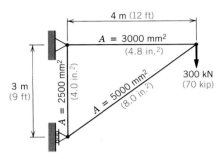

PROBLEM 10-45

10-46 Determine the stiffness matrix for the given steel structure.

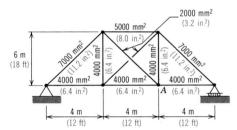

PROBLEM 10-46

10-47 Determine the stiffness matrix for the given steel structure.

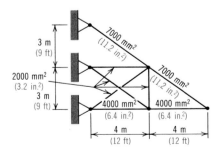

PROBLEM 10-47

10-48 to 10-51 Use Castigliano's second theorem to determine the vertical deflection at point *A*. Consider the bending strain energy only. The beams are all S510 × 128 (S20 × 86) and are 4 m (12 ft) long.

PROBLEM 10-48

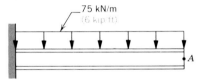

PROBLEM 10-49

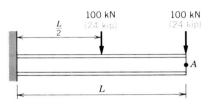

PROBLEM 10-50

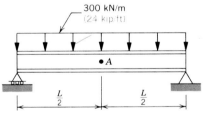

PROBLEM 10-51

10-52 For the simply supported beam loaded at the third point, determine the deflection under the load.

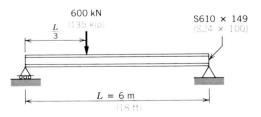

PROBLEM 10-52

10-53 Determine the deflection at midspan of the beam in Prob. 10-52.

10-54 Using Castigliano's second theorem, determine the vertical displacement at *A* in Prob. 10-30.

10-55 Rework Prob. 10-30 using Castigliano's second theorem.

10-56 Rework Prob. 10-16 using Castigliano's second theorem.

10-57 In Prob. 10-16, use Castigliano's second theorem to determine the deflection at *B* due to load *P* applied at *A*.

10-58 For each frame determine the horizontal displacement at point *A* due to the given load. In all cases the steel section is a W760 × 314 (W30 × 210).

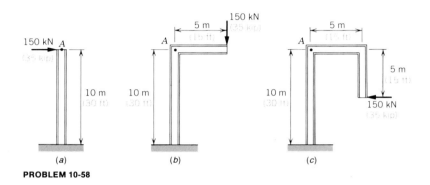

(a)   (b)   (c)

PROBLEM 10-58

10-59 Determine the horizontal displacement at the end of the rod in Prob. 10-20.

10-60 Determine the horizontal displacement at the end of the rod in Prob. 10-21.

10-61 Determine the load on the roller at midspan.

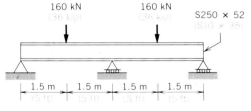

PROBLEM 10-61

10-62 Use Castigliano's second theorem to determine the force on the roller at the left end of the beam.

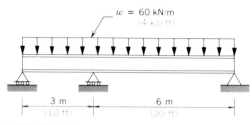

PROBLEM 10-62

10-63 Use Castigliano's second theorem to determine the force on the roller supporting the end of the given cantilever beam.

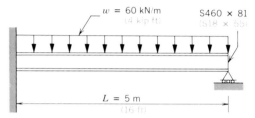

PROBLEM 10-63

10-64 Before the load is applied to the beam given in Prob. 10-63, there is a space of 20 mm (0.8 in.) between the end of the beam and the roller. Determine the force on the roller when the distributed load is applied.

10-65 Determine the horizontal deflection in the frame

    a. At the point of application of the load.
    b. At the roller.

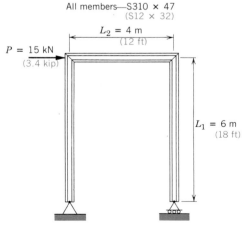

PROBLEM 10-65

10-66 Determine the horizontal deflection at the point of application of the load in the frame given in Prob. 10-65 when the roller is replaced by a pinned support.

10-67 Calculate the maximum bending stress.

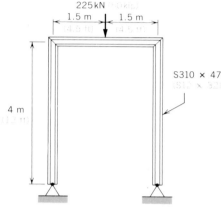

PROBLEM 10-67

10-68 Use Castigliano's second theorem to solve Prob. 10-31.

10-69 Use Castigliano's second theorem in Prob. 10-46 to find the vertical deflection at A due to a vertical load of 500 kN (110 kip) applied at A.

10-70 Solve Prob. 10-39 using Castigliano's second theorem.

10-71 Determine the rotation of the left end of the beam in Prob. 10-52.

10-72 Calculate the rotation at the end of the cantilevers.

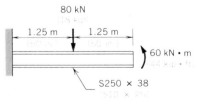

PROBLEM 10-72 (*a*)

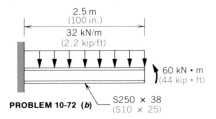

PROBLEM 10-72 (b) — S250 × 38 (S10 × 25)

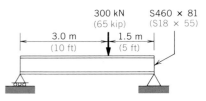

PROBLEM 10-74

10-73 Calculate the vertical displacement at the end of the cantilevers given in Prob. 10-72.

10-74 Calculate the rotation at

a. The left end.

b. The right end due to the given concentrated load.

10-75 Calculate the deflection in the y-direction at the point A and the rotation about the x-axis at point B due to the given load.

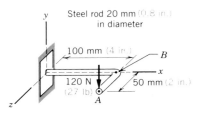

PROBLEM 10-75

Challenging Problems

10-A The vertical members A-C and E-G are made of four L200 × 200 × 30 (L8 × 8 × 1⅛) welded together to form box sections. The other members consist of pairs of L75 × 75 × 10 (L3 × 3 × ⅜) and can be considered to be pin-connected at A, B, D, E, and F. Determine

a. The lateral deflection at A.

b. The maximum bending stress in the vertical members due to the horizontal load at A.

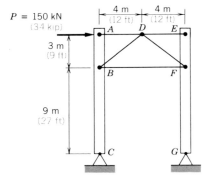

PROBLEM 10-A

10-B A hollow rectangular steel section measures 120 by 200 mm (4.8 by 8.0 in.) on the outside and has an 8-mm (0.32-in.) wall. The member is 4 m (12 ft) long

and has one end built-in. Calculate the rotation of the other end when a torque load of 40 kN·m (30 kip·ft) is applied.

10-C In the given position there is no force in the spring. Plot PE as a function of D for D ranging from −200 to 1000 mm (−8 to 40 in.). From the curve find the stable and unstable equilibrium positions when

a. W = 9 kN (2 kip).

b. W = 18 kN (4 kip).

c. W = 27 kN (6 kip).

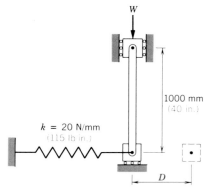

PROBLEM 10-C

10-D Calculate the stress in member *A-B*.

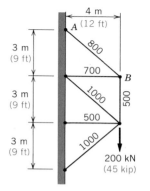

	mm²	(in.²)
	800	(1.3)
	700	(1.1)
	1000	(1.6)
	500	(0.8)

PROBLEM 10-D

10-E Calculate the maximum bending stress.

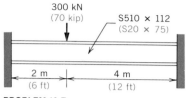

PROBLEM 10-E

11

PRESSURE
VESSELS

11-1 INTRODUCTION

The design methods given in Chapter 1 are adequate for determining the required thickness for cylindrical pressure vessels and pipes in a great many practical cases. Stress formulas based on (1-7) have been used to design the great bulk of the pressure vessels and piping now in existence. If hemispherical end caps having the same thickness as the shell are to be used to complete the vessel, no more formulas are required to design reasonably good pressure vessels. This is because, as we shall see later, such a hemispherical end cap would be stressed to about half the level of the shell; hence the design of the cap would be very conservative.

In reality the hemispherical end cap, or *head*, of Fig. 11-1a is seldom used, instead shapes as shown in Fig. 11-1b and c are commonly employed. The ellipsoidal head can be derived from the spherical head by foreshortening all the axial coordinates by a fixed factor so that the profile is changed from a circle to an ellipse. The profile of the torospherical head consists of circular arcs. The radius near the edge, known as the "knuckle radius," is small while the radius at the center is large. This profile is rotated about the vessel centerline to generate the head. The knuckle curve sweeps out a portion of a torus while the long-radius curve sweeps out a portion of a sphere, hence the name *torospherical* head.

For a practical comparison consider three heads that have the same thickness as the cylindrical part of the shell. The hemisphere is at a disadvantage because it requires more material and makes the vessel longer than necessary. The ellipsoidal head may seem to be preferable to the torospherical head because its profile can be described in a single mathematical equation. However, that equation is of such a nature that lengthy computations are required for the stress analysis of an ellipsoidal head.

All heads are formed by pressing red-hot disks, called blanks, between pairs of dies shaped to match the inside and outside surfaces of the required head. The dies for hemispherical heads are very simple to produce and to check for wear while those for the torospherical heads are only slightly more complicated. The dies for ellipsoidal heads present the die maker with some very difficult problems; when the head thickness is taken into account the surface profiles are no longer ellipses.

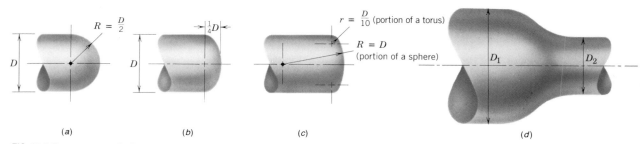

FIG. 11-1 Pressure vessel shapes commonly used in practice. (*a*) Hemispherical head. (*b*) Ellipsoidal head. (*c*) Torospherical head. (*d*) Transition piece.

When all factors are taken into consideration it is easy to understand why the torospherical head is almost always selected by pressure-vessel designers.

The fundamental difference between the vessels that can be analyzed by the equations of Chapter 1 and the vessels shown in Fig. 11-1 is that in the latter group the shell is curved in two directions. We will now derive the general equation, which relates stress to pressure in a shell having curvature in two directions, and we will see that the formulas of Chapter 1 are special cases of the general shell equation.

11-2 GENERAL FORMULA FOR SHELL STRESS

Consider a location, such as point P on the vessel shown in Fig. 11-2, at which the stresses due to pressure are to be determined. The vessel resembles a football but is intended to represent a vessel of any shape. The chosen location, P, is on top of the vessel, but the point could be anywhere on the surface, the important feature of the location being only that there is curvature in two directions. The point chosen is convenient in that the standard views show the curvature in both directions without distortion. The results would be the same, but the development much more lengthy, if we had chosen a point at a more general location such as P'.

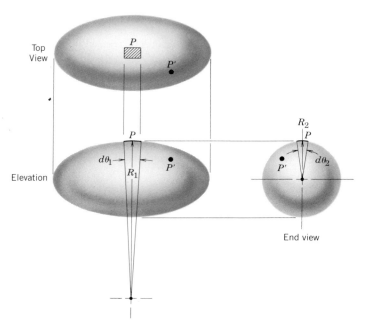

FIG. 11-2 Pressure vessel showing element in the neighborhood of P for derivation of the general shell equation.

Taking an element from the shell, in the neighborhood of P, as a free body, we have the dimensions and stresses as shown in Fig. 11-3a. Note that no provision is made for transverse shear stress. The tacit assumption that the shear stress is zero has implications that will be discussed later. The forces on the element resulting from stress and pressure are shown in Fig. 11-3b. The pressure p acting on the wall puts a radial compressive stress of that magnitude into the metal at the inner surface of the wall. The radial stress diminishes to zero at the outer surface of the wall in a manner that we are not prepared to determine at this stage.

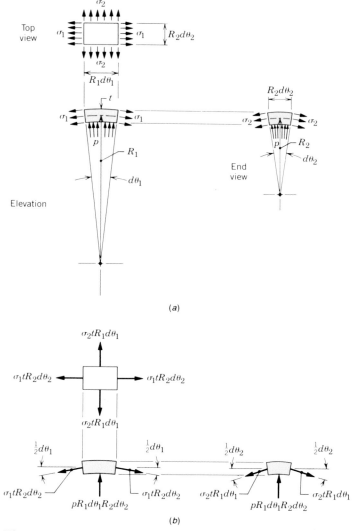

FIG. 11-3 Pressure vessel element considered for derivation of the general shell equation. (a) Dimensions and stresses. (b) Forces acting on the surfaces of the element.

Since in a thin-walled vessel the radial stress is small relative to σ_1 and σ_2, it will not usually be taken into account. Thus a rather complicated stress state is reduced to a simple biaxial state for design purposes.

As in our previous treatment of shells, stresses σ_1 and σ_2 are taken as being constant through the wall thickness. The stresses will vary significantly if the wall is thick, but for thin walls the distribution is substantially constant. Thus we are again developing *thin-wall* theory.

For the upward forces in Fig. 11-3b to be in equilibrium:

$$pR_1\,d\theta_1\,R_2\,d\theta_2 - \sigma_1 tR_2\,d\theta_2 \sin\left(\tfrac{1}{2}\,d\theta_1\right) - \sigma_1 tR_2\,d\theta_2 \sin\left(\tfrac{1}{2}\,d\theta_1\right)$$

$$- \sigma_2 tR_1\,d\theta_1 \sin\left(\tfrac{1}{2}\,d\theta_2\right) - \sigma_2 tR_1\,d\theta_1 \sin\left(\tfrac{1}{2}\,d\theta_2\right) = 0$$

Since the sine of a small angle is approximately equal to the angle, the equation can be simplified to

$$\sigma_1 tR_2\,d\theta_2\,d\theta_1 + \sigma_2 tR_1\,d\theta_1\,d\theta_2 = pR_1\,d\theta_1\,R_2\,d\theta_2 \qquad \textbf{(11-1)}$$

We will find it useful, in interpreting some of the unexpected results that we will encounter later, to examine the terms in (11-1) for physical significance. The term on the right is the force that is exerted by pressure acting on the surface of the element and so is the load on the element. The load can be considered to be carried partly by stress σ_1 and partly by stress σ_2, the portion of the load carried by σ_1 being the first term on the left, the portion carried by stress σ_2 being the second term. It is obvious that (11-1) can be simplified by dividing both sides by $d\theta_1$ and $d\theta_2$. We will go further and divide all terms by $R_1\,d\theta_1\,R_2\,d\theta_2$ to give

$$\boxed{\dfrac{\sigma_1 t}{R_1} + \dfrac{\sigma_2 t}{R_2} = p} \qquad \textbf{(11-2)}$$

Equation (11-2) is referred to as the *general formula for shell stress.*

Although *p* is not, strictly speaking, a load, we will still find it useful to refer to it as the load on the shell, and likewise to refer to the first term in (11-2) as the part of the load carried by σ_1, and the second term as the part carried by σ_2.

If we attempt to solve for stresses in a given shell containing fluid under a known pressure, we can in general not obtain a solution. We find that one equation is, as we might expect, not sufficient to evaluate both stresses. The equation is valid and useful, but we must find one stress by some other means. When the shell has an axis of symmetry, which is usually the case in pressure vessels, one stress can always be found by the method illustrated in the following example, and then the second stress determined by substitution into (11-2).

Example 11-1

A pressure vessel wall is 15 mm thick and the pressure inside the vessel is 10 000 kPa. If the end closure is as shown in Fig. 11-4a, find the stresses in the plane of the diagram and normal to the plane at point P.

Solution.   As indicated in Fig. 11-4b, the radius of curvature of the element in the plane of the figure is

$$R_1 = 300 \text{ mm}$$

Normal to the plane of the figure, the radius of curvature is not so obvious. By making use of your knowledge of descriptive and analytical geometry, you will be able to confirm that it is the length dimensioned as R_2 in Fig. 11-4b. By trigonometry we obtain

$$R_2 = 300 - \frac{100}{\cos 30°} = 184.5 \text{ mm}$$

Substituting into (11-2) yields

$$\frac{\sigma_1 \times 15}{300} + \frac{\sigma_2 \times 15}{184.5} = 10.0 \tag{1}$$

This is a valid equation but it contains two unknown stresses; hence we must determine one stress by another method. To solve for σ_1, consider the free body of Fig. 11-4c consisting of the shell to the right of a transverse plane through P and the fluid inside that portion of the shell.

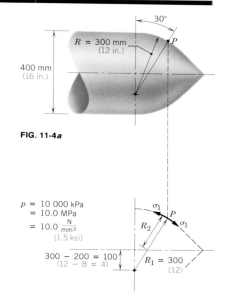

FIG. 11-4a

$p = 10\ 000 \text{ kPa}$
$\quad = 10.0 \text{ MPa}$
$\quad = 10.0 \dfrac{\text{N}}{\text{mm}^2}$
$\quad\quad (1.5 \text{ ksi})$

$300 - 200 = 100$
$(12 - 8 = 4)$

$R_1 = 300$
(12)

FIG. 11-4b

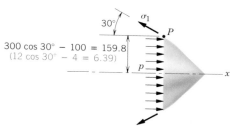

$300 \cos 30° - 100 = 159.8$
$(12 \cos 30° - 4 = 6.39)$

FIG. 11-4c

$$\sum F_x = 0$$

$$\pi(159.8)^2 p - \sigma_1 \times 15$$

$$\times 2\pi \times 159.8 \times \cos 30° = 0$$

$$\sigma_1 = \frac{\pi(159.8)^2\, 10.0}{15 \times 2\pi \times 159.8 \times \cos 30°} \quad \frac{\text{mm}^2 \cdot \text{N/mm}^2}{\text{mm} \cdot \text{mm}}$$

$$= 61.5 \ \frac{\text{N}}{\text{mm}^2}$$

Substituting into (1) gives

$$\frac{61.5 \times 15}{300} + \frac{\sigma_2 \times 15}{184.5} = 10 \tag{2}$$

$$3.08 + 0.0813\sigma_2 = 10$$

$$0.0813\sigma_2 = 10 - 3.08 = 6.92 \tag{3}$$

$$\sigma_2 = 85.1 \ \frac{\text{N}}{\text{mm}^2}$$

Notice that the pressure load (10.0 N/mm^2) is distributed between both load carrying terms in the ratio of (3.08/10 = 0.308 =) 31% to σ_1 and (6.92/10 = 0.692 =) 69% to σ_2, and also that, as we might have expected, both stresses are positive and hence tensile. If the value of R_1(= 300) in **(2)** had been much smaller, as it would be in a typical torospherical head, the first term would have been much larger, say, 30. This would have made the right-hand side of **(3)**, (10 − 30 = −20), negative. The resulting negative sign attached to σ_2 would indicate a compressive stress. A compressive stress in a vessel containing fluid under pressure seems at first to be absurd, and a natural reaction would be to look for the error in the calculations and in the theory. However, no error will be found; a careful examination using your engineering instinct and intuition will lead you to conclude that the first impression of the tensile action was wrong and there is indeed compressive stress in the circumferential direction. The assignment problems offer you an opportunity to solve cases in which there is compressive circumferential stress.

Example 11-1

A pressure vessel wall is 0.6 in. thick and the pressure inside the vessel is 1.5 ksi. If the end closure is as shown in Fig. 11-4a, find the stresses in the plane of the diagram and normal to the plane at point P.

Solution.   As indicated in Fig. 11-4b, the radius of curvature of the element in the plane of the figure is

$$R_1 = 12 \text{ in.}$$

Normal to the plane of the figure, the radius of curvature is not so obvious. By making use of your knowledge of descriptive and analytical geometry, you will be able to confirm that it is the length dimensioned as R_2 in Fig. 11-4b. By trigonometry we obtain

$$R_2 = 12 - \frac{4}{\cos 30^\circ} = 7.38 \text{ in.}$$

Substituting into (11-2) yields

$$\frac{\sigma_1 \times 0.6}{12} + \frac{\sigma_2 \times 0.6}{7.38} = 1.5 \tag{1}$$

This is a valid equation but it contains two unknown stresses; hence we must determine one stress by another method. To solve for σ_1, consider the free body of Fig. 11-4c consisting of the shell to the right of a transverse plane through P and the fluid inside that portion of the shell.

$$\sum F_x = 0$$

$$\pi(6.39)^2 p - \sigma_1 \times 0.6$$

$$\times 2\pi \times 6.39 \times \cos 30^\circ = 0$$

$$\sigma_1 = \frac{\pi(6.39)^2 1.5}{0.6 \times 2\pi \times 6.39 \times \cos 30^\circ} \frac{\text{in.}^2 \cdot \text{kip/in.}^2}{\text{in.} \cdot \text{in.}}$$

$$= 9.22 \text{ kip/in.}^2$$

Substituting into **(1)**

$$\frac{9.22 \times 0.6}{12} + \frac{\sigma_2 \times 0.6}{7.38} = 1.5 \qquad \textbf{(2)}$$

$$0.46 + 0.0813\sigma_2 = 1.5$$

$$0.0813\sigma_2 = 1.5 - 0.46 = 1.04 \qquad \textbf{(3)}$$

$$\sigma_2 = \textbf{12.8 kip/in.}^2$$

Notice that the pressure load (1.5 ksi) is distributed between both load carrying terms in the ratio of $(0.46/1.5 = 0.307 =)$ 31% to σ_1 and $(1.04/1.5 = 0.693 =)$ 69% to σ_2, and also that, as we might have expected, both stresses are positive and hence tensile. If the value of $R_1(= 12)$ in **(2)** had been much smaller, as it would be in a typical torospherical head, the first term would have been much larger, say, 4. This would have made the right hand side of **(3)**, $(1.5 - 4 = -2.5)$, negative. The resulting negative sign attached to σ_2 would indicate a compressive stress. A compressive stress in a vessel containing fluid under pressure seems at first to be absurd, and a natural reaction would be to look for the error in the calculations and in the theory. However, no error will be found; a careful examination using your engineering instinct and intuition will lead you to conclude that the first impression of the tensile action was wrong and there is indeed compressive stress in the circumferential direction. The assignment problems offer you an opportunity to solve cases in which there is compressive circumferential stress.

In the derivation of the shell formula the element in Fig. 11-3 was taken without shear stress on the edges. Recalling the relationship between the shear and bending moment in beams, we know that a beam with zero shear must be subjected to a constant bending moment or, conversely, a beam with a constant bending moment of any magnitude, including zero, is subjected to zero shear. In terms of the pressure vessel wall we can say that if there is no bending moment in the wall, there is no shear. If the wall had been made of a material that could not sustain a bending moment, such as a thin membrane, then the shear assumption would have been valid. The analysis that led to (11-2) can then be described as *membrane theory* and the stresses described as *membrane stresses*. Note that membrane theory is not to be confused with the membrane analogy discussed previously in Chapter 7 in connection with torsion.

Membrane theory enables us to design many practical pressure vessels because generally we are dealing with cases where the wall thickness is very small relative to the other dimensions and hence the behavior is substantially that of a membrane. When the wall is thick, as it is in some practical cases, the flexural rigidity may be substantial and the stresses may deviate significantly from those given by membrane theory. The source of the stress deviation can be seen in a pressure vessel with a torospherical head. If we consider the cylindrical shell as not attached to the head, its diameter will increase due to strain in the material. An unattached toroidal region of the head would decrease in diameter because of the large compressive circumferential stress. Since, in practice, the torospherical head would be welded to the cylindrical shell, the head is not free to contract, and a substantial amount of shear and bending results in the region of the junction. This action significantly alters the stress patterns and greatly reduces the compressive stresses that are produced by membrane action. As a result, in practice, heads that appear to have excessive stress when analyzed by membrane theory may in fact be quite safe. In such cases the design must be verified by experimental evidence or by analysis using numerical methods that are beyond the scope of this book.

We will now consider two special applications of the general shell equation (11-2). If it is used in the case of a cylindrical shell, Fig. 11-5, the equation becomes

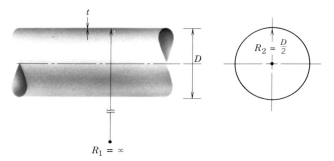

FIG. 11-5 Cylindrical shell.

$$\frac{\sigma_1 t}{\infty} + \frac{\sigma_2 t}{D/2} = p$$

Because of zero curvature in the longitudinal direction, the radius of curvature R_1 is infinite and the first term becomes zero. This is an indication that the longitudinal stress, irrespective of its magnitude, is incapable of carrying any of the pressure load. The equation can then be solved for the remaining stress, giving

$$\sigma_2 = \frac{Dp}{2t}$$

This stress is in the circumferential direction and is seen to be identical to that obtained by (1-7).

As the second special case consider a spherical shell, or a segment of a spherical shell, having radius R_s. From symmetry σ_1 and σ_2 are identical and will be represented by σ_s. Then (11-2) becomes

$$\frac{\sigma_s t}{R_s} + \frac{\sigma_s t}{R_s} = p$$

$$\sigma_s = \frac{R_s p}{2t} \qquad\qquad \textbf{(11-3)}$$

In terms of the diameter of a sphere

$$\sigma_s = \frac{D_s p}{4t} \qquad\qquad \textbf{(11-4)}$$

When compared with a cylinder of the same material and with equal diameters and thicknesses, it is seen that a sphere can safely contain fluid at twice the pressure. This can be accounted for by noting that the equal curvature in two directions causes the pressure load to be equally distributed between the stress in both directions, thus doubling the load-carrying capacity relative to the cylinder.

Pressure vessels of the type that we have been considering normally contain fluid at a pressure that is greater than the surrounding pressure. The stresses are then mainly tensile and, if they are less than the allowable stress, the vessel is safe. If, however, the vessel is subjected to external pressure, as for example the hull of a submarine, the membrane stresses become compressive and the vessel could fail by buckling. Just as axially loaded members are usually substantially weaker in compression than in tension, a pressure vessel subjected to external pressure may fail by buckling at a very low value of stress. Buckling can occur when a vessel is subjected to external pressure or when internal pressure acts on the convex side of an element of a vessel. To illustrate the latter case, imagine the heads shown in Fig. 11-1(a), (b), and (c) to be reversed so that they are turned inward with the fluid pressure acting on the convex surface. The buckling strength of pressure vessels is very difficult to determine and is beyond the scope of our present studies.

Problems 11-1 to 11-9

11-3 MEMBRANE SHELL THEORY AND MEMBRANE ANALOGY

In Chapter 7 we made use of the membrane analogy as an intuitive method for determining the regions of high shear stress in members subjected to torque. Although governed by the same equation, that is, (11-2), the problem to be solved in the analysis of a shell is quite different from that of the analogy.

Shell theory is used to determine stresses in a shell that has given curvatures. It is assumed that the curvatures do not change substantially when the pressure is applied and stresses are obtained by considerations of equilibrium.

Solutions by the membrane analogy are based on a membrane, either a thin sheet of rubber or a soap film, that is stretched so that it is under uniform stress before the pressure load is applied. The stress is assumed to be unaltered by the application of the pressure load, an exact assumption in the case of the soap film and a reasonably good assumption in the case of the rubber membrane provided that the deflection is small. Then the shell equation (11-2), when applied to the membrane used in the membrane analogy, becomes

$$\frac{\sigma t}{R_1} + \frac{\sigma t}{R_2} = p \tag{11-5}$$

where R_1 and R_2 are the unknowns. In the soap film, σt is replaced by twice the surface tension. But that is immaterial, the important point still being that the radii are the unknowns. A typical application is shown in Fig. 11-6 where the deflection of the membrane is required in order to calculate the shear stress in a shaft having a rectangular cross section. Since the membrane deflection is better defined by w, the displacement normal to the x-y plane, we can substitute

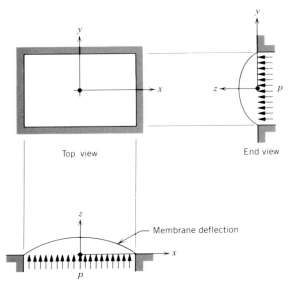

Top view

End view

Membrane deflection

Elevation

FIG. 11-6

$$-\frac{\partial^2 w}{\partial x^2} = \frac{1}{R_1} \tag{11-6}$$

and

$$-\frac{\partial^2 w}{\partial y^2} = \frac{1}{R_2} \tag{11-7}$$

giving the equilibrium equation

$$\frac{\partial^2 w}{\partial x^2} + \frac{\partial^2 w}{\partial y^2} = -\frac{p}{\sigma t} \tag{11-8}$$

Equations (11-6) and (11-7) are valid only when $\partial w/\partial x$ and $\partial w/\partial y$ are small. However, this is not a new restriction, since it was previously stated that the membrane analogy requires that the deflection be small.

Although membrane deflection has been discussed as though it is a useful experimental technique, it is not often used because of practical difficulties. It is, as mentioned earlier, a useful way of visualizing shear stress distribution and of locating regions of high stress. It is also useful in that it gives us (11-8), which can be solved analytically for some simple cases. In more advanced cases it can be solved by some numerical method such as replacing the differentials by finite difference approximations. Although numerical solutions are becoming increasingly important, the techniques will not be dealt with here.

11-4   FLAT PLATES SUPPORTING A UNIFORM PRESSURE

For practical reasons, designers are often obliged to use flat plates to support uniformly distributed loads. Examples of such cases are shown in Fig. 11-7. The flat elements have infinite radii of curvature in both directions; hence both terms on the left side of (11-2) are zero. This means that the flat plate is incapable of supporting a normal pressure load by membrane stresses. The plate must then support the load by bending stresses, in a manner similar to a beam, and must be designed so that the bending stresses are not excessive.

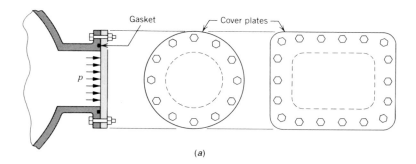

(a)

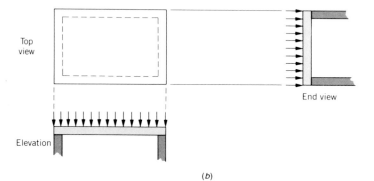

(b)

FIG. 11-7 Examples of flat plates supporting uniform pressure. (a) Circular or rectangular cover plate over an opening in a pressure vessel. (b) Floor slab.

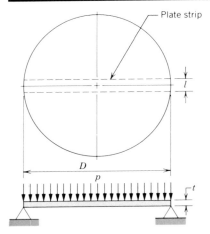

FIG. 11-8 Circular plate with simply supported edges.

The relationships between the load, dimensions, and stress for various plate elements can be derived from theories that will not be presented here; only the final results will be given.

Consider first a circular plate with simply supported edges subjected to a uniform pressure load as shown in Fig. 11-8. Without an adequate theory we might be inclined to take a diametral strip of uniform width and treat it as a beam. The maximum moment would then be

$$M_{\text{max(strip)}} = \tfrac{1}{8} p l D^2$$

and the maximum bending stress

$$\sigma_{\text{max(strip)}} = \frac{M_{\text{max}}}{\tfrac{1}{6} l t^2} = \frac{3}{4} p \left(\frac{D}{t} \right)^2$$

This method ignores the support that would be given by a strip at right angles to the one chosen. The stresses and strains in these strips are coupled by the Poisson effect, and so we should not be surprised to find that the actual stress is less than that given above and that stress is in some way dependent on Poisson's ratio. Well-established theory leads to the conclusion that the maximum bending stress for a simply supported disk occurs at the center and is given by

$$\sigma_{\text{max}} = \frac{3 + v}{8} \sigma_{\text{max(strip)}}$$

$$= \frac{3(3 + v)}{32} p \left(\frac{D}{t} \right)^2 \tag{11-9}$$

When a circular plate has built-in edges we would expect, from our experience with built-in beams, the maximum bending stress to occur at the support. The maximum bending stress does, in fact, occur at the edge support and is given by

$$\sigma_{\text{max}} = \frac{1}{4} \sigma_{\text{max(strip)}}$$

$$= \frac{3}{16} p \left(\frac{D}{t} \right)^2 \tag{11-10}$$

It is interesting to note that in this case the stress is independent of Poisson's ratio.

A rectangular plate can be imagined to carry the load by the beam action of strips running laterally and longitudinally in the plate as shown in Fig. 11-9.

Since the transverse strip is shorter, it will be more rigid and will carry a greater portion of the load than the longitudinal strip. For an infinitely long plate, all of the load will be carried by strips running laterally. The maximum bending stress will then be

$$\sigma_{\text{max(strip)}} = \frac{3}{4} p \left(\frac{a}{t} \right)^2 \tag{11-11}$$

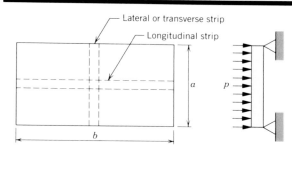

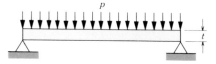

FIG. 11-9 Rectangular plate with simply supported edges.

When the plate has a finite length, the stress will be reduced from that given by (11-11) by a factor C. Then

$$\sigma_{\max} = C \times \sigma_{\max(\text{strip})}$$

$$= C \times \frac{3}{4} p \left(\frac{a}{t} \right)^2 \qquad \textbf{(11-12)}$$

Values for C can be taken from the curves in Fig. 11-10 for either simply supported or built-in edges. For simply supported edges, C varies over a wide range depending mainly on the length-to-width ratio and also slightly on Poisson's ratio. Note that for most metals the curve for $v = 0.3$ can be used. For other values of Poisson's ratio, which is always between 0.0 and 0.5, interpolation can be used to get the required C. If Poisson's ratio is not known, the curve for $v = 0.5$ can be used, thus introducing a safe-side error that in most cases will be quite small. For simply supported plates the maximum stress occurs on the surface at the center of the plate and is in the direction of the short span.

When the edges are built-in, the maximum stress is independent of Poisson's ratio; hence a single curve is sufficient for C rather than a family of curves. This is consistent with circular disks where Poisson's ratio was also not a factor when the edge was built-in. The maximum stress in built-in plates occurs on the surface at the center of the longer sides and is in the direction of the short span.

Comparing the curves in Fig. 11-10 reveals that there is an advantage in building-in the edges of very long plates; with everything else remaining the same, the plate with built-in edges can carry 50% more load. For shorter plates the added capacity is not as great, and for square plates the capacity may actually be reduced by building-in the edges. This is a rather curious behavior, since we have seen many cases of beams that had their capacity increased by building-in the ends; the

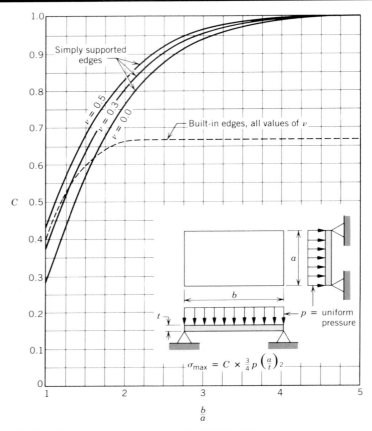

FIG. 11-10 Maximum bending stress in a rectangular plate.

circular disk formulas (11-9) and (11-10) also show a substantial increase in load capacity.

Problems 11-10 to 11-17

PROBLEMS

11-1 The given shell is 12 mm (0.5 in.) thick and contains fluid at a pressure of 1500 kPa (220 psi). Determine the stresses at *A*, *B*, *C*, *D*, *E*, and *F*.

> **Note:** Do not ignore the message that (11-2) will attempt to convey to you when you apply it to point *F*.

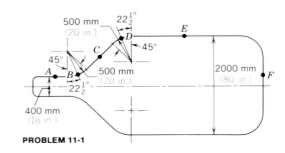

PROBLEM 11-1

11-2 Determine the stresses at points *A*, *B*, and *C* in the given pressure vessel due to an internal pressure of 1200 kPa (175 psi). The thickness of the material is 6 mm (0.24 in.).

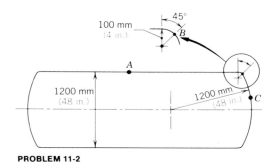

PROBLEM 11-2

11-3 A stainless steel pressure vessel has 5-mm-thick (0.2-in.-thick) walls. It consists of a cylindrical portion, 2 m (6 ft) long and 0.5 m (20 in.) outside diameter, and hemispherical end caps. When a pressure of 8000 kPa (1200 psi) is applied to the inside of the vessel, determine the change in volume of

a. The cylindrical portion of the vessel.
b. The complete vessel.

11-4 When the pressure is applied to the vessel in Prob. 11-3, the wall becomes thinner. Calculate the change in thickness:

a. In the cylindrical portion.
b. In the end caps.

11-5 Pressure in the given stainless steel tank is to be measured by strain gages attached to the surface at *A* and *B*. Calculate the pressure when the measured strains, in μm/m (μin./in.), are

a. $\varepsilon_A = 927$ $\varepsilon_B = 218$
b. $\varepsilon_A = 887$ $\varepsilon_B = 195$ (not reliable)
c. $\varepsilon_A = 724$ (not reliable) $\varepsilon_B = 338$
d. $\varepsilon_A = 819$ $\varepsilon_B = 307$

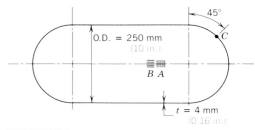

PROBLEM 11-5

11-6 A delta strain gage rosette is attached to the vessel in Prob. 11-5 at *C*. Calculate the strain in each element of the gage when the vessel is subjected to a pressure of 9600 kPa (1400 psi).

11-7 A tank having a mean diameter of 300 mm (12 in.) and a 5-mm-thick (0.2-in.-thick) wall contains oxygen at a pressure of 10 000 kPa (1450 psi). The heads are torospherical with a knuckle radius of 30 mm (1.2 in.) and a spherical radius of 200 mm (8 in.). In order to provide a base for the tank to stand on, one head is made concave as shown in the diagram. Calculate the membrane stresses at *A*, *B*, *C*, *D*, and *E*.

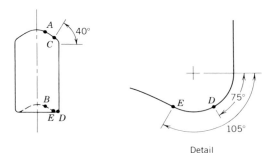

Detail

PROBLEM 11-7

11-8 A conical head is being considered for a pressure vessel to contain fluid at 1800 kPa (260 psi). If the allowable stress is 200 MPa (29 ksi), determine the required thickness of the head. Disregard the reinforcing ring that would be required at the junction between the cylinder and cone.

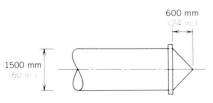

PROBLEM 11-8

11-9 The removable head for a vacuum chamber consists of a ring having a 15 by 15 mm (0.6 × 0.6 in.) cross section and a segment of a sphere having a radius of 2000 mm (80 in.) and a thickness of 0.8 mm (0.032 in.). Calculate:

a. The stress in the spherical segment.
b. The circumferential stress in the ring.

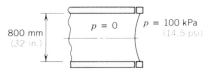

PROBLEM 11-9

11-10 A cylindrical pressure vessel has a diameter of 200 cm (80 in.) and contains steam at 7000 kPa (1000 psi). The allowable stress for the steel in all parts of the vessel is 150 MPa (22 ksi).

a. Determine the minimum required thicknesses t_c, t_s, and t_d.
b. Compare the amount of material in the disk with the amount in the hemispherical head.

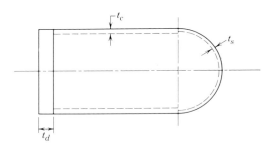

PROBLEM 11-10

11-11 A disk is to be used to close a circular hole in a pressure vessel wall. The pressure inside the vessel is 5000 kPa (725 psi) and the allowable stress in the disk is 130 MPa (19 ksi). Determine the required thickness of the disk if it is

a. Supported on a soft gasket.
b. Welded into place.

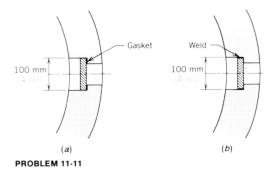

PROBLEM 11-11

11-12 A circular cover plate similar to that shown in Fig. 11-7a is to be designed. Fluid in the vessel is at 1200 kPa (175 psi), the O-ring gasket diameter is 360 mm (15 in.), the bolt-circle diameter is 400 mm (16.6 in.), and the allowable stress is 120 MPa (17.4 ksi). Determine the required thickness of the cover plate.

> **Note:** You will have to use your engineering judgment in deciding on the edge conditions to use and also in determining the effective diameter of the disk.

11-13 A vessel containing fluid at 300 kPa (45 psi) is to be fitted with a viewing port having an area of 40 000 mm² (64 in.²). Glass 15 mm (0.6 in.) thick is to be used for the window and three different shapes are being considered:

a. Circular — 225 mm (9 in.) in diameter.
b. Square — 200 mm × 200 mm (8 in. × 8 in.).
c. Rectangular — 400 mm × 100 mm (16 in. × 4 in.).

> The edge is supported on a soft gasket and hence is simply supported. Calculate the maximum bending stress in each of the proposed ports.

11-14 Calculate the stresses in the glass in the various ports described in Prob. 11-13 if some means can be found to build-in the edges.

11-15 In order to give a view of the sea bottom, a glass bottom is being built into a boat. The draft of the fully loaded boat is 0.6 m (24 in.). The glass that is available is 12 mm (0.5 in.) thick and has an allowable bending stress of 7 MPa (1.0 ksi). The edges of the glass will be simply supported. Determine the maximum viewing area that can be provided if the shape of the glass is

a. Square.
b. Rectangular with a length-to-width ratio of 2.

11-16 A window pane is 1500 × 2500 mm (60 × 100 in.) and is supported in a soft rubber gasket on all four sides. The glass is subjected to a wind load of 1.20 kPa (0.17 psi) and has an allowable stress of 7 MPa (1.0 ksi). Determine the required thickness of the glass.

11-17 For the window pane of Prob. 11-16, what thickness of glass would be required if a method could be devised for securely clamping the edges?

Challenging Problems

11-A The given vessel contains fluid at a pressure of 3000 kPa (435 psi). The wall thickness is 25 mm (1 in.) in all parts and the cylindrical part has a diameter of 1 m (40 in.). At points along the line *A-B* calculate the membrane stresses and draw curves showing the stress profiles.

failure most likely to occur if the material is not strong enough?

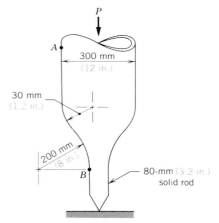

PROBLEM 11-B

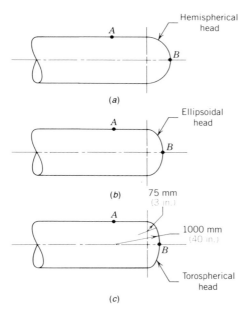

PROBLEM 11-A

11-B A pile tip as shown is being assessed to determine its practicability. The wall is 8 mm (0.32 in.) thick in the tapered part.

The worst loading occurs when the tip is against bedrock and the pile driver exerts its full force, which is 800 kN (180 kip). Consider the driving force and the reaction at the tip to be the only loads.

Determine the membrane stresses at points along the line *A-B* and plot stress profiles. Where is

11-C A conical hopper is to be filled with Portland cement having a density of 1500 kg/m³ (94 lb/ft³). The cement may be treated as having the properties of a fluid. If the allowable stress for the hopper steel is 180 MPa (26 ksi), determine the minimum safe thickness for the plate.

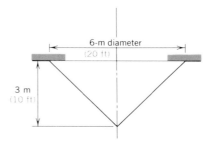

PROBLEM 11-C

11-D The allowable stress for the reinforcing ring material in Prob. 11-8 is 200 MPa (29 ksi). Determine the cross-sectional area required for the ring.

11-E In order to design tires for a piece of earth-moving equipment, it is necessary to determine stresses in the given torus. Calculate the membrane stresses at points *A*, *B*, and *C* if the wall is 20 mm (0.8 in.) thick and the internal pressure is 400 kPa (60 psi).

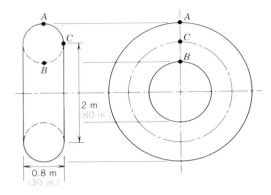

PROBLEM 11-E

DESIGN FOR CYCLIC LOAD

12-1 INTRODUCTION

For the design of members in which buckling is not a possible mode of failure, we have developed a simple yet satisfactory procedure. We determine the stress in a proposed member due to the loads and compare the maximum stress with an allowable stress for the material that has been selected for the member. The allowable stress is determined by testing to destruction specimens of the same material in a tensile testing machine. For reliable results many specimens would be tested and the results averaged, but it is important to note that the strength of each specimen is obtained from a *single* application of load. Then, strictly speaking, the allowable stress based on this data is pertinent only to cases in which the load is applied once during the lifetime of the member. That is, when there is a *static load*. The main members of large bridges and other structures can be designed for static load, but in a great many other practical cases the load cycles or at least fluctuates in magnitude. Static design was applied successfully in the early days of civil engineering but, with the advent of machines, parts that were designed for static loads were found to be failing after being in service for a period of time. Such failures were observed, for instance, in railroad locomotive axles that had been properly designed for static conditions but failed in service after many cycles of load application. Because the material seemed to weaken in service, the failure was attributed to "*fatigue.*"

When fatigue was first recognized, compensation was provided by increasing the safety factor, using some crude rule of thumb, which meant that in some cases parts were ultraconservative while in others failure still occurred. Further increases in the safety factor would have made machines excessively massive and if applied to aircraft would have resulted in machines that were too heavy to fly. The aircraft industry provided the motivation and resources for tests and studies that have revealed many of the secrets of the nature of fatigue and the procedures by which parts can be designed to carry cyclic loads with an acceptable degree of reliability. Today when a failure does occur it is usually due to faulty fatigue analysis because all aspects of fatigue behavior are still not known or because the designer was not fully conversant with the existing body of knowledge. Failure may also be due to the actual members not being exactly the same as the designer had intended, perhaps because of poorly written specifications or inadequate inspection.

As we will see later in this chapter, the design process becomes more complex when fatigue is taken into consideration. Today a designer could spend many years absorbing the knowledge that already exists and thereafter devote full attention to just keeping abreast of current developments as presented in technical publications dealing with the art of fatigue design. Obviously only a small fraction of those engaged in design can keep fully informed on developments in fatigue, but all designers need a basic knowledge of this important design consideration. Although the treatment of fatigue in this chapter will be restricted to a few fundamental yet practical cases, it will provide a good base for more comprehensive studies for those who decide to follow the subject further.

12-2 THE NATURE OF FATIGUE FAILURE

If we were to test a number of models made of a material with known tensile properties, we could predict quite accurately the tensile load that would cause failure in a single-load application. However, if a smaller load were applied and removed repeatedly, it is quite likely that failure would occur after a number of load applications. Such a *fatigue failure*, under repeated or *cyclic loading*, would exhibit some unexpected characteristics. We are accustomed to test specimens of ductile materials elongating and necking down in a conspicuous manner before breaking, thereby giving warning of impending failure. However, fatigue failure in ductile materials occurs without warning; no obvious elongation or necking can be seen. There is an absence of the familiar cup and cone, the fatigue failure surface being normal to the tensile stress. Because of the similarity in behavior and appearance to brittle materials tested under static load, fatigue was at one time referred to as *brittle failure* and was thought to have been caused by the material losing its ductility. Although the failure exhibits some of the characteristics of a brittle material, the material does, in fact, retain its ductility.

The surface of a broken shaft shown in Fig. 12-1 is typical of a fatigue failure. The failure invariably starts at some discontinuity in the material

FIG. 12-1 Fatigue failure of a shaft.

such as a sharp reentrant angle, a tool or punch mark, or a flaw in the material. Stress near the discontinuity is larger than in the surrounding material; hence local yielding may occur at the discontinuity although the bulk of the material is well below the yield stress. During the first few cycles of plastic flow the atoms on each side of the slip plane form new bonds without losing strength in a measurable degree. However, under continued cycling, microscopic cracks will be generated along the slip planes. If we were able to observe these microscopic cracks, we would find them on the plane of the maximum shear stress, that is, on planes at 45° to the axis of the applied uniaxial tension and compression. Once such a crack has been initiated, the stress concentration at the end of the crack promotes further growth, and an appreciable crack may develop in the plane of maximum shear as the cycling continues.

The nucleus of the crack in the shaft of Fig. 12-1 can be seen at the extreme left of the cross section, and there is a portion of the surface that slopes at 45° to the shaft axis. The slope is not obvious in the photograph.

After the initial fatigue crack has grown to an appreciable size, the failure process enters a second phase. In this phase the crack continues to spread but travels in a plane normal to the maximum tensile stress. The crack does not progress with each load cycle, but rather pauses for a number of cycles, as though waiting for the metal in the highly stressed zone ahead of the crack to be weakened by repeated stresses, and then makes a perceptible advance in a step. The stepwise advance of the crack can be seen on the fracture surface in the *"oyster-shell"* or *"beach"* markings, which have the appearance of waves emanating from the region of initial failure. They are characteristic of fatigue failure and are useful in that they indicate the location of the material flaw, or the design defect, that was responsible for initiating the failure.

During the course of the crack development the fracture surfaces are rubbed together and somewhat smoothed. This gives a rubbed texture to the surfaces that is useful in identifying fatigue failure.

When the uncracked area of the cross section has been reduced in size to the point where it can no longer carry the load, sudden and complete failure takes place. This region can be seen at the right side of the shaft in Fig. 12-1. Because fatigue failure occurs without clearly visible signs of distress prior to complete rupture, such failures can be catastrophic and lead to a large loss of life as well as major economic loss. To guard against failure, it is important for you to become acquainted with the major factors that influence fatigue as outlined in the following sections of this chapter.

12-3 RELATIONSHIP BETWEEN STRENGTH AND NUMBER OF CYCLES

Imagine a simple tensile test in which a load, less than the ultimate, is applied and released cyclically. A large number of specimens are subjected to this test with a different load magnitude for each case. The data from such tests, when plotted, would give points as shown in Fig. 12-2.

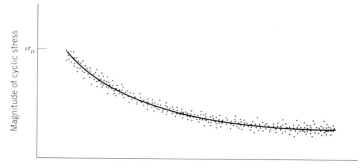

FIG. 12-2 Cyclic loading leads to a reduction in strength.

A considerable amount of scatter would be observed, but a curve drawn through the points will show some definite trends.

The test results show that the strength under cyclic load, that is, the *fatigue strength*, is generally less than the ultimate stress, σ_u. The number of cycles required to cause failure increases as the stress is decreased, and finally a stress level is reached below which failure will not occur regardless of how many cycles are applied.

Since some of the specimens will require a million or more cycles to cause failure, gathering sufficient data to draw a stress versus the number of cycles to failure curve is a costly undertaking. If we do not take special care, the scatter in the points will be large and this will increase the number of points required. Since it is impractical to load a specimen a million times in an ordinary tensile test machine, an entirely different method is used to gather fatigue data.

The standard fatigue testing machine is shown diagramatically in Fig. 12-3. It is a very simple device that quickly applies a cyclic stress of accurately known value to a rotating beam. The stress that is of interest is the maximum stress which is applied to the surface fibers in the central region of the test specimen. Any fiber in this region goes through a stress cycle from maximum tension to an equal compression once for each rotation of the shaft. When the specimen breaks, the motor shuts off and a counter reveals the number of cycles required to cause failure. This stress cycle is not the same as that which was applied in the previous tests using the tensile machine, so the results would differ, but the stress-number relationship would still be of the same general form as that of Fig. 12-2. Since fatigue data are usually obtained from rotating-beam tests, only data from such tests will be considered in this chapter. When the actual stress cycle is different, for example, when the mean stress is not zero, some method of predicting cycles to failure must be employed. More about that later.

Let us imagine that we have a new material and that we want to display its fatigue characteristics by drawing the stress-number curve. We will imagine that a large number of standard test beams have been made and are to be tested. Such an undertaking would require a large number of test machines and would consume several months of time. If we were experienced in this work, we would know that the most useful

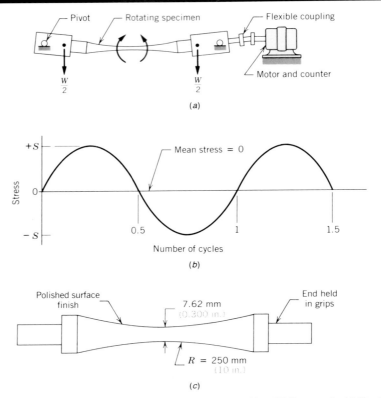

FIG. 12-3 Fatigue test. (*a*) Rotating beam fatigue test machine. (*b*) Stress cycle. (*c*) Standard specimen.

curve is obtained by plotting the logarithm of stress, S, against the logarithm of the number of cycles to failure, N. The laws of nature relating S to N are too complex to be expressed by a simple mathematical formula. To get a working relationship for design purposes, we approximate the S-N curve by one that is composed of the straight lines shown in Fig. 12-4. Experience shows that for steels the straight lines can be taken through three critical points as follows:

1. For one cycle the stress to cause failure is σ_u, which is consistent with our definition of the static ultimate strength.
2. Based on empirical evidence, at 10^3 cycles the stress can be taken as $0.9\sigma_u$.
3. At 10^6 cycles and beyond, the stress is constant. This is an important stress and is designated, S_N', the *endurance limit*.

Nonferrous metals do not exhibit a leveling of the curve and hence have no definite endurance limit. Note that for a very wide range of steels the value of the endurance limit is close to one-half the ultimate stress. If we assume that this applies to all steels, we can construct the complete S-N curve from simple static tensile test data when actual fatigue test data are not available. Designers commonly accept the simplified curve, with $S_N' = 0.5\sigma_u$, when the expense of conducting a series of fatigue tests is not warranted by the importance of the job.

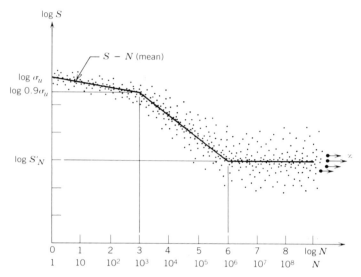

FIG. 12-4 Typical test data for steel.

12-4   STRENGTH REDUCTION FACTOR: RELIABILITY

The idealized S-N curve in Fig. 12-5 is drawn through points that have a considerable scatter. The scatter is the result of the statistical nature of fatigue and persists in spite of the great care taken to make the test specimens identical in every detail. If we used the mean curve in Fig. 12-5 to give a design stress for a particular number of cycles of a rotating shaft, there is only a 50% chance that the shaft would survive the required number of cycles. The S-N (mean) curve in Fig. 12-5 should then be designated as "50% reliable."

At 1 million cycles the statistical spread of the stress values can be taken as Gaussian with a standard deviation of 8%. Under these conditions, if a design stress that is 8% less than the value given by the curve is used, the probability of a part's surviving for 10^6 cycles is 0.841 or we could say that such a part is 84.1% reliable. For most designs this is not acceptable. However, if the distribution is, in fact, Gaussian, no amount of stress reduction will produce a design that is 100% reliable. It is customary to compromise by reducing the stress by three standard deviations, that is, by 24%. Thus a stress of $0.76S_N'$, which gives a theoretical reliability of 99.9%, is considered to produce a safe design. In actual fact the distribution is not truly Gaussian and the reliability is even greater than 99.9%.

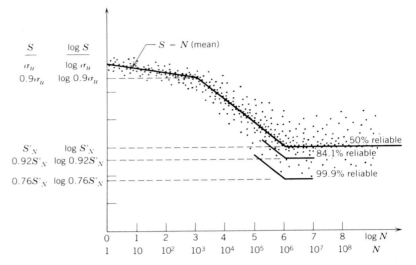

FIG. 12-5 Idealized S-N curve for steel.

Studies of the spread of data show that the standard deviation is less than 8% for smaller numbers of cycles, but for the sake of simplicity all data will be taken as having a standard deviation of 8%. This gives design curve B in Fig. 12-6, which provides an adequate factor of safety if the material and conditions under which the designed part operates are identical with those of the test specimen.

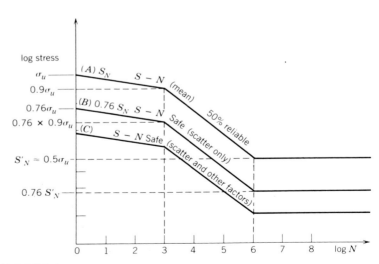

FIG. 12-6 S-N design curve.

The factor having the value 0.76, which was used to reduce the mean stress to a reliable design value, will be designated C_R, and is the first of several *strength reduction factors* that we will encounter. Thus the strength reduction factor for *reliability* is

$$C_R = 0.76 \qquad \textbf{(12-1)}$$

and the allowable stress for a design that is to have a safe life of N cycles is

$$\sigma_a = C_R S_N \qquad \textbf{(12-2)}$$

12-5 MATHEMATICAL RELATIONSHIP BETWEEN STRENGTH AND NUMBER OF CYCLES

Equation (12-2) gives an allowable stress for cyclic design, as equation (1-2) did for static design, and it may seem that we are near the end of our treatment of fatigue. However, there are further factors that we would not anticipate from our work in static design. For example, in static design the material strength varies with size but to such a small extent that the test specimen strength can be used to design parts that are very much larger than the specimen. When the load is cyclic, size has a significant influence on strength and cannot be ignored.

Furthermore, stress concentration, around bolt holes and at the bottoms of surface irregularities could be ignored in static design for ductile materials. However, stress concentration has a marked effect on fatigue strength and hence must be taken into account in the design of parts subjected to cyclic load.

When these other factors are taken into consideration, the safe line will be lowered still more, as shown by curve C in Fig. 12-6. In order to designate the stress at the critical points on the safe curve, we will use symbols S_{10^3} and S_{10^6} to indicate the safe stress at 1000 cycles and 1 000 000 cycles, respectively.

In a great many applications the number of cycles is over one million and then only the value of S_{10^6} is significant. At the other end of the range, that is, between one cycle and a thousand cycles, the strength reduction is quite small and it is common practice to use static design procedures with perhaps some increase in the safety factor to compensate for the cyclic nature of the load. Between one thousand and one million cycles we could obtain S_N by plotting the S-N curve, but to save time it is preferable to use the straight-line formula

$$S_N = S_{10^6} \left(\frac{S_{10^3}}{S_{10^6}} \right)^{\left(\frac{6 - \log N}{3} \right)} \qquad \textbf{(12-3)}$$

Or if we need to know the safe number of cycles for a given stress, the formula is

$$N = 10^{\left(6 - 3\frac{\log S_N/S_{10^6}}{\log S_{10^3}/S_{10^6}}\right)}$$ **(12-4)**

12-6 STRENGTH REDUCTION FACTOR: MEAN STRESS

Rotating beam tests give information about the strength of the test material when a complete reversal of stress takes place, that is, for a stress cycle such as that of Fig. 12-7a. The stress has a *mean value* of zero and consequently only the *amplitude*, S_A, is pertinent.

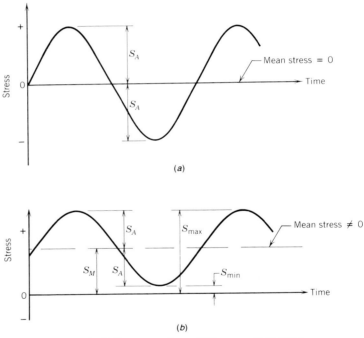

FIG. 12-7 Stress cycle. (*a*) Zero mean stress. (*b*) Non-zero mean stress.

Any cyclic stress can be decomposed into two parts; a constant mean stress, S_M, and a superimposed fluctuating stress of amplitude, S_A. This, the general load cycle case, is shown in Fig. 12-7b.

When the mean stress is zero, the general case becomes the particular case that we have been considering and the allowable amplitude for any given number of cycles can be predicted by methods that are now familiar. Introducing a nonzero mean stress will, as might be expected, reduce the allowable amplitude. To determine the new safe value for the cyclic component of stress, we will consider the relationship between the maximum stress S_{max}, minimum stress S_{min}, and mean stress S_M. These stresses are related by curves A-C and B-C in Fig. 12-8a, and we will now establish these curves, as far as possible, by using our present knowledge.

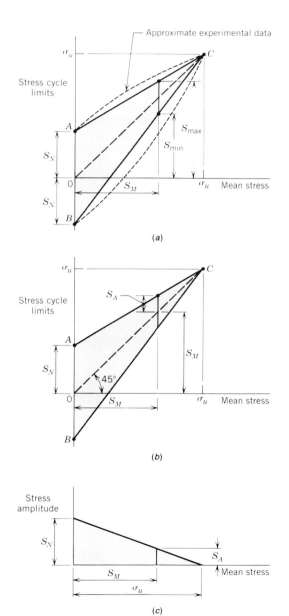

FIG. 12-8 (*a*) Relationship between stress cycle limits and mean stress. (*b*) Relationship between stress amplitude and mean stress. (*c*) Goodman diagram.

First, when the mean stress is zero we have the complete reversal of stress that is used in the standard fatigue test. Then S_{max} and S_{min} both have the magnitude of S_N and we can plot points A and B, thus determining one point on each curve.

At the other extreme, if the mean stress is equal to the ultimate stress, σ_u, the specimen can withstand no further load and hence the corresponding amplitude in zero. Then, S_{max} and S_{min} are the same and are equal to σ_u; thus both curves pass through point C. Experiments to establish other points on the curves A-C and B-C would show the usual scatter of the points but, when averaged, would give the general shape of the broken curved lines in Fig. 12-8a. We will take the lines as being straight from A to C and B to C and thus introduce an error on the safe side.

Because C was plotted with both coordinates equal to σ_u, the line O-C, in Fig. 12-8b, slopes upward at $45°$. The vertical distance to a typical point on O-C is S_M, and the additional vertical distance to the line A-C is S_A. Then the triangle A-O-C in Fig. 12-8b is all that is necessary to determine the value of S_A for given S_M, S_N, and σ_u.

The same relationships exist in the simplified diagram shown in Fig. 12-8c. This diagram was devised by Goodman, and the straight line on the diagram is referred to as the *Goodman line*. By drawing the *Goodman diagram*, we set up a relationship between S_A and S_M based on the known properties of the material, S_N and σ_u. In many practical problems the mean stress S_M is known and S_A is to be found. In this case it is not really necessary to draw the diagram, the equation for the straight line,

$$S_A = \left(1 - \frac{S_M}{\sigma_u}\right) S_N \qquad \text{(12-5)}$$

being much more convenient.

We can regard the term in brackets in (12-5) as being a strength reduction factor that takes into account the effect of mean stress in reducing the allowable amplitude of the cyclic component. We will refer to this factor as C_M, the strength reduction factor that takes into account the detrimental effect of a superimposed *mean stress*. Then

$$C_M = 1 - \frac{S_M}{\sigma_u} \qquad \text{(12-6)}$$

When both reliability and mean stress are taken into account, the critical values on the *S-N* design curve become:

$$S_{10^3} = C_R \times C_M \times 0.9\sigma_u \qquad \text{(12-7)}$$

and

$$S_{10^6} = C_R \times C_M \times S_N' \qquad \text{(12-8)}$$

When C_M has been calculated and substituted into (12-7) and (12-8), the whole effect of mean stress has been taken into account and we need not consider it again. The S_{10^3} and S_{10^6} thus obtained are the completely reversing stresses that can be safely applied for 10^3 and 10^6 cycles, respectively. When these S_{10^3} and S_{10^6} values and an appropriate N are substituted into (12-3), the S_N thus obtained is the completely reversing stress that can be superimposed on the constant stress and safely cycled N times.

We will use (12-6) in all of our exercises and problems to take the constant component of stress into account. However, before leaving this topic we will take a brief look at some other commonly accepted practices.

The *modified Goodman line*, in Fig. 12-9*a*, ensures that the maximum stress will not exceed the yield strength of the material. The modified line is rather cumbersome to use, since it must be expressed as two equations. In practice the additional line seldom controls the design, so the best procedure is to design on the basis of the Goodman line and then check to make sure that the stresses do not represent a point on the broken part of the Goodman line.

The *Gerber line*, in Fig. 12-9*b*, follows the experimental points more closely than the Goodman line. However, being curved, it is more difficult to work with. The Gerber line is more widely used in Europe than in North America.

The *Soderberg line*, as shown in Fig. 12-9*c*, has the virtue of simplicity but is somewhat more conservative than the Goodman line. When compared with the experimental data, it seems that the Goodman line errs sufficiently on the safe side and that the Soderberg line is ultraconservative.

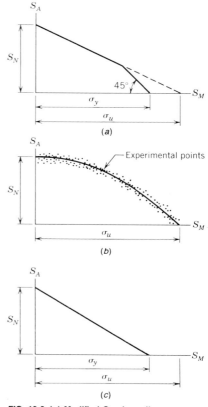

FIG. 12-9 (*a*) Modified Goodman line. (*b*) Gerber line. (*c*) Soderberg line.

Example 12-1

A polished rotating steel shaft, 200 mm long and 6 mm in diameter, is simply supported at the ends and carries a lateral load at midspan. The shaft is also subjected to a steady axial tensile load of 8 kN along the shaft centerline.

The ultimate strength of the steel is 900 MPa and its endurance limit is 500 MPa. If the shaft is to turn 100 000 revolutions in the service life of the machine, determine the magnitude of the lateral load that the shaft can carry safely.

Solution. Mean stress, $S_M = \dfrac{8000}{(\pi/4)6^2} \dfrac{\text{N}}{\text{mm}^2}$

$$= 283 \text{ MPa}$$

Ultimate strength, $\sigma_u = 900$ MPa   (given)

$$C_M = 1 - \frac{S_M}{\sigma_u} \quad \text{(12-6)}$$

$$= 1 - \frac{283}{900} = 0.69$$

$$C_R = 0.76 \qquad (12\text{-}1)$$

$$S_N' = 500 \text{ MPa} \qquad (\text{given})$$

$$S_{10^3} = C_R C_M 0.9\sigma_u \qquad (12\text{-}7)$$

$$= 0.76 \times 0.69 \times 0.9 \times 900 = 425 \text{ MPa}$$

$$S_{10^6} = C_R C_M S_N' \qquad (12\text{-}8)$$

$$= 0.76 \times 0.69 \times 500 = 262 \text{ MPa}$$

For a life of 100 000 cycles

$$\sigma_a = S_N = S_{10^5} = S_{10^6}\left(\frac{S_{10^3}}{S_{10^6}}\right)^{(6-5)/3} \qquad (12\text{-}3)$$

$$= 262\left(\frac{425}{262}\right)^{1/3}$$

$$= 262 \times 1.18 = 308 \text{ MPa}$$

The stress-load relationship for the shaft shown in Fig. 12-10 is:

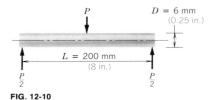

FIG. 12-10

$$\text{bending stress} = \frac{Mc}{I} = \frac{(P/2)(L/2)(D/2)}{(\pi/64)D^4} = \frac{8}{\pi}\frac{PL}{D^3}$$

For a safe design,

$$\frac{8}{\pi}\frac{PL}{D^3} \leqslant \sigma_a$$

$$P \leqslant \frac{\pi}{8}\frac{D^3\sigma_a}{L}$$

$$= \frac{\pi}{8}\frac{6^3 \times 308}{200}\frac{\text{mm}^3 \cdot \text{N/mm}^2}{\text{mm}}$$

$$= 131 \text{ N}$$

The safe lateral load is 131 N.

Example 12-1

A polished rotating steel shaft, 8 in. long and 0.25 in. in diameter, is simply supported at the ends and carries a lateral load at midspan. The shaft is also subjected to a steady axial tensile load of 2 kip along the shaft centerline.

The ultimate strength of the steel is 130 ksi and its endurance limit is 70 ksi. If the shaft is to turn 100 000 revolutions in the service life of the machine, determine the magnitude of the lateral load that the shaft can carry safely.

Solution. Mean stress, $S_M = \dfrac{2}{(\pi/4)0.25^2} \dfrac{\text{kip}}{\text{in.}^2}$

$$= 40.7 \text{ kip/in.}^2$$

Ultimate strength, $\sigma_u = 130$ kip/in.^{2} (given)

$$C_M = 1 - \frac{S_M}{\sigma_u} \quad (12\text{-}6)$$

$$= 1 - \frac{40.7}{130} = 0.69$$

$$C_R = 0.76 \quad (12\text{-}1)$$

$$S_N' = 70 \text{ kip/in.}^2 \quad \text{(given)}$$

$$S_{10^3} = C_R C_M 0.9 \sigma_u \quad (12\text{-}7)$$

$$= 0.76 \times 0.69 \times 0.9 \times 130 = 61.4 \text{ kip/in.}^2$$

$$S_{10^6} = C_R C_M S_N' \quad (12\text{-}8)$$

$$= 0.76 \times 0.69 \times 70 = 36.7 \text{ kip/in.}^2$$

For a life of 100,000 cycles

$$\sigma_a = S_N = S_{10^5} = S_{10^6} \left(\frac{S_{10^3}}{S_{10^6}} \right)^{(6-5)/3} \quad (12\text{-}3)$$

$$= 36.7 \left(\frac{61.4}{36.7} \right)^{1/3}$$

$$= 36.7 \times 1.19 = 43.7 \text{ kip/in.}^2$$

The stress-load relationship for the shaft shown in Fig. 12-10 is:

$$\text{bending stress} = \frac{Mc}{I} = \frac{(P/2)(L/2)(D/2)}{(\pi/64)D^4} = \frac{8}{\pi} \frac{PL}{D^3}$$

For a safe design,

$$\frac{8}{\pi} \frac{PL}{D^3} \leqslant \sigma_a$$

$$P \leqslant \frac{\pi}{8} \frac{D^3 \sigma_a}{L}$$

$$= \frac{\pi}{8} \frac{0.25^3 \times 43.7}{8} \frac{\text{in.}^3 \cdot \text{kip/in.}^2}{\text{in.}}$$

$$= 0.034 \text{ kip} = 34 \text{ lb}$$

The safe lateral load is 34 lb.

Problems 12-1 to 12-6

12-7 STRENGTH REDUCTION FACTOR: SIZE

We will now consider the effect of size on fatigue strength. Static tests on specimens having a variety of sizes, but made of the same material, show that there is some reduction in the ultimate stress as the specimen size increases. This reduction is quite small and is usually ignored by designers unless the part is exceptionally large. A decrease in fatigue strength is also observed, but in this case the reduction is not negligible.

The reduction can be accounted for by imagining two rotating beam tests. In one test the standard specimen is used while in the other the specimen has the same proportions as the standard specimen but is scaled up, let us say by a factor of two. If both tests are run at the same maximum stress and we treat the highly stressed region of the member as being that volume of material that is stressed to 95% or more of the maximum stress, the volume of highly stressed material in the large specimen is eight times that of the standard specimen. Then the probability of significant flaws occurring in the highly stressed region is greater in the large specimen and we expect to find that its fatigue strength is lower.

The problem in design is that we must design on the basis of data from standard test specimens and adjust the strength to compensate for a reduced strength due to size. This is done by another strength reduction factor, C_S, which takes into account the size of the part relative to the test specimen.

There are many empirical formulas for C_S, for example,

$$C_S = \left(\frac{V}{V_0} \right)^{-0.034}$$

(12-9)

where V_0 is the volume of material in the standard specimen that is stressed to 95% or more of the maximum and V is the volume of the part being designed that has stress in the same range.

Determining V_0 is a considerable undertaking but could be done once and would never have to be repeated. However, calculating V would usually be a lengthy, and perhaps impossible, task and would have to be repeated whenever dimensions or loads were altered. The value of the formula is that it points out the significance of the volume of the highly stressed material that should be kept in mind during the practice of the art of fatigue design.

In exercises and problems we will use a simpler and less accurate formula but one that is widely accepted. For circular cross sections,

$$C_S = 1 - \frac{(D - 7.62)}{380} \qquad \text{for } 7.62 < D < 150 \text{ mm}$$

(12-10)

where D is the diameter in millimetres.

When a shaft has the same diameter as the test specimen ($D = 7.62$ mm), there is no reduction in strength due to size. For larger sizes the factor decreases linearly with D and would become zero for a shaft having a diameter of 388 mm. This is, of course, absurd and is the reason for the upper limit on D in the formula. The formula would allow for additional strength in rods that are smaller than the standard specimen. Although such rods may in fact be stronger, it is good design practice to not utilize this extra strength by taking

$$\boxed{C_S = 1 \qquad \text{for } D \leqslant 7.62} \qquad \textbf{(12-11)}$$

In exercises and problems we will use a simpler and less accurate formula but one that is widely accepted. For circular cross sections,

$$\boxed{C_S = 1 - \frac{(D - 0.3)}{15} \qquad \text{for } 0.3 < D < 6 \text{ in.}} \qquad \textbf{(12-10)}$$

where D is diameter in inches.

When a shaft has the same diameter as the test specimen ($D = 0.3$ in.) there is no reduction in strength due to size. For larger sizes the factor decreases linearly with D and would become zero for a shaft having a diameter of 15.3 in. This is, of course, absurd and is the reason for the upper limit on D in the formula. The formula would allow for additional strength in rods that are smaller than the standard specimen. Although such rods may in fact be stronger, it is good design practice not to utilize this extra strength by taking

$$\boxed{C_S = 1 \qquad \text{for } D \leqslant 0.3 \text{ in.}} \qquad \textbf{(12-11)}$$

When the cross section is not circular, the usual practice is to use, in place of D, the dimension of the section that is normal to the neutral axis.

Returning now to the *S-N* curves for design in Fig. 12-6, we will further modify the "safe" line to take into account the size factor, C_S. As already mentioned, size has a negligible effect on the static ultimate strength and hence will not influence the left end of the curve. Its full influence is felt on S_N' and consequently on S_{10^6}. To take the new factor into account, we need to modify (12-8) to

$$S_{10^6} = C_R C_M C_S S_N'$$

Between the extremes there is S_{10^3}, which is influenced to a small extent by C_S. Most designers choose to ignore this influence and to evaluate S_{10^3} by (12-7).

12-8 STRENGTH REDUCTION FACTOR: FINISH

All cases that have been considered up to this point have specified *polished surfaces* so that the surface would be similar to that of the standard test specimen. Since polishing is an expensive process, in practice it is rarely used on machine parts; therefore, to be realistic we must be capable of making allowance for the effect of surface finish on fatigue strength.

If we were to make a series of tests on specimens of the same material but with different surface finishes, we would find that the endurance limit was strongly dependent on the finish. To take into account the reduction in fatigue strength due to *surface finish*, we will introduce another strength reduction factor, C_F.

Tests to determine C_F show that it depends on the ultimate strength of the material as well as on the texture of the surface. Since a rough surface introduces depressions that cause stress concentration, a reduction in strength is to be expected when the surface is not polished. However, the influence of material strength on what seems to be a purely geometric consideration is not expected and, although it can be accounted for, we will accept it without explanation. Values of C_F for our problems will be taken from Fig. 12-11.[*] It is interesting to note the tendency for C_F to decrease with increasing ultimate strength. This means that some of the gain in load-carrying capacity that may have been anticipated by using a higher-strength material is lost because of a reduction in C_F. From the curves it is evident that high-strength materials can benefit most from a fine surface finish. This accounts for the practice of specifying finer finishes for such materials.

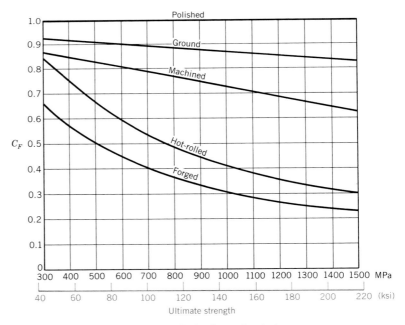

FIG. 12-11 C_F, surface finish strength reduction factors for steel.

[*] Plotted from data obtained from V. M. Faires *Design of Machine Elements*, 4th ed., Macmillan, New York, 1965.

For ductile materials the static strength is not appreciably influenced by the surface finish and the S_{10^3} strength is affected by an insignificant amount. Consequently, we can continue to calculate S_{10^3} by (12-7) but must include a surface finish reduction factor to the equation for S_{10^6}. Thus the equation becomes

$$S_{10^6} = C_R C_M C_S C_F S_N'$$ **(12-12)**

Example 12-2

An element of a leaf spring is made of hot-rolled steel having an ultimate strength of 1000 MPa. The leaf is 1.40 m long, 80 mm wide, and 12 mm thick. As indicated in Fig. 12-12, it is simply supported at the ends and carries a load at the center that cycles between zero and 2.2 kN. Determine the safe number of load cycles.

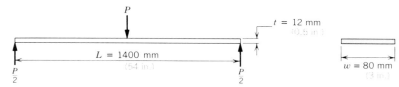

FIG. 12-12

Solution

$$M_{\max} = \frac{1}{4} PL$$

$$\sigma_{\max} = \frac{M_{\max} c}{I} = \frac{\frac{1}{4} PL(t/2)}{\frac{1}{12} wt^3} = \frac{3}{2} \frac{PL}{wt^2}$$

$$= \frac{3 \times 2200 \times 1400}{2 \times 80 \times 12^2} \frac{\text{N} \cdot \text{mm}}{\text{mm} \cdot \text{mm}^2}$$

$$= 401 \text{ MPa}$$

$$S_A = S_M = \frac{\sigma_{\max}}{2} = \frac{401}{2} = 201 \text{ MPa}$$

$$C_R = 0.76 \qquad (12\text{-}1)$$

$$C_M = 1 - \frac{S_M}{\sigma_u} \qquad (12\text{-}6)$$

$$= 1 - \frac{201}{1000} = 0.80$$

$$S_{10^3} = C_R C_M 0.9 \sigma_u \qquad (12\text{-}7)$$

$$= 0.76 \times 0.80 \times 0.9 \times 1000 = 547 \text{ MPa}$$

$$C_S = 1 - \frac{D - 7.62}{380} \qquad (12\text{-}10)$$

$$= 1 - \frac{t - 7.62}{380} = 1 - \frac{12 - 7.62}{380}$$

$$= 0.99$$

$$C_F = 0.42 \qquad \text{(Fig. 12-11)}$$

$$S_{10^6} = C_R C_M C_S C_F S_N' \qquad (12\text{-}12)$$

$$= 0.76 \times 0.80 \times 0.99 \times 0.42 \times \frac{1000}{2} = 126 \text{ MPa}$$

$$S_N = S_A = 201$$

$$N = 10^{\left(6 - 3\frac{\log S_N/S_{10^6}}{\log S_{10^3}/S_{10^6}}\right)} \qquad (12\text{-}4)$$

$$= 10^{\left(6 - 3\frac{\log 201/126}{\log 547/126}\right)}$$

$$= 10^{5.05} = 112\,000$$

The safe number of load cycles is 112 000.

Example 12-2

An element of a leaf spring is made of hot-rolled steel having an ultimate strength of 145 ksi. The leaf is 4.5 ft long, 3 in. wide, and 0.5 in. thick. As indicated in Fig. 12-12, it is simply supported at the ends and carries a load at the center that cycles between zero and 0.5 kip. Determine the safe number of load cycles.

Solution

$$M_{\text{max}} = \frac{1}{4} PL$$

$$\sigma_{\text{max}} = \frac{M_{\text{max}} c}{I} = \frac{\frac{1}{4} PL(t/2)}{\frac{1}{12} wt^3} = \frac{3}{2} \frac{PL}{wt^2}$$

$$= \frac{3}{2} \frac{0.5 \times 54}{3 \times 0.5^2} \frac{\text{kip} \cdot \text{in.}}{\text{in.} \cdot \text{in.}^2}$$

$$= 54 \text{ kip/in.}^2$$

$$S_A = S_M = \frac{\sigma_{\text{max}}}{2} = \frac{54}{2} = 27 \text{ kip/in.}^2$$

$$C_R = 0.76 \qquad (12\text{-}1)$$

$$C_M = 1 - \frac{S_M}{\sigma_u} \qquad (12\text{-}6)$$

$$= 1 - \frac{27}{145} = 0.81$$

$$S_{10^3} = C_R C_M 0.9 \sigma_u \qquad (12\text{-}7)$$

$$= 0.76 \times 0.81 \times 0.9 \times 145 = 80.3 \text{ kip/in.}^2$$

$$C_S = 1 - \frac{D - 0.3}{15} \qquad (12\text{-}10)$$

$$= 1 - \frac{t - 0.3}{15} = 1 - \frac{5 - 0.3}{15}$$

$$= 0.99$$

$$C_F = 0.42 \qquad (\text{Fig. } 12\text{-}11)$$

$$S_{10^6} = C_R C_M C_S C_F S_N' \qquad (12\text{-}12)$$

$$= 0.76 \times 0.81 \times 0.99 \times 0.42 \times \frac{145}{2} = 18.6 \text{ kip/in.}^2$$

$$S_N = S_A = 27.0$$

$$N = 10^{\left(6 - 3 \frac{\log S_N / S_{10^6}}{\log S_{10^3} / S_{10^6}}\right)} \qquad (12\text{-}4)$$

$$= 10^{\left(6 - 3 \frac{\log 27.0/18.6}{\log 80.3/18.6}\right)}$$

$$= 10^{5.236} = 172,000$$

The safe number of load cycles is 172,000.

Problems 12-7 to 12-8

12-9 STRENGTH REDUCTION FACTOR: CONCENTRATION OF STRESS

We now come to the most difficult of all the fatigue strength reduction factors, that is C_C, the factor that takes into account the stress concentration at holes, notches, and abrupt changes in cross section. The effect of stress concentration on the design of parts made of brittle and ductile materials was discussed in Chapter 1. At that time it was pointed out that the stress concentration factor is ignored when the material is ductile but is applied when the material is brittle. In effect, for static cases we used a stress concentration factor that was either 1.0 or its theoretical value depending upon the ductility of the material. For cyclic stress the stress concentration factor lies in the range between these two extreme values. The value that applies within that range depends upon a property of the material that we have not yet encountered.

A plate subjected to cyclic load would have its strength substantially reduced by a small hole regardless of whether the plate material was ductile or brittle. A fatigue crack would start in the region of highest stress and spread across the plate, the sharp end of the crack providing a very high stress concentration and making it possible for the crack to enter regions where the initial stress was very low. We would thus expect the introduction of the small hole to reduce the fatigue strength of the plate by a factor of three; that is, that the strength reduction factor for *stress concentration*,

$$C_C = \frac{1}{3} \tag{12-13a}$$

or, in general,

$$C_C = \frac{1}{K} \tag{12-13b}$$

where K is the stress concentration factor.

Experiments with plates of various materials subjected to cyclic load would show that the effect of the hole on fatigue is not solely shape-dependent, as implied by (12-13), but also depends upon the material. This could be stated by saying that some materials are more sensitive to stress concentration or more *notch-sensitive* than others.

Although we did not mention it at the time, we encountered a manifestation of *notch sensitivity* in the curves for C_F in Fig. 12-11. These curves show that for equal stress concentration at the bottoms of grooves or notches in a given surface, the strength reduction varies with the strength of the material. We see that the high-strength materials are more influenced by the surface roughness than the low-strength materials. Experiments with plates having holes, or any other configuration that causes stress concentration, would show the same tendency for high-strength materials to be more notch-sensitive than low-strength materials.

In order to deal with stress concentration and notch sensitivity, we will follow standard practice and replace (12-13b) with

$$C_C = \frac{1}{K_f} \tag{12-14}$$

where K_f is the *fatigue stress concentration factor*. This factor depends upon both the geometry of the part and the notch sensitivity of the material. To distinguish between the two stress concentration factors, we will designate the *theoretical* or *static* stress concentration factor as K_t.

The two stress concentration factors are related through

$$K_f = 1 + q(K_t - 1) \tag{12-15}$$

where q is the *notch sensitivity factor*.

The factor q can range from 0 to 1. A material that is not at all influenced by a stress concentration would have a q equal to zero, which makes the fatigue stress concentration factor equal to unity. This means that notches have no effect on the fatigue life of parts made of material having zero notch sensitivity. Low-grade cast iron is quite insensitive to notches because the material has so many built-in stress raisers, in the form of blow-holes, graphite deposits, and other flaws, that additional holes and notches have little influence on fatigue life. Consequently, some engineers use $q = 0$ when designing for low-strength cast iron and $q = 0.2$ for other grades of cast iron.

The notch sensitivity of steel depends upon its ultimate strength and also on the sharpness of the notch in the region of the highest stress. These parameters are related by the curves in Fig. 12-13,* which give q for known steel strength and radius of curvature at the point of highest stress. Inspection of these curves shows that some of the gain made by paying a premium price for high-strength material is lost in the larger value of the notch sensitivity factor. This can be offset by designing to avoid sharp reentrant angles, especially in high-strength materials.

The theoretical stress concentration factor, K_t, is difficult to determine because of the infinite variety of shapes and proportions that are encountered in practice. For given dimensions, K_t can be determined by experimental measurement, finite element calculations, or theoretical analysis. All of these methods are costly and time-consuming and most designers rely on published results in the form of tables and charts of stress concentration factors. Quite frequently, the shape or proportions of the part being designed do not exactly match those for which K_t is known. The designer must then use the available data as a guide to select a value that seems reasonable but one that tends to err on the side of safety. This is part of the art of engineering design.

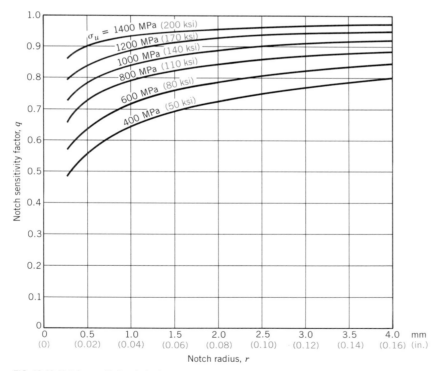

FIG. 12-13 Notch sensitivity of steel.

*Plotted from data obtained from Deutschman, Michels, and Wilson, *Machine Design Theory and Practice*, Macmillan, New York, 1975.

Figure 12-14a to i gives stress concentration factors for shapes and loads that are encountered in the problems.* These are adequate for academic problems, but for practical designs the factors should be taken from one or more of the numerous books that list stress concentration factors for a great variety of shapes and loads.

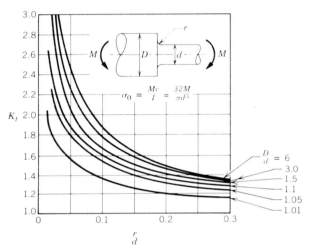

FIG. 12-14a

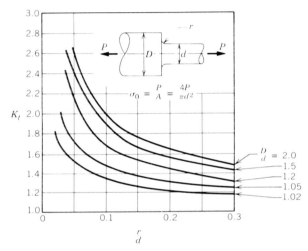

FIG. 12-14b

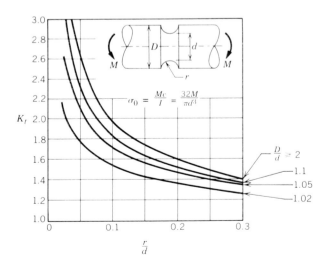

FIG. 12-14c

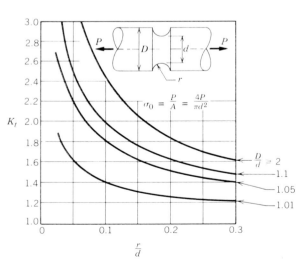

FIG. 12-14d

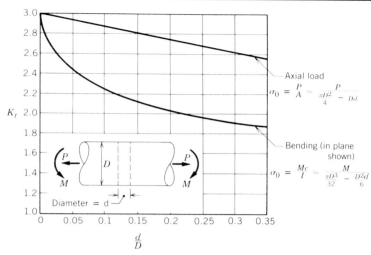

FIG. 12-14e

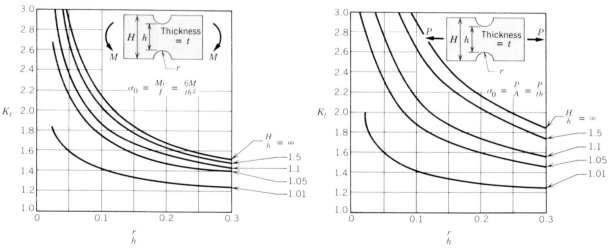

FIG. 12-14f

FIG. 12-14g

*Figure 12-14a to g was plotted from data obtained from R. E. Peterson *Stress Concentration Design Factors*, Wiley, New York, 1953, and *Stress Concentration Factors*, Wiley, New York, 1974. Figure 12-14h and i was plotted from formulas given in Roark and Young *Formulas for Stress and Strain*, McGraw-Hill, New York, 1975.

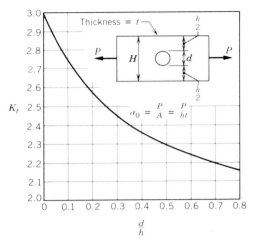

FIG. 12-14*h*

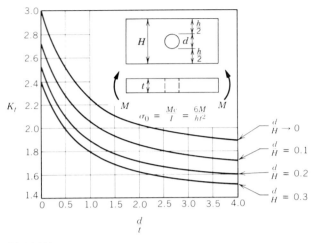

FIG. 12-14*i*

12-10 STRENGTH REDUCTION FACTOR: OTHER CONDITIONS

In addition to the strength reduction factors that have been given, there is a catchall factor, C_O, which is used to take into account *other* conditions that have an influence on fatigue. For example, if a part is plated, the fatigue strength is reduced and it is appropriate to use $C_O = 0.8$. At first this may seem to be wrong, since a smooth-plated surface would appear to have no irregularities to cause surface stress concentration. However, the plating material frequently has very fine cracks that concentrate stress and hence reduce fatigue strength.

Any production process, such as forging, hot rolling, or heat treatment, that exposes the surface of steel to air at a high temperature will cause the surface to become decarburized. This locally changes the chemical composition of the material and hence its strength and also causes a residual tensile stress. Both of these effects contribute to a reduction in fatigue strength. There is evidence that this can cause as much as a 75% reduction in strength, that is, $C_O = 0.25$ may be required if the steel is severely decarburized. Case hardening has the opposite effect and increases fatigue life by making the surface material stronger and also by causing a residual compressive stress. Whether caused by changing carbon content, quenching, or other processes, allowance for the residual surface stresses is made through C_O. Since these stresses are usually impossible to calculate, their influence is taken into account by an estimate of C_O based on experience.

Temperature over 70°C tends to reduce fatigue life; some designers use

$$C_o = \frac{345}{275 + T} \qquad \text{for } T > 70°C \qquad \textbf{(12-16)}$$

to allow for this temperature effect.

Temperature over 160 F tends to reduce fatigue life; some designers use

$$C_o = \frac{620}{460 + T} \quad \text{for } T > 160 \text{ F}$$ (12-16)

to allow for this temperature effect.

Corrosion also has a very strong influence on fatigue life. For example, high-strength steel in salt water can loose 90% of its fatigue strength.

In most cases the effects that determine C_o have little or no influence on the low-cycle end of the S-N curve. Consequently, C_o is usually omitted from the factors that determine S_{10^3}.

12-11 SUMMARY OF EQUATIONS AND STRENGTH REDUCTION FACTORS

Rewriting the strength equations to include all fatigue strength reduction factors, we obtain the final form as

$$S_{10^3} = C_R C_M 0.9 \sigma_u$$ (12-17)

$$S_{10^6} = C_R C_M C_S C_F C_C C_O S_N'$$ (12-18)

where

C_R = reliability reduction factor
C_M = mean stress reduction factor
C_S = size reduction factor
C_F = surface finish reduction factor
C_C = stress concentration reduction factor
C_O = reduction factor for other effects

Example 12-3

The stepped shaft of Fig. 12-15 is made of AISI 4340 steel and is machined all over. The shaft rotates while the direction of the load remains fixed. Determine the safe number of revolutions for the shaft.

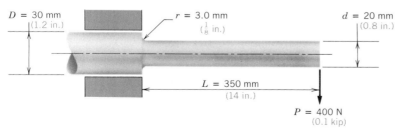

$D = 30$ mm (1.2 in.) $r = 3.0$ mm ($\frac{1}{8}$ in.) $d = 20$ mm (0.8 in.)

$L = 350$ mm (14 in.)

$P = 400$ N (0.1 kip)

FIG. 12-15

Solution. The material properties are

$$\sigma_u = 690 \text{ MPa} \qquad \text{(Table A)}$$

$$S_N' = \frac{\sigma_u}{2} \qquad \text{(assumed)}$$

$$= \frac{690}{2} = 345 \text{ MPa}$$

and the strength reduction factors:

$$C_R = 0.76 \qquad \text{(12-1)}$$

$$C_M = 1.0 \qquad \text{(12-6) with } S_M = 0$$

$$C_S = 1 - \frac{(D - 7.62)}{380} = 0.97 \qquad \text{(12-10)}$$

$$C_F = 0.79 \qquad \text{(Fig. 12-11)}$$

The stress concentration is

$$\frac{r}{d} = \frac{3.0}{20} = 0.15$$

$$\frac{D}{d} = \frac{30}{20} = 1.5$$

$$K_t = 1.52 \qquad \text{(Fig. 12-14}a\text{)}$$

$$q = 0.85 \qquad \text{(Fig. 12-13)}$$

$$K_f = 1 + q(K_t - 1) \qquad \text{(12-15)}$$

$$= 1 + 0.85(1.52 - 1) = 1.44$$

$$C_C = \frac{1}{K_f} \qquad \text{(12-14)}$$

$$= \frac{1}{1.44} = 0.69$$

$$C_O = 1.0$$

The strength-number relations are

$$S_{10^3} = C_R C_M 0.9 \sigma_u \qquad \text{(12-17)}$$

$$= 0.76 \times 1.0 \times 0.9 \times 690 = 472 \text{ MPa}$$

$$S_{10^6} = C_R C_M C_S C_F C_C C_O S_N' \qquad \text{(12-18)}$$

$$= 0.76 \times 1.0 \times 0.97 \times 0.79 \times 0.69 \times 1.0 \times 345$$

$$= 139 \text{ MPa}$$

The shaft life is

$$S_N = \sigma_{\max} = \frac{Mc}{I}$$

$$= \frac{PL(d/2)}{(\pi/64)d^4} = \frac{32}{\pi}\frac{PL}{d^3}$$

$$= \frac{32}{\pi}\frac{400 \times 350}{20^3}\frac{\text{N} \cdot \text{mm}}{\text{mm}^3}$$

$$= 178 \text{ MPa}$$

$$N = 10^{\left(6 - 3\frac{\log S_N/S_{10^6}}{\log S_{10^3}/S_{10^6}}\right)} \qquad (12\text{-}4)$$

$$= 10^{\left(6 - 3\frac{\log 178/139}{\log 472/139}\right)}$$

$$= 250\ 000$$

The shaft, therefore, has **a safe life of 250 000 revolutions.**

Example 12-3

The stepped shaft of Fig. 12-15 is made of AISI 4340 steel and is machined all over. The shaft rotates while the direction of the load remains fixed. Determine the safe number of revolutions for the shaft.

Solution.   The material properties are

$$\sigma_u = 100 \text{ ksi} \qquad \text{(Table A)}$$

$$S_N' = \frac{\sigma_u}{2} \qquad \text{(assumed)}$$

$$= \frac{100}{2} = 50 \text{ ksi}$$

and the strength reduction factors:

$$C_R = 0.76 \qquad (12\text{-}1)$$

$$C_M = 1.0 \qquad (12\text{-}6) \text{ with } S_M = 0$$

$$C_S = 1 - \frac{D - 0.3}{15} = 0.94 \qquad (12\text{-}10)$$

$$C_F = 0.79 \qquad (\text{Fig. } 12\text{-}11)$$

The stress concentration is

$$\frac{r}{d} = \frac{0.125}{0.8} = 0.16$$

$$\frac{D}{d} = \frac{1.2}{0.8} = 1.5$$

$$K_t = 1.5 \qquad (\text{Fig. } 12\text{-}14a)$$

$$q = 0.85 \quad \text{(Fig. 12-13)}$$

$$K_f = 1 + q(K_t - 1) \quad (12\text{-}15)$$

$$= 1 + 0.85(1.5 - 1) = 1.43$$

$$C_C = \frac{1}{K_f} \quad (12\text{-}14)$$

$$= \frac{1}{1.43} = 0.70$$

$$C_O = 1.0$$

The strength-number relations are

$$S_{10^3} = C_R C_M 0.9\sigma_u \quad (12\text{-}17)$$

$$= 0.76 \times 1.0 \times 0.9 \times 100 = 68.4 \text{ kip/in.}^2$$

$$S_{10^6} = C_R C_M C_S C_F C_C C_O S_N' \quad (12\text{-}18)$$

$$= 0.76 \times 1.0 \times 0.94 \times 0.79 \times 0.70 \times 1.0 \times 50$$

$$= 19.8 \text{ kip/in.}^2$$

The shaft life is

$$S_N = \sigma_{\max} = \frac{Mc}{I}$$

$$= \frac{PL(d/2)}{(\pi/64)d^4} = \frac{32}{\pi} \frac{PL}{d^3}$$

$$= \frac{32}{\pi} \frac{0.1 \times 14}{0.8^3} \frac{\text{kip} \cdot \text{in.}}{\text{in.}^3}$$

$$= 27.9 \text{ kip/in.}^2$$

$$N = 10^{\left(6 - 3\frac{\log S_N/S_{10^6}}{\log S_{10^3}/S_{10^6}}\right)} \quad (12\text{-}4)$$

$$= 10^{\left(6 - 3\frac{\log 27.9/19.8}{\log 68.4/19.8}\right)}$$

$$= 148,000$$

The shaft, therefore, has **a safe life of 148,000 revolutions.**

Example 12-4

Determine the safe number of revolutions for the shaft in Example 12-3 if an additional tensile load of 65 kN is applied along the axis of the shaft.

Solution. The constant component of stress, or the mean stress, S_M, in the reduced part of the shaft is

$$S_M = \frac{65\ 000}{(\pi/4)20^2} \frac{N}{mm^2}$$

$$= 207\ MPa$$

and the additional strength reduction factor is

$$C_M = 1 - \frac{S_M}{\sigma_u} \qquad (12\text{-}6)$$

$$= 1 - \frac{207}{690} = 0.70$$

Note that the stress concentration factor, which could have been obtained from Fig. 12-14a, was not applied to S_M. In this case it is correct to ignore the stress concentration because we are dealing with the constant component of stress and concentration is not significant when the stress is static.

Although the axial load alters the strength reduction factor for mean stress, all others remain as in Example 12-13. This gives strength-number relations:

$$S_{10^3} = 0.76 \times 0.70 \times 0.9 \times 690 = 330\ MPa$$

$$S_{10^6} = 0.76 \times 0.70 \times 0.97 \times 0.79 \times 0.69 \times 1.0 \times 345$$

$$= 97\ MPa$$

The shaft life is

$$N = 10^{\left(6 - 3\frac{\log 178/97}{\log 330/97}\right)}$$

$$= 32\ 000$$

With the addition of the axial load, the shaft has **a safe life of only 32 000 revolutions.**

Example 12-4

Determine the safe number of revolutions for the shaft in Example 12-3 if an additional tensile load of 15 kip is applied along the axis of the shaft.

Solution.   The constant component of stress, or the mean stress, S_M, in the reduced part of the shaft is

$$S_M = \frac{15}{(\pi/4)0.8^2} \frac{kip}{in.^2}$$

$$= 29.8\ kip/in.^2$$

and the additional strength reduction factor is

$$C_M = 1 - \frac{S_M}{\sigma_u} \qquad (12\text{-}6)$$

$$= 1 - \frac{29.8}{100} = 0.70$$

Note that the stress concentration factor, which could have been obtained from Fig. 12-14a, was not applied to S_M. In this case it is correct to ignore the stress concentration because we are dealing with the constant component of stress and concentration is not significant when the stress is static.

Although the axial load alters the strength reduction factor for mean stress, all others remain as in Example 12-3. This gives strength-number relations:

$$S_{10^3} = 0.76 \times 0.70 \times 0.9 \times 100 = 47.9 \text{ kip/in.}^2$$

$$S_{10^6} = 0.76 \times 0.70 \times 0.94 \times 0.79 \times 0.70 \times 1.0 \times 50$$

$$= 13.8 \text{ kip/in.}^2$$

The shaft life is

$$N = 10^{\left(6 - 3\frac{\log 27.9/13.8}{\log 47.9/13.8}\right)}$$

$$= 20,000$$

With the addition of the axial load, the shaft has **a safe life of only 20,000 revolutions.**

Problems 12-9 to 12-20

12-12 LOADS OF VARYING AMPLITUDE

The cases that have been dealt with up to this point have all had loads that cycled between limits that remain fixed for the life of the part. In many applications the loads change significantly with time. For example, the suspension system of a motor vehicle is subjected to a great variety of load amplitudes as the vehicle travels over roads having different degrees of roughness. Similarly, the frames of aircraft are subjected to many different loads depending on air turbulence. Also the loads on the connecting rods of internal combustion engines will change as the power output varies.

The load on an aircraft is of a random nature, but studies of strain gage data gathered from many flights show that loads can be predicted for design purposes. The load-time curves are broken into harmonics and simplified to predict that during a given time interval there will be n_1 cycles of load L_1, n_2 cycles of load L_2, and so on. Of course, all cycles of load L_1 are not applied sequentially within one time period followed by all cycles of load L_2 and then all of L_3, and so on. In reality a few cycles of one load will be applied followed by a few cycles of another load. There is evidence to suggest that the safety of a part is not materially influenced by the sequence of load application. Consequently, in design we treat the load as though L_1 is applied the full number of cycles n_1 followed by L_2 applied for n_2 cycles, and so on.

To test for safety, we consider each load system separately and calculate the number of cycles of the load that could be carried safely. Thus we find that L_1 could be carried safely for N_1 cycles, L_2 for N_2 cycles,

L_3 for N_3 cycles, and so on. The fact that, in reality, L_1 is applied only n_1 times means that if it were the only load, the fractional portion of the life of the part given by n_1/N_1 has been consumed. Or we may say that fatigue damage has occurred to the extent of n_1/N_1 of the life of the part.

In treating load L_2, an undamaged part is considered and the safe number of cycles N_2 is calculated. The application of n_2 cycles will damage the part to the extent of n_2/N_2 of its life. Similarly, the damage done by L_3 is n_3/N_3 of the life of the part.

When all loads are considered to act sequentially on the same part, the fatigue damage will be cumulative and the part will be safe if

$$\frac{n_1}{N_1} + \frac{n_2}{N_2} + \frac{n_3}{N_3} + \ldots \leqslant 1 \qquad \text{(12-19)}$$

This is known as the *accumulative fatigue damage formula* or *Miner's rule*. Some designers make allowance for the possibility of additional damage, due to interaction between the loads, by changing the right-hand side to a value less than 1, for example, 0.8.

Problems 12-21 to 12-24

12-13 SUMMARY

In this chapter we have studied a simple approach to a very complex subject. The subject has been made complex by the physical nature of fatigue failure; it is impossible to reduce it to a design procedure that is as simple as that of some other engineering designs such as beams in bending. There are a large number of parameters influencing fatigue life, and in fatigue experiments it is often difficult to keep all other parameters constant while one is varied to determine its influence on fatigue. Consequently, there is often a wide scatter in experimental data; to such an extent that it may obscure the relationship that is being determined. This often makes it impractical to establish exact mathematical formulas and, as you have seen, we frequently made assumptions that were known to be approximate. These assumptions were required because of the lack of data or because we wanted to keep the design procedures from becoming inordinately complicated. For example, we assumed that several fatigue strength reduction factors do not influence S_{10^3}; we also assumed that strength reduces linearly with diameter although we know that this is an oversimplification.

The methods presented here have enabled you to become acquainted with the main factors that are taken into account in designing for fatigue. These methods are widely used but are not universally accepted; therefore, you should expect to find other methods in use when you enter practice and you should be prepared to adopt the methods that are recommended by your more experienced colleagues. Above all you should not think that you are now ready to do fatigue design; a long apprenticeship under experienced designers will be required before you are ready to practice this branch of the art of engineering.

PROBLEMS

12-1 The given polished shaft has an overhanging end that supports a lateral load. The shaft turns at 50 rpm and is part of a machine that is being designed for a life of 10 000 hours. The steel has an ultimate strength of 1200 MPa (180 ksi).

    a. Determine the magnitude of the load *P* that can be safely carried by the shaft.

    b. If an axial load of 4 kN (900 lb) is applied in addition to the load determined in part (a), what is the safe life of the shaft in hours?

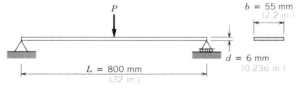

PROBLEM 12-1

12-2 An element of a leaf spring can be treated as a simply supported beam with a load at the center as shown in the figure. The steel in the spring has an ultimate strength of 1000 MPa (145 ksi). The surface may be taken as being polished (this assumption is not very realistic, but you are not yet ready to deal with parts having any other surface finish). A downward load of 1.3 kN (290 lb) is placed on the spring and then released. How many times can this cycle be applied safely?

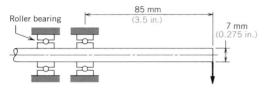

PROBLEM 12-2

12-3 What is the maximum load *P* that can be applied in Prob. 12-2 if the spring is to have an infinite life?

12-4 The given shaft is made of AISI 4063 and has a polished surface. It carries an axial tensile load of 5 kN (1.1 kip) and a transverse load. Determine the allowable transverse load if the shaft rotates 50 000 revolutions during its service life.

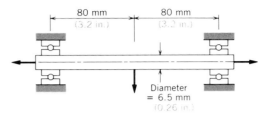

PROBLEM 12-4

12-5 The given cantilever has a polished surface and is made of AISI 4063. The load cycles between 700 N (160 lb) downward and 350 kN (80 lb) upward. Determine the safe number of cycles.

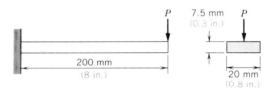

PROBLEM 12-5

12-6 A polished steel shaft, having an ultimate strength of 600 MPa (90 ksi) and an endurance limit of 350 MPa (50 ksi), passes through three bearings in the frame of a machine as shown in the figure. The manufacturing tolerances are such that the center bearing may be laterally displaced as much as 0.50 mm (0.02 in.) from its theoretically correct position. Assuming that the other loads on the shaft are negligible and that the center bearing has the maximum misalignment, what is the safe life of the shaft in revolutions?

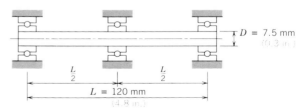

PROBLEM 12-6

12-7 If the spring in Prob. 12-2 had been machined rather than polished, how many load cycles could it carry safely?

12-8 A 50-mm (2-in.) shaft is made of AISI 4340 hot rolled steel. The shaft turns at 80 rpm and is part of a machine being designed for a service life of 10 000 hours. The shaft is 1.5 m (60 in.) long and is supported by bearings at the ends and center. How much misalignment can be allowed in the center bearing? Assume that the stress is due to misalignment only. Note that this problem would be much more realistic, but very hard to solve, if the shaft also carried a torque load.

12-9 A 25-mm-diameter (1.0-in.-diameter) rod in a machine is made of steel having an ultimate strength of 1200 MPa (180 ksi). A circumferential retaining-ring groove is machined into the surface of the shaft. The groove is 1.6 mm (0.064 in.) wide, has straight sides with a semicircular bottom and a total depth of 1.8 mm (0.072 in.). The rod operates at a temperature of 100°C (212°F) and carries an axial load that cycles between 20 kN (4.5 kip) compression and 100 kN (22.5 kip) tension. How many load cycles can the rod carry safely?

12-10 A pump rod is subjected to an axial load that alternates between 15-kN (3.4-kip) tension and 15-kN (3.4-kip) compression. The rod is made of AISI 1020 steel. Determine the safe number of load cycles if

a. The rod surface is in the as-forged state.
b. The rod is machined all over.

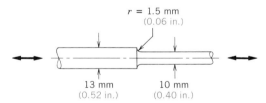

PROBLEM 12-10

12-11 The rod given in Prob. 12-10 carries a load that fluctuates between zero- and 15-kN (3.4-kip) tension. Determine the safe number of cycles for conditions (a) and (b).

12-12 The spring leaf given in Example 12-2 is altered by having a hole drilled at the location of the load *P*. The hole has a diameter of 8 mm (0.32 in.). So that the net section will not be reduced, the width is increased from 80 to 88 mm (3.00 to 3.32 in.). Determine the safe numbers of load cycles.

12-13 The given shaft is machined to operate in a sleeve bearing. The shaft has a semicircular oil groove and is subjected to a completely reversed bending moment of 40 kN·m (30 kip·ft) at the oil-grove location. The shaft is made from AISI 4063 steel and is machined all over. Determine the safe life of the shaft in hours if the shaft rotates at 2 rpm.

PROBLEM 12-13

12-14 A 50-mm (2-in.) shaft is made of AISI 1045 steel and has a 2-mm (0.08-in.) transverse hole drilled through its center. While rotating at 20 rpm, it is subjected to a bending moment of 700 N·m (520 lb·ft) in the region of the hole. Determine the safe life of the shaft in hours:

a. If the surface of the shaft is in the as-forged state.
b. If the shaft is forged to the diameter given above and then the surface, to a depth of 2 mm (0.08 in.), is removed by machining.

12-15 A 75-mm (3-in.) shaft is made of AISI 4063 steel and is subjected to an alternating load. The shaft must be modified in order to prevent a roller bearing from moving axially along the shaft. Two alternatives are being considered:

(i) Use a snap ring which would require that a circumferential groove be machined. The groove would be 4 mm (0.16 in.) wide and 3 mm (0.12 in.) deep with a semicircular bottom.
(ii) Pin a collar to the shaft. A transverse hole 4 mm (0.16 in.) in diameter would be required for the pin.

Which of the above alternatives will weaken the shaft by the least amount?

a. When the load is a bending moment.
b. When the load is an axial tension and compression force.

12-16 A shaft, made of steel having an ultimate strength of 1100 MPa (160 ksi), is 1.2 m (48 in.) long and is machined to a diameter of 24 mm (1 in.). It is supported at the ends in self-aligning ball bearings and carries a lateral load at midlength. A 4-mm (0.16-in.) hole is drilled through the center of the shaft at 500 mm (20 in.) from one end. Determine the safe load if the shaft is to have an indefinite life.

12-17 The given machine part is made of AISI 1090 and is machined all over. It carries an axial load that cycles between 100-kN (22-kip) tension and 700-kN (154-kip) tension. Determine the safe number of cycles.

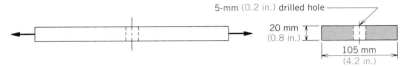

PROBLEM 12-17

12-18 The given rod is made of AISI 4063 steel and carries an axial load only. How many load cycles can the rod carry safely under the following conditions?

a. The load cycles between 100-kN (22-kip) tension and 100-kN (22-kip) compression.
b. The load cycles between 100-kN (22-kip) tension and 50-kN (11-kip) compression.

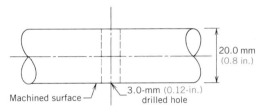

PROBLEM 12-18

12-19 The given rotating shaft carries a downward load of 10 kN (2.2 kip). The shaft material has an ultimate strength of 1200 MPa (174 ksi) and an endurance limit of 700 MPa (100 ksi). All surfaces are machined. Determine the safe number of rotations for the shaft when

a. The downward force is the only load.
b. There is also an axial tensile force of 200 kN (45 kip).

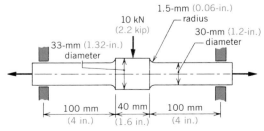

PROBLEM 12-19

12-20 Calculate the safe number of vertical load cycles for the notched cantilever when the load cycles between

a. A downward force of 25 kN (5.5 kip) and an upward force of the same magnitude.
b. Zero force and an upward force of 25 kN (5.5 kip).

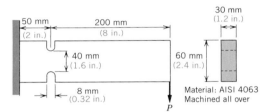

PROBLEM 12-20

12-21 The given shaft is machined all over and is made of steel having an ultimate strength of 1000 MPa (145 ksi). While supporting a load of 250 N (56 lb), it rotates 30 000 revolutions. The load then changes to 400 N (90 lb). For how many revolutions can it safely carry the increased load?

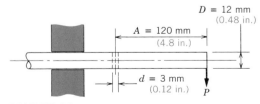

PROBLEM 12-21

12-22 The given notched plate is machined all over and is made of AISI 1020. It is designed to carry an alternating axial load of 20 kN (4.5 kip).

a. Determine the safe number of cycles for the design load.

b. Through an error the part is subjected to double the design load for 2000 cycles. By how many cycles will this reduce the safe life of the part?

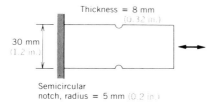

Thickness = 8 mm
(0.32 in.)

30 mm
(1.2 in.)

Semicircular
notch, radius = 5 mm (0.2 in.)

PROBLEM 12-22

12-23 A part being designed for a punch press is subjected to an alternating axial load. It has a rectangular section 25 × 5 mm (1 × 0.2 in.) and has a 5-mm (0.2-in.) drilled hole as in Fig. 12-14h. The part is made of AISI 1090, is machined all over, and operates at a temperature of 150°C (300°F).

Because of different thicknesses of material being punched, the part is subjected to a variety of loads. During one typical hour of operating the following loads are anticipated:

	Load (kN)	(kip)	Number of Applications
(i)	10	(2.2)	100
(ii)	30	(6.6)	5
(iii)	50	(11.0)	1

a. Determine the safe life of the parts in hours.
b. Determine the factor by which the life will be altered if the material that produces load (iii) is no longer punched.
c. Determine the safe life if only load (i) is applied.

d. If the machine is going to be used to punch one thickness of material, determine the load for an indefinite life.

12-24 A motor vehicle has the given stub axle made of steel having an ultimate strength of 1050 MPa (152 ksi). The surface in the region of high stress is machined and ground. The axle does not rotate as the wheel turns; consequently, the stress is substantially constant as long as the vehicle is traveling on a smooth road. Under these conditions the force, F, is 11.0 kN (2420 lb).

Whenever the wheel strikes an obstruction, the force, F, can be considered to increase by a certain amount and then decrease by the same amount before returning to its constant level. For design purposes the wheel may be considered to encounter two types of bumps, designated minor and major. A minor bump, which can be expected once every 2 kilometres, imposes a transient force of 5 kN (1120 lb). A major bump occurs with one-tenth the frequency and results in a transient force of 8 kN (1800 lb). Determine the safe life of the axle in kilometres.

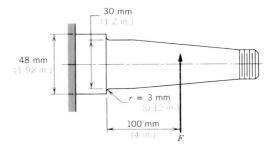

30 mm
(1.2 in.)

48 mm
(1.92 in.)

r = 3 mm
(0.12 in.)

100 mm
(4 in.)

F

PROBLEM 12-24

APPENDIX A

TABLES (SI UNITS)

Table A   Physical Properties of Commonly Used Engineering Materials

Table B   Properties of Rolled Structural Steel Shapes

        *B-1 W-Shapes**
        *B-2 S-Shapes**
        *B-3 C-Shapes**
        *B-4 Angles, Equal Legs**
        *B-5 Angles, Unequal Legs**

Table C   Properties of Steel Pipe*

Table D   Properties of Steel Rods

Table E   Properties of Reinforcing Bars

Table F   Properties of Timber

Table G   Formulas for Beam Deflections

Table H   Properties of Areas

Table I   SI Units

*Reprinted from the *Handbook of Steel Construction*, Third Edition, with kind permission of the Canadian Institute of Steel Construction.

TABLE A Physical Properties of Commonly Used Engineering Materials

Note: Material properties depend upon conditions that are not referred to in this table. For example, strength may depend upon rolling temperature (steel), moisture content (wood), age (concrete), and so on. Therefore, the values given below, while adequate for textbook exercises, should not be used in engineering design.

Material	Yield Stress T or C (MPa)	Ultimate Stress T or C (MPa)	Typical Allowable Stress T or C (MPa)	Shear (MPa)	Young's Modulus E (MPa)	Poisson's Ratio ν	Shear Modulus G (MPa)	Density (kg/m³)	Coeff. of Expansion α [μm/(m·°C)]
Structural Steel									
Low-strength	±230	+380	±140	100	200000	0.30	77000	7850	12
Medium-strength	±310	+450	±190	130	200000	0.30	77000	7850	12
High-strength	±410	+620	±250	170	200000	0.30	77000	7850	12
Wrought Steel									
AISI 1020	±250	+450			200000	0.30	77000	7850	12
AISI 1045	±410	+660			200000	0.30	77000	7850	12
AISI 1090	±460	+840			200000	0.30	77000	7850	12
AISI 4340	±470	+690			200000	0.30	77000	7850	12
AISI 4063	±1100	+1240			200000	0.30	77000	7850	12
AISI 8760	±1380	+1500			200000	0.30	77000	7850	12
Stainless Steel									
AISI 302	±260	+620			190000	0.30	73000	7700	12
AISI 431	±650	+860			190000	0.30	73000	7700	12
Reinforcing Steel									
Grade 300	±300	+450	±120		200000	0.30	77000	7850	12
Grade 350	±350	+550	±140		200000	0.30	77000	7850	12
Grade 400	±400	+600	±160		200000	0.30	77000	7850	12
Gray Cast Iron		+150			90000	0.20	38000	7600	11
		−600							
Aluminum Alloys									
1100-0	±35	+90			70000	0.33	26000	2800	23
6061-T6	±270	+310	±140		70000	0.33	26000	2800	23
7075-T6	±490	+560			70000	0.33	26000	2800	23
Magnesium Alloy	±160	+230			45000	0.32	17000	1800	25
Copper-Base Alloys									
Phosphor bronze	±510	+560			110000	0.35	41000	8200	21
Yellow brass	±410	+510			100000	0.35	37000	8600	19
Concrete									
Low-strength		−20	−9	0.4	21000	0.15		2400	11
High-strength		−35	−16	0.6	28000	0.15		2400	11
Wood									
Douglas fir		+50	±9[a]	0.9	12000			550	
			−6[b]						
Western hemlock		+42	±7[a]	0.7	10000			480	
			−5[b]						
Eastern spruce		+38	±7[a]	0.6	8000			450	
			−5[b]						
Glass		+30	+7		70000	0.17	30000	2600	15
		−300							
Rubber		+25				0.5		970	

[a] Tension or compression stress in beams.
[b] Compression stress parallel to grain in short columns.

TABLE B-1 W-Shapes

Note: Under "Designation" the first number is the nominal depth of the section in millimetres; the second number is the mass in kilograms per metre of length.

I = moment of inertia
r = radius of gyration
S = section modulus

Designation (mm) (kg/m)	Area (mm^2)	Depth (mm)	Flange Width (mm)	Flange Thick. (mm)	Web Thick. (mm)	I_x ($10^6 \, mm^4$)	r_x (mm)	S_x ($10^3 \, mm^3$)	I_y ($10^6 \, mm^4$)	r_y (mm)
W920 × 446	57000	933	423	43	24	8470	385	18200	540	97
× 387	49300	921	420	37	21	7180	382	15600	453	96
× 342	43600	912	418	32	19	6250	379	13700	390	95
× 289	36800	927	308	32	19	5040	370	10900	156	65
× 253	32300	919	306	28	17	4370	368	9520	134	64
× 223	28600	911	304	24	16	3770	363	8270	112	63
W840 × 329	42000	862	401	32	20	5350	357	12400	349	91
× 193	24700	840	292	22	15	2780	335	6630	90	61
W760 × 314	40100	786	384	33	20	4270	326	10900	316	89
× 257	32800	773	381	27	17	3420	323	8840	250	87
× 185	23500	766	267	24	15	2230	308	5820	75	57
× 161	20400	758	266	19	14	1860	302	4900	61	55
W690 × 140	17800	684	254	19	12	1360	276	3980	52	54
W610 × 195	24900	622	327	24	15	1680	260	5400	142	76
× 155	19700	611	324	19	13	1290	256	4220	108	74
× 125	15900	612	229	20	12	985	249	3220	39	50
× 101	13000	603	228	15	10	764	242	2530	30	48
W530 × 92	11800	533	209	16	10	552	216	2070	24	45
× 74	9520	529	166	14	10	411	208	1550	10	33
W460 × 82	10400	460	191	16	10	370	189	1610	19	42
× 67	8680	454	190	13	8	300	186	1320	15	41
× 52	6630	450	152	11	8	212	179	943	6	31
W410 × 132	16800	425	263	22	13	538	179	2530	67	63
× 100	12700	415	260	17	10	398	177	1920	50	62
× 60	7580	407	178	13	8	216	169	1060	12	40
× 39	4990	399	140	9	6	127	160	634	4	29
W360 × 509	64900	446	416	63	39	2050	178	9170	754	108
× 347	44200	407	404	44	27	1250	168	6140	481	104
× 196	25000	372	374	26	16	636	159	3420	229	96
× 147	18800	360	370	20	12	463	157	2570	167	94
W310 × 79	10100	306	254	15	9	177	132	1160	40	63
× 33	4180	313	102	11	7	65	125	415	2	21
W250 × 115	14600	269	259	22	14	189	114	1410	64	66
× 89	11400	260	256	17	11	143	112	1100	48	65
× 67	8550	257	204	16	9	104	110	806	22	51
× 33	4170	258	146	9	6	49	108	379	5	34
× 18	2270	251	101	5	5	22	99	179	1	20
W200 × 59	7560	210	205	14	9	61	90	582	20	52
× 42	5310	205	166	12	7	41	88	399	9	41
× 27	3390	207	133	8	6	26	87	249	3	31

TABLE B-2   S-Shapes

Note: Under "Designation" the first number is the nominal depth of the section in millimetres, the second number is the mass in kilograms per metre of length.

I = moment of inertia
r = radius of gyration
S = section modulus

Designation (mm)(kg/m)	Area (mm^2)	Depth (mm)	Flange Width (mm)	Flange Thick. (mm)	Web Thick. (mm)	I_x (10^6 mm^4)	r_x (mm)	S_x (10^3 mm^3)	I_y (10^6 mm^4)	r_y (mm)
S610 × 180	22900	622	204	28	20	1310	239	4220	34.7	39
× 158	20100	622	200	28	16	1220	246	3940	32.4	40
× 149	18900	610	184	22	19	996	230	3270	20.1	33
× 134	17100	610	181	22	16	939	234	3080	18.9	33
× 119	15200	610	178	22	13	879	240	2880	17.9	34
S510 × 143	18300	516	183	23	20	702	196	2720	21.1	34
× 128	16400	516	179	23	17	660	201	2560	19.6	35
× 112	14200	508	162	20	16	532	194	2090	12.5	30
× 98	12500	508	159	20	13	497	199	1960	11.7	31
S460 × 104	13300	457	159	18	18	387	171	1690	10.3	28
× 81	10400	457	152	18	12	335	179	1470	8.8	29
S380 × 74	9500	381	143	16	14	203	146	1060	6.6	26
× 64	8150	381	140	16	10	187	151	980	6.1	27
S310 × 74	9470	305	139	17	17	127	116	833	6.6	26
× 61	7730	305	133	17	12	113	121	744	5.7	27
× 52	6650	305	129	14	11	95.8	120	629	4.2	25
× 47	6040	305	127	14	9	91.1	123	597	3.9	26
S250 × 52	6660	254	126	12	15	61.6	96	485	3.6	23
× 38	4820	254	118	12	8	51.4	103	405	2.8	24
S200 × 34	4370	203	106	11	11	27.0	79	266	1.8	20
× 27	3500	203	102	11	7	24.0	83	237	1.6	21
S180 × 30	3800	178	98	10	11	17.8	68	200	1.3	19
× 23	2910	178	93	10	6	15.4	73	173	1.1	20
S150 × 26	3270	152	91	9	12	10.9	58	144	0.98	17
× 19	2370	152	85	9	6	9.19	62	121	0.78	18
S130 × 22	2790	127	83	8	12	6.33	48	100	0.69	16
× 15	1890	127	76	8	5	5.12	52	80.6	0.51	16
S100 × 14	1800	102	71	7	8	2.85	40	55.8	0.38	15
× 11	1450	102	68	7	5	2.55	42	50.1	0.32	15
S75 × 11	1430	76	64	7	9	1.22	29	32.0	0.25	13
× 8	1070	76	59	7	4	1.05	31	27.4	0.19	13

TABLE B-3   C-Shapes

Note: Under "Designation" the first number is the nominal depth of the section in millimetres, the second number is the mass in kilograms per metre of length.

I = moment of inertia
r = radius of gyration
S = section modulus

Designation (mm)(kg/m)	Area (mm^2)	Depth (mm)	Flange Width (mm)	Flange Thick. (mm)	Web Thick. (mm)	I_x $(10^6 mm^4)$	r_x (mm)	S_x $(10^3 mm^3)$	C_x (mm)	I_y $(10^6 mm^4)$	r_y (mm)
C380 × 74	9480	381	94	16	18	168	133	881	20.3	4.60	22.0
× 60	7570	381	89	16	13	145	138	760	19.7	3.84	22.5
× 50	6430	381	86	16	10	131	143	687	20.0	3.39	23.0
C310 × 45	5690	305	81	13	13	67.3	109	442	17.0	2.12	19.3
× 37	4720	305	77	13	10	59.9	113	393	17.1	1.85	19.8
× 31	3920	305	77	13	7	53.5	117	351	17.5	1.59	20.1
C250 × 45	5670	254	76	11	17	42.8	86.9	337	16.3	1.60	16.8
× 37	4750	254	73	11	13	37.9	89.3	299	15.7	1.40	17.2
× 30	3780	254	69	11	10	32.7	93.0	257	15.3	1.16	17.5
× 23	2880	254	65	11	6	27.8	98.2	219	15.9	0.92	17.9
C230 × 30	3800	229	67	10	11	25.5	81.9	222	14.8	1.01	16.3
× 22	2840	229	63	10	7	21.3	86.6	186	14.9	0.81	16.8
× 20	2530	229	61	10	6	19.8	88.5	173	15.1	0.72	16.8
C200 × 28	3560	203	64	10	12	18.2	71.5	180	14.4	0.82	15.2
× 21	2600	203	59	10	8	14.9	75.7	147	14.0	0.63	15.5
× 17	2170	203	57	10	6	13.5	78.9	133	14.5	0.54	15.8
C180 × 22	2780	178	58	9	11	11.3	63.8	127	13.5	0.57	14.3
× 18	2310	178	55	9	8	10.0	65.8	113	13.2	0.48	14.4
× 15	1850	178	53	9	5	8.86	69.2	99	13.8	0.41	14.8
C150 × 19	2450	152	54	9	11	7.12	53.9	93.7	12.9	0.43	13.2
× 16	1980	152	51	9	8	6.22	56.0	81.9	12.6	0.35	13.3
× 12	1540	152	48	9	5	5.45	59.0	70.6	12.8	0.28	13.5
C130 × 17	2190	127	52	8	12	4.36	44.6	68.7	12.9	0.35	12.6
× 13	1700	127	47	8	8	3.66	46.4	57.6	11.9	0.25	12.2
× 10	1260	127	44	8	5	3.09	49.5	48.6	12.2	0.20	12.4
C100 × 11	1370	102	43	8	8	1.91	37.3	37.4	11.5	0.17	11.3
× 9	1190	102	42	8	6	1.77	38.6	34.6	11.6	0.16	11.5
× 8	1020	102	40	8	5	1.61	39.7	31.6	11.6	0.13	11.4
C75 × 9	1120	76	40	7	9	0.85	27.5	22.3	11.4	0.12	10.5
× 7	933	76	37	7	7	0.75	28.3	19.7	10.8	0.10	10.1
× 6	763	76	35	7	4	0.67	29.6	17.6	10.9	0.08	10.1

TABLE B-4 Angles, Equal Legs

Note: The three dimensions given under "Designation" are the lengths of the legs and the thickness of the legs.

I = moment of inertia
r = radius of gyration
z refers to the axis about which the moment of inertia is a minimum.

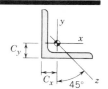

Designation (mm)(mm)(mm)	Mass (kg/m)	Area (mm²)	$C_x = C_y$ (mm)	$I_x = I_y$ (10^3 mm⁴)	$r_x = r_y$ (mm)	I_z (10^3 mm⁴)	r_z (mm)
L200 × 200 × 30	87.1	11100	60.9	40300	60.3	16900	39.0
× 25	73.6	9380	59.2	34800	60.9	14300	39.1
× 20	59.7	7600	57.4	28800	61.6	11700	39.3
× 16	48.2	6140	55.9	23700	62.1	9580	39.5
× 13	39.5	5030	54.8	19700	62.6	9930	39.7
× 10	30.6	3900	53.7	15500	63.0	6210	39.9
L150 × 150 × 20	44.0	5600	44.8	11600	45.5	4810	29.3
× 16	35.7	4540	43.4	9630	46.0	3920	29.4
× 13	29.3	3730	42.3	8050	46.4	3270	29.6
× 10	22.8	2900	41.2	6370	46.9	2580	29.8
L125 × 125 × 16	29.4	3740	37.1	5410	38.0	2230	24.4
× 13	24.2	3080	36.0	4540	38.4	1850	24.5
× 10	18.8	2400	34.9	3620	38.8	1460	24.7
× 8	15.2	1940	34.2	2960	39.1	1190	24.8
L100 × 100 × 16	23.1	2940	30.8	2650	30.0	1120	19.5
× 13	19.1	2430	29.8	2240	30.4	924	19.5
× 10	14.9	1900	28.7	1800	30.8	737	19.7
× 8	12.1	1540	28.0	1480	31.1	604	19.8
× 6	9.14	1160	27.2	1140	31.3	457	19.9
L90 × 90 × 13	17.0	2170	27.2	1600	27.2	672	17.6
× 10	13.3	1700	26.2	1290	27.6	526	17.6
× 8	10.8	1380	25.5	1070	27.8	432	17.7
× 6	8.20	1040	24.7	826	28.1	333	17.9
L75 × 75 × 13	14.0	1780	23.5	892	22.4	379	14.6
× 10	11.0	1400	22.4	725	22.8	298	14.6
× 8	8.92	1140	21.7	602	23.0	246	14.7
× 6	6.78	864	21.0	469	23.3	189	14.8
× 5	5.69	725	20.6	398	23.4	161	14.9
L65 × 65 × 10	9.42	1200	19.9	459	19.6	194	12.7
× 8	7.66	976	19.2	383	19.8	157	12.7
× 6	5.84	744	18.5	300	20.1	122	12.8
× 5	4.91	625	18.1	255	20.2	104	12.9
L55 × 55 × 10	7.85	1000	17.4	268	16.4	114	10.7
× 8	6.41	816	16.7	225	16.6	93	10.7
× 6	4.90	624	16.0	177	16.9	73	10.8
× 5	4.12	525	15.6	152	17.0	61	10.8
× 4	3.33	424	15.2	125	17.1	50	10.9
× 3	2.52	321	14.9	96	17.3	39	11.0
L45 × 45 × 8	5.15	656	14.2	118	13.4	50	8.76
× 6	3.96	506	13.6	94	13.7	39	8.79
× 5	3.34	425	13.1	81	13.8	33	8.82
× 4	2.70	344	12.7	67	13.9	27	8.87
× 3	2.05	261	12.4	52	14.1	21	8.93
L35 × 35 × 6	3.01	384	10.9	42	10.5	18	6.81
× 5	2.55	325	10.6	36	10.6	15	6.83
× 4	2.07	264	10.2	30	10.7	12	6.86
× 3	1.58	201	9.9	24	10.8	10	6.91
L25 × 25 × 5	1.77	225	8.1	12	7.4	5	4.87
× 4	1.44	184	7.7	10	7.5	4	4.87
× 3	1.11	141	7.4	8	7.6	3	4.89

TABLE B-5   Angles, Unequal Legs

Note: The three dimensions given under "Designation" are the lengths of the legs and the thickness of the legs.

I = moment of inertia
r = radius of gyration
z refers to the axis about which the moment of inertia is a minimum.

Designation (mm)(mm)(mm)	Mass (kg/m)	Area (mm²)	C_x (mm)	C_y (mm)	I_x (10^3 mm⁴)	r_x (mm)	I_y (10^3 mm⁴)	r_y (mm)	I_z (10^3 mm⁴)	r_z (mm)	α (degrees)
L200 × 150 × 25	63.8	8120	41.3	66.3	31600	62.3	15100	43.2	8310	32.0	28.5
× 20	51.8	6600	39.5	64.5	26200	63.0	12700	43.8	6800	32.1	28.8
× 16	42.0	5340	38.1	63.1	21600	63.5	10500	44.3	5570	32.3	29.0
× 13	34.4	4380	37.0	62.0	17900	64.0	8770	44.7	4630	32.5	29.1
L200 × 100 × 20	44.0	5600	24.3	74.3	22600	63.6	3840	26.2	2540	21.3	14.4
× 16	35.7	4540	22.8	72.8	18700	64.2	3220	26.6	2080	21.4	14.7
× 13	29.3	3730	21.7	71.7	15600	64.6	2720	27.0	1740	21.6	14.9
× 10	22.8	2900	20.5	70.5	12300	65.1	2714	21.8	1380	15.2	
L150 × 100 × 16	29.4	3740	25.9	50.9	8400	47.4	3000	28.3	1740	21.6	23.5
× 13	24.2	3080	24.9	49.9	7030	47.8	2530	28.7	1450	21.7	23.7
× 10	18.8	2400	23.8	48.7	5580	48.2	2030	29.1	1150	21.9	24.0
× 8	15.2	1940	23.0	48.0	4550	48.5	1670	29.3	939	22.0	24.1
L125 × 90 × 16	25.0	3180	24.7	42.2	4840	39.0	2090	25.6	1170	19.2	26.5
× 13	20.6	2630	23.7	41.2	4070	39.4	1770	26.0	980	19.3	26.8
× 10	16.1	2050	22.6	40.1	3250	39.8	1420	26.4	780	19.5	27.1
× 8	13.0	1660	21.8	39.3	2660	40.1	1180	26.6	638	19.6	27.2
L125 × 75 × 13	19.1	2430	18.9	43.9	3820	39.6	1040	20.7	638	16.2	19.6
× 10	14.9	1900	17.8	42.8	3050	40.0	841	21.0	505	16.3	20.0
× 8	12.1	1540	17.1	42.1	2500	40.3	697	21.3	414	16.4	20.2
× 6	9.14	1160	16.3	41.3	1920	40.6	542	21.6	320	16.6	20.4
L100 × 90 × 13	18.1	2300	26.1	31.1	2170	30.7	1660	26.8	779	18.4	38.5
× 10	14.1	1800	25.0	30.0	1740	31.1	1330	27.2	616	18.5	38.7
× 8	11.4	1460	24.3	29.3	1430	31.4	1100	27.5	505	18.6	38.7
× 6	8.67	1100	23.5	28.5	1110	31.7	853	27.8	385	18.7	38.8
L100 × 75 × 13	16.5	2110	20.9	33.4	2040	31.1	976	21.5	540	16.0	28.4
× 10	13.0	1650	19.8	32.3	1640	31.5	791	21.9	428	16.1	28.8
× 8	10.5	1340	19.0	31.5	1350	31.8	656	22.2	352	16.2	29.0
× 6	7.96	1010	18.3	30.8	1040	32.1	511	22.4	268	16.3	29.2
L90 × 75 × 13	15.5	1980	21.8	29.3	1510	27.6	946	21.9	482	15.6	33.9
× 10	12.2	1550	20.7	28.2	1220	28.0	767	22.2	382	15.7	34.2
× 8	9.86	1260	20.0	27.5	1010	28.3	636	22.5	315	15.8	34.3
× 6	7.49	954	19.3	26.8	779	28.6	495	22.8	241	15.9	34.5
× 5	6.28	800	18.9	26.4	660	28.7	421	22.9	205	16.0	34.6
L90 × 65 × 10	11.4	1450	17.3	29.8	1160	28.3	507	18.7	280	13.9	26.8
× 8	9.23	1180	16.6	29.1	958	28.5	422	18.9	231	14.0	27.1
× 6	7.02	894	15.9	28.4	743	28.8	330	19.2	180	14.2	27.4
× 5	5.89	750	15.5	28.0	629	29.0	281	19.4	151	14.2	27.5
L80 × 60 × 10	10.2	1300	16.5	26.5	808	24.9	388	17.3	213	12.8	28.5
× 8	8.29	1060	15.8	25.8	670	25.2	324	17.5	176	12.9	28.8
× 6	6.31	804	15.1	25.1	522	25.5	254	17.8	136	13.0	29.0
× 5	5.30	675	14.7	24.7	443	25.6	217	17.9	114	13.0	29.2
L75 × 50 × 8	7.35	936	13.0	25.5	525	23.7	187	14.1	109	10.8	23.5
× 6	5.60	714	12.2	24.7	410	24.0	148	14.4	85	10.9	23.8
× 5	4.71	600	11.9	24.4	349	24.1	127	14.5	71	10.9	24.0
L65 × 50 × 8	6.72	856	13.8	21.3	351	20.2	180	14.5	96	10.6	29.8
× 6	5.13	654	13.1	20.6	275	20.5	142	14.7	75	10.7	30.1
× 5	4.32	550	12.7	20.2	235	20.7	122	14.9	64	10.8	30.2
× 4	3.49	444	12.4	19.9	192	20.8	100	15.0	52	10.8	30.4

TABLE B5 *Continued*

Designation (mm)(mm)(mm)	Mass (kg/m)	Area (mm²)	C_x (mm)	C_y (mm)	I_x (10³ mm⁴)	r_x (mm)	I_y (10³ mm⁴)	r_y (mm)	I_z (10³ mm⁴)	r_z (mm)	α (degrees)
L55 × 35 × 6	3.96	504	9.0	19.0	152	17.4	48	9.77	29	7.55	21.6
× 5	3.34	425	8.7	18.7	130	17.5	41	9.89	24	7.59	21.9
× 4	2.70	344	8.3	18.3	107	17.7	34	10.0	20	7.65	22.1
× 3	2.05	261	7.9	17.9	83	17.8	27	10.2	16	7.72	22.3
L45 × 30 × 6	3.25	414	8.2	15.7	82	14.0	29	8.35	17	6.44	23.1
× 5	2.75	350	7.9	15.4	70	14.2	25	8.46	15	6.47	23.4
× 4	2.23	284	7.5	15.0	58	14.3	21	8.58	12	6.51	23.7
× 3	1.70	216	7.1	14.6	45	14.5	16	8.72	9	6.57	24.0

TABLE C **Properties of Steel Pipe**

Nominal Diameter (mm)	Outside Diameter (mm)	Inside Diameter (mm)	Wall Thickness (mm)	Area (mm²)	I (10⁶ mm⁴)	r (mm)	Mass (kg/m)
25	33.4	26.64	3.38	319	0.0364	10.7	2.50
38	48.3	40.94	3.68	516	0.129	15.8	4.05
50	60.3	52.48	3.91	693	0.277	20.0	5.43
63	73.0	62.68	5.16	1100	0.637	24.1	8.62
75	88.9	77.92	5.49	1440	1.26	29.6	11.3
88	101.6	90.12	5.74	1730	1.99	34.0	13.6
100	114.3	102.26	6.02	2050	3.01	38.3	16.1
125	141.3	128.20	6.55	2770	6.31	47.7	21.9
150	168.3	154.08	7.11	3600	11.7	57.0	28.2
200	219.1	202.74	8.18	5420	30.2	74.6	42.5
250	273.1	254.56	9.27	7680	66.9	93.3	60.2
300	323.9	304.84	9.53	9410	116	111	73.8

TABLE D **Properties of Steel Rods**

I = moment of inertia with respect to a diameter
r = radius of gyration with respect to a diameter
S = section modulus

Diameter (mm)	Area (mm²)	I (mm⁴)	r (mm)	S (mm³)	Mass (kg/m)
6	28.3	64	1.50	21	0.222
10	78.3	491	2.50	98	0.617
14	154	1890	3.50	269	1.21
16	201	3220	4.00	402	1.58
20	314	7850	5.00	785	2.47
25	491	19200	6.25	1530	3.85
30	707	39800	7.50	2650	5.55
35	962	73700	8.75	4210	7.55
40	1260	126000	10.00	6283	9.86
45	1590	201000	11.25	8950	12.5
50	1960	307000	12.50	12300	15.4
60	2830	636000	15.00	21200	22.2
75	4420	1550000	18.75	41400	34.7
90	6360	3220000	22.50	71600	49.9
100	7850	4910000	25.00	98200	61.7

TABLE E   Properties of Reinforcing Bars

Note: The number given under "Designation" is the nominal diameter in millimetres.

Designation	Diameter (mm)	Area (mm²)	Mass (kg/m)
10	11.3	100	0.79
15	16.0	200	1.57
20	19.5	300	2.36
25	25.2	500	3.93
30	29.9	700	5.50
35	35.7	1000	7.85
45	43.7	1500	11.8
55	56.4	2500	19.6

TABLE F   Properties of Timber

I = moment of inertia
S = section modulus

Dimensions		Area	I_x	S_x
b (mm)	h (mm)	(10^3 mm^2)	(10^6 mm^4)	(10^3 mm^3)
50	100	5.0	4.17	83
50	150	7.5	14.1	188
50	200	10.0	33.3	333
50	250	12.5	65.1	521
50	300	15.0	112.5	750
50	400	20.0	266.7	1333
100	100	10.0	8.33	167
100	150	15.0	28.1	375
100	200	20.0	66.7	667
100	250	25.0	130.2	1042
100	300	30.0	225.0	1500
100	400	40.0	533.3	2667
150	150	22.5	42.2	562
150	200	30.0	100.0	1000
150	250	37.5	195.3	1563
150	300	45.0	337.5	2250
150	400	60.0	800.0	4000
200	200	40.0	133.3	1333
200	250	50.0	260.4	2083
200	300	60.0	450.0	3000
200	400	80.0	1067.0	5333
250	250	62.5	325.5	2604
250	300	75.0	562.5	3750
250	400	100.0	1333.0	6667
300	300	90.0	675.0	4500
300	400	120.0	1600.0	8000

TABLE G   Formulas for Beam Deflections

Case Number	Load and End Conditions	End Rotation	Maximum Deflection
1		$\theta = \dfrac{1}{16}\dfrac{PL^2}{EI}$	$\delta = \dfrac{1}{48}\dfrac{PL^3}{EI}$
2		$\theta = \dfrac{1}{24}\dfrac{wL^3}{EI}$	$\delta = \dfrac{5}{384}\dfrac{wL^4}{EI}$
3		$\theta = \dfrac{1}{2}\dfrac{PL^2}{EI}$	$\delta = \dfrac{1}{3}\dfrac{PL^3}{EI}$
4		$\theta = \dfrac{1}{6}\dfrac{wL^3}{EI}$	$\delta = \dfrac{1}{8}\dfrac{wL^4}{EI}$
5		$\theta = 0$	$\delta = \dfrac{1}{192}\dfrac{PL^3}{EI}$
6		$\theta = 0$	$\delta = \dfrac{1}{384}\dfrac{wL^4}{EI}$
7		$\theta = \dfrac{ML}{EI}$	$\delta = \dfrac{1}{2}\dfrac{ML^2}{EI}$
8		$\theta_1 = \dfrac{1}{6}\dfrac{ML}{EI}$ At left end $\theta_2 = \dfrac{1}{3}\dfrac{ML}{EI}$ At right end	$\delta = \dfrac{1}{9\sqrt{3}}\dfrac{ML^2}{EI}$ At $x = \dfrac{1}{\sqrt{3}}L$
9		$\theta = \dfrac{1}{4}\dfrac{ML}{EI}$	$\delta = \dfrac{1}{27}\dfrac{ML^2}{EI}$ At $x = \dfrac{2}{3}L$

TABLE H **Properties of Areas**

Case Number	Shape	Properties
1 General area (integration formulas)	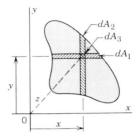	$I_x = \int y^2\, dA_1$ $I_x = \int\int y^2\, dA_3$ $r_x = \sqrt{I_x/A}$ $I_y = \int x^2\, dA_2$ $I_y = \int\int x^2\, dA_3$ $r_y = \sqrt{I_y/A}$ $I_{xy} = \int\int xy\, dA_3$ $J_0 = \int\int z^2\, dA_3$ $\quad = I_x + I_y$
2 General area (Parallel-axis theorem)	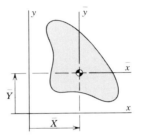	$I_x = I_{\bar{x}} + A\bar{Y}^2$ $r_x = \sqrt{r_{\bar{x}}^2 + \bar{Y}^2}$ $I_y = I_{\bar{y}} + A\bar{X}^2$ $r_y = \sqrt{r_{\bar{y}}^2 + \bar{X}^2}$ $I_{xy} = I_{\bar{x}\,\bar{y}} + A\bar{X}\bar{Y}$
3 Symmetrical area	Area is symmetrical about either the $\bar{x}$- or $\bar{y}$-axis 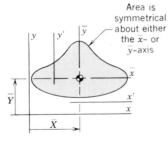	$I_{\bar{x}\,\bar{y}} = 0$ $I_{x\,\bar{y}} = 0$ $I_{\bar{x}\,y'} = 0$ $I_{xy} = A\bar{X}\bar{Y}$
4 Rectangle	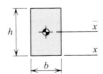	$A = bh$ $I_{\bar{x}} = \dfrac{1}{12}bh^3$ $r_{\bar{x}} = \dfrac{1}{\sqrt{12}}h$ $I_x = \dfrac{1}{3}bh^3$ $r_x = \dfrac{1}{\sqrt{3}}h$

TABLE H *Continued*

Case Number	Shape	Properties

5
Triangle

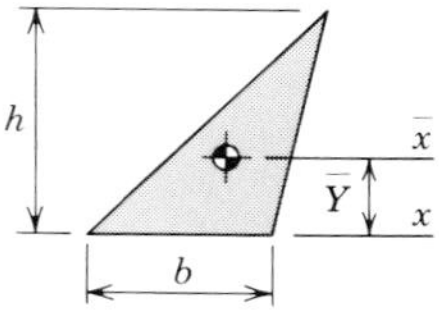

$A = \dfrac{1}{2}bh$

$\bar{Y} = \dfrac{1}{3}h$

$I_{\bar{x}} = \dfrac{1}{36}bh^3$

$r_{\bar{x}} = \dfrac{1}{3\sqrt{2}}h$

$I_x = \dfrac{1}{12}bh^3$

$r_x = \dfrac{1}{\sqrt{6}}h$

6
Circle

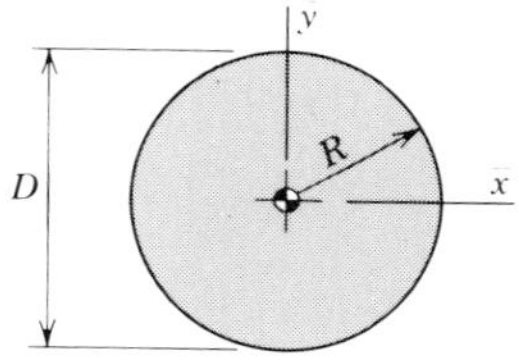

$A = \dfrac{\pi}{4}D^2 = \pi R^2$

$I_{\bar{x}} = \dfrac{\pi}{64}D^4 = \dfrac{\pi}{4}R^4$

$r_{\bar{x}} = \dfrac{1}{4}D = \dfrac{1}{2}R$

$J_{\bar{o}} = \dfrac{\pi}{32}D^4 = \dfrac{\pi}{2}R^4$

7
Ellipse

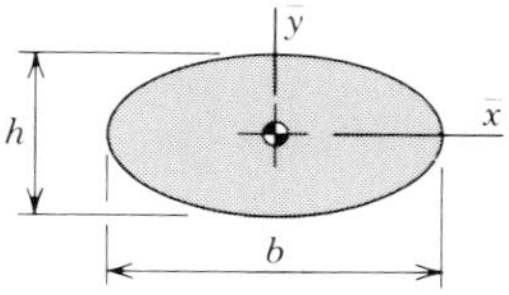

$A = \dfrac{\pi}{4}bh$

$I_{\bar{x}} = \dfrac{\pi}{64}bh^3$

$r_{\bar{x}} = \dfrac{1}{4}h$

$I_{\bar{y}} = \dfrac{\pi}{64}b^3h$

$r_{\bar{y}} = \dfrac{1}{4}b$

8
Hollow circle

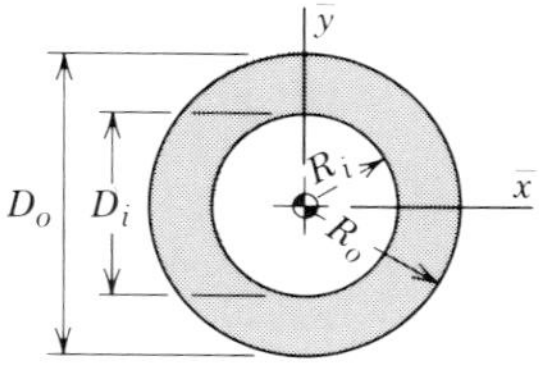

$A = \dfrac{\pi}{4}(D_o^2 - D_i^2) = \pi(R_o^2 - R_i^2)$

$I_{\bar{x}} = \dfrac{\pi}{64}(D_o^4 - D_i^4) = \dfrac{\pi}{4}(R_o^4 - R_i^4)$

$r_{\bar{x}} = \dfrac{1}{4}\sqrt{D_o^2 + D_i^2} = \dfrac{1}{2}\sqrt{R_o^2 + R_i^2}$

$J_{\bar{o}} = \dfrac{\pi}{32}(D_o^4 - D_i^4) = \dfrac{\pi}{2}(R_o^4 - R_i^4)$

TABLE H *Continued*

Case Number	Shape	Properties
9 Hollow circle (approximation) (thin tube)	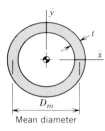 D_m Mean diameter	$A = \pi D_m t$ $I_{\bar{x}} = \dfrac{\pi}{8}[1 + (t/D_m)^2]\, D_m^3 t$ $\simeq \dfrac{\pi}{8} D_m^3 t$ (error $< 1\%$ when $t < \frac{1}{10}D_m$) $r_{\bar{x}} = \dfrac{1}{2\sqrt{2}}\sqrt{1 + (t/D_m)^2}\; D_m$ $\simeq \dfrac{1}{2\sqrt{2}} D_m$ (error $< 0.4\%$ when $t < \frac{1}{10}D_m$) $J_{\bar{o}} = \dfrac{\pi}{4}[1 + (t/D_m)^2]\, D_m^3 t$ $\simeq \dfrac{\pi}{4} D_m^3 t$ (error $< 1\%$ when $t < \frac{1}{10}D_m$)
10 Semicircle	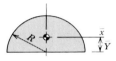	$A = \dfrac{\pi}{2} R^2$ $\bar{Y} = \dfrac{4}{3\pi} R = 0.4244 R$ $I_{\bar{x}} = \left(\dfrac{\pi}{8} - \dfrac{8}{9\pi}\right) R^4 = 0.1098\, R^4$ $r_{\bar{x}} = \sqrt{\dfrac{1}{4} - \dfrac{16}{9\pi^2}}\; R = 0.2643 R$
11 Hollow semicircle (approximation) (half of thin tube)	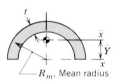 R_m, Mean radius	$A = \pi R_m t$ $\bar{Y} \simeq \dfrac{2}{\pi} R_m$ (error $< 0.4\%$ when $t < \frac{1}{5}R_m$) $I_x \simeq \dfrac{\pi}{2} R_m^3 t$ (error $< 1\%$ when $t < \frac{1}{5}R_m$) $I_{\bar{x}} \simeq \left(\dfrac{\pi}{2} - \dfrac{4}{\pi}\right) R_m^3 t = 0.2976 R_m^3 t$ (error $< 2.4\%$ when $t < \frac{1}{5}R_m$) (error $< 0.6\%$ when $t < \frac{1}{10}R_m$)

TABLE I   SI Units

Units Commonly Used in Engineering:

Quantity	Name of Unit	Symbol
Time	second	s
Length	metre	m
	millimetre	mm
Mass	kilogram	kg
	tonne	t (= 1000 kg)
Force	newton	N
Pressure	pascal	Pa ($= N/m^2$)
	kilopascal	kPa
Stress	megapascal	MPa ($= N/mm^2$)
Work, energy	joule	J ($= N \cdot m$)
Power	watt	W ($= J/s = N \cdot m/s$)
Temperature	degrees Celsius	°C

Prefixes:

Prefix	Symbol	Multiplication Factor	
giga	G	1 000 000 000	$= 10^9$
mega	M	1 000 000	$= 10^6$
kilo	k	1 000	$= 10^3$
milli	m	0.001	$= 10^{-3}$
micro	μ	0.000 001	$= 10^{-6}$

Dimensional Equivalents:

1 inch = 25.40 mm   1 m = 39.37 inches
1 foot = 304.8 mm   1 m = 3.281 feet
1 mile = 1.609 km   1 km = 0.6214 mile
1 pound (force) = 4.448 N   1 N = 0.2248 pounds (force)
1 pound (mass) = 0.4536 kg   1 kg = 2.205 pounds (mass)
1 kip/foot = 14.59 kN/m   1 kN/m = 0.0685 kip/foot
1 kip/square inch = 6.895 MPa   1 MPa = 0.1450 kip/square inch
1 foot-pound = 1.356 J   1 J = 0.738 foot-pound
1 horsepower = 746 W   1 kW = 1.341 horsepower

$$g = 32.2 \frac{ft}{s^2} = 9.81 \frac{m}{s^2}$$

Gravity (standard) exerts a force of 9.81 N on a mass of 1 kg.

APPENDIX B

TABLES (IMPERIAL UNITS)

Table A   Physical Properties of Commonly Used Engineering Materials

Table B   Properties of Rolled Structural Steel Shapes

        B-1 W-Shapes*
        B-2 S-Shapes*
        B-3 C-Shapes*
        B-4 Angles, Equal Legs†
        B-5 Angles, Unequal Legs†

Table C   Properties of Steel Pipe†

Table D   Properties of Steel Rods

Table E   Properties of Reinforcing Bars

Table F   Properties of Timber

Table G   Formulas for Beam Deflections

Table H   Properties of Areas

*Reprinted from the *Handbook of Steel Construction*, Third Edition, with kind permission of the Canadian Institute of Steel Construction. The Imperial unit designations are no longer used in Canada.

†Reprinted from the *Handbook of Steel Construction*, Second Edition, with kind permission of the Canadian Institute of Steel Construction.

TABLE A   Physical Properties of Commonly Used Engineering Materials

Note: Material properties depend upon conditions which are not referred to in this table. For example, strength may depend upon rolling temperature (steel), moisture content (wood), age (concrete), etc. Therefore, the values given below, while adequate for textbook exercises, should not be used in engineering design.

Material	Yield Stress T or C (ksi)	Ultimate Stress T or C (ksi)	Typical Allowable Stress T or C (ksi)	Shear (ksi)	Young's Modulus E (ksi)	Poisson's Ratio (ν)	Shear Modulus G (ksi)	Density (lb/in.^{3})	Coefficient of Expansion α [μin./(in.°F)]
Structural Steel									
Low-strength	± 33	$+55$	± 20	14	29000	0.3	11000	0.283	6.5
Medium-strength	± 45	$+65$	± 28	19	29000	0.3	11000	0.283	6.5
High-strength	± 60	$+90$	± 35	25	29000	0.3	11000	0.283	6.5
Wrought Steel									
AISI 1020	± 36	$+65$			29000	0.3	11000	0.283	6.5
AISI 1045	± 60	$+96$			29000	0.3	11000	0.283	6.5
AISI 1090	± 67	$+122$			29000	0.3	11000	0.283	6.5
AISI 4340	± 68	$+100$			29000	0.3	11000	0.283	6.5
AISI 4063	± 160	$+180$			29000	0.3	11000	0.283	6.5
AISI 8760	± 200	$+218$			29000	0.3	11000	0.283	6.5
Stainless Steel									
AISI 302	± 38	$+90$			27000	0.3	10000	0.278	6.5
AISI 431	± 94	$+125$			27000	0.3	10000	0.278	6.5
Reinforcing Steel									
Grade 40	± 40	$+60$	± 20		29000	0.3	11000	0.283	6.5
Grade 50	± 50	$+80$	± 20		29000	0.3	11000	0.283	6.5
Grade 60	± 60	$+90$	± 24		29000	0.3	11000	0.283	6.5
Gray Cast Iron		$+22$ -87			13000	0.2	5000	0.274	6.1
Aluminum Alloys									
1100–0	± 5	$+13$			10000	0.33	4000	0.101	12.8
6061–T6	± 39	$+45$	± 20		10000	0.33	4000	0.101	12.8
7075–T6	± 71	$+81$			10000	0.33	4000	0.101	12.8
Magnesium Alloy	± 23	$+33$			6500	0.32	2500	0.065	14.0
Copper-Base Alloys									
Phosphor bronze	± 74	$+81$			16000	0.35	6000	0.296	12.0
Yellow brass	± 60	$+74$			15000	0.35	5500	0.310	10.0
Concrete									
Low-strength		-2.9	-1.3	0.06	3000	0.15		0.087	6.1
High-strength		-5.1	-2.3	0.09	4000	0.15		0.087	6.1
Wood									
Douglas fir		$+7.2$	$\pm 1.3^a$ -0.9^b	0.13	1700			0.020	
Western hemlock		$+6.1$	$\pm 1.0^a$ -0.7^b	0.10	1400			0.017	
Eastern spruce		$+5.5$	$\pm 1.0^a$ -0.7^b	0.09	1200			0.016	
Glass		$+4.3$ -43.5	$+1.0$		10000	0.17	4300	0.094	8.3
Rubber		$+3.6$				0.5		0.035	

[a] Tension or Compression stress in beams.
[b] Compression stress parallel to grain in short columns.

TABLE B-1 W-Shapes

Note: Under "Designation" the first number is the nominal depth of the section in inches, the second number is the mass in pounds per foot of length.

I = moment of inertia
r = radius of gyration
S = section modulus

Designation (in.) (lb/ft)	Area (in.^{2})	Depth (in.)	Flange Width (in.)	Flange Thick. (in.)	Web Thick. (in.)	I_x (in.^{4})	r_x (in.)	S_x (in.^{3})	I_y (in.^{4})	r_y (in.)
W36 × 300	88.3	36.74	16.65	1.68	0.945	20300	15.2	1110	1300	3.84
× 260	76.5	36.26	16.55	1.44	0.840	17300	15.0	954	1090	3.77
× 230	67.6	35.90	16.47	1.26	0.760	15000	14.9	836	940	3.73
× 194	57.0	36.49	12.12	1.26	0.765	12100	14.6	663	375	2.56
× 170	50.0	36.17	12.03	1.10	0.680	10500	14.5	581	320	2.53
× 150	44.2	35.85	11.98	0.94	0.625	9040	14.3	504	270	2.47
W33 × 220	65.0	33.93	15.81	1.27	0.775	12800	14.0	754	840	3.59
× 120	38.3	33.30	11.53	0.96	0.605	6710	13.2	406	218	2.39
W30 × 210	62.0	30.94	15.10	1.31	0.775	10300	12.9	666	757	3.49
× 172	50.8	30.44	14.98	1.06	0.655	8200	12.7	539	598	3.43
× 124	36.5	30.17	10.52	0.93	0.585	5360	12.1	355	181	2.23
× 108	31.7	29.83	10.48	0.76	0.545	4470	11.9	300	146	2.15
W27 × 94	27.7	26.92	9.99	0.74	0.490	3270	10.9	243	124	2.12
W24 × 130	38.5	24.48	12.85	0.96	0.605	4020	10.2	328	340	2.97
× 100	30.6	24.06	12.75	0.75	0.500	3100	10.1	258	259	2.91
× 84	24.7	24.10	9.02	0.77	0.470	2370	9.80	197	94.4	1.95
× 68	20.1	23.73	8.97	0.58	0.415	1830	9.54	154	70.4	1.87
W21 × 62	18.3	20.99	8.24	0.62	0.400	1330	8.53	127	57.5	1.77
× 49	14.7	20.83	6.53	0.54	0.380	984	8.18	94.4	24.9	1.30
W18 × 55	16.2	18.11	7.53	0.63	0.390	890	7.41	98.3	44.9	1.66
× 45	13.4	17.86	7.48	0.50	0.335	719	7.33	80.5	34.8	1.61
× 35	10.3	17.70	6.00	0.42	0.300	510	7.04	57.6	15.3	1.22
W16 × 88	26.2	16.75	10.36	0.88	0.525	1300	7.04	155	163	2.49
× 71	19.7	16.33	10.23	0.66	0.395	954	6.96	177	119	2.46
× 40	11.8	16.01	6.99	0.50	0.305	518	6.63	64.7	28.9	1.56
× 26	7.68	15.69	5.50	0.34	0.250	301	6.26	38.4	9.6	1.12
W14 × 342	101.0	17.54	16.36	2.47	1.54	4900	6.97	559	1810	4.23
× 237	68.5	16.04	15.89	1.72	1.07	3010	6.63	375	1150	4.10
× 136	38.8	14.66	14.73	1.03	0.645	1530	6.28	209	548	3.76
× 95	29.1	14.16	14.56	0.78	0.485	1110	6.18	157	402	3.72
W12 × 53	15.6	12.06	9.99	0.58	0.345	425	5.22	70.5	95.8	2.48
× 22	6.48	12.31	4.03	0.42	0.260	156	4.91	25.3	4.7	0.85
W10 × 77	22.6	10.60	10.19	0.87	0.530	455	4.49	85.8	154	2.61
× 60	17.6	10.22	10.08	0.68	0.420	341	4.40	66.7	116	2.57
× 45	13.3	10.10	8.02	0.62	0.350	248	4.32	49.1	53.4	2.00
× 21	6.49	10.17	5.75	0.36	0.240	118	4.26	23.2	11.4	1.33
× 11	3.54	9.87	3.96	0.21	0.190	54	3.90	10.9	2.2	0.78
W8 × 40	11.7	8.25	8.07	0.56	0.360	146	3.53	35.4	49.1	2.05
× 28	8.25	8.06	6.53	0.46	0.285	98	3.45	24.3	21.7	1.62
× 17	5.26	8.14	5.25	0.33	0.230	62	3.43	15.2	8.0	1.23

TABLE B-2 S-Shapes

Note: Under "Designation" the first number is the nominal depth of the section in inches, the second number is the mass in pounds per foot of length.

  I = moment of inertia
  r = radius of gyration
  S = section modulus

Designation (in.)(lb/ft)	Area (in.²)	Depth (in.)	Flange Width (in.)	Flange Thick. (in.)	Web Thick. (in.)	I_x (in.⁴)	r_x (in.)	S_x (in.³)	I_y (in.⁴)	r_y (in.)
S24 × 121	35.6	24.5	8.05	1.09	0.800	3160	9.42	258	83.9	1.54
× 106	31.2	24.5	7.87	1.09	0.620	2940	9.71	240	77.7	1.58
× 100	29.4	24.0	7.24	0.870	0.745	2390	9.02	199	48.2	1.28
× 90	26.5	24.0	7.13	0.870	0.625	2250	9.21	188	45.4	1.31
× 80	23.5	24.0	7.00	0.870	0.500	2110	9.48	176	42.8	1.35
S20 × 96	28.3	20.3	7.20	0.920	0.800	1680	7.70	166	50.6	1.34
× 86	25.4	20.3	7.06	0.920	0.660	1580	7.89	156	47.3	1.36
× 75	22.0	20.0	6.39	0.795	0.635	1280	7.63	128	30.2	1.17
× 66	19.4	20.0	6.26	0.795	0.505	1190	7.83	119	28.1	1.20
S18 × 70	20.6	18.0	6.25	0.691	0.711	928	6.71	103	24.5	1.09
× 55	16.1	18.0	6.00	0.691	0.461	807	7.08	89.7	21.2	1.15
S15 × 50	14.7	15.0	5.64	0.622	0.550	487	5.76	64.9	15.9	1.04
× 43	12.6	15.0	5.50	0.622	0.411	448	5.96	59.7	14.6	1.08
S12 × 50	14.7	12.0	5.48	0.659	0.687	306	4.56	51.0	15.9	1.04
× 41	12.0	12.0	5.25	0.659	0.462	273	4.77	45.5	13.8	1.07
× 35	10.3	12.0	5.08	0.544	0.428	230	4.73	38.3	10.0	0.985
× 32	9.37	12.0	5.00	0.544	0.350	219	4.83	36.5	9.49	1.01
S10 × 35	10.3	10.0	4.94	0.491	0.594	147	3.78	29.4	8.45	0.906
× 25	7.47	10.0	4.66	0.491	0.311	124	4.07	24.8	6.88	0.960
S8 × 23	6.77	8.0	4.17	0.425	0.441	64.9	3.10	16.2	4.35	0.802
× 18	5.41	8.0	4.00	0.425	0.271	57.7	3.27	14.4	3.77	0.835
S7 × 20	5.89	7.0	3.86	0.392	0.450	42.5	2.69	12.1	3.20	0.737
× 15	4.50	7.0	3.66	0.392	0.252	36.8	2.86	10.5	2.67	0.770
S6 × 17	5.07	6.0	3.56	0.359	0.465	26.4	2.28	8.80	2.33	0.678
× 12	3.67	6.0	3.33	0.359	0.232	22.2	2.46	7.40	1.85	0.710
S5 × 15	4.34	5.0	3.28	0.326	0.494	15.2	1.87	6.08	1.68	0.622
× 10	2.94	5.0	3.00	0.326	0.214	12.3	2.05	4.92	1.23	0.647
S4 × 10	2.80	4.0	2.80	0.293	0.326	6.8	1.56	3.40	0.91	0.570
× 8	2.26	4.0	2.66	0.293	0.193	6.1	1.64	3.04	0.77	0.584
S3 × 8	2.21	3.0	2.51	0.260	0.349	2.9	1.15	1.95	0.59	0.517
× 6	1.67	3.0	2.33	0.260	0.170	2.5	1.23	1.69	0.46	0.525

TABLE B-3   C-Shapes

Note: Under "Designation" the first number is the nominal depth of the section in inches, the second number is the mass in pounds per foot of length.

I = moment of inertia
r = radius of gyration
S = section modulus

Designation (in.)(lb/ft)	Area (in.^{2})	Depth (in.)	Flange Width (in.)	Flange Thick. (in.)	Web Thick. (in.)	I_x (in.^{4})	r_x (in.)	S_x (in.^{3})	C_x (in.)	I_y (in.^{4})	r_y (in.)
C15 × 50	14.7	15.0	3.72	0.650	0.716	404	5.24	53.9	0.802	11.2	0.873
× 40	11.8	15.0	3.52	0.650	0.520	349	5.44	46.5	0.782	9.35	0.890
× 34	9.97	15.0	3.40	0.650	0.400	315	5.62	42.0	0.792	8.24	0.909
C12 × 30	8.82	12.0	3.17	0.501	0.510	162	4.29	27.0	0.676	5.18	0.766
× 25	7.35	12.0	3.05	0.501	0.387	144	4.43	24.0	0.676	4.51	0.783
× 21	6.09	12.0	2.94	0.501	0.282	129.	4.60	21.5	0.700	3.92	0.802
C10 × 30	8.82	10.0	3.03	0.436	0.673	103	3.42	20.6	0.650	3.97	0.671
× 25	7.35	10.0	2.89	0.436	0.526	91.2	3.52	18.2	0.619	3.38	0.678
× 20	5.88	10.0	2.74	0.436	0.379	79.0	3.67	15.8	0.608	2.83	0.694
× 15	4.49	10.0	2.60	0.436	0.240	67.4	3.87	13.5	0.636	2.30	0.716
C9 × 20	5.88	9.0	2.65	0.413	0.448	61.0	3.22	13.6	0.584	2.44	0.644
× 15	4.41	9.0	2.48	0.413	0.285	51.1	3.40	11.4	0.589	1.94	0.663
× 13	3.94	9.0	2.43	0.413	0.233	47.9	3.49	10.6	0.603	1.78	0.672
C8 × 19	5.51	8.0	2.53	0.390	0.487	44.0	2.83	11.0	0.567	2.00	0.602
× 14	4.04	8.0	2.34	0.390	0.303	36.2	2.99	9.05	0.556	1.54	0.617
× 12	3.38	8.0	2.26	0.390	0.220	32.6	3.11	8.15	0.574	1.33	0.627
C7 × 15	4.33	7.0	2.30	0.366	0.419	27.3	2.51	7.80	0.533	1.39	0.567
× 12	3.60	7.0	2.19	0.366	0.314	24.3	2.60	6.94	0.527	1.18	0.573
× 10	2.87	7.0	2.09	0.366	0.210	21.3	2.72	6.09	0.543	0.978	0.584
C6 × 13	3.83	6.0	2.16	0.343	0.437	17.4	2.13	5.80	0.516	1.06	0.526
× 10	3.09	6.0	2.03	0.343	0.314	15.2	2.22	5.07	0.501	0.874	0.532
× 8	2.40	6.0	1.92	0.343	0.200	13.1	2.34	4.37	0.514	0.700	0.540
C5 × 12	3.38	5.0	2.03	0.320	0.472	10.4	1.75	4.16	0.506	0.815	0.491
× 9	2.65	5.0	1.89	0.320	0.325	8.9	1.83	3.56	0.480	0.639	0.491
× 7	1.97	5.0	1.75	0.320	0.190	7.5	1.95	3.00	0.486	0.484	0.496
C4 × 7	2.13	4.0	1.72	0.296	0.321	4.6	1.47	2.29	0.461	0.437	0.453
× 6	1.84	4.0	1.65	0.296	0.247	4.2	1.51	2.10	0.456	0.377	0.453
× 5	1.59	4.0	1.58	0.296	0.184	3.9	1.56	1.93	0.460	0.323	0.451
C3 × 6	1.76	3.0	1.60	0.273	0.356	2.1	1.09	1.39	0.457	0.309	0.419
× 5	1.47	3.0	1.50	0.273	0.258	1.9	1.12	1.24	0.440	0.250	0.412
× 4	1.21	3.0	1.41	0.273	0.170	1.7	1.17	1.11	0.439	0.199	0.406

TABLE B-4   Angles, Equal Legs

Note: The three dimensions given under "Designation" are the lengths of the legs and the thickness of the legs.

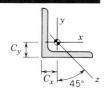

I = moment of inertia
r = radius of gyration
z refers to the axis about which the moment of inertia is a minimum

Designation (in.)(in.)(in.)	Mass (lb/ft)	Area (in.²)	$C_x = C_y$ (in.)	$I_x = I_y$ (in.⁴)	$r_x = r_y$ (in.)	I_z (in.⁴)	r_z (in.)
L8 × 8 × 1⅛	56.9	16.7	2.41	98.0	2.42	40.6	1.56
× 1	51.0	15.0	2.37	89.0	2.44	36.5	1.56
× ¾	38.9	11.4	2.28	69.7	2.47	28.1	1.57
× ⅝	32.7	9.61	2.23	59.4	2.49	24.0	1.58
× ½	26.4	7.75	2.19	48.6	2.50	19.6	1.59
× ⅜	19.9	5.86	2.14	37.3	2.52	15.0	1.60
L6 × 6 × ¾	28.7	8.44	1.78	28.2	1.83	11.6	1.17
× ⅝	24.2	7.11	1.73	24.2	1.84	9.9	1.18
× ½	19.6	5.75	1.68	19.9	1.86	8.0	1.18
× ⅜	14.9	4.36	1.64	15.4	1.88	6.2	1.19
L5 × 5 × ⅝	20.0	5.86	1.48	13.6	1.52	5.6	0.98
× ½	16.2	4.75	1.43	11.3	1.54	4.6	0.98
× ⅜	12.3	3.61	1.39	8.7	1.56	3.5	0.99
× 5⁄16	10.3	3.03	1.37	7.4	1.57	3.0	0.99
L4 × 4 × ⅝	15.7	4.61	1.23	6.7	1.20	2.8	0.78
× ½	12.8	3.75	1.18	5.6	1.22	2.3	0.78
× ⅜	9.8	2.86	1.14	4.4	1.23	1.8	0.79
× 5⁄16	8.2	2.40	1.12	3.7	1.24	1.5	0.79
× ¼	6.6	1.94	1.09	3.0	1.25	1.2	0.80
L3½ × 3½ × ½	11.1	3.25	1.06	3.6	1.06	1.5	0.68
× ⅜	8.5	2.48	1.01	2.9	1.07	1.2	0.69
× 5⁄16	7.2	2.09	0.99	2.5	1.08	0.99	0.69
× ¼	5.8	1.69	0.97	2.0	1.09	0.80	0.69
L3 × 3 × ½	9.4	2.75	0.93	2.2	0.90	0.93	0.58
× ⅜	7.2	2.11	0.89	1.8	0.91	0.71	0.58
× 5⁄16	6.1	1.78	0.87	1.5	0.92	0.62	0.59
× ¼	4.9	1.44	0.84	1.2	0.93	0.50	0.59
× 3⁄16	3.7	1.09	0.82	0.96	0.94	0.38	0.59
L2½ × 2½ × ⅜	5.9	1.73	0.76	0.98	0.75	0.42	0.49
× 5⁄16	5.0	1.47	0.74	0.85	0.76	0.35	0.49
× ¼	4.1	1.19	0.72	0.70	0.77	0.29	0.49
× 3⁄16	3.07	0.90	0.69	0.55	0.78	0.22	0.49
L2 × 2 × ⅜	4.70	1.36	0.64	0.48	0.59	0.21	0.39
× 5⁄16	3.92	1.15	0.61	0.42	0.60	0.17	0.39
× ¼	3.19	0.94	0.59	0.35	0.61	0.14	0.39
× 3⁄16	2.44	0.71	0.57	0.27	0.62	0.11	0.39
× 5⁄32	2.04	0.60	0.56	0.23	0.62	0.094	0.39
× ⅛	1.65	0.48	0.55	0.19	0.63	0.076	0.40
L1¾ × 1¾ × 5⁄16	3.39	1.00	0.55	0.27	0.52	0.120	0.34
× ¼	2.77	0.81	0.53	0.23	0.53	0.094	0.34
× 3⁄16	2.12	0.62	0.51	0.18	0.54	0.072	0.34
× 5⁄32	1.77	0.52	0.50	0.15	0.54	0.062	0.34
× ⅛	1.44	0.42	0.48	0.13	0.55	0.051	0.35
L1⅜ × 1⅜ × ¼	2.12	0.62	0.43	0.105	0.41	0.045	0.27
× 3⁄16	1.63	0.48	0.41	0.083	0.42	0.035	0.27
× 5⁄32	1.38	0.41	0.40	0.072	0.42	0.029	0.27
× ⅛	1.11	0.33	0.39	0.059	0.42	0.024	0.27
L1 × 1 × 3⁄16	1.16	0.34	0.32	0.030	0.30	0.013	0.19
× 5⁄32	0.98	0.29	0.31	0.026	0.30	0.011	0.19
× ⅛	0.80	0.23	0.30	0.022	0.30	0.009	0.20

TABLE B-5   Angles, Unequal Legs

Note: The three dimensions given under "Designation" are the lengths of the legs and the thickness of the legs.

I = moment of inertia
r = radius of gyration
z refers to the axis about which the moment of inert inertia is a minimum

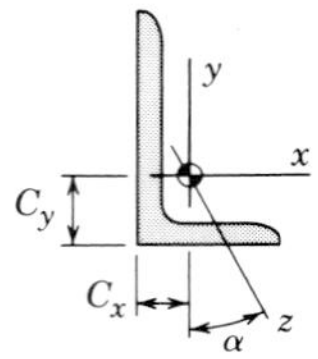

Designation (in.)(in.)(in.)	Mass (lb/ft)	Area (in.²)	C_x (in.)	C_y (in.)	I_x (in.⁴)	r_x (in.)	I_y (in.⁴)	r_y (in.)	I_z (in.⁴)	r_z (in.)	α (degrees)
L8 × 6 × 1	44.2	13.00	1.65	2.65	80.8	2.49	38.8	1.73	21.3	1.28	28.5
× ¾	33.8	9.94	1.56	2.56	63.4	2.53	30.7	1.76	16.5	1.29	28.8
× ⅝	28.5	8.36	1.52	2.52	54.1	2.54	26.3	1.77	13.9	1.29	29.0
× ½	23.0	6.75	1.47	2.47	44.3	2.56	21.7	1.79	11.4	1.30	29.2
L8 × 4 × ¾	28.7	8.44	0.95	2.95	54.9	2.55	9.4	1.05	6.10	0.85	14.5
× ⅝	24.2	7.11	0.91	2.91	46.9	2.57	8.1	1.07	5.26	0.86	14.7
× ½	19.6	5.75	0.86	2.86	38.5	2.59	6.7	1.08	4.25	0.86	15.0
× ⅜	14.8	4.35	0.81	2.81	29.6	2.60	5.3	1.09	3.32	0.87	15.2
L6 × 4 × ⅝	20.0	5.86	1.03	2.03	21.1	1.90	7.5	1.13	4.33	0.86	23.5
× ½	16.2	4.75	0.99	1.99	17.4	1.91	6.3	1.15	3.60	0.87	23.8
× ⅜	12.3	3.61	0.94	1.94	13.5	1.93	4.9	1.17	2.80	0.88	24.0
× ⁵⁄₁₆	10.3	3.03	0.92	1.92	11.4	1.94	4.2	1.17	2.35	0.88	24.2
L5 × 3½ × ⅝	16.8	4.92	0.95	1.70	12.0	1.56	4.8	0.99	2.77	0.75	25.3
× ½	13.6	4.00	0.91	1.66	10.0	1.58	4.0	1.01	2.25	0.75	25.6
× ⅜	10.4	3.05	0.86	1.61	7.8	1.60	3.2	1.02	1.76	0.76	25.9
× ⁵⁄₁₆	8.7	2.56	0.84	1.59	6.6	1.61	2.7	1.03	1.48	0.76	26.0
L5 × 3 × ½	12.8	3.75	0.75	1.75	9.5	1.59	2.6	0.83	1.58	0.65	19.6
× ⅜	9.8	2.86	0.70	1.70	7.4	1.61	2.0	0.84	1.21	0.65	20.0
× ⁵⁄₁₆	8.2	2.40	0.68	1.68	6.3	1.61	1.7	0.85	1.05	0.66	20.2
× ¼	6.6	1.94	0.66	1.66	5.1	1.62	1.4	0.86	0.85	0.66	20.4
L4 × 3½ × ½	11.9	3.50	1.00	1.25	5.3	1.23	3.8	1.04	1.81	0.72	36.9
× ⅜	9.1	2.67	0.96	1.21	4.2	1.25	3.0	1.06	1.42	0.73	37.0
× ⁵⁄₁₆	7.7	2.25	0.93	1.18	3.6	1.26	2.6	1.07	1.20	0.73	37.1
× ¼	6.2	1.81	0.91	1.16	2.9	1.27	2.1	1.07	0.96	0.73	37.2
L4 × 3 × ½	11.1	3.25	0.83	1.33	5.0	1.25	2.4	0.86	1.33	0.64	28.5
× ⅜	8.5	2.48	0.78	1.28	4.0	1.26	1.9	0.88	1.02	0.64	28.8
× ⁵⁄₁₆	7.2	2.09	0.76	1.26	3.4	1.27	1.6	0.89	0.88	0.65	29.0
× ¼	5.8	1.69	0.74	1.24	2.8	1.28	1.4	0.90	0.71	0.65	29.2
L3½ × 3 × ½	10.2	3.00	0.88	1.13	3.5	1.07	2.3	0.88	1.15	0.62	35.5
× ⅜	7.9	2.30	0.83	1.08	2.7	1.09	1.8	0.90	0.88	0.62	35.8
× ⁵⁄₁₆	6.6	1.93	0.81	1.06	2.3	1.10	1.6	0.90	0.77	0.63	35.9
× ¼	5.4	1.56	0.79	1.04	1.9	1.11	1.3	0.91	0.62	0.63	36.0
× ³⁄₁₆	4.0	1.18	0.76	1.01	1.5	1.11	1.0	0.92	0.47	0.63	36.1
L3½ × 2½ × ⅜	7.2	2.11	0.66	1.16	2.6	1.10	1.1	0.72	0.62	0.54	26.4
× ⁵⁄₁₆	6.1	1.78	0.63	1.14	2.2	1.11	0.94	0.73	0.52	0.54	26.6
× ¼	4.9	1.44	0.61	1.11	1.8	1.12	0.78	0.74	0.42	0.54	26.8
× ³⁄₁₆	3.7	1.08	0.59	1.09	1.4	1.12	0.6	0.74	0.32	0.54	27.0
L3 × 2½ × ⅜	6.6	1.92	0.71	0.96	1.7	0.93	1.0	0.74	0.52	0.52	34.0
× ⁵⁄₁₆	5.6	1.62	0.68	0.93	1.4	0.94	0.90	0.74	0.46	0.53	34.2
× ¼	4.5	1.31	0.66	0.91	1.2	0.95	0.74	0.75	0.37	0.53	34.4
× ³⁄₁₆	3.9	1.00	0.64	0.89	0.91	0.95	0.58	0.76	0.28	0.53	34.6
L3 × 2 × ⁵⁄₁₆	5.0	1.47	0.52	1.02	1.3	0.95	0.47	0.57	0.27	0.43	23.5
× ¼	4.1	1.19	0.49	0.99	1.1	0.95	0.39	0.57	0.22	0.43	23.8
× ³⁄₁₆	3.7	0.90	0.47	0.97	0.84	0.97	0.31	0.58	0.17	0.44	24.0
L2½ × 2 × ⁵⁄₁₆	4.5	1.31	0.56	0.81	0.79	0.78	0.45	0.58	0.23	0.42	31.8
× ¼	3.62	1.06	0.54	0.79	0.65	0.78	0.37	0.59	0.19	0.42	32.0
× ³⁄₁₆	2.75	0.81	0.51	0.76	0.51	0.79	0.29	0.60	0.15	0.43	32.3
× ⁵⁄₃₂	2.30	0.67	0.50	0.75	0.43	0.79	0.24	0.60	0.12	0.42	32.4

TABLE B-5   *Continued*

Designation (in.)(in.)(in.)	Mass (lb/ft)	Area (in.²)	C_x (in.)	C_y (in.)	I_x (in.⁴)	r_x (in.)	I_y (in.⁴)	r_y (in.)	I_z (in.⁴)	r_z (in.)	α (degrees)
L2 × 1½ × ¼	2.77	0.81	0.41	0.66	0.32	0.62	0.15	0.43	0.082	0.32	28.5
× $\frac{3}{16}$	2.12	0.62	0.39	0.64	0.25	0.63	0.12	0.44	0.064	0.32	28.8
× $\frac{5}{32}$	1.77	0.52	0.38	0.63	0.21	0.64	0.102	0.44	0.054	0.32	29.0
× ⅛	1.44	0.42	0.37	0.62	0.17	0.64	0.085	0.45	0.044	0.33	29.2
L1¾ × 1¼ × ¼	2.34	0.69	0.35	0.60	0.20	0.54	0.085	0.35	0.049	0.27	25.9
× $\frac{3}{16}$	1.80	0.53	0.33	0.58	0.16	0.55	0.068	0.36	0.038	0.27	26.4
× $\frac{5}{32}$	1.51	0.44	0.32	0.57	0.14	0.56	0.058	0.36	0.032	0.27	26.6
× ⅛	1.23	0.36	0.31	0.56	0.11	0.56	0.049	0.37	0.026	0.27	26.8

TABLE C   Properties of Steel Pipes

Nominal Diameter (in.)	Outside Diameter (in.)	Inside Diameter (in.)	Wall Thickness (in.)	Area (in.²)	I (in.⁴)	r (in.)	Mass (lb/ft)
1	1.315	1.049	0.133	0.494	0.087	0.420	1.68
1½	1.900	1.610	0.145	0.799	0.310	0.623	2.72
2	2.375	2.067	0.154	1.08	0.666	0.787	3.65
2½	2.875	2.469	0.203	1.70	1.53	0.947	5.79
3	3.500	3.068	0.216	2.23	3.02	1.16	7.58
3½	4.000	3.548	0.226	2.68	4.79	1.34	9.11
4	4.500	4.026	0.237	3.17	7.23	1.51	10.79
5	5.563	5.047	0.258	4.30	15.2	1.88	14.62
6	6.625	6.065	0.280	5.58	28.1	2.25	18.97
8	8.625	7.981	0.322	8.40	72.5	2.94	28.55
10	10.750	10.020	0.365	11.91	161.	3.67	40.48
12	12.750	12.000	0.375	14.58	279.	4.38	49.56

TABLE D   Properties of Steel Rods

I = moment of inertia with respect to a diameter
r = radius of gyration with respect to a diameter
S = section modulus

Diameter (in.)	Area (in.²)	I (in.⁴)	r (in.)	S (in.³)	Mass (lb/ft)
¼	0.049	0.00019	0.0625	0.00153	0.166
⅜	0.110	0.00097	0.0938	0.00518	0.375
½	0.196	0.00307	0.125	0.0123	0.667
⅝	0.307	0.00749	0.156	0.0240	1.04
¾	0.442	0.0155	0.187	0.0414	1.50
⅞	0.601	0.0288	0.219	0.0658	2.04
1	0.785	0.0491	0.250	0.0982	2.67
1¼	1.23	0.120	0.312	0.192	4.17
1½	1.77	0.248	0.375	0.331	6.00
1¾	2.41	0.460	0.437	0.526	8.17
2	3.14	0.785	0.500	0.785	10.7
2½	4.91	1.92	0.625	1.53	16.7
3	7.07	3.98	0.750	2.65	24.0
3½	9.62	7.37	0.875	4.21	32.7
4	12.57	12.57	1.000	6.28	42.7

TABLE E   Properties of Reinforcing Bars

Note: The number given under "Designation" is the diameter in eighths of an inch.

Designation	Diameter (in.)	Area (in.^{2})	Mass (lb/ft)
# 4	0.500	0.196	0.667
# 5	0.625	0.307	1.04
# 6	0.750	0.442	1.50
# 8	1.000	0.785	2.67
# 10	1.250	1.23	4.17
# 11	1.375	1.48	5.04
# 14	1.750	2.41	8.17
# 18	2.250	3.98	13.50

TABLE F   Properties of Timber

I = moment of inertia
S = section modulus

Dimensions		Area (in.^{2})	I_x (in.^{4})	S_x (in.^{3})
b (in.)	h (in.)			
2 × 4		8.0	10.7	5.3
2 × 6		12.0	36.0	12.0
2 × 8		16.0	85.3	21.3
2 × 10		20.0	166.7	33.3
2 × 12		24.0	288.0	48.0
2 × 16		32.0	682.7	85.3
4 × 4		16.0	21.3	10.6
4 × 6		24.0	72.0	24.0
4 × 8		32.0	170.7	42.7
4 × 10		40.0	333.3	66.7
4 × 12		48.0	576.0	96.0
4 × 16		64.0	1365.3	170.7
6 × 6		36.0	108.0	36.0
6 × 8		48.0	256.0	64.0
6 × 10		60.0	500.0	100.0
6 × 12		72.0	864.0	144.0
6 × 16		96.0	2048.0	256.0
8 × 8		64.0	341.3	85.3
8 × 10		80.0	666.7	133.3
8 × 12		96.0	1152.0	192.0
8 × 16		128.0	2730.7	341.3
10 × 10		100.0	833.3	166.7
10 × 12		120.0	1440.0	240.0
10 × 14		140.0	2286.7	326.7
12 × 12		144.0	1728.0	288.0
12 × 14		168.0	2744.0	392.0

TABLE G   Formulas for Beam Deflections

Case Number	Load and End Conditions	End Rotation	Maximum Deflection
1		$\theta = \dfrac{1}{16}\dfrac{PL^2}{EI}$	$\delta = \dfrac{1}{48}\dfrac{PL^3}{EI}$
2		$\theta = \dfrac{1}{24}\dfrac{wL^3}{EI}$	$\delta = \dfrac{5}{384}\dfrac{wL^4}{EI}$
3		$\theta = \dfrac{1}{2}\dfrac{PL^2}{EI}$	$\delta = \dfrac{1}{3}\dfrac{PL^3}{EI}$
4		$\theta = \dfrac{1}{6}\dfrac{wL^3}{EI}$	$\delta = \dfrac{1}{8}\dfrac{wL^4}{EI}$
5		$\theta = 0$	$\delta = \dfrac{1}{192}\dfrac{PL^3}{EI}$
6		$\theta = 0$	$\delta = \dfrac{1}{384}\dfrac{wL^4}{EI}$
7		$\theta = \dfrac{ML}{EI}$	$\delta = \dfrac{1}{2}\dfrac{ML^2}{EI}$
8		$\theta_1 = \dfrac{1}{6}\dfrac{ML}{EI}$ At left end $\theta_2 = \dfrac{1}{3}\dfrac{ML}{EI}$ At right end	$\delta = \dfrac{1}{9\sqrt{3}}\dfrac{ML^2}{EI}$ At $x = \dfrac{1}{\sqrt{3}}L$
9		$\theta = \dfrac{1}{4}\dfrac{ML}{EI}$	$\delta = \dfrac{1}{27}\dfrac{ML^2}{EI}$ At $x = \dfrac{2}{3}L$

TABLE H Properties of Areas

Case Number	Shape	Properties
1 General area (integration formulas)	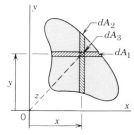	$I_x = \int y^2\, dA_1$ $I_x = \int\int y^2\, dA_3$ $r_x = \sqrt{I_x/A}$ $I_y = \int x^2\, dA_2$ $I_y = \int\int x^2\, dA_3$ $r_y = \sqrt{I_y/A}$ $I_{xy} = \int\int xy\, dA_3$ $J_0 = \int\int z^2\, dA_3$ $\quad = I_x + I_y$
2 General area (Parallel-axis theorem)	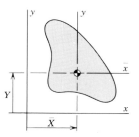	$I_x = I_{\bar{x}} + A\bar{Y}^2$ $r_x = \sqrt{r_{\bar{x}}^2 + \bar{Y}^2}$ $I_y = I_{\bar{y}} + A\bar{X}^2$ $r_y = \sqrt{r_{\bar{y}}^2 + \bar{X}^2}$ $I_{xy} = I_{\bar{x}\bar{y}} + A\bar{X}\bar{Y}$
3 Symmetrical area	Area is symmetrical about either the x- or y-axis 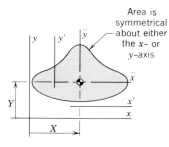	$I_{\bar{x}\bar{y}} = 0$ $I_{x\bar{y}} = 0$ $I_{\bar{x}y} = 0$ $I_{xy} = A\bar{X}\bar{Y}$
4 Rectangle	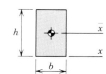	$A = bh$ $I_{\bar{x}} = \dfrac{1}{12}bh^3$ $r_{\bar{x}} = \dfrac{1}{\sqrt{12}}h$ $I_x = \dfrac{1}{3}bh^3$ $r_x = \dfrac{1}{\sqrt{3}}h$

TABLE H *Continued*

Case Number	Shape	Properties
5 Triangle		$A = \dfrac{1}{2}bh$ $\bar{Y} = \dfrac{1}{3}h$ $I_{\bar{x}} = \dfrac{1}{36}bh^3$ $r_{\bar{x}} = \dfrac{1}{3\sqrt{2}}h$ $I_x = \dfrac{1}{12}bh^3$ $r_x = \dfrac{1}{\sqrt{6}}h$
6 Circle		$A = \dfrac{\pi}{4}D^2 = \pi R^2$ $I_{\bar{x}} = \dfrac{\pi}{64}D^4 = \dfrac{\pi}{4}R^4$ $r_{\bar{x}} = \dfrac{1}{4}D = \dfrac{1}{2}R$ $J_o = \dfrac{\pi}{32}D^4 = \dfrac{\pi}{2}R^4$
7 Ellipse		$A = \dfrac{\pi}{4}bh$ $I_{\bar{x}} = \dfrac{\pi}{64}bh^3$ $r_{\bar{x}} = \dfrac{1}{4}h$ $I_{\bar{y}} = \dfrac{\pi}{64}b^3h$ $r_{\bar{y}} = \dfrac{1}{4}b$
8 Hollow circle		$A = \dfrac{\pi}{4}(D_o^2 - D_i^2) = \pi(R_o^2 - R_i^2)$ $I_{\bar{x}} = \dfrac{\pi}{64}(D_o^4 - D_i^4) = \dfrac{\pi}{4}(R_o^4 - R_i^4)$ $r_{\bar{x}} = \dfrac{1}{4}\sqrt{D_o^2 + D_i^2} = \dfrac{1}{2}\sqrt{R_o^2 + R_i^2}$ $J_o = \dfrac{\pi}{32}(D_o^4 - D_i^4) = \dfrac{\pi}{2}(R_o^4 - R_i^4)$

TABLE H *Continued*

Case Number	Shape	Properties
9 Hollow circle (approximation) (thin tube)	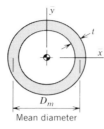 D_m Mean diameter	$A = \pi D_m t$ $I_{\bar{x}} = \dfrac{\pi}{8}[1 + (t/D_m)^2]\, D_m^3 t$ $\simeq \dfrac{\pi}{8} D_m^3 t \quad$ (error < 1% when $t < \frac{1}{10} D_m$) $r_{\bar{x}} = \dfrac{1}{2\sqrt{2}}\sqrt{1 + (t/D_m)^2}\; D_m$ $\simeq \dfrac{1}{2\sqrt{2}} D_m \quad$ (error < 0.4% when $t < \frac{1}{10} D_m$) $J_{\bar{o}} = \dfrac{\pi}{4}[1 + (t/D_m)^2]\, D_m^3 t$ $\simeq \dfrac{\pi}{4} D_m^3 t \quad$ (error < 1% when $t < \frac{1}{10} D_m$)
10 Semicircle	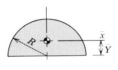	$A = \dfrac{\pi}{2} R^2$ $\bar{Y} = \dfrac{4}{3\pi} R = 0.4244R$ $I_{\bar{x}} = \left(\dfrac{\pi}{8} - \dfrac{8}{9\pi}\right) R^4 = 0.1098R^4$ $r_{\bar{x}} = \sqrt{\dfrac{1}{4} - \dfrac{16}{9\pi^2}}\; R = 0.2643R$
11 Hollow semicircle (approximation) (half of thin tube)	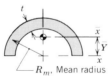 R_m, Mean radius	$A = \pi R_m t$ $\bar{Y} \simeq \dfrac{2}{\pi} R_m \quad$ (error < 0.4% when $t < \frac{1}{5} R_m$) $I_x \simeq \dfrac{\pi}{2} R_m^3 t \quad$ (error < 1% when $t < \frac{1}{5} R_m$) $I_{\bar{x}} \simeq \left(\dfrac{\pi}{2} - \dfrac{4}{\pi}\right) R_m^3 t = 0.2976 R_m^3 t$ (error < 2.4% when $T < \frac{1}{5} R_m$) (error < 0.6% when $t < \frac{1}{10} R_m$)

PHOTO CREDITS

INDEX

Accumulative fatigue damage, 573
Allowable stress:
  bearing, 12
  bending, 107
  bolts in shear, 107
  definition of, 4–5
  test to determine, 47–53
  typical values, Table A
Aluminum column design formulas, 208, 213
Angle of twist, 297, 308–309
Angles, properties for design:
  equal legs, Table B-4
  unequal legs, Table B-5
Anisotropic material, 56
Area:
  centroid, 96
  moment, 96, 120
  moment of inertia, 97–107, 348–352
  polar moment of inertia, 299
  product of inertia, 348–355
  properties of, Table H
Area-moment, see Moment-area
Axial load:
  concentric, elongation due to, 56–58
  eccentric, stress due to, 108–112

Beam:
  bending stress, 97–100
    equation, 97
    maximum, 100
  composite, 368–374
  concrete, 91
  crippling, 367
  curved, 125–136, 179–184
  definition, 91
  deflection, 153–185
    by direct energy method, 457–461
    formulas for, Table G
    by moment-area method, 175–178
    by Rayleigh-Ritz method, 473–478
    by singularity function, 165–169
    by superposition, 172–174
    due to shear, 457–461

design, 107
differential relationships, 171
flexure equation, 155, 159
inelastic bending, 379–386
local buckling, 363–367
nonlinear bending, 374–379
plastic design, 386–390
plastic hinge, 383–385
reinforced concrete, shear, 137
rotation-moment relationship, 97, 152
shear center, 341–342, 358–365
shear failure in, 113–116
statically indeterminate, 157–165
strain energy:
  bending, 450–451
  shear, 451–452
stress by:
  approximate method, 28–38
  exact method, 91–100, 112–125
transformed cross-section, 370
unsymmetrical bending, 342–348, 355–358
Beam-column, 422–435
  concentrated load, 428–435
  distributed load, 422–428
Bearing stress, 12
Bending, inelastic, 339–386
Bending, nonlinear, 374–379
Bending moment:
  diagram, 17–27
  equation, 18–24
  peak and maximum, 161
  rate of change, 25
  sign convention, 17
Bending stress equiation, 97
Bending stress in beam, 92–100
Bending stress in plate, 533–536
Biaxial stress, see Plane stress
Bolted end connections, 10–13
Brittle failure, 543
Brittle material, 53, 60
Buckling:
  critical load, 196–200
  due to eccentric load, 112

energy considerations, 203–205
Euler equation, 200, 205, 214
in flange of beam, 136–137
in test specimen, 54
introduction to, 195
in web due to shear, 236
see also Column
Bulk modulus, 274

Castigliano's first theorem, 478–491
Castigliano's second theorem, 491–511
Centroid, 96, 105
Centroidal axis, 4
Channel, see C-Shape
Coil spring, see Helical spring
Column:
  aluminum, design formulas, 208, 213
  beam-column, 422–435
  crippling, 436
  critical stress, 205
  design, 436–437
  design formulas, 206–214
    limitations, 208
  eccentrically loaded, 412–422
  effective length, 211–213
  effective length factor, 214–217, 401
  end conditions, 211–214
  end constraint, partial, 401–412
  equilibrium, stable and unstable, 204
  Euler buckling equation, 200, 205, 214
  Euler buckling load, 401
  failure, mode of, 195
  ideal, 196, 203
  local buckling, 435–436
  secant formula, 414
  sidesway, 432–433
  slenderness ratio, 204–206
  stability, 204
  steel, design formulas, 207, 213
  timber, design formulas, 208, 213
  unbraced frame, 431–435

Composite beam, 368–374
Compression member, *see* Column
Concrete beam, 91
Constants of integration, 154
Contraflexure, point of, 211
Cover plate, 91
 shear stress in rivets, 123–125
Crippling, beam, 367
Crippling, column, 436
Critical buckling load, 196–200
Critical stress in column, 205
Cross section, transformed, 370
C-Shapes, properties for design, Table
 B-3
Curved beam, 125–136, 179–184
Cyclic loading, *see* Fatigue

Deflection:
 beam, 153–185
 due to shear, 457–461
 formulas for, Table G
 Rayleigh-Ritz method, 473–478
 see also Displacement
Deformation due to axial load, 56–68
Deformation in truss member, 62
Degrees of freedom, 468
Density of engineering materials, Table
 A
Design, column, 206–214, 436–437
Design cycle, 2
Design stress, *see* Stress, allowable
Diagonal torsion and compression, 236
Direct energy method, 453–464
Displacement by:
 Castigliano's first theorem, 476–483
 direct energy method, 453–464
 Rayleigh-Ritz method, 473–478
 stiffness matrix, 484–491
Double shear, 11
Ductility, 52, 60
Dummy load, 497–502

Eccentric load, stress, 108–112
Eccentrically loaded column, 412–422
Effective length, column, 211–213
Effective length factor, 214–217, 401
Elastic constants:
 bulk modulus, 274
 Lamé's constant, 274
 modulus of elasticity, 54
 Poisson's ratio, 269–271
 shear modulus, 272–273
 Young's modulus, 54
Elastic material, linear, 54
Elastic material, nonlinear, 54

Elastic-perfectly plastic, 60, 379
Elongation, permanent, 52
End conditions, columns, 211–214,
 401–412
End connections, bolted, 10–13
End connections, welded, 13–17
Endurance Limit, 546
Energy, *see* Strain energy
Engineering stress, 52
Equilibrium configuration, by potential
 energy, 464–472
Euler's buckling equation, 200, 205,
 214, 401
Euler, Leonhard, 196
Expansion coefficient, values of, Table
 A
Extensometer, 49

Factor of safety, *see* Safety factor
Failure mechanism, cleavage, 227
Failure mechanism, shear, 228
Failure, shear, beam, 113–116
Failure theory, 246–263
 brittle material, 262
 distortion strain energy, 259
 maximum principal strain, 250
 maximum principal stress, 247
 maximum shear stress, 253
 total strain energy, 257
Failure under uniaxial stress, 47, 48,
 52, 53
Fatigue:
 accumulative damage, 573
 endurance limit, 546
 Gerber line, 553
 Goodman line, 551–553
 loads of varying amplitude, 572–573
 Miner's rule, 573
 nature of failure, 543–563
 notch sensitivity, 562–563
 Soderberg line, 553
 strength-number relationship,
  544–547, 549
 strength reduction factor:
  concentration of stress, 561–566
  corrosion, 567
  finish, 557–558
  mean stress, 550–555
  reliability, 547–549
  size, 556–557
  temperature, 566–567
 stress concentration factor, 562
Fillet weld, 14
Flexure equation for beam, 153
Force, longitudinal, due to shear, 118

Gage length, 49
Gerber line, 553
Goodman line, 551–553
Groove in shaft, stress concentration,
 317–318
Gusset plate, 10–11, 13–14
Gyration, radius of, 205

Helical spring:
 axial elongation, 312–313
 shear stress, 314
Hinge, plastic, 383–385
Homogeneous material, 56
Hooke's law, 54
Hoop stress, *see* Stress, pressure
 vessel, circumferential
Horizontal shear stress, 116, 119

I-Beam, *see* S-Shape
Idealized material, stress-strain
 relationship, 379–380
Impact loading, 52
Inelastic bending, 379–386
Internal force, 4
Isotropic material, 56

Keyway in shaft, stress concentration,
 317–318

Lamé's constant, 274
Lateral strain, 53, 270
Linearly elastic material, 54
Local buckling:
 beam, 365–367
 column, 435, 436
Longitudinal force due to shear, 118

Membrane analogy for torsion,
 314–331
Membrane theory, pressure vessel
 stress, 529, 531–532
Miner's rule, 573
Modular ratio, 370
Modulus of elasticity, 54
Modulus:
 secant, 55
 tangent, 55
Mohr's circle for:
 moment and product of inertia,
  351–355
 strain, 279
 stress, biaxial, 232–243
 stress, triaxial, 244–245
Moment, plastic, 383–386
Moment, yield, 386

Moment-area method, deflection by, 175–178
Moment of area, 96, 120
Moment of area, second, 97
Moment of inertia, 100–107
  definition, 97
  examples, 98, 101–103
  formulas for, 97
  parallel axis theorem, 104
  polar, 299
  principal, 352
Moment-rotation relationship, see Rotation-moment relationship

Net section, 11
Neutral axis, 93
  location, 96, 385
Neutral surface, 93
Nonlinear bending, 374–379
Nonlinearly elastic material, 54
Normal stress, 12
Notch sensitivity, 562–563

Offset method for yield stress, 55

Panel point, 10–11
Parallel axis theorem:
  moment of inertia, 104
  product of inertia, 353
Permanent set, 50
Physical properties of materials, Table A
Pipe, properties for design, Table C
Plane stress, 229
Plastic design, 386–391
Plastic flow, effect on stress concentration, 61–62
Plastic hinge, 383–385
Plastic moment, bending, 383–386
Plastic moment, torque, 331–333
Plate girder, 91
  shear, 139
Plates, bending stress, 533–536
Poisson's ratio, 269–271
  engineering materials, value of, Table A
Potential energy:
  equilibrium configuration by, 464–472
  statically indeterminate systems, 472
Power transmitted, rotating shaft, 305–308
Pressure vessel:
  cylindrical, stresses, 38–47
  end caps, 523
  spherical, stress, 530–531

stress formula, 524–531
stress, membrane theory, 529, 531–532
Principal axes, inertia, 352–355
Principal strains from rosette data, 285–289
Principal stress, 233–235
Principal stress equations, 240
Product of inertia, 348–355
  parallel axis theorem, 353
Properties for design:
  areas, Table H
  beam deflection, Table G
  physical, materials, Table A
  reinforcing bars, Table E
  steel pipe, Table C
  steel rods, Table D
  structural shapes, Table B
  timber, Table F
Proportional limit, 51

Radius of gyration, 205
Rate of change of bending moment, 25
Rayleigh-Ritz method, beam deflection by, 473–478
Reinforced concrete beam:
  shear failure, 115, 137
  T-beam, 92, 114
Reinforcing bars, properties for design, Table E
Reduction in area, 52
Redundant member, 63
Redundant support, 62
Repeated loading, see Fatigue
Residual stress, 73
Rods, steel, properties for design, Table D
Rotation-moment relationship:
  beam, 97, 152
  torsion, 297, 308–309
Rupture modulus, 377, 379

Safety factor, 5, 47, 53, 107
Secant formula for column, 414
Secant modulus, 55
Section modulus, definition, 108
Shaft, rotating, power transmitted, 305–308
Shape factor:
  bending, 386
  torsion, 331
Shear:
  center, 341–342, 358–365
  double, 11

diagram, 17–27
equation for beam, 20–22
failure in beam, 113–116
flow, flanged section, 360–365
force:
  sign convention, 19
  summation rule, 24
modulus definition, 272–273
modulus, values of, Table A
plate girder, 137
reinforced concrete beam, 137
stress:
  allowable, bolts, 12
  beam, 112–125
  bolt, 11–12
  cover-plate rivets, 123–125
  helical spring, 314
  horizontal, 116, 119
  maximum, 240
  sign convention, 228
  torsion, 322–326
  vertical, 116, 119
Sidesway, 432–433
Sign convention:
  bending moment, 17
  shear force, 19
Singularity function, 165–169
  properties, 166–167
Slenderness ratio, 204–206
Soderberg line, 553
S-Shape, properties for design, Table B-2
Stability:
  column, 204
  potential energy, 465
Statically indeterminate beam, 157–165
Statically indeterminate system:
  Castigliano's second, 502–507
  introduction, 63–70
  minimum potential energy, 472
Steel column, design formulas, 207, 213
Step function, see Singularity function
Stiffness matrix, 485
Strain:
  definition, 49
  elastic, 50
  hardening, 51
  inelastic, 50–51
  lateral, 53, 270
  Mohr's circle for, 279
  tapered bar, 50
  thermal, 70

Strain energy:
  bar uniformly stressed, 449
  beam:
    bending, 450–451
    shear, 451–452
  density, 447
    normal stress, 447–448
    shear stress, 448
    stretched spring, 445, 446
Strain gage:
  electric resistance, 275–277,
    281–289
  mechanical, 48
  rosette:
    delta, 286
    rectangular, 282, 286
Strain-stress relationships, see Stress-
    strain relationships
Strength reduction factors, see Fatigue
Stress:
  allowable, 4–5, 47–53, 107
  allowable, typical values, Table A
  beam:
    approximate method, 28–38
    exact method, 92–100
  bearing, 12
  bending, 92–100
  bending, maximum, 100
  biaxial, see Plane stress
  composite beam, 371–374
  concentration, 58–62
    cyclic load, 62
    effect of plastic flow, 61–62
    factor, 59–62, 561–566
    keyway in circular shaft,
      317–318
    torsion, 317–318, 321
  design, see Allowable stress
  eccentric load, 108–112
  engineering, 52
  horizontal shear, 116
  inclined plane, 224, 238
  Mohr's circle for, 232–243
  normal, 12

plane, 229
plate in bending, 533–536
pressure vessel:
  circumferential, 38–41
  longitudinal, 41–42, 44, 526
  torsional, 45–46
principal, 233–235, 240
residual, 73
rod due to torsion, 295, 308
shear, 11–12
shear in beam, 112–125
spherical pressure vessel, 530–531
strain gage data, from, 281–284
thermal, 70–73
torsional, see Torsion
triaxial, 244
true, 52
ultimate, 51
  engineering materials, value of,
    Table A
uniaxial, 53
vertical shear, 116
yield, 51
  engineering materials, value of,
    Table A
Stress-strain relationship, 48–58
  biaxial, 271–275
  idealized material, 379–380
  triaxial, 274
  uniaxial, 269–271
Superposition, deflection by, 172–174
Superposition, method of, 109

Tangent modulus, 55
Testing machine, tensile, 49
Testing machine, fatigue, 546
Theory of failure, see Failure theories
Thermal expansion, coefficient of, 70,
    Table A
Thermal stress, 70–73
Thermal strain, 70
Timber:
  allowable stress, Table A
  beam, shear failure, 113–114, 116

column, design formulas, 208, 213
properties for design, Table F
Torsion:
  angle of twist, 297, 308–309, 319
  fully plastic, 331
  inelastic, 326–334
  membrane analogy, 314–331
  non-circular section, 318–331
  rectangular section, 318–319
  shape factor, 331
  stress:
    cylindrical rod, 295–308
    thin-walled tube, 45–46
    structural shape, 320–322
    thin-walled closed section,
      322–326
Transformed cross-section, 370
Triaxial stress, 244
Triaxial stress-strain relationship, 274
True stress, 52
Truss member deformation, 62

Ultimate stress, 51, 53
  engineering materials, value of, Table
    A
Uniaxial stress, 53
Unsymmetrical bending, 342–348,
    355–358

Vertical shear stress, 116, 119

Weld, fillet, 14
Wide-flange, see W-Shape
W-Shapes, properties for design, Table
    B-1

Yield moment, 386
Yield stress, 51
  engineering materials, value of, Table
    A
  offset method, 55
Young's modulus, 54
  engineering materials, value of, Table
    A